Methods in Enzymology

Volume 165

Microbial Toxins: Tools in Enzymology

EDITED BY

Sidney Harshman

DEPARTMENT OF MICROBIOLOGY
VANDERBILT UNIVERSITY SCHOOL OF MEDICINE
NASHVILLE, TENNESSEE

ACADEMIC PRESS, INC.
Harcourt Brace Jovanovich, Publishers
San Diego New York Berkeley Boston
London Sydney Tokyo Toronto

ACADEMIC PRESS, INC.
San Diego, California 92101

United Kingdom Edition published by
ACADEMIC PRESS, INC. (LONDON) LTD.
24-28 Oval Road, London NW1 7DX

LIBRARY OF CONGRESS CATALOG CARD NUMBER: 54-9110

ISBN 0-12-182066-1 (alk. paper)

PRINTED IN THE UNITED STATES OF AMERICA
88 89 90 91 9 8 7 6 5 4 3 2 1

Table of Contents

Section I. Preparation of Toxins

A. Gram-Positive Cocci

B. Gram-Positive Bacilli

C. Gram-Negative Cocci

D. Gram-Negative Bacilli

E. *Vibrios*

F. Immunotoxins

Section II. Assay of Toxins

Addenda

Contributors to Volume 165

Article numbers are in parentheses following the names of contributors.
Affiliations listed are current.

JOSEPH E. ALOUF (8, 9, 10, 14, 41), *Unité des Antigènes Bactériens (UA CNRS 557), Institute Pasteur-Unite Associée, 75724 Paris Cedex 15, France*

JOHN P. ARBUTHNOTT (5, 46, 48), *Department of Microbiology, Moyne Institute, Trinity College, Dublin 2, Ireland*

CHRISTOPHER J. BAILEY (5), *Department of Biochemistry, Moyne Institute, Trinity College, Dublin 2, Ireland*

JOSEPH T. BARBIERI (11), *Department of Microbiology, Medical College of Wisconsin, Milwaukee, Wisconsin 53226*

MERLIN S. BERGDOLL (44, 45), *Food Research Institute, University of Wisconsin-Madison, Madison, Wisconsin 53706*

ALAN W. BERNHEIMER (30), *Professor Emeritus—Microbiology, New York University School of Medicine, New York, New York 10016*

SUCHARIT BHAKDI (40), *Institute of Medical Microbiology, University of Giessen, D-6300 Giessen, Federal Republic of Germany*

T. HARRY BIRKBECK (3), *Department of Microbiology, Anderson College, The University of Glasgow, Glasgow G11 6NU, Scotland*

DEBRA A. BLOMSTER-HAUTAMAA (6), *Department of Microbiology, University of Minnesota, Minneapolis, Minnesota 55455*

G. A. BOHACH (20, 43), *Department of Microbiology, Medical School, University of Minnesota, Minneapolis, Minnesota 55455*

J. THOMAS BUCKLEY (27), *Department of Biochemistry and Microbiology, University of Victoria, Victoria, British Columbia V8W 2Y2, Canada*

STEPHEN F. CARROLL (11, 31), *Department of Protein Chemistry, XOMA Corporation, Berkeley, California 94710*

PAUL CASSIDY (1), *Cardiac Muscle Research Laboratory, Boston University School of Medicine, Boston, Massachusetts 02118*

S. J. CAVALIERI (20), *Department of Medical Microbiology, Creighton University School of Medicine, Omaha, Nebraska 68178*

JENIFER COBURN (35), *Sackler School of Biomedical Sciences, Tufts University, Department of Microbiology, Boston, Massachusetts 02111*

R. JOHN COLLIER (11, 31), *Department of Microbiology and Molecular Genetics and the Shipley Institute of Medicine, Harvard Medical School, Boston, Massachusetts 02115*

LARRY W. DANIEL (42), *Department of Biochemistry, Bowman Gray School of Medicine of Wake Forest University, Winston-Salem, North Carolina 27103*

JOYCE C. S. DE AZAVEDO (5, 46, 48), *Department of Microbiology, Moyne Institute, Trinity College, Dublin 2, Ireland*

ARTHUR DONOHUE-ROLFE (22, 33, 36), *Division of Geographic Medicine, Department of Medicine, Tufts University School of Medicine, Boston, Massachusetts 02111*

JOHN H. FREER (3), *Department of Microbiology, Anderson College, The University of Glasgow, Glasgow G11 6NU, Scotland*

ROSWITHA FÜSSLE (40), *Institute of Medical Microbiology, University of Giessen, D-6300 Giessen, Federal Republic of Germany*

CHRISTIANE GEOFFROY (8, 10), *Unité des Antigènes Bactériens (UA CNRS 557), Institut Pasteur, 75724 Paris Cedex 15, France*

D. MICHAEL GILL (34, 35), *Department of Molecular Biology and Microbiology, Tufts University Schools of Medicine, Dental Medicine, and Veterinary Medicine, Boston, Massachusetts 02111*

LARRY D. GRAY (25), *Department of Microbiology and Immunology, Bowman Gray School of Medicine, Winston-Salem, North Carolina 27103*

RICHARD N. GREENBERG (19), *Department of Internal Medicine, Division of Infectious Diseases, Saint Louis University School of Medicine, Saint Louis, Missouri 63104*

SIDNEY HARSHMAN (1), *Department of Microbiology, Vanderbilt University School of Medicine, Nashville, Tennessee 37232*

S. PETER HOWARD (27), *Centre de Biochimie et de Biologie Moléculaire, Centre National de la Recherche Scientifique, 31 Ch. J. Auguier, 13402 Marseille, France*

JOHN J. IANDOLO (7), *Division of Biology, Kansas State University, Manhattan, Kansas 66506*

MARY JACEWICZ (22, 33), *Division of Geographic Medicine and Infectious Diseases, Department of Medicine, New England Medical Center, Boston, Massachusetts 02111*

COLETTE JOLIVET-REYNAUD (13, 41), *Unité des Antigènes Bactériens (UA CNRS 557), Institut Pasteur, 75724 Paris Cedex 15, France*

ANNE V. KANE (22), *Grasp Center, Department of Medicine, New England Medical Center, Boston, Massachusetts 02111*

IWAO KATO (4), *Second Department of Microbiology, Chiba University School of Medicine, 1-8-1, Inohana, Chiba 280, Japan*

MARTHA KENNEDY (42), *Department of Biochemistry, Bowman Gray School of Medicine of Wake Forest University, Winston-Salem, North Carolina 27103*

GERALD T. KEUSCH (22, 33, 36), *Department of Medicine, Division of Geographic Medicine, Tufts University School of Medicine, Boston, Massachusetts 02111*

LYNN KING (42), *Department of Biochemistry, Bowman Gray School of Medicine of Wake Forest University, Winston-Salem, North Carolina 27103*

MAHENDRA H. KOTHARY (25), *Department of Microbiology and Immunology, Bowman Gray School of Medicine, Winston-Salem, North Carolina 27103*

KENNETH J. KOZAK (21), *Department of Microbiology and Molecular Genetics, University of Cincinnati College of Medicine, Cincinatti, Ohio 45267-0524*

ARNOLD S. KREGER (25), *Department of Microbiology and Immunology, Bowman Gray School of Medicine, Winston-Salem, North Carolina 27103*

STEPHEN H. LEPPLA (16), *U.S. Army Medical Research Institute of Infectious Diseases, Fort Detrick, Frederick, Maryland 21701-5011*

JONATHAN J. LIPMAN (37, 38), *Division of Nephrology, Departments of Medicine and Surgery, Vanderbilt University School of Medicine, Nashville, Tennessee 37232*

ASA LJUNGH (28), *Department of Medical Microbiology, University of Lund, S-223 62 Lund, Sweden*

CATHERINE LORIDAN (9), *Unité des Antigènes Bactériens (UA CNRS 557), Institute Pasteur, 75724 Paris Cedex 15, France*

ROGER N. LUCKEN (48), *Viral Vaccine and Monoclonal Antibody Production, Wellcome Biotech, Beckenham, Kent BR33BS, England*

BRUCE A. MCCLANE (15), *Department of Microbiology, Biochemistry and Molecular Biology, University of Pittsburgh School of Medicine, Pittsburgh, Pennsylvania 15261*

JAMES L. MCDONEL (15), *Department of Biology, Indiana University at South Bend, South Bend, Indiana 46616*

JOHN J. MEKALANOS (24), *Department of Microbiology and Molecular Genetics, Harvard Medical School, Boston, Massachusetts 02115*

JOHN L. MIDDLEBROOK (12), *Department of Toxinology, U.S. Army Medical Research Institute of Infectious Diseases, Fort Detrick, Frederick, Maryland 21701-5011*

CESARE MONTECUCCO (49), *Centro C.N.R. Biomembrane and Dipartimento di Scienze Biomediche, Universitá di Padova, Via Trieste 35131, Padova, Italy*

THOMAS C. MONTIE (23), *Department of Microbiology, University of Tennessee, Knoxville, Tennessee 37996-0845*

HERVÉ MOREAU (14, 41), *Unité des Antigènes Bactériens (UA CNRS 557), Institut Pasteur, 75724 Paris Cedex 15, France*

MASATOSHI NODA (4), *Department of Bacterial Infection, Institute of Medical Science, University of Tokyo, 4-6-1, Shirokanedai, Minato-ku, Toyko 108, Japan*

THOMAS N. OELTMANN (29, 36), *Oncology Division, Department of Medicine, Vanderbilt University Medical School, Nashville, Tennessee 37232*

V. Y. PERERA (18), *Clinical Division, Bio-Rad Laboratories, Hercules, California 94547*

ANDREW G. PLAUT (17), *Division of Gastroenterology, Department of Medicine, Tufts-New England Medical Center, Boston, Massachusetts 02111*

MICHEL PLOMMET (2), *Department of Animal Pathology, Institut National de la Recherche Agronomique (INRA), Centre de Tours-Nouzilly, 37380 Nouzilly, France*

JEFFRY REIDLER (52), *Westinghouse Electric Co., Waltz Mill Site, Madison, PA 15663-0286*

JOHN P. ROBINSON (52, 13), *Department of Microbiology, Vanderbilt University, Nashville, TN 37232*

ABDUL M. K. SAEED (19), *Department of Epidemiology, School of Public Health and Community Medicine, University of Washington, Seattle, Washington 98195*

CATHARINE B. SAELINGER (21, 32), *Department of Microbiology and Molecular Genetics, University of Cincinnati College of Medicine, Cincinnati, Ohio 46267 0524*

P. M. SCHLIEVERT (6, 43, 47), *Department of Microbiology, University of Minnesota, Minneapolis, Minnesota 55455*

JAMES J. SCHMIDT (12), *Department of Toxinology, U.S. Army Medical Research Institute of Infectious Diseases, Fort Detrick, Frederick, Maryland 21701 5011*

LANCE L. SIMPSON (12), *Departments of Medicine and Pharmacology, Jefferson Medical College, Philadelphia, Pennsylvania 19107*

I. S. SNYDER (20), *Department of Microbiology and Immunology, West Virginia University Medical Center, Morgantown, West Virginia 26506*

NANCY SUGG (1), *Department of Microbiology, Vanderbilt University School of Medicine, Nashville, Tennessee 37232*

ANDREAS SZIEGOLEIT (40), *Institute of Medical Microbiology, University of Giessen, D-6300 Giessen, Federal Republic of Germany*

YOSHIFUMI TAKEDA (26), *The Institute of Medical Science, The University of Tokyo, 4-6-1, Shirokanedai Minato-ku, Tokyo 108, Japan*

MONICA THELESTAM (39), *Department of Bacteriology, Karolinska Institutet, S-104 01 Stockholm, Sweden*

JØRGEN TRANUM-JENSEN (40, 50, 51), *Anatomy Institute C, University of Copenhagen, The Panum Institute, DK-2200 Copenhagen N, Denmark*

RODNEY K. TWETEN (7), *Department of Microbiology and Immunology, University of Oklahoma Health Sciences Center, Oklahoma City, Oklahoma 73190*

TORKEL WADSTRÖM (28), *Department of Medical Microbiology, University of Lund, S-223 62 Lund, Sweden*

RONALD G. WILEY (29, 36), *Neurology Department, VAMC, and Vanderbilt University School of Medicine, Nashville, Tennessee 37212*

MARILYN WOOLKALIS (34, 35), *Department of Pharmacology, University of Pennsylvania School of Medicine, Philadelphia, Pennsylvania 19104*

Preface

Microbial toxins are being used as precise tools to dissect biochemical pathways and to elucidate complex chemical structures. They have been exploited to define GTP-binding proteins, to characterize lipid structures in membranes, and to selectively permeabilize cells, to mention but a few applications. This volume is a convenient source both for methods of preparing a variety of microbial toxins and for their assay.

The book has been divided into two sections. Methods for the preparation of various toxins are reported in the first section. Emphasis is on simplicity of the procedure and the purity of biological activity rather than on physical chemical purity. For convenience, the toxins have been grouped under the classification of the organisms that produce them. Included are examples of hemolysins, proteases, protein synthesis inhibitors, ADP-ribosyltransferases, lipases, enterotoxins, neurotoxins, and the construction of a toxin hybrid molecule. The second section is devoted to descriptions of different ways of measuring the biochemical or physiological activities of the various toxins. It begins with a general procedure for the assay of hemolysins followed by more specific methods for assaying inhibition of protein synthesis, ADP-ribosylating activity, neurotoxic actions, membrane permeabilization, lipases, enterotoxins, and special assays for TSST-1. The section concludes with a description for photolabeling of membrane-penetrating toxins, the electron microscopic study of toxins, and the analysis of two-dimensional crystals of toxins.

I would like to express my appreciation to the contributors for their participation in this venture and for generously sharing their expertise. I thank also the competent and courteous staff of Academic Press for their efforts and advice.

Sidney Harshman

METHODS IN ENZYMOLOGY

EDITED BY

Sidney P. Colowick and Nathan O. Kaplan

VANDERBILT UNIVERSITY
SCHOOL OF MEDICINE
NASHVILLE, TENNESSEE

DEPARTMENT OF CHEMISTRY
UNIVERSITY OF CALIFORNIA
AT SAN DIEGO
LA JOLLA, CALIFORNIA

I. Preparation and Assay of Enzymes
II. Preparation and Assay of Enzymes
III. Preparation and Assay of Substrates
IV. Special Techniques for the Enzymologist
V. Preparation and Assay of Enzymes
VI. Preparation and Assay of Enzymes (*Continued*)
Preparation and Assay of Substrates
Special Techniques
VII. Cumulative Subject Index

METHODS IN ENZYMOLOGY

EDITORS-IN-CHIEF

Sidney P. Colowick and Nathan O. Kaplan

VOLUME VIII. Complex Carbohydrates
Edited by ELIZABETH F. NEUFELD AND VICTOR GINSBURG

VOLUME IX. Carbohydrate Metabolism
Edited by WILLIS A. WOOD

VOLUME X. Oxidation and Phosphorylation
Edited by RONALD W. ESTABROOK AND MAYNARD E. PULLMAN

VOLUME XI. Enzyme Structure
Edited by C. H. W. HIRS

VOLUME XII. Nucleic Acids (Parts A and B)
Edited by LAWRENCE GROSSMAN AND KIVIE MOLDAVE

VOLUME XIII. Citric Acid Cycle
Edited by J. M. LOWENSTEIN

VOLUME XIV. Lipids
Edited by J. M. LOWENSTEIN

VOLUME XV. Steroids and Terpenoids
Edited by RAYMOND B. CLAYTON

VOLUME XVI. Fast Reactions
Edited by KENNETH KUSTIN

VOLUME XVII. Metabolism of Amino Acids and Amines (Parts A and B)
Edited by HERBERT TABOR AND CELIA WHITE TABOR

VOLUME XVIII. Vitamins and Coenzymes (Parts A, B, and C)
Edited by DONALD B. MCCORMICK AND LEMUEL D. WRIGHT

VOLUME XIX. Proteolytic Enzymes
Edited by GERTRUDE E. PERLMANN AND LASZLO LORAND

VOLUME XX. Nucleic Acids and Protein Synthesis (Part C)
Edited by KIVIE MOLDAVE AND LAWRENCE GROSSMAN

VOLUME XXI. Nucleic Acids (Part D)
Edited by LAWRENCE GROSSMAN AND KIVIE MOLDAVE

VOLUME XXII. Enzyme Purification and Related Techniques
Edited by WILLIAM B. JAKOBY

VOLUME XXIII. Photosynthesis (Part A)
Edited by ANTHONY SAN PIETRO

VOLUME XXIV. Photosynthesis and Nitrogen Fixation (Part B)
Edited by ANTHONY SAN PIETRO

VOLUME XXV. Enzyme Structure (Part B)
Edited by C. H. W. HIRS AND SERGE N. TIMASHEFF

VOLUME XXVI. Enzyme Structure (Part C)
Edited by C. H. W. HIRS AND SERGE N. TIMASHEFF

VOLUME XXVII. Enzyme Structure (Part D)
Edited by C. H. W. HIRS AND SERGE N. TIMASHEFF

VOLUME XXVIII. Complex Carbohydrates (Part B)
Edited by VICTOR GINSBURG

VOLUME XXIX. Nucleic Acids and Protein Synthesis (Part E)
Edited by LAWRENCE GROSSMAN AND KIVIE MOLDAVE

VOLUME XXX. Nucleic Acids and Protein Synthesis (Part F)
Edited by KIVIE MOLDAVE AND LAWRENCE GROSSMAN

VOLUME XXXI. Biomembranes (Part A)
Edited by SIDNEY FLEISCHER AND LESTER PACKER

VOLUME XLVI. Affinity Labeling
Edited by WILLIAM B. JAKOBY AND MEIR WILCHEK

VOLUME XLVII. Enzyme Structure (Part E)
Edited by C. H. W. HIRS AND SERGE N. TIMASHEFF

VOLUME XLVIII. Enzyme Structure (Part F)
Edited by C. H. W. HIRS AND SERGE N. TIMASHEFF

VOLUME XLIX. Enzyme Structure (Part G)
Edited by C. H. W. HIRS AND SERGE N. TIMASHEFF

VOLUME L. Complex Carbohydrates (Part C)
Edited by VICTOR GINSBURG

VOLUME LI. Purine and Pyrimidine Nucleotide Metabolism
Edited by PATRICIA A. HOFFEE AND MARY ELLEN JONES

VOLUME LII. Biomembranes (Part C: Biological Oxidations)
Edited by SIDNEY FLEISCHER AND LESTER PACKER

VOLUME LIII. Biomembranes (Part D: Biological Oxidations)
Edited by SIDNEY FLEISCHER AND LESTER PACKER

VOLUME LIV. Biomembranes (Part E: Biological Oxidations)
Edited by SIDNEY FLEISCHER AND LESTER PACKER

VOLUME LV. Biomembranes (Part F: Bioenergetics)
Edited by SIDNEY FLEISCHER AND LESTER PACKER

VOLUME LVI. Biomembranes (Part G: Bioenergetics)
Edited by SIDNEY FLEISCHER AND LESTER PACKER

VOLUME LVII. Bioluminescence and Chemiluminescence
Edited by MARLENE A. DELUCA

VOLUME LVIII. Cell Culture
Edited by WILLIAM B. JAKOBY AND IRA PASTAN

VOLUME LIX. Nucleic Acids and Protein Synthesis (Part G)
Edited by KIVIE MOLDAVE AND LAWRENCE GROSSMAN

VOLUME LX. Nucleic Acids and Protein Synthesis (Part H)
Edited by KIVIE MOLDAVE AND LAWRENCE GROSSMAN

VOLUME 61. Enzyme Structure (Part H)
Edited by C. H. W. HIRS AND SERGE N. TIMASHEFF

VOLUME 62. Vitamins and Coenzymes (Part D)
Edited by DONALD B. McCORMICK AND LEMUEL D. WRIGHT

VOLUME 63. Enzyme Kinetics and Mechanism (Part A: Initial Rate and Inhibitor Methods)
Edited by DANIEL L. PURICH

VOLUME 64. Enzyme Kinetics and Mechanism (Part B: Isotopic Probes and Complex Enzyme Systems)
Edited by DANIEL L. PURICH

VOLUME 65. Nucleic Acids (Part I)
Edited by LAWRENCE GROSSMAN AND KIVIE MOLDAVE

VOLUME 66. Vitamins and Coenzymes (Part E)
Edited by DONALD B. McCORMICK AND LEMUEL D. WRIGHT

VOLUME 67. Vitamins and Coenzymes (Part F)
Edited by DONALD B. McCORMICK AND LEMUEL D. WRIGHT

VOLUME 68. Recombinant DNA
Edited by RAY WU

VOLUME 69. Photosynthesis and Nitrogen Fixation (Part C)
Edited by ANTHONY SAN PIETRO

VOLUME 70. Immunochemical Techniques (Part A)
Edited by HELEN VAN VUNAKIS AND JOHN J. LANGONE

VOLUME 71. Lipids (Part C)
Edited by JOHN M. LOWENSTEIN

VOLUME 72. Lipids (Part D)
Edited by JOHN M. LOWENSTEIN

VOLUME 73. Immunochemical Techniques (Part B)
Edited by JOHN J. LANGONE AND HELEN VAN VUNAKIS

VOLUME 74. Immunochemical Techniques (Part C)
Edited by JOHN J. LANGONE AND HELEN VAN VUNAKIS

VOLUME 75. Cumulative Subject Index Volumes XXXI, XXXII, XXXIV–LX
Edited by EDWARD A. DENNIS AND MARTHA G. DENNIS

VOLUME 76. Hemoglobins
Edited by ERALDO ANTONINI, LUIGI ROSSI-BERNARDI, AND EMILIA CHIANCONE

VOLUME 77. Detoxication and Drug Metabolism
Edited by WILLIAM B. JAKOBY

VOLUME 78. Interferons (Part A)
Edited by SIDNEY PESTKA

VOLUME 79. Interferons (Part B)
Edited by SIDNEY PESTKA

VOLUME 80. Proteolytic Enzymes (Part C)
Edited by LASZLO LORAND

VOLUME 81. Biomembranes (Part H: Visual Pigments and Purple Membranes, I)
Edited by LESTER PACKER

VOLUME 82. Structural and Contractile Proteins (Part A: Extracellular Matrix)
Edited by LEON W. CUNNINGHAM AND DIXIE W. FREDERIKSEN

VOLUME 83. Complex Carbohydrates (Part D)
Edited by VICTOR GINSBURG

VOLUME 84. Immunochemical Techniques (Part D: Selected Immunoassays)
Edited by JOHN J. LANGONE AND HELEN VAN VUNAKIS

VOLUME 85. Structural and Contractile Proteins (Part B: The Contractile Apparatus and the Cytoskeleton)
Edited by DIXIE W. FREDERIKSEN AND LEON W. CUNNINGHAM

VOLUME 86. Prostaglandins and Arachidonate Metabolites
Edited by WILLIAM E. M. LANDS AND WILLIAM L. SMITH

VOLUME 87. Enzyme Kinetics and Mechanism (Part C: Intermediates, Stereochemistry, and Rate Studies)
Edited by DANIEL L. PURICH

VOLUME 88. Biomembranes (Part I: Visual Pigments and Purple Membranes, II)
Edited by LESTER PACKER

VOLUME 89. Carbohydrate Metabolism (Part D)
Edited by WILLIS A. WOOD

VOLUME 90. Carbohydrate Metabolism (Part E)
Edited by WILLIS A. WOOD

VOLUME 91. Enzyme Structure (Part I)
Edited by C. H. W. HIRS AND SERGE N. TIMASHEFF

VOLUME 92. Immunochemical Techniques (Part E: Monoclonal Antibodies and General Immunoassay Methods)
Edited by JOHN J. LANGONE AND HELEN VAN VUNAKIS

VOLUME 93. Immunochemical Techniques (Part F: Conventional Antibodies, Fc Receptors, and Cytotoxicity)
Edited by JOHN J. LANGONE AND HELEN VAN VUNAKIS

VOLUME 94. Polyamines
Edited by HERBERT TABOR AND CELIA WHITE TABOR

VOLUME 95. Cumulative Subject Index Volumes 61–74, 76–80
Edited by EDWARD A. DENNIS AND MARTHA G. DENNIS

VOLUME 96. Biomembranes [Part J: Membrane Biogenesis: Assembly and Targeting (General Methods; Eukaryotes)]
Edited by SIDNEY FLEISCHER AND BECCA FLEISCHER

VOLUME 97. Biomembranes [Part K: Membrane Biogenesis: Assembly and Targeting (Prokaryotes, Mitochondria, and Chloroplasts)]
Edited by SIDNEY FLEISCHER AND BECCA FLEISCHER

VOLUME 98. Biomembranes (Part L: Membrane Biogenesis: Processing and Recycling)
Edited by SIDNEY FLEISCHER AND BECCA FLEISCHER

VOLUME 99. Hormone Action (Part F: Protein Kinases)
Edited by JACKIE D. CORBIN AND JOEL G. HARDMAN

VOLUME 100. Recombinant DNA (Part B)
Edited by RAY WU, LAWRENCE GROSSMAN, AND KIVIE MOLDAVE

VOLUME 101. Recombinant DNA (Part C)
Edited by RAY WU, LAWRENCE GROSSMAN, AND KIVIE MOLDAVE

VOLUME 102. Hormone Action (Part G: Calmodulin and Calcium-Binding Proteins)
Edited by ANTHONY R. MEANS AND BERT W. O'MALLEY

VOLUME 103. Hormone Action (Part H: Neuroendocrine Peptides)
Edited by P. MICHAEL CONN

VOLUME 104. Enzyme Purification and Related Techniques (Part C)
Edited by WILLIAM B. JAKOBY

VOLUME 105. Oxygen Radicals in Biological Systems
Edited by LESTER PACKER

VOLUME 106. Posttranslational Modifications (Part A)
Edited by FINN WOLD AND KIVIE MOLDAVE

VOLUME 107. Posttranslational Modifications (Part B)
Edited by FINN WOLD AND KIVIE MOLDAVE

VOLUME 108. Immunochemical Techniques (Part G: Separation and Characterization of Lymphoid Cells)
Edited by GIOVANNI DI SABATO, JOHN J. LANGONE, AND HELEN VAN VUNAKIS

VOLUME 109. Hormone Action (Part I: Peptide Hormones)
Edited by LUTZ BIRNBAUMER AND BERT W. O'MALLEY

VOLUME 110. Steroids and Isoprenoids (Part A)
Edited by JOHN H. LAW AND HANS C. RILLING

VOLUME 111. Steroids and Isoprenoids (Part B)
Edited by JOHN H. LAW AND HANS C. RILLING

VOLUME 112. Drug and Enzyme Targeting (Part A)
Edited by KENNETH J. WIDDER AND RALPH GREEN

VOLUME 113. Glutamate, Glutamine, Glutathione, and Related Compounds
Edited by ALTON MEISTER

VOLUME 114. Diffraction Methods for Biological Macromolecules (Part A)
Edited by HAROLD W. WYCKOFF, C. H. W. HIRS, AND SERGE N. TIMASHEFF

VOLUME 115. Diffraction Methods for Biological Macromolecules (Part B)
Edited by HAROLD W. WYCKOFF, C. H. W. HIRS, AND SERGE N. TIMASHEFF

VOLUME 116. Immunochemical Techniques (Part H: Effectors and Mediators of Lymphoid Cell Functions)
Edited by GIOVANNI DI SABATO, JOHN J. LANGONE, AND HELEN VAN VUNAKIS

VOLUME 117. Enzyme Structure (Part J)
Edited by C. H. W. HIRS AND SERGE N. TIMASHEFF

VOLUME 118. Plant Molecular Biology
Edited by ARTHUR WEISSBACH AND HERBERT WEISSBACH

VOLUME 119. Interferons (Part C)
Edited by SIDNEY PESTKA

VOLUME 120. Cumulative Subject Index Volumes 81–94, 96–101

VOLUME 121. Immunochemical Techniques (Part I: Hybridoma Technology and Monoclonal Antibodies)
Edited by JOHN J. LANGONE AND HELEN VAN VUNAKIS

VOLUME 122. Vitamins and Coenzymes (Part G)
Edited by FRANK CHYTIL AND DONALD B. MCCORMICK

VOLUME 123. Vitamins and Coenzymes (Part H)
Edited by FRANK CHYTIL AND DONALD B. MCCORMICK

VOLUME 124. Hormone Action (Part J: Neuroendocrine Peptides)
Edited by P. MICHAEL CONN

VOLUME 125. Biomembranes (Part M: Transport in Bacteria, Mitochondria, and Chloroplasts: General Approaches and Transport Systems)
Edited by SIDNEY FLEISCHER AND BECCA FLEISCHER

VOLUME 126. Biomembranes (Part N: Transport in Bacteria, Mitochondria, and Chloroplasts: Protonmotive Force)
Edited by SIDNEY FLEISCHER AND BECCA FLEISCHER

VOLUME 127. Biomembranes (Part O: Protons and Water: Structure and Translocation)
Edited by LESTER PACKER

VOLUME 128. Plasma Lipoproteins (Part A: Preparation, Structure, and Molecular Biology)
Edited by JERE P. SEGREST AND JOHN J. ALBERS

VOLUME 129. Plasma Lipoproteins (Part B: Characterization, Cell Biology, and Metabolism)
Edited by JOHN J. ALBERS AND JERE P. SEGREST

VOLUME 130. Enzyme Structure (Part K)
Edited by C. H. W. HIRS AND SERGE N. TIMASHEFF

VOLUME 131. Enzyme Structure (Part L)
Edited by C. H. W. HIRS AND SERGE N. TIMASHEFF

VOLUME 132. Immunochemical Techniques (Part J: Phagocytosis and Cell-Mediated Cytotoxicity)
Edited by GIOVANNI DI SABATO AND JOHANNES EVERSE

VOLUME 133. Bioluminescence and Chemiluminescence (Part B)
Edited by MARLENE DELUCA AND WILLIAM D. MCELROY

VOLUME 134. Structural and Contractile Proteins (Part C: The Contractile Apparatus and the Cytoskeleton)
Edited by RICHARD B. VALLEE

VOLUME 135. Immobilized Enzymes and Cells (Part B)
Edited by KLAUS MOSBACH

VOLUME 136. Immobilized Enzymes and Cells (Part C)
Edited by KLAUS MOSBACH

VOLUME 137. Immobilized Enzymes and Cells (Part D)
Edited by KLAUS MOSBACH

VOLUME 138. Complex Carbohydrates (Part E)
Edited by VICTOR GINSBURG

VOLUME 139. Cellular Regulators (Part A: Calcium- and Calmodulin-Binding Proteins)
Edited by ANTHONY R. MEANS AND P. MICHAEL CONN

VOLUME 140. Cumulative Subject Index Volumes 102–119, 121–134

VOLUME 141. Cellular Regulators (Part B: Calcium and Lipids)
Edited by P. MICHAEL CONN AND ANTHONY R. MEANS

VOLUME 142. Metabolism of Aromatic Amino Acids and Amines
Edited by SEYMOUR KAUFMAN

VOLUME 143. Sulfur and Sulfur Amino Acids
Edited by WILLIAM B. JAKOBY AND OWEN GRIFFITH

VOLUME 144. Structural and Contractile Proteins (Part D: Extracellular Matrix)
Edited by LEON W. CUNNINGHAM

VOLUME 145. Structural and Contractile Proteins (Part E: Extracellular Matrix)
Edited by LEON W. CUNNINGHAM

VOLUME 146. Peptide Growth Factors (Part A)
Edited by DAVID BARNES AND DAVID A. SIRBASKU

VOLUME 147. Peptide Growth Factors (Part B)
Edited by DAVID BARNES AND DAVID A. SIRBASKU

VOLUME 148. Plant Cell Membranes
Edited by LESTER PACKER AND ROLAND DOUCE

VOLUME 149. Drug and Enzyme Targeting (Part B)
Edited by RALPH GREEN AND KENNETH J. WIDDER

VOLUME 150. Immunochemical Techniques (Part K: *In Vitro* Models of B and T Cell Functions and Lymphoid Cell Receptors)
Edited by GIOVANNI DI SABATO

VOLUME 151. Molecular Genetics of Mammalian Cells
Edited by MICHAEL M. GOTTESMAN

VOLUME 152. Guide to Molecular Cloning Techniques
Edited by SHELBY L. BERGER AND ALAN R. KIMMEL

VOLUME 153. Recombinant DNA (Part D)
Edited by RAY WU AND LAWRENCE GROSSMAN

VOLUME 154. Recombinant DNA (Part E)
Edited by RAY WU AND LAWRENCE GROSSMAN

VOLUME 155. Recombinant DNA (Part F)
Edited by RAY WU

VOLUME 156. Biomembranes (Part P: ATP-Driven Pumps and Related Transport: The Na,K-Pump)
Edited by SIDNEY FLEISCHER AND BECCA FLEISCHER

VOLUME 157. Biomembranes (Part Q: ATP-Driven Pumps and Related Transport: Calcium, Proton, and Potassium Pumps)
Edited by SIDNEY FLEISCHER AND BECCA FLEISCHER

VOLUME 158. Metalloproteins (Part A)
Edited by JAMES F. RIORDAN AND BERT L. VALLEE

VOLUME 159. Initiation and Termination of Cyclic Nucleotide Action
Edited by JACKIE D. CORBIN AND ROGER A. JOHNSON

VOLUME 160. Biomass (Part A: Cellulose and Hemicellulose)
Edited by WILLIS A. WOOD AND SCOTT T. KELLOGG

VOLUME 161. Biomass (Part B: Lignin, Pectin, and Chitin)
Edited by WILLIS A. WOOD AND SCOTT T. KELLOGG

VOLUME 162. Immunochemical Techniques (Part L: Chemotaxis and Inflammation)
Edited by GIOVANNI DI SABATO

VOLUME 163. Immunochemical Techniques (Part M: Chemotaxis and Inflammation)
Edited by GIOVANNI DI SABATO

VOLUME 164. Ribosomes
Edited by HARRY F. NOLLER, JR., AND KIVIE MOLDAVE

VOLUME 165. Microbial Toxins: Tools for Enzymology
Edited by SIDNEY HARSHMAN

VOLUME 166. Branched-Chain Amino Acids (in preparation)
Edited by ROBERT HARRIS AND JOHN R. SOKATCH

VOLUME 167. Cyanobacteria (in preparation)
Edited by LESTER PACKER AND ALEXANDER N. GLAZER

VOLUME 168. Hormone Action (Part K: Neuroendocrine Peptides) (in preparation)
Edited by P. MICHAEL CONN

VOLUME 169. Platelets: Receptors, Adhesion, Secretion (Part A) (in preparation)
Edited by JACEK HAWIGER

Section I

Preparation of Toxins

A. Gram-Positive Cocci
Articles 1 through 10

B. Gram-Positive Bacilli
Articles 11 through 16

C. Gram-Negative Cocci
Articles 17 and 18

D. Gram-Negative Bacilli
Articles 19 through 23

E. *Vibrios*
Articles 24 through 28

F. Immunotoxins
Article 29

[1] Preparation and Purification of Staphylococcal Alpha Toxin

By SIDNEY HARSHMAN, NANCY SUGG, and PAUL CASSIDY

Staphylococcal alpha toxin is an extracellular protein produced by most strains of pathogenic *Staphylococcus aureus*. It is selectively hemolytic with a marked preference for rabbit red blood cells. It induces dermonecrosis, spastic muscle paralysis, and it is lethal for laboratory animals. These properties have been reviewed by Thelestam and Blomqvist.[1a] Various methods of preparation of alpha toxin have been compared in a review by Freer and Arbuthnot[1] and most recently a purification method using ion-exchange chromatograph has been described.[2a] The alpha toxin genome has been cloned and sequenced and the complete amino acid sequence deduced therefrom.[2] Earlier purification procedures suffered from incomplete separations of alpha toxin from delta toxin.[1] The definitive method of purification employs preparative gel electrophoresis, which yields a major band of alpha toxin that is stable with respect to both size and charge.[3]

The method of purification of alpha toxin described here is based on selective adsorption and elution from glass-pore beads.[4] The glass-pore bead method has the advantage of being both simple and rapid and giving high yields of alpha toxin which are free of other toxin products and identical in properties to the major form of the toxin obtained by the preparative gel electrophoresis procedure.[3]

Preparation of Medium

The medium of choice is the yeast extract dialysate medium described by Bernheimer and Schwartz.[5] However, with the glass-pore bead method described here, we have found it possible to use undialyzed yeast

[1a] M. Thelestam and L. Blomqvist, *Toxicon* **26,** 151 (1988).
[1] J. H. Freer and J. P. Arbuthnott, *Pharmacol. Ther.* **19,** 55 (1983).
[2a] I. Lind, G. Ahnert-Hilger, G. Fuchs, and M. Gratzl, *Analyt. Biochem.* **164,** 84 (1987).
[2] G. S. Gray and M. Kehoe, *Infect. Immun.* **46,** 615 (1984).
[3] H. R. Six and S. Harshman, *Biochemistry* **12,** 2672 (1973).
[4] P. Cassidy and S. Harshman, *Infect. Immun.* **13,** 982 (1976).
[5] A. W. Bernheimer and L. L. Schwartz, *J. Gen. Microbiol.* **30,** 455 (1963).

extract without disturbing the purification procedure. For 1 liter of medium:

Yeast extract, 25 g
Acid-hydrolyzed casein, 20 g
Glucose, 2.5 g
Thiamin, 0.13 mg
Nicotinic acid, 1.2 mg

The components are dissolved and the pH adjusted to 7.5 with 1 to 2 ml of 10 *N* NaOH. It is convenient to distribute the medium in 500-ml portions into 2-liter Erlenmeyer flasks before sterilizing by autoclaving. A starter flask is prepared at this time containing 50 ml of medium in a 500-ml Erlenmeyer flask.

Selection of Stock Organisms

The organism routinely used for the production of staphylococcal alpha toxin is the Wood 46 strain of *Staphylococcus aureus*. Variation in the Wood 46 strain exists. It is, therefore, prudent to check the stock culture for alpha toxin production. Organisms that, at confluence, produce titers of 500 to 1500 hemolytic units of toxin per milliliter of culture fluid are acceptable. Stocks of suitable strains of *Staphylococcus aureus* are prepared by growing the organisms to stationary phase and adding an equal volume of sterile glycerol. One-milliliter aliquots of the mixture, stored at −70°, remain stable and viable for several years.

Preparation of Alpha Toxin

The starter culture is inoculated with 0.5 ml of stock culture and incubated with aeration at 37° for 24 hr. It is convenient to use a model G-25 New Brunswick incubator shaker (New Brunswick Scientific Company, Edison, NJ) with the platform speed set at 200 rpm. The production flasks, containing 500 ml of medium, are inoculated with 5.0 ml of the 24-hr culture and incubation, with shaking, is continued for 18 hr. The cultures are chilled on ice and all reagents from this point forward are maintained at 0–10°. For small batches of 1 to 2 liters, it is convenient to remove the cells by centrifugation. For batches of 3 liters or more, it is more rapid to remove the cells by membrane filtration (Durapore filter type HVLP-filter code porosity 0.5 μm, Millipore Corporation, Bedford, MA).

Glass-Pore Column Chromatography

Reagents needed for chromatography on glass-pore beads are as follow:

1. Culture supernatant adjusted to pH 6.8 with NaOH
2. 0.01 *M* $KHPO_4$, pH 6.8
3. 1.0 *M* $KHPO_4$, pH 7.5
4. Glass-pore beads, CPG-10, 120/200 mesh, mean pore diameter 340 Å (Electronucleonics, Inc., Fairfield, NJ)

The glass-pore beads are washed with a 10% (v/v) solution of hypochlorous acid; commercial Clorox is suitable. This is followed by copious rinses with distilled water and, finally, the beads are equilibrated with 0.01 *M* $KHPO_4$ buffer at pH 6.8. Care should be taken in stirring the glass-pore beads so as to minimize any grinding action, which pulverizes the beads and reduces the flow rate through the glass-pore beads column. After several uses, one should decant any very fine glass powder during the distilled water wash steps. The glass-pore beads are then poured into a column and the culture supernatant, adjusted to pH 6.8, is applied. Flow rates of 10–20 ml/min are acceptable. A column of 3 × 9 cm will retain the toxin from 1 liter of culture fluid; a column of 5 × 20 cm is routinely used for 3-liter batches of culture fluid.

In order to monitor the operation of the glass-pore bead column, it is convenient to set up a rapid hemolytic assay (see chapter [30] in this volume for details of the standard hemolytic assay). The culture supernatant fluid is sampled at 5-, 10-, and 20-μl volumes which are each mixed with 0.5 ml of 1% suspension of rabbit red blood cells in phosphate-buffered saline, 1% bovine serum albumin [0.14 *M* NaCl, 0.01 *M* $NaPO_4$, pH 7.2, 0.1% (w/v) bovine serum albumin, PBS-A]. The mixtures are incubated at 37° for 10 min and observed for sparkling (complete) hemolysis. The smallest volume giving complete hemolysis is then selected to test each aliquot of effluent from the glass-pore bead column. During the application of the culture supernatant, effluent fractions of 100 ml each are collected and tested for hemolytic activity using the rapid assay described above. Saturation of the capacity of the glass-pore bead column to retain toxin is detected by the appearance of hemolytic activity in the effluent fractions.

The column is next washed with 0.01 *M* $KHPO_4$ buffer, pH 6.8. Effluent fractions of 50-ml volumes are collected and washing continued until no more tan color is eluted from the column. This is usually achieved by three column volumes. The fractions eluted with 0.01 *M* $KHPO_4$ buffer, pH 6.8, should be devoid of hemolytic activity.

The bound alpha toxin is eluted with 1.0 *M* $KHPO_4$ buffer, pH 7.5. Eluate fractions of 5- to 10-ml volumes are collected, commencing after the first column volume. Fractions positive in the rapid hemolytic assay are combined and the solution brought to 90% saturation by the addition of solid $(NH_4)_2SO_4$ (66.2 g/100 ml of solution). The solution is stored overnight in the cold room.

Sephadex G-75 Chromatography

Elution of the glass-pore bead column, with 1 *M* $KHPO_4$ buffer, pH 7.5, coelutes a small amount of tan pigment derived from the culture medium. This material is inert and is not detectable in Coomassie blue-stained polyacrylamide gels. However, it can be removed from the preparation of alpha toxin by chromatography on Sephadex G-75, where the tan pigment elutes after the bulk of the hemolytic activity. Chromatography on Sephadex G-75 also separates any small amount of the hexamer form of alpha toxin, which may have been present in culture supernatant and coeluted from the glass-pore bead column.

In harvesting the alpha toxin from the ammonium sulfate suspension it is convenient to first remove the inorganic salt formed by the mixing of $(NH_4)_2SO_4$ with the $KHPO_4$ buffer. The salt is packed by low-speed centrifugation (3000 rpm, 5 min, 5°) and the alpha toxin recovered from the fluid phase by high-speed centrifugation (10,000 rpm, 15 min, 5°). The tan pellet, containing the alpha toxin, is dissolved in 5–10 ml of 100 m*M* $NaHPO_4$ buffer, pH 7.2, and applied to a Sephadex G-75 column (2.6 × 25 cm) equilibrated with the same buffer. The column is developed with 100 m*M* phosphate buffer, pH 7.2, at a flow rate of 1.0 ml/min, and fractions of 5.0 ml are collected. The fractions containing hemolytic activity are pooled and the alpha toxin precipitated by the addition of solid $(NH_4)_2SO_4$ (66.2 g/100 ml of solution) as before. The fraction eluting at the void volume contains the hexamer form of the toxin, which is not hemolytic. It may be collected by $(NH_4)_2SO_4$ precipitation as described above for the alpha toxin monomer. The tan material, derived from the growth medium, elutes after the alpha toxin and these fractions are usually discarded.

After overnight storage in the cold room, the alpha toxin is recovered from the ammonium sulfate solution by centrifugation (10,000 rpm, 10 min, 5°) and the white precipitate obtained is dissolved in 2 to 5 ml of 100 m*M* $NaHPO_4$ buffer, pH 7.2. The final solution should contain 2–10 mg of alpha toxin per milliliter of buffer. At concentrations above 2 mg of protein per milliliter, the toxin can be stored at −70° for periods up to 2 years without detectable loss of hemolytic titer or significant accumulation of the hexamer form of the toxin.

Purity of the Toxin

Two criteria of purity of the alpha toxin preparation are routinely assessed: (1) the specific hemolytic activity and (2) the physical homogeneity as revealed by acrylamide gel electrophoresis. The hemolytic assay, using fresh washed rabbit red blood cells, is performed as described by Bernheimer and Schwartz[5]; see chapter [30] in this volume. Specific activities range from 20,000 to 30,000 hemolytic units per milligram of alpha toxin. Protein is determined by either the Lowry[6] or Bradford[7] methods using BSA as the standard or more conveniently, spectrophotometrically, by the absorbance at 280 nm using an extinction coefficient of 1.1 for 1 mg/ml/cm path length.

Polyacrylamide gel electrophoresis using the buffer system described by Laemmli[8] and 7.5% acrylamide permits the detection of both the monomer and hexamer forms of alpha toxin. It is important to avoid temperatures above 60° when preparing the samples for electrophoresis, if the hexamer form is to be preserved. The hexamer form of alpha toxin dissociates to the monomer form when heated at 100° in buffers containing sodium dodecyl sulfate.[9]

Concluding Remarks

The glass-pore bead procedure for preparing staphylococci alpha toxin described above is simple and rapid. It is completed in 3 days' time and recoveries of 10–20 mg of pure alpha toxin per liter of culture medium are usual. The only significant problem encountered is that some isolates of Wood 46 strain of *Staphylococcus aureus* produce excessive amounts of proteases which cleave the monomer into inactive half-molecules. These inactive half-molecules, which are coeluted with the alpha toxin molecule from the glass-pore bead column, are readily detected following acrylamide gel electrophoresis by the usual staining or by immunochemical methods. To some extent, they can be removed during the chromatography on Sephadex G-75. It is more effective, however, to reduce the growth time of the culture from 18 to 16 or 15 hr, thereby reducing the time of exposure of the alpha toxin to the proteases. The reduction in yield of alpha toxin, by the shortened culture time, is usually less than 30%.

[6] O. H. Lowry, N. J. Rosebrough, A. L. Farr, and R. J. Randall, *J. Biol. Chem.* **193,** 265 (1951).

[7] M. M. Bradford, *Analyt. Biochem.* **72,** 248 (1976).

[8] U. K. Laemmli, *Nature (London)* **227,** 680 (1970).

[9] P. Cassidy and S. Harshman, *Biochemistry* **18,** 232 (1979).

[2] Preparation and Purification of γ-Hemolysin of Staphylococci

By MICHEL PLOMMET

Overview

In 1938, Smith and Price[1] first described γ-hemolysin (Hγ), a toxin of *Staphylococcus aureus*. The existence of this toxin was subsequently confirmed by several authors (see reviews in Refs. 2–5).

The most important stages were as follows: (1) isolation of the *S. aureus* 5R strain,[6] which produces Hγ free from the other α-, β-, and δ-hemolysins; (2) separation of the two hemolysins which are active against human red blood cells (HRBC), i.e., Hγ and δ-hemolysin (Hδ), by heating (Hγ is thermolabile at 60°, 10 min) or by dissolving in ethanol (Hδ is soluble),[7] and (3) isolation of the two HγI and HγII components by purification on hydroxyapatite (HA). The components are inactive separately, but hemolytic and toxic in the mouse when combined.[8,9] This finding was rapidly confirmed[10] by determination of the molecular weights: HγI (M_r 29,000, p*I* 9.8), HγII (M_r 26,000, p*I* 9.9). It is somewhat surprising that this method of separating the I and II components on HA was not followed up by other authors[11,12] using the 5R strain, and this has led to some confusion. These questions have been discussed.[2–5,13] However, the separation of the I and II components is of obvious importance in the investigation of the mode of action of Hγ on cell membranes. It is also possible

[1] M. L. Smith and S. A. Price, *J. Pathol. Bacteriol.* **47,** 579 (1938).

[2] A. C. McCartney and J. P. Arbuthnott, *in* "Bacterial Toxins and Cell Membranes" (J. Jeljaszewicz and T. Wadström, eds.), p. 113. Academic Press, London, 1978.

[3] M. Rogolsky, *Microbiol. Rev.* **43,** 320 (1979).

[4] R. Möllby, *in* "Staphylococci and Staphylococcal Infections" (C. S. F. Easmon and C. Adlam, eds.), Vol. 2, p. 619. Academic Press, London, 1983.

[5] T. Wadström, *in* "Staphylococci and Staphylococcal Infections" (C. S. F. Easmon and C. Adlam, eds.), Vol. 2, p. 685. Academic Press, London, 1983.

[6] D. D. Smith, *J. Pathol. Bacteriol.* **84,** 359 (1962).

[7] A. W. Jackson and R. M. Little, *Can. J. Microbiol.* **4,** 453 (1958).

[8] F. Guyonnet, M. Plommet, and C. Bouillanne, *C. R. Hebd. Seances Acad. Sci., Ser. D.* **267,** 1180 (1968).

[9] F. Guyonnet, *Ann. Rech. Vet.* **1,** 155 (1970).

[10] A. G. Taylor and A. W. Bernheimer, *Infec. Immun.* **10,** 54 (1974).

[11] R. Möllby and T. Wadström, *Infect. Immun.* **3,** 633 (1971).

[12] H. B. Fackrell and G. M. Wiseman, *J. Gen. Microbiol.* **92,** 11 (1976).

[13] G. M. Wiseman, *Bacteriol. Rev.* **39,** 317 (1975).

to neutralize each component specifically within a mixture, using the appropriate antisera.[14]

The effects of Hγ on cells and membranes have been investigated by two teams[15-17] (see reviews in Refs. 2 and 4). γ-Hemolysin is produced by about half the strains isolated from infectious processes[4,8] and could be associated with strains isolated from Toxic Shock Syndrome.[17a] Similarly, patients suffering from staphylococcal infections present anti-Hγ antibodies,[18-21] which implies that the toxin is produced *in vivo* and that it may be involved in the virulence of the bacterium. γ-Hemolysin is toxic toward the mouse, with an LD_{50} ip of 7000 hemolytic units (HU).[9,22] It is also toxic toward the rabbit[22] and the guinea pig.[17] The two components are not toxic separately, but only when combined, following a synergistic law which is slightly different from that governing their hemolytic activity.[9] Death is accompanied by acute stasis in the liver, spleen, and kidneys, with images of cell lesions.[9,17]

After separating the HγI from HγII on HA, the authors developed a method for production in a liquid medium[23,24] and then described a simple preparation method using HA during successive stages of purification and immunization.[14] It is this method, known in the authors' laboratory as hydroxyapatite connection, which is described here. It was inspired by the work of Tisclius *et al.*[25] and that of Bernardi[26] on the purification of proteins on HA and also by the work of Raynaud and Relyveld[27] on the purification of the diphtheria toxin.

Crude hemolysin (HγI + HγII) has been used in pathology both to study its role in experimental infection (unpublished, nonsignificant per-

[14] F. Guyonnet and M. Plommet, *Ann. Inst. Pasteur, Paris* **118,** 19 (1970).
[15] M. Thelestam, R. Möllby, and T. Wadström, *Infect. Immun.* **8,** 938 (1973).
[16] B. Petrini and R. Möllby, *Infect. Immun.* **31,** 952 (1981).
[17] H. B. Fackrell and G. M. Wiseman, *J. Gen. Microbiol.* **92,** 1 (1976).
[17a] E. C. Carlson, *J. Infec. Dis.* **154,** 186 (1986).
[18] A. G. Taylor and M. Plommet, *J. Clin. Pathol.* **26,** 409 (1973).
[19] A. G. Taylor, J. Cook, W. J. Fincham, and F. J. C. Millard, *J. Clin. Pathol.* **28,** 284 (1975).
[20] E. Borderon, B. Glorion, P. Burdin, J. C. Borderon, and J. Castaing, *Med. Mal. Infect.* **7,** 14 (1977).
[21] P. Queneau, E. Lejeune, C. Veysseyre, M. Titoulet, M. P. Vallat, and M. Carraz, *Rev. Rhum.* **45,** 325 (1978).
[22] T. Wadström and R. Möllby, *Toxicon* **10,** 511 (1972).
[23] M. Plommet and C. Bouillanne, *Ann. Biol. Anim., Biochim., Biophys.* **6,** 529 (1966).
[24] G. Bézard and M. Plommet, *Ann. Rech. Vet.* **4,** 355 (1973).
[25] A. Tiselius, S. Hjerten, and O. Levin, *Arch. Biochem. Biophys.* **65,** 132 (1956).
[26] G. Bernardi, this series, Vol. 22 [29], p. 325 (1971).
[27] M. Raynaud and E. H. Relyveld, *C. R. Hebd. Seances Acad. Sci., Ser. D* **242,** 424 (1956).

sonal results and Ref. 28) and for diagnosis and prognosis by monitoring the antitoxin titer in patients suffering from deep staphylococcal infections.[18–21]

Titration of γ-Hemolysin

One hemolytic unit is the quantity of the hemolysin required to hemolyze 50% of a 1% suspension of HRBC in 0.7 ml, according to the original method of Jackson and Little[29] developed for titration of δ-hemolysin. This method, slightly modified, is also suitable for use for Hγ. It is accurate, and distinguishes between hemolysis due to Hδ from that due to Hγ from the slope of the dose–response curve.[14] The titers of Hγ and Hδ are approximately additive, and so it is possible to estimate the amount of each present in a mixture by titrating before and after heating (60° for 10 min), since this process eliminates all gamma activity.

A range of volumes, e.g., 0.05 to 0.5 ml, of hemolysin diluted in a pH 6.8 buffer (BSS: KH_2PO_4, 1.029 g; $Na_2HPO_4 \cdot 12H_2O$, 4.02 g; NaCl, 8 g; H_2O, 1000 ml) are placed in test tubes, which are then filled to 3 ml. The standard HRBC suspension (0.5 ml) is then added, and the tubes incubated for 37° for 20 min. Short-time centrifugation (3000 *g* for 2 min) separates the supernatant. One milliliter of the supernatant is diluted in 6 ml of a 0.1% solution of Na_2CO_3. The optical density (OD) determined at 541 nm is plotted on log/probit graph paper and the dose required to produce 50% hemolysis (OD = 0.25) read from the graph. This dose contains 5 HU.

The standard HRBC suspension is prepared from human blood collected, washed three times with Alsever's solution, and diluted in BSS to about 7% of cell pellet. The suspension is then adjusted so that 1 ml diluted in 49 ml of the Na_2CO_3 solution has an OD of 0.5 at 541 nm.

This standard suspension can be kept for 1 week. Depending on the blood used, variations on the order of 20% may be recorded. According to some authors, it is better to use type O Rh^+ red cells,[10] but the present author believes that these variations are individual rather than related to the blood group.

HγI and HγII are assayed in the same manner by the addition of an excess (50 HU) of the absent component.

The titer values reported have relative significance only, since they are dependent on the experimental conditions. It would be a simple matter to

[28] M. Kurek, K. Pryjma, S. Bartkowski, and P. B. Heczko, *Med. Microbiol. Immunol.* **163,** 61 (1977).

[29] A. W. Jackson and R. M. Little, *Can. J. Microbiol.* **4,** 435 (1958).

develop a seroneutralization assay method using two antisera—anti-HγI and anti-HγII—obtained from rabbits.[14] This would provide a stable reference against which the results from different laboratories could be compared.

Production of γ-Hemolysin

After initially adopting a laborious method of production on cellophane, which is also used by other authors,[12] the present authors gradually developed a method of production in liquid medium, initially in small volumes of CCY medium,[23] then in an enriched medium in a fermenter.[24] The production parameters have also been analyzed and presented recently by Möllby.[4] After many attempts and failures, the authors are of the opinion that success is dependent on three factors: (1) the choice of bacterial clone selected; (2) the culture medium, and, in particular, the yeast extract dialysate; and (3) aeration by a mixture of 80/20 O_2/CO_2.

1. Maintenance and Preparation of the S. aureus 5R Strain

Although other strains could have been used, the authors actually used the 5R strain originally isolated by Smith.[6] This strain, of human origin, naturally in rough phase (Fig. 1), produces γ-hemolysin almost exclusively, but presents a high level of spontaneous mutations involving the loss of Hγ production, or, at a low level, acquisition of the capacity of secreting α- and β-hemolysins. Since the genetic determination of

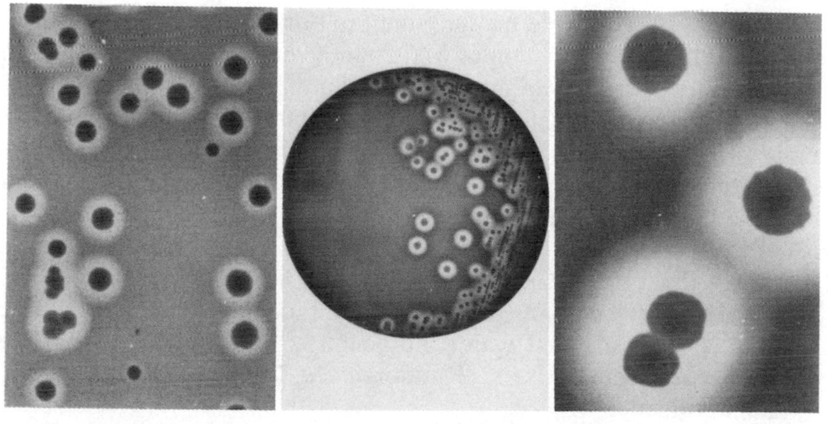

FIG. 1. *Staphylococcus aureus* strain 5R on human blood agar after a 1-day incubation on agarose (left) or a 2-day incubation on noble agar (incubation at 37° with CO_2, 10%). Note the rough phase indicated by the irregular edge of colonies. Rough colonies do not disaggregate on flooding.

γ-hemolysin secretion was unknown until recent molecular cloning of Hγ gene in *Escherichia coli*,[29a] it is essential to maintain the strain very carefully to ensure that negative mutants are not selected.

The strain, stored in lyophilized state, is first cultured in broth for 4 hr and then isolated on a nutrient 1% agar medium containing washed human red blood cells. The medium should contain 3 ml of HRBC pellet/100 ml. Since hemolysin is inhibited by agar, an uncharged or low-charge agar should be used,[10,11,14] such as agarose (Industrie Biologique Française) or noble agar (Difco). After incubating for 24 hr at 37° in an atmosphere of 10–20% CO_2 colonies develop which are rough in appearance and possess γ-type hemolytic properties (clear hemolysis with a blurred edge). These colonies are harvested and recultured four or five times on the same medium. It should be noted that type β mutant colonies (incomplete hemolysis, clear edge) do occur from time to time among colonies of the 5R strain. These mutants can be detected by isolation on sheep's blood or, more simply, on the human red cell medium by means of flooding, after incubation, with a 4% solution of formaldehyde. The strain has to be cloned until pure colonies are obtained. After a final passage, a tube containing 2.5% heart infusion broth medium (Difco) is liberally seeded and incubated overnight. This culture can be kept for 1 month before use as the initial culture for production, or it may be harvested and lyophilized. After storage for 1 month, the screening for hemolytic colonies on HRBC should be repeated.

2. *Culture Medium*

Commercially available media, such as brain heart infusion (Difco), can be used for the production of toxins on a very small scale. However, it is necessary to make up a special medium in order to obtain larger amounts. We initially used CCY medium,[30] but gradually were able to simplify and improve this medium by eliminating some ingredients and adding others. The EGY medium used is as follows: peptone Evans (Evans Medical, Ltd., Liverpool), 10 g (if this is not available, casamino acids, Difco, can be used, but gives less satisfactory results); sodium β-glycerophosphate, 20 g; 60% sodium lactate, 16.3 ml; $Na_2HPO_4 \cdot 12H_2O$, 6.25 g; KH_2PO_4, 0.4 g; 0.32% citric acid, 2 ml; depigmented yeast extract dialysate, 200 ml; distilled water, sufficient to make 1000 ml, pH 6.9; antifoam agent as required (e.g., Rhodorsil 426, Rhône Poulenc, Paris, 0.4 ml); and sterilized at 120°.

[29a] M. Mulvey and J. P. Arbuthnott, *in* "Bacterial Protein Toxins" (P. Falmagne, ed.), p. 287. G. Fisher, Stuttgart (1986).

[30] G. P. Gladstone and W. E. van Heynigen, *Br. J. Exp. Pathol.* **38,** 123 (1957).

Despite several attempts involving most of the known constituents of the dialysate (with the exception of the amino acids), it was not possible to replace this with pure factors. This led the authors to rid the dialysate of some of its components, notably the pigments, by adsorption on HA, in order to simplify the subsequent stages of the purification process.

The yeast extract dialysate is prepared as follows: yeast extract (Oxoid), 120 g; and water, 85 ml, are dialyzed in 80-cm 8/32 tubes (Visking). These tubes, which are kept under pressure, are then placed inside 7 × 100 cm glass tubes containing about 800 ml of distilled water. The tubes are then shaken slowly for 24 hr at 4° and the dialysate collected. Four liters of dialysate is taken and to it is added 500 g of hydroxyapatite (Bio-Rad, Richmond, CA), which has been washed three times with a 0.001 *M*, pH 6.8 potassium phosphate (KP) buffer and shaken for 30 min. The depigmented dialysate is collected by decantation and can then be stored in frozen form. The hydroxyapatite can be regenerated by washing in 1 *M*, pH 6.8 KP buffer, followed by boiling for 15 min in distilled water, and, finally, by three successive washings in 0.001 *M* buffer.

3. *Culture*

The culture can be carried out in a toxin vessel (Fig. 2, for a volume of 1 liter) or in a fermenter. The authors used a 20-liter fermenter for a volume of 10 liters.

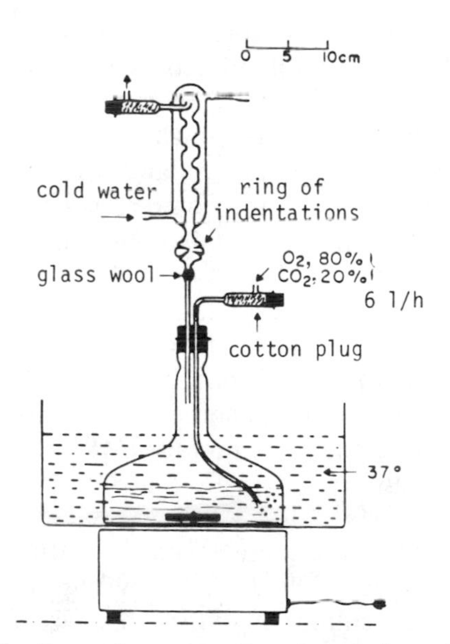

FIG. 2. Small vessel apparatus for γ-hemolysin production.

The strain harvested from the agar tube is put into a stage of exponential growth by two passages over EGY medium (50 ml) in a 250-ml conical flask at 37° shaken by vigorous rotation at 400 rpm. After 4 hr, 10% of the first culture is transferred into the second culture in another conical flask containing 100 ml. One to 2% of this second culture is, in turn, transferred to the fermenter in the medium at 37°, previously aerated with the O_2/CO_2, 80/20 mixture.

For cultures in a flask, aeration is provided by moderate speed rotation (about 700 rpm). The strain is harvested after culture for 16 hr when the titer is 200–300 HU.

For cultures in the fermenter, the pH is regulated to around 6.9–7.0 by the addition of 1 *M* NaOH. By the end of the culture, the pH has risen to 7.4–7.8. There are two phases of aeration: up to 5 hr the O_2/CO_2 mixture flow rate is 1 liter min^{-1}, with rotary shaking at 100 rpm. After 5 hr, the flow is increased to 2 liters and the rotation to 300 rpm. An appropriate amount of antifoam is added using a suitable apparatus. After 16 hr of culture, the bacterial concentration reaches $3–4 \times 10^{10}$ and the hemolytic titer 600 to 1500 HU/ml. The culture medium is then cooled to 4° and the harvest collected by decantation in the tank followed by centrifuging to remove the bacterial cells.

Harvesting the Crude γ-Hemolysin (Components I and II)

The hemolysin is separated from the culture medium by means of batch adsorption. The medium is diluted by the addition of 2 vol of distilled water and then warmed to 20°. One gram of HA washed in 0.001 *M*, pH 6.8 KP buffer is then added per 10,000 U and the mixture shaken for 30 min. After decantation, the supernatant is checked for any residual hemolysin by testing separately for components I and II. If necessary, more hydroxyapatite is added. After adsorption, the HA is placed in a 10-liter beaker and washed three times with 0.2 *M*, pH 6.8 buffer, shaking slowly. The HA is then transferred to a 45 × 8 cm column and the hemolysin (components I and II) obtained by means of elution by washing the column with 0.7 *M*, pH 6.8 KP buffer. The initial fractions, which do not contain toxin, are discarded. The hemolytic eluate (about 3 liters) is then concentrated over a Diaflo membrane (MW 10, Amicon Corp.), dialyzed against a 0.1 *M* buffer, and finally centrifuged to extract the precipitate formed during dialysis.

The yield from 10 liters of culture is $5–10 \times 10^6$ HU in 150 ml. This crude toxin contains about 1500 HU/mg Lowry protein. It does not contain substances from the culture medium, which are eliminated during the

0.001 *M* adsorption, nor dialysis pigments, nor most of the proteins excreted by the staphylococci, which are eliminated during the 0.2 *M* washing. This crude toxin can be stored in either the frozen or lyophilized state.

Separation of Components I and II

Components I and II are separated by adsorption and elution on HA. The column (45 × 8 cm) is loaded with crude hemolysin at 0.1 *M*, washed at 0.2 *M*, and then eluted at 0.3 *M* to obtain component I and at 0.5 *M* to obtain component II. The two components are then concentrated on Diaflo and dialyzed against distilled water. They can then be stored at −30° or lyophilized. The mixture of both components has an activity of 4000 HU/mg protein.

Concluding Remarks

Since this method for producing and separating components I and II was described in detail in 1973[24] and adopted by other authors,[10] the present authors have produced many batches of crude hemolysin or of components I and II with consistent results. Most of this hemolysin was intended for use in the assay of antistaphylolysins in the serum of human patients or else for tests of the antitoxic immunity in experimental staphylococcal infections. Consequently, there was no need to pursue greater purity, such as that described by Fackrell and Wiseman,[12] who reported activity of 10^5 HU/mg protein for total hemolysin (I and II not separated). The availability of a supernatant with a high titer (1500 HU) does, however, constitute an obvious advantage for subsequent purification, since in this study the authors obtained an activity of 4000 HU/mg protein, with an overall purification coefficient of 100, whereas Fackrell had to purify 2700 times to obtain an activity of 10^5 HU. It is obvious, however, that for the investigation of the mode of action of hemolysin on the cell membrane, there is a major advantage in being able to obtain the two components separately and to have available the corresponding, specific antibodies.

To the best of the authors' knowledge, there has been no notable progress in the production of γ-hemolysin, with the exception of the work of Möllby.[4] The authors' prolonged investigation of optimal production conditions[24] leads them to think that further progress is possible along the following lines: (1) improved regulation of aeration in the fermenter; (2) identification of the active factor(s) in yeast extract. This factor could be an amino acid, like, for instance, histidine in the case of α-hemolysin[31]; (3) the genetic investigation of the synthesis and excretion of the two components of hemolysin. Once the data are available, it should be possible for the production of hemolysin to progress beyond its present empirical state.

[31] A. B. Dalen, *J. Gen. Microbiol.* **74,** 53 (1973).

[3] Purification and Assay of Staphylococcal δ-Lysin

By T. HARRY BIRKBECK and JOHN H. FREER

The δ-lysin (also known as δ-hemolysin, delta toxin) of *Staphylococcus aureus* is a surface-active polypeptide[1,2] which causes membrane-permeability changes in a wide range of cells and cell organelles.[3] Two immunologically distinct forms have been described, one produced by *S. aureus* strains isolated from humans and a variant form produced by strains of *S. aureus* of canine origin.[4] Comparison of the amino acid sequences of the lysins from strain 186X (human origin) with that of strain CN7450 (canine origin) shows 9 changes in the 26 amino acids of the peptide[5,6] but conservation of charge distribution and both peptides have similar biological properties. A similar polypeptide produced by *Staphylococcus haemolyticus* is immunologically partially related to, and shares many properties with, the above lysins and may be a third form of δ-lysin.[7]

Strains Used for Production of δ-Lysin

For production of δ-lysin, strains of *S. aureus* deficient in production of α-lysin (alpha toxin) have generally been employed; we prefer strain

[1] G. Colacicco, M. K. Basu, A. R. Buckelew, and A. W. Bernheimer, *Biochim. Biophys. Acta* **465,** 378 (1977).
[2] M. Bhakoo, T. H. Birkbeck, and J. H. Freer, *Can. J. Biochem. Cell Biol.* **63,** 1 (1984).
[3] A. S. Kreger, K. S. Kim, F. Zaboretzky, and A. W. Bernheimer, *Infect. Immun.* **3,** 449 (1971).
[4] W. H. Turner, *Infect. Immun.* **20,** 485 (1978).
[5] J. E. Fitton, A. Dell, and W. V. Shaw, *FEBS Lett.* **115,** 209 (1980).
[6] J. E. Fitton, D. F. Hunt, J. Marasco, J. Shabanowitz, S. Winston, and A. Dell, *FEBS Lett.* **169,** 25 (1984).
[7] T. H. Birkbeck, R. J. Bisaillon, and R. Beaudet, *in* "Bacterial Protein Toxins" (P. Falmagne, J. E. Alouf, F. J. Fehrenbach, M. Thelestam, and J. Jeljascewicz, eds.). Fischer-Verlag, Stuttgart, Federal Republic of Germany, 1986.

NCTC 10345, a mutant of the Wood 46 strain which is similar to that used by Kreger *et al.*[3] but strain 186X[8] is also recommended.

Preparation of Media

Yeast extract diffusate medium[9] is prepared from yeast extract diffusate, 3200 ml (see below); Bacto casamino acids (Difco), 64 g; glucose, 3 g; nicotinic acid, 3.7 mg; aneurine hydrochloride, 0.4 mg. The pH value should be adjusted to 7.1 with 1 *N* NaOH before dispensing the medium in 500-ml amounts in 2-liter flanged Erlenmeyer flasks and autoclaving at 15 $lb/in.^2$ for 15 min. The yeast extract diffusate is prepared by dissolving 400 g yeast extract (Difco) in 1 liter distilled water by steaming and, after cooling, pouring the solution into two 50-cm lengths of 7-cm Visking dialysis tubing (Scientific Instrument Centre, London) previously soaked in 70% ethanol to minimize contamination. Each dialysis sac is immersed in 1600 ml distilled water in 5-liter beakers and stirred for 48 hr at 4°. The dialysis sacs and contents are discarded and the two batches of diffusate adjusted in volume to 3200 ml with distilled water.

Production of Crude Culture Supernatants Containing δ-Lysin

Staphylococcus aureus NCTC 10345 is inoculated onto a horse blood agar plate [10% (v/v) whole blood, washed three times in sterile 0.85% saline] and after incubation for 24 hr at 37°, bacteria from an isolated colony which shows a large hemolytic zone are inoculated into 50 ml yeast extract diffusate medium in a flanged 250-ml Erlenmeyer flask which is then incubated on a rotary shaker for 8 hr at 37°. Six 2-liter flasks of prewarmed medium are inoculated with 10 ml of the *S. aureus* culture and the flasks incubated for 18 hr at 37° on a rotary shaker operating at 135 rpm. Excessive foaming may be prevented by addition of a suitable antifoam solution (Silicolapse 5000, ICI, or Antifoam A, Sigma).

The culture purity should be confirmed by microscopy and the cultures pooled and centrifuged at 17,000 *g* to remove bacteria to yield approximately 3 liters of crude culture supernatant for purification of δ-lysin.

Fractionation of Culture Supernatants and Isolation of δ-Lysin

Purification with Hydroxylapatite.[3] Substantial quantities of δ-lysin may readily be prepared by the method of Kreger *et al.*,[3] which involves

[8] N. G. Heatley, *J. Gen. Microbiol.* **69,** 269 (1971).

[9] A. W. Bernheimer and L. L. Schwartz, *J. Gen. Microbiol.* **30,** 455 (1963).

batch adsorption of lysin to hydroxylapatite, elution of adsorbed medium components and contaminants by washing with phosphate buffers, pH 6.8, of up to 0.4 *M*, and elution of the δ-lysin with 1 *M* phosphate buffer, pH 7.4.

Preparation of Hydroxylapatite. We have not found commercial sources of hydroxylapatite to be suitable for purification of δ-lysin from some strains of *S. aureus* and prefer to prepare hydroxylapatite in the laboratory. Calcium phosphate exists in several crystalline forms, including brushite and hydroxylapatite, and Tiselius *et al.*[10] showed that brushite can be readily converted into hydroxylapatite by heating in phosphate buffers.

Brushite is prepared by allowing 2 liters of 0.5 *M* $CaCl_2$ and 2 liters of 0.5 *M* Na_2HPO_4 to drip at equal flow rates (6 ml/min) into a glass beaker with stirring. After settling of the brushite the supernatant is removed by decantation and the precipitate washed four times with distilled water by decantation before being resuspended in 4 liters of distilled water. After addition of 100 ml 10 *M* NaOH, the brushite suspension is boiled and stirred for 1 hr, allowed to cool, and the fine particles removed by decantation. The precipitate is washed a further four times in distilled water by decantation, resuspension in 3 liters of 0.01 *M* potassium phosphate buffer, pH 6.8, and heating to boiling. Following decantation of the supernatant, fresh 0.01 *M* phosphate buffer is added and the suspension is boiled for 5 min. The cycle of decantation and addition of fresh buffer is repeated and the suspension boiled and stirred for 15 min, first in 0.01 *M* buffer and then twice in 0.001 *M* phosphate buffer, to convert the brushite to hydroxylapatite. The suspension may be stored indefinitely at 4° in 0.001 *M* phosphate buffer and a small sample of the suspension should be dried to establish the dry weight of hydroxylapatite per milliliter of suspension.

Reaction with Hydroxylapatite. All procedures are done at 4° with centrifugation at 300 *g* for 10 min unless otherwise indicated. Hydroxylapatite is added to the crude culture supernatant (20 g/liter) and the mixture stirred for 3 hr and allowed to settle overnight. The supernatant is discarded following decantation and the hydroxylapatite is washed five times by stirring with 2 liters of 0.01 *M* phosphate buffer, pH 6.8, with centrifugation at each stage to sediment the hydroxylapatite. To remove weakly bound proteins the hydroxylapatite is stirred six times with 500 ml 0.4 *M* phosphate buffer, pH 6.8, for 30 min each time with centrifugation at each stage to sediment the hydroxylapatite. The washings are discarded and the delta-lysin is eluted in six cycles of stirring the hydroxylapatite with

[10] A. Tiselius, S. Hjerten, and O. Levin, *Arch. Biochem. Biophys.* **65,** 132 (1956).

100 ml 1 *M* potassium phosphate buffer, pH 7.4, each time with centrifugation to sediment the hydroxylapatite. After pooling the 1 *M* phosphate buffer washes containing δ-lysin, traces of hydroxylapatite are removed by centrifugation at 12,000 *g* for 10 min at 4°. If lyophilized lysin is required, salts are removed by exhaustive dialysis against water and any "insoluble" δ-lysin is removed by centrifugation at 10,000 *g* for 20 min before lyophilization. If a higher yield of lysin is desired then only traces of insoluble δ-lysin are formed if the eluate from hydroxylapatite is dialyzed exhaustively against Dulbecco A PBS (in grams per liter: NaCl, 8; KCl, 0.2; Na_2HPO_4, 1.15; KH_2PO_4, 0.2; pH 7.3) and the solution is concentrated 20× in an Amicon ultrafiltration cell (PM10 membrane) before clarification by centrifugation at 50,000 *g* for 30 min and storage at −20°.

Lysin prepared by the hydroxylapatite method contains traces of material derived from the yeast extract dialysate medium, which react in the *Limulus* amoebocyte lysate (LAL) assay for endotoxins[11] and these may be extracted with dilute trichloroacetic acid (TCA) to yield active lysin, which does not cause LAL clotting at concentrations of up to 1 mg/ml. Equal volumes of ice-cold lysin solution and 1.0 *M* TCA are mixed and the resulting precipitate collected by centrifugation at 4° and washed once with 1 vol of ice-cold 0.5 *M* trichloroacetic acid. The precipitate is dissolved in distilled water, dialyzed thoroughly against distilled water, and lyophilized to yield a white powder.

Purification by Solvent Transfer (Heatley)[3]

δ-Lysin is an amphipathic molecule which dissolves both in aqueous buffers and in chloroform/methanol (2:1, v/v); this was exploited by Heatley, who devised a simple and efficient purification scheme based on the transfer of the lysin between the two phases formed with chloroform/methanol/water, the lysin partitioning predominantly in the upper phase at acid pH values and in the lower phase at neutral pH values.

Solvents. Analytical grade solvents should be used and approximately 400 ml chloroform, 175 ml methanol, and 625 ml acetone are required to process 3 liters of crude culture supernatant. In addition, 50 ml 0.1 *M* and 120 ml 0.01 *M* phosphate buffer, pH 6.8, are used.

Procedure. Crude culture supernatant (3 liters) is heated to 60° with stirring for 1 hr, cooled to 20° and the precipitate removed by centrifugation at 16,000 *g* for 45 min at 20°. The pH value of the supernatant is adjusted to 5.3 by addition of 5 *M* H_3PO_4, chloroform is added to a final concentration of 1.5% (v/v), and the solution stirred overnight at room

[11] T. H. Birkbeck, *Toxicon* **20,** 253 (1982).

temperature. The lysin is precipitated at this stage and the precipitate is collected by centrifugation at 16,000 *g* for 45 min at 4° and washed once by resuspension in chloroform-saturated 0.01 *M* phosphate buffer. After centrifugation, the precipitate is dissolved in 270 ml chloroform/methanol (2 : 1, v/v) and held at 4° overnight. Any precipitate is removed by centrifugation and the clear supernatant is filtered (Whatman No. 1 filter paper) into a separating funnel; 50 ml 0.1 *M* phosphate buffer, pH 6.8, is added and the solutions vigorously mixed. After separation of the phases, the lower phase is removed and the upper phase extracted a second time with 170 ml chloroform/methanol/water (86 : 14 : 1, v/v). The lower phase is removed, combined in a 2-liter separating funnel with the lower phase from the first extraction, and extracted five times with 20 ml acidified chloroform/methanol/water (3 : 48 : 47, v/v) containing 100, 300, 200, 200 and 200 μl of 1 *M* H_3PO_4 in successive washes. The upper phases are retained each time during transfer to a second 2-liter separating funnel and lysin is precipitated from each upper phase by addition of 125 ml of acetone. The white precipitates obtained are recovered by centrifugation, dissolved in distilled water, and lyophilized to yield a white powder.

Analysis of Purified δ-Lysin

From 3 liters of crude culture supernatant, approximately 1.0 g of purified lysin should be obtained, with a specific activity of 160–320 HU_{50}/mg against human or horse erythrocytes and up to 750 HU_{50}/mg against cod erythrocytes.

The ultraviolet absorption spectrum ($A_{280}^{1\%} = 11.4$) shows a minimum at 250 nm and a maximum at 273 nm with a secondary peak at 292 nm which is characteristic of tryptophan, the amino acid predominating in the ultraviolet spectrum of NCTC 10345 and 186X δ-lysin.

On analysis by sodium dodecyl sulfate–polyacrylamide gel electrophoresis (SDS–PAGE; Laemmli[12]) δ-lysin migrates at, or in advance of, the dye front in 11% acrylamide gels but in 15% gels yields a diffuse band with an apparent M_r 6000, approximately twice the value found from the amino acid sequence. The solubility of δ-lysin in organic solvents leads to poor fixation in SDS–PAGE gels if methanol/acetic acid mixtures are used and a preliminary fixation step for 30 min in 10% glutaraldehyde is required for adequate fixation. Lysin prepared by the solvent transfer method shows a single band on SDS–PAGE but heavy loading of lysin purified using hydroxylapatite shows traces of a contaminating protein of M_r 32,000. The identity of this protein is unknown.

[12] U. K. Laemmli, *Nature (London)* **227,** 680 (1970).

Assay of Staphylococcal δ-Lysin

Unlike the other lysins of *S. aureus,* the δ-lysin is of relatively low specific lytic activity yet is active against a wide range of cells, subcellular organelles, and artificial membranes.[13] Its membrane-damaging activity is conveniently assayed against erythrocytes by measuring the release of hemoglobin (hemolysis) as a result of membrane damage. Although erythrocytes from a wide range of species are sensitive to the action of δ-lysin, those most sensitive are from marine fish, particularly the mackerel (*Scomber scombrus*). Erythrocytes from gadoid fish such as the Atlantic cod (*Gadus morhua*) are also sensitive indicators and are more easily obtained. However, for most laboratories, human or horse erythrocytes provide a convenient and satisfactory assay system for δ-lysin.

Details are given here for the titration of δ-lysin against human or horse erythrocytes; additional information on the collection, storage, and use of marine fish erythrocytes for the titration of δ-lysin has been published.[14,15]

Preparation of Erythrocyte Suspensions

Source of Human Blood. (**Warning:** The possibility that human blood of unknown origin may carry infectious agents such as hepatitis or AIDS viruses should be noted and appropriate precautions taken.)

Citrated or heparinized blood may be used. It can be obtained fresh or as outdated transfusion stock. For one or two titrations, sufficient fresh blood can be obtained by swabbing the thumb with 70% ethanol and allowing it to dry before piercing the skin with a sterile lancet. About 0.5 ml of blood can be collected via a glass capillary and transferred directly into a microcentrifuge tube containing 0.5 ml sterile 3.8% (w/v) sodium citrate.

Source of Horse Blood. Defibrinated horse blood can be obtained from commercial suppliers. It has a shelf life of 1–2 weeks at 4°.

Erythrocyte Washing Procedure. Erythrocytes are sedimented by centrifugation at 700 *g* for 10 min and washed three times by resuspension in 10 vol of phosphate-buffered saline (PBS) (see formulation given previously) and centrifugation at 700 *g* for 10 min. Small volumes (< 1 ml) may be washed more rapidly by centrifugation in a microcentrifuge capable of running at relatively low speed. If there is visible lysis of the erythrocytes

[13] J. H. Freer and J. P. Arbuthnott, *in* "Pharmacology of Bacterial Toxins" (F. Dorner and J. Drews, eds.), pp. 581–633. Pergamon, Oxford, England, 1986.

[14] L.-P. Chao and T. H. Birkbeck, *J. Fish Biol.* **13,** 483 (1978).

[15] L.-P. Chao and T. H. Birkbeck, *J. Med. Microbiol.* **11,** 303 (1978).

during washing (hemoglobin in the washing fluid) then the blood should be discarded and fresh samples obtained. Thorough washing of the erythrocyte suspension is essential, since residual serum lipoproteins may partially neutralize the lytic activity of the δ-lysin.

Preparation of Standardized Erythrocyte Suspensions. Standardized suspensions of erythrocytes are prepared by adding 0.9 ml of packed washed erythrocytes to 100 ml of PBS and adjusting the final concentration of erythrocytes such that a 5.0 ml sample of the suspension, after complete lysis by the addition of ~1.0 mg saponin, gives an absorbance value at 545 nm of 0.8 ± 0.02 in a spectrophotometer with cuvettes of 5-mm light path. Under these conditions the cell concentration is approximately 0.7% (v/v).

Doubling Dilution Hemolytic Titrations. Serial doubling dilutions of δ-lysin are made in 50-μl volumes of PBS in a microtiter tray using 50-μl diluters. Controls without toxin are included. To each well of the tray is added 50 μl erythrocyte suspension before the contents of the wells are mixed by tapping the sides of the tray. The tray is then covered and incubated at 37° for 30 min. The degree of lysis is then assessed visually and the dilution of lysin which causes a release of 50% of the hemoglobin is termed one hemolytic unit.

Calculation of Specific Hemolytic Activity. One hemolytic unit (HU_{50}) is defined as the amount of lysin which causes the release of 50% of the hemoglobin from the cells in 1 ml of a 0.35% suspension of erythrocytes. Thus, for a 1 mg/ml solution of δ-lysin giving an end point of 1/320 in the above assay, the potency would be 320 HU_{50}/mg.

[4] Purification and Crystallization of Staphylococcal Leukocidin

By MASATOSHI NODA and IWAO KATO

Staphylococcal leukocidin, known to be important in the pathogenicity of certain staphylococcal diseases,[1,2] consists of two protein components (S and F) that act synergistically to induce cytotoxic changes in

[1] A. M. Woodin, *Ann. N.Y. Acad. Sci.* **128,** 152 (1965).

[2] K. Wenk and H. Blobel, *Zentralbl. Bakteriol., Parasitenk., Infektionskr. Hyg., Abt. 1: Orig., Reihe A* **213,** 479 (1970).

human and rabbit polymorphonuclear leukocytes.[3–5] These components when tested individually are not cytotoxic. The S and F components are preferentially bound and inactivated by G_{M1} ganglioside and phosphatidylcholine, respectively.[6] Specific binding of the S component to G_{M1} ganglioside on the cell membrane of leukocidin-sensitive cells induces activation of phospholipid methyltransferases[7] and phospholipase A_2[8] and an increase in the number of F component-binding sites.[9] This increased number of binding sites for F component is dependent on the specific enzymatic action of S component and is correlated with increased phospholipid methylation and phospholipase A_2 activation. F component, bound to leukocidin-sensitive cells in the presence of S component, rapidly caused degradation of the cell membrane resulting in cell lysis. This degradation of cell membrane is associated with stimulation of ouabain-insensitive NA^+,K^+-ATPase activity and inhibition of cyclic AMP-dependent protein kinase.[8]

Purification of Leukocidin

The V8 strain of *Staphylococcus aureus* (American Type Culture Collection No. 27733) is believed to produce more leukocidin than most staphylococcal strains. The cocci are grown at 37° for 22 hr on an enriched medium (Table I)[10] in a reciprocating shaker (120 cpm). The culture is then centrifuged at 11,000 *g* for 20 min at 4°, and the clear supernatant fluid (~5000 ml, step I) is purified at 4° by the following procedure (Scheme 1). $ZnCl_2$ (3.7 *M*) is added dropwise to the supernatant (pH 6.5) until a final concentration of 75 m*M* is reached. After 30 min at 4°, a precipitate is formed which is collected by centrifugation for 15 min at 8000 *g*. The pellet is dissolved gradually in 0.4 *M* sodium phosphate buffer (pH 6.5). To remove metal ions, the solution is dialyzed against 0.05 *M* sodium acetate buffer (pH 5.2) containing 0.2 *M* NaCl (buffer A). Solid $(NH_4)_2SO_4$ is then added until saturation is reached, and the solution allowed to stand overnight. The precipitate is collected by centrifugation at 15,000 *g* for 20 min, dissolved in a small volume of buffer A, and then

[3] A. M. Woodin, *Biochem. J.* **75,** 158 (1960).
[4] H. Soboll, A. Ito, W. Schaeg, and H. Blobel, *Zentralbl. Bakteriol., Parasitenk., Infektionskr. Hyg., Abt. 1: Orig., Reihe A* **224,** 184 (1973).
[5] M. Noda, T. Hirayama, I. Kato, and F. Matsuda, *Biochim. Biophys. Acta* **633,** 33 (1980).
[6] M. Noda, I. Kato, T. Hirayama, and F. Matsuda, *Infect. Immun.* **29,** 678 (1980).
[7] M. Noda, T. Hirayama, F. Matsuda, and I. Kato, *Infect. Immun.* **50,** 142 (1985).
[8] M. Noda, I. Kato, T. Hirayama, and F. Matsuda, *Infect. Immun.* **35,** 38 (1982).
[9] M. Noda, I. Kato, F. Matsuda, and T. Hirayama, *Infect. Immun.* **34,** 362 (1981).
[10] G. P. Gladstone and W. E. van Heyningen, *Br. J. Exp. Pathol.* **38,** 123 (1957).

TABLE I
COMPOSITION OF GROWTH MEDIUM (pH 7.4)

Component	Concentration
Yeast extract	25 g
Casamino acid	20 g
Sodium glycerophosphate	20 g
$Na_2HPO_4 \cdot 12H_2O$	6.25 g
KH_2PO_4	400 mg
$MgSO_4 \cdot 7H_2O$	20 mg
$MnSO_4 \cdot 4H_2O$	10 mg
Sodium lactate, 50%	19.8 ml
$FeSO_4 \cdot 7H_2O$, 0.32% (w/v), plus citric acid, 0.32% (w/v)	2 ml
Distilled water to total	1 liter

dialyzed against buffer A. The dialysate (315×10^5 U, step II) is applied to a column (5×90 cm) of carboxymethyl-Sephadex C-50 equilibrated with buffer A. The fractions containing the highest leukocidin activity are eluted with 0.05 *M* sodium acetate buffer (pH 5.2) containing 1.2 *M* NaCl. The active fractions are pooled (254×10^5 U, step III), saturated with $(NH_4)_2SO_4$ by addition of the salt, and allowed to stand overnight. The precipitate is collected by centrifugation, dissolved in buffer A, and then dialyzed against buffer A. Further purification is achieved by applying the leukocidin preparation to a column (2.5×85 cm) of carboxymethyl-Sephadex C-50 equilibrated with buffer A and elution with a linear gradient from 0.2 to 1.2 *M* NaCl in 0.05 *M* sodium acetate buffer (pH 5.2). The fractions with F (fast) component activity, eluted at about 0.45 *M* NaCl (230×10^5 U, step IV), and S component activity, eluted at about 0.9 *M* NaCl (235×10^5 U, step IV) (Fig. 1), are pooled and concentrated by dialyzing against saturated $(NH_4)_2SO_4$. Each concentrated solution is applied to a column (2.5×85 cm) of Sephadex G-100 equilibrated with buffer A. The fractions with F or S component activity (F component, 217 $\times 10^5$ U or S component, 231×10^5 U, step V) are concentrated by dialyzing against saturated $(NH_4)_2SO_4$, and then dialyzed against 0.05 *M* veronal buffer (pH 8.6). After dialysis each component is purified by zone electrophoresis on starch [12 mA for 20 hr (vessel, $1.5 \times 3 \times 40$ cm)][11] using 0.05 *M* veronal buffer (pH 8.6). Both the F and S (Fig. 2) fractions show a single protein peak associated with leukocidin activity. Fractions with highest activity (F component, 210×10^5 U or S component, 209×10^5 U, step VI) are dialyzed against buffer A and pooled for crystalliza-

[11] I. H. Fine and L. A. Costello, this series, Vol. 6, p. 958.

Step

I. Culture filtrate (Step I, Table II)

II. $ZnCl_2$ precipitation at 4°C, pH 6.5

centrifugation (8000 × g, 15 min)

ppt | sup

dissolved in 0.4 M sodium phosphate buffer (pH 6.5)

centrifugation (10 000 × g, 20 min)

ppt | sup

dialyzed against 0.05 M acetate buffer (pH 5.2) containing 0.2 M NaCl [buffer A]

concentrated by saturated $(NH_4)_2SO_4$ solution (pH 7.0) [SAS]

centrifugation (15 000 × g, 20 min)

ppt | sup

dialyzed against buffer A (Step II, Table II)

III. Carboxymethyl-Sephadex C-50 column chromatography

eluted stepwise

Effluent with buffer A

Effluent with 0.05 M acetate buffer (pH 5.2) containing 1.2 M NaCl (Step III, Table II)

Solution concentrated by SAS

IV. Carboxymethyl-Sephadex C-50 column chromatography

eluted with linear gradient (0.2—1.2 M NaCl in 0.05 M acetate buffer, pH 5.2)

F fraction (eluted with about 0.45 M NaCl concentration) (Step IV, Table II) | S fraction (eluted with about 0.9 M NaCl concentration) (Step IV, Table II)

concentrated by dialyzing against SAS

V. Sephadex G-100 gel filtration

eluted buffer A (Step V, Table II)

Solution concentrated by dialyzing against SAS

dialyzed against 0.05 M veronal buffer (pH 8.6)

VI. Starch zone-electrophoresis in 0.05 M veronal buffer (pH 8.6)

Purified F component (Step VI, Table II) | Purified S component (Step VI, Table II)

SCHEME 1. Purification procedures for leukocidin.

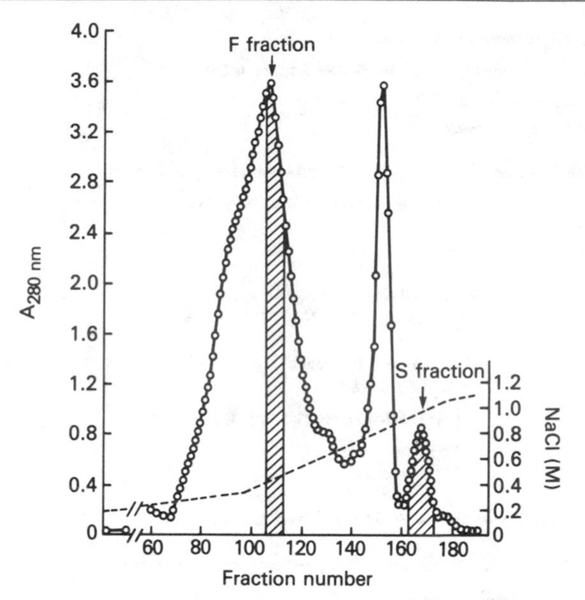

FIG. 1. The gradient elution pattern of leukocidin from CM-Sephadex C-50. CM-Sephadex C-50 column (2.5 × 85 cm) was equilibrated with 0.05 *M* sodium acetate buffer, pH 5.2, containing 0.2 *M* NaCl. The F and S fractions of leukocidin were eluted by 0.05 *M* sodium acetate buffer, pH 5.2, containing a linear gradient of NaCl concentration (0.2–1.2 *M*). Leukocidin activity of the F and S fractions was determined by the microscopic slide adhesion method.[10] Aliquots of 3 ml were measured for $A_{280\ nm}$ (○—○) and NaCl concentration (– – –) from each fraction (8 ml). The individual diagonally hatched areas of the F fraction (fraction Nos. 107–113) and S fraction (fraction Nos. 163–178) were applied to the gel filtration in step V.

tion. The recovery and degree of purification at each step are shown in Table II. The yields of the highly purified F and S components of leukocidin are about 2 mg/liter of culture filtrate.

Crystallization of the F and S Components of Leukocidin

The F and S components of leukocidin are crystallized by dialyzing against a saturated $(NH_4)_2SO_4$ solution, pH 7.0, at 4° for 16 hr. Slight opalescence appears in 6 hr, along with white precipitate on the bottom of the cellulose tube. The crystalline precipitate formed is collected by gentle centrifugation at 4° and dissolved with the same volume of chilled 0.05 *M* sodium acetate buffer (pH 5.2) containing 0.2 *M* NaCl. The crystals are highly soluble in the buffer. Recrystallization of the two components is repeated twice more by dialyzing the component solution against a large volume of a 95% saturated $(NH_4)_2SO_4$ solution at 4°. Microscopic examinations of the white precipitates of the purified F and S components

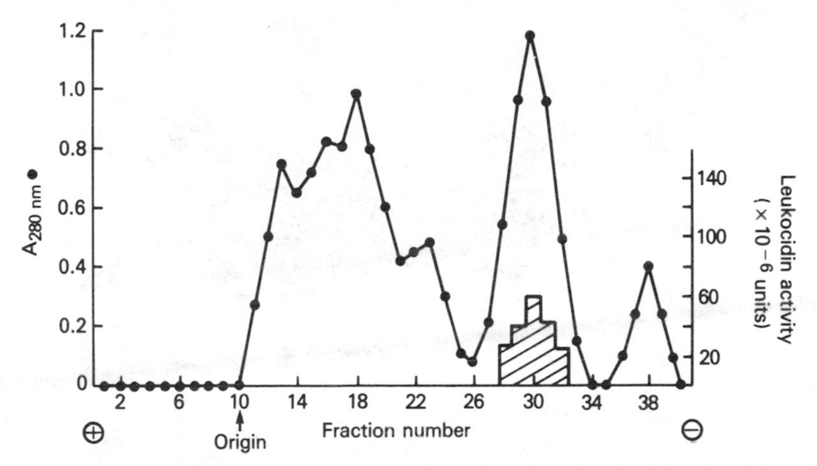

FIG. 2. Starch zone electrophoresis of the S fraction. Sample, the S fraction (42 mg) in step V; buffer, 0.05 M veronal buffer, pH 8.6; vessel, 1.5 × 3 × 40 cm; current, 12 mA for 20 hr at 4°. The diagonally hatched area depicts leukocidin activity. Each fraction (2.5 ml) was measured for $A_{280\ nm}$ (●—●).

TABLE II
PREPARATION OF LEUKOCIDIN

Step in purification	Volume (ml)	Total protein (mg)	Total activity (1×10^{-5}U)	Specific activity (1×10^{-5} U/mg protein)	Recovery of activity (%)	Recovery of protein (%)
I. Culture filtrate	5,000	13,840	346	0.025	100	100
II. $ZnCl_2$ precipitation	300	6,330	315	0.050	91.1	45.7
III. 1.2 M NaCl effluent	1,500	1,420	254	0.180	73.4	10.3
IV. Gradient elution						
F component	50	138	230[a]	1.67	66.5	1.0
S component	80	47.1	235[a]	4.99	67.9	0.34
V. Gel filtration						
F component	45	39.9	217[a]	5.44	62.7	0.29
S component	40	42.0	231[a]	5.50	66.8	0.30
VI. Zone electrophoresis						
F component	15	12.0	210[a]	17.50	60.7	0.09
S component	10	11.0	209[a]	19.00	60.4	0.08

[a] For the determination of the activity of each fraction, the F (or S) component (6 ng) was added into the serially diluted S (or F) fraction solution in 20 μl of phosphate-buffered saline (pH 7.2) containing 0.5% gelatin, since each fraction itself has no or little leukocidin activity.

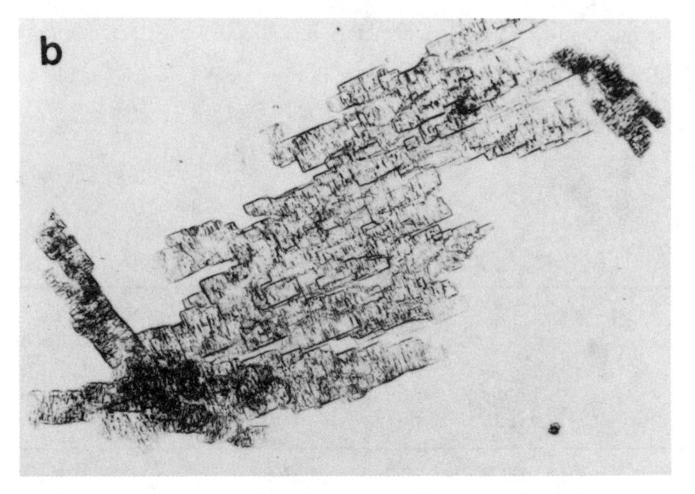

FIG. 3. Crystals of S component (a) and F component (b) after recrystallization three times at 4°. ×200.

of leukocidin in the dialysis bag reveals crystals in the form of square plates and very fine needles, respectively (Fig. 3).

Determination of Leukocidin Activity

Polymorphonuclear leukocytes are prepared from rabbit peripheral blood with isokinetic gradient of Ficoll[12] and washed with Hanks' solu-

[12] T. G. Pretlow II and D. E. Luberoff, *Immunology* **24,** 85 (1973).

tion. The purity of polymophonuclear leukocyte suspensions averages 98% as judged by examination of Giemsa-stained smears. Viability averages 99% as assayed by nigrosine dye exclusion. Aliquots of 10^6 cells (10 μl) are placed on glass slides. Since the F and S fractions individually have little or no leukocidin activity, to quantify the components of leukocidin, 6 ng of the S (or F) component of leukocidin is added to serially diluted fractions containing the other component. To standardize leukocidin, the microscopic slide adhesion method[10] is used. Serial dilutions (10 μl) of each component of leukocidin in 0.1 *M* phosphate-buffered saline (pH 7.2) containing 0.5% gelatin are incubated at 37° for 10 min with the other component (6 ng) on a glass slide (total volume, 20 μl) in a moist chamber. After incubation, morphological changes in a slide field (about 1000 cells) are observed by phase-contrast microscopy. The end point is the smallest amount of leukocidin causing about 100% morphological changes in a standard polymorphonuclear leukocyte suspension (10^6 cells), the number of units in a leukocidin preparation being numerically equal to the dilution at the end point. When 500 pg of purified F component and various amounts (100–700 pg) of purified S component are incubated with suspensions of polymorphonuclear leukocytes (10^6 cells) on the glass slide (total volume, 20 μl), optimal destruction of all polymorphonuclear leukocytes is obtained at 500 pg of S component (Fig. 4). A linear relationship between polymorphonuclear leukocyte destruction (percentage) and protein concentration is obtained in the range of 300–500 pg. Results similar to those of Fig. 4 are obtained when 500 pg of S component is incubated with various amounts (100–700 pg) of F component and polymorphonuclear leukocytes (10^6 cells). Optimal destruction of all polymorphonuclear leukocytes is obtained at 500 pg of F component. One unit of each component of leukocidin is obtained at 500 pg. When a suspension of rabbit polymorphonuclear leukocytes (10^6 cells) is

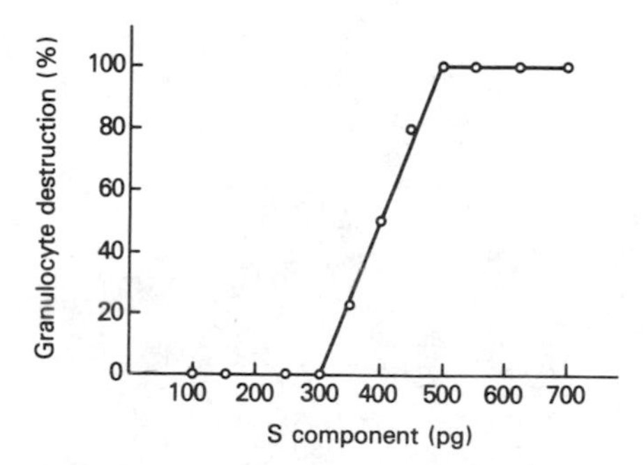

FIG. 4. Tritration of the optimum leukocidic dose of the F and S components. The mixtures (20 μl) of 500 pg F component and various amounts (100–700 pg) of S component were incubated with the suspensions of granulocytes (10^6 cells) for 10 min at 37°.

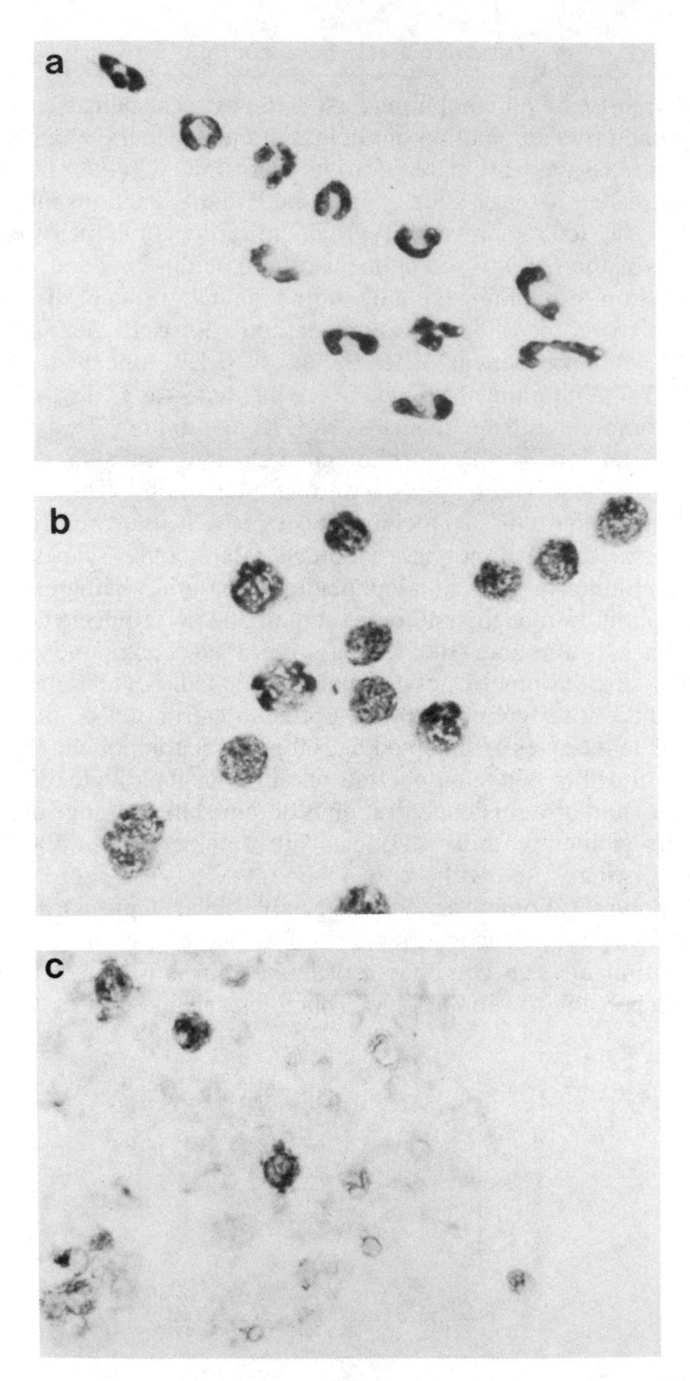

FIG. 5. Morphological changes in rabbit polymorphonuclear leukocytes caused by staphylococcal leukocidin. (a) Before exposure to leukocidin; (b) after exposure to leukocidin for 2 min; (c) after exposure to leukocidin for 10 min. Cells were stained with Giemsa.

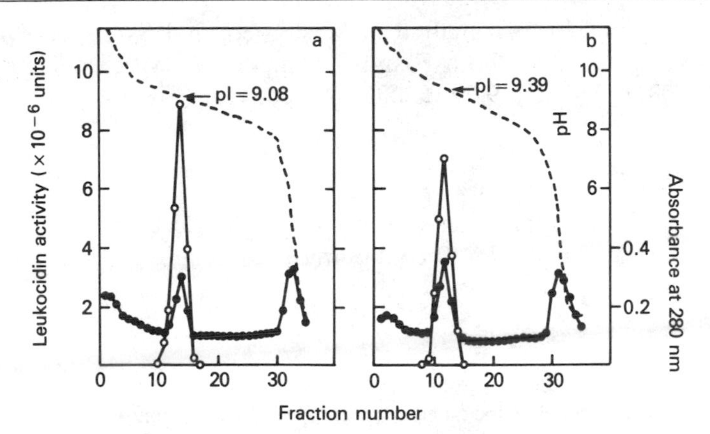

FIG. 6. Isolectric focusing of the crystallized F component (a) and S component (b) of leukocidin. ●—●, $A_{280\ nm}$; ○—○, leukocidin activity; – – – –, pH.

incubated with 500 pg of both components of leukocidin, it is revealed by light microscopy that, after 2 min of the incubation, the polymorphonuclear leukocytes become round and somewhat swollen; the lobulated nuclei eventually become spherical. The terminal event in leukocidin action is cytoplasmic degranulation, rupture of nuclei, and complete cell lysis. These effects are fully developed after 10 min of incubation (Fig. 5).

Leukocidin activity is determined also by assaying for ^{86}Rb release from ^{86}Rb-labeled polymorphonuclear leukocytes. Similar data are obtained by examining morphological changes of polymorphonuclear leukocytes. The study of leukocidin effects is confined to the first 10 min of intoxication; a maximal effect results from treating 10^6 polymorphonuclear leukocytes per 20 μl with 500 pg of both components of leukocidin.

Criteria of Purity

The F and S components crystallize in the form of plates and needles, respectively. The amino-terminal residues of both components are alanine. The crystallized components migrate as single bands on the SDS–polyacrylamide gels, which are analyzed by the methods of Shapiro *et al.*[13] and Dunker and Rueckert.[14] Molecular weights of crystallized F and S components of leukocidin determined by the method of Weber and Osborn[15] from relative mobilities of marker proteins are 32,000 and 31,000, respectively. To further establish purity, the isoelectric point (p*I*)

[13] A. K. Shapiro, E. Vinnela, and J. V. Maizel, Jr., *Biochem. Biophys. Res. Commun.* **28,** 815 (1967).

[14] A. K. Dunker and R. R. Rueckert, *J. Biol. Chem.* **244,** 5072 (1969).

[15] K. Weber and M. Osborn, *J. Biol. Chem.* **244,** 4406 (1969).

of each is determined by the method of Vesterberg and Svensson.[16] A single protein is evident for both F and S components with isoelectric points of p*I* 9.08 ± 0.05 and 9.39 ± 0.05, respectively (Fig. 6).

Storage of Leukocidin

The purified F and S components of leukocidin stored at −80° in 0.1 *M* phosphate-buffered saline (pH 7.2) containing 0.5% gelatin are stable for at least 3 months.

Acknowledgment

We thank Joel Moss for reading the manuscript and for suggesting improvements.

[16] O. Vesterberg and H. Svensson, *Acta Chem. Scand.* **20,** 820 (1966).

[5] Purification of Epidermolytic Toxin of *Staphylococcus aureus*

By JOYCE C. S. DE AZAVEDO, CHRISTOPHER J. BAILEY, and JOHN P. ARBUTHNOTT

Two immunologically distinct serotypes of epidermolytic toxin (ET) have been identified in a number of laboratories and are termed ETA and ETB; they are described in recent reviews.[1,2]

Strains of staphylococci are capable of producing either or both serotypes of ET. Although the prevalence of ET production is higher among phage group II strains, strains belonging to other phage groups are also capable of producing either serotype.[3–5]

Production of ET by an individual strain may vary considerably depending on the growth medium and culture conditions. The following procedure for production and purification of ET has been successfully

[1] M. Rogolsky, *Microbiol. Rev.* **43,** 320 (1979).
[2] J. H. Freer and J. P. Arbuthnott, *Pharmacol. Ther.* **19,** 55 (1983).
[3] I. Kondo, S. Sakurai, and Y. Sarai, *Infect. Immun.* **8,** 156 (1974).
[4] J. P. Arbuthnott and B. Billcliffe, *J. Med. Microbiol.* **9,** 191 (1976).
[5] J. de Azavedo and J. P. Arbuthnott, *J. Med. Microbiol.* **14,** 341 (1981).

used by the authors for a number of years and gives a reasonable yield of highly purified protein.[6-8]

Recommended Purification Procedure

Growth Medium. A yeast diffusate medium is prepared as follows[9]:

400 g yeast extract (Oxoid)
64 g casamino acids (Oxoid)
8 g glucose (BDH, Analar)
3.7 mg nicotinic acid (BDH, Analar)
0.4 mg thiamin (BDH, Analar)
38.4 g $Na_2HPO_4 \cdot 12H_2O$
7.14 g NaCl

Dissolve the yeast extract in 1 liter distilled H_2O in a steam bath for 15–20 min. Transfer to dialysis tubing (Visking tubing, 5-cm diameter) and dialyze against 3200 ml distilled H_2O for 48–72 hr. Discard dialysis sacs and their contents. Add the remaining ingredients to the diffusate, adjust pH to 7.8 with NaOH, and restore volume to 3200 ml with distilled H_2O. Sterilize by autoclaving at 15 lb for 30 min.

Inoculum. A single colony of the producer strain is inoculated into 25 ml yeast diffusate medium contained in a 250-ml flanged Erlenmeyer flask and incubated at 37° for 18–20 hr on an orbital shaker at 150 rpm. Six 2-liter flanged Erlenmeyer flasks, each containing 500 ml medium, are each inoculated with 1 ml of the overnight culture and incubated at 37° with shaking (150 rpm) for 20–40 hr. Under these conditions, maximum yields of ETA are produced between 20 and 24 hr and ETB between 36 and 40 hr. Cultures are spun at 20,000 *g* in a Sorvall RC4 centrifuge at 4° for 20 min and the supernatants are pooled and tested for the presence of ET. Yields may vary from 100 to 1000 μg/ml ET with different batches of medium and with different strains. Care should be taken to choose a strain which elaborates a single serotype since several strains may produce both serotypes of ET.

Assay. Various methods of assay are available, which are presented in Chapter [46] of this volume. In the following purification procedure, Ouchterlony immunodiffusion against specific sera is used for detection

[6] J. P. Arbuthnott, B. Billcliffe, and W. D. Thompson, *FEBS Lett.* **46,** 92 (1976).
[7] C. J. Bailey, J. de Azavedo, and J. P. Arbuthnott, *Biochim. Biophys. Acta* **624,** 111 (1980).
[8] C. J. Bailey, S. R. Martin, and P. B. Bayley, *Biochem. J.* **203,** 779 (1982).
[9] A. W. Bernheimer and L. L. Schwartz, *J. Gen. Microbiol.* **30,** 445 (1963).

and estimation of toxin. Doubling dilutions (5 μl) of test sample and specific anti-ET (5 μl) are added to wells (2.5-mm diameter) punched in agarose gels and allowed to diffuse at room temperature overnight. ET at concentrations of 60 μg/ml and above give a precipitin line.

Concentration of Culture Supernatant. Culture supernatant is cooled to 4° and $(NH_4)_2SO_4$ is slowly added to 90% saturation (650 g/liter). The precipitate is stirred overnight at 4°, harvested by filtration through Celite (Hyflo-Super-Cel filter aid; Koch Light Laboratories) in a Büchner funnel, and resuspended in a small volume of 1% (w/v) glycine (150–300 ml). Culture supernatant is thus concentrated 10- to 20-fold.

Preparative Isoelectric Focusing. Concentrated culture supernatant is dialyzed against 1% (w/v) glycine to remove excess ammonium sulfate. Between 400 and 800 mg protein is applied to a 400 ml (LKB 8100-2) preparative isoelectric focusing column. Initial separation is effected using a broad pH gradient (pH 3.5–9.5). The pH gradient is generated using 1% (w/v) ampholine carrier ampholytes (LKB) and is stabilized using a sucrose density gradient (5–55%, w/v). Focusing is carried out until a final potential of 800 V is reached (approximately 72 hr). Fractions (4–8 ml) are collected and the pH, optical density at 280 nm, and ET level of each fraction are monitored. Fractions with the highest ET activity (pH range 6.7–7.2) are pooled, concentrated if necessary, and dialyzed against 1% glycine. Pooled peak fractions from one or two broad gradient columns are refocused on a 110 ml column (LKB 8100-1) using a narrow pH gradient (pH 6–8). Focusing, using a small column, is generally achieved within 48 hr. The p*I* values of ETA and ETB, respectively, are 7.0 and 6.9 and the fractions with the highest activity are found in this region. Generally, between 3 and 5 mg of highly purified ET are recoverable from 1 liter of culture supernate.

Criteria of Purity. The pure material has a specific activity corresponding to a minimal dose of about 0.4 μg of protein (i.e., 2500 mouse U/mg in the neonatal mouse assay) and runs as a single polypeptide band on SDS–polyacrylamide gel electrophoresis even when the gels are overloaded.[7] The second isoelectric focusing step may be shown to remove trace impurities of hemolysin, coagulase, staphylokinase, lipase, DNase, and gelatinase activities.[7] For some purposes, purification as far as the first isoelectric focusing step is adequate; for the most demanding uses, e.g., monospecific antibody production, two stages of isoelectric focusing are required.

Storage of the Purified Toxin. After isoelectric focusing, the purified toxins may be held at −20° for periods of up to 1 month without marked loss of activity. However, ETA solution becomes turbid over longer periods of storage in frozen solution, although this effect has not been correlated with loss of potency. Solutions of ETB accumulate peptide fragments and lose activity[10] over longer term storage at −20°. Furthermore, the presence in the preparations of ampholytes, sucrose, and traces of other material may be unacceptable. To avoid problems, fractions, after isoelectric purification, may be buffer exchanged into 0.1 *M* NH_4HCO_3 on Sephadex G-50, dispensed into suitable ampoules, and freeze dried.[8] At −20° lyophilized material appears to be indefinitely (>4 years) stable.

Other Purification Procedures. A large number of workers[11–15] have used ion-exchange chromatography to purify concentrates of crude supernatant from the culture mixture. Final purification, if not obtained by an ion-exchange step,[16] can make use of hydroxylapatite chromatography[13,14] or isoelectric focusing.[6,7,17,18] The latter step has the advantage of high resolution, but is slow and costly. More recent experience suggests that chromatofocusing has some advantages over isoelectric focusing.[15] None of the published procedures can adequately resolve mixtures of ETA and ETB; the technique of affinity chromatography should overcome this problem.

During purification, other forms of toxin in low relative amounts have been observed by many workers.[6,12,15,16] At least some of the heterogeneity on isoelectric focusing has been shown to be artifactual,[19] resulting from either the discrete binding of ampholines or from other effects. At the present time there seems to be no reason to assume other than that the observed heterogeneity is artifactual in origin; forms other than ETA and ETB have not declared themselves convincingly.

Properties of the Purified Toxins

The two serotypes are identical in their action and potency. They induce splitting of the epidermis of man, monkeys, golden hamsters, and mice[20,21] at the level of the stratum granulosum.[22]

[10] I. Kondo, S. Sakurai, and Y. Sarai, *Zentralbl. Bakteriol, Parasitenk., Infektionskr. Hyg., Abt. 1, Suppl.* **5,** 489 (1976).
[11] F. A. Kapral and M. M. Miller, *Infect. Immun.* **4,** 541 (1971).
[12] I. S. Kondo, S. Sakurai, and Y. Sarai, *Infect. Immun.* **8,** 156 (1973).
[13] A. D. Johnson, J. F. Metzer, and Y. Spero, *Infect. Immun.* **12,** 1206 (1975).
[14] A. D. Johnson, L. Spero, J. S. Cades, and B. T. de Cicco, *Infect. Immun.* **24,** 679 (1979).
[15] Y. Piedmont and H. Monteil, *FEMS Lett.* **17,** 191 (1983).
[16] L. Dimoad and K. D. Wuepper, *Infect. Immun.* **13,** 627 (1976).
[17] M. E. Melish, L. A. Glasgow, and M. D. Turner, *J. Infect. Dis.* **125,** 129 (1972).
[18] B. B. Wiley, L. A. Glasgow, and M. Rogolsky, *Zentralbl. Bakteriol., Parasitenk., Infektionskr. Hyg., Abt. 1, Suppl.* **5,** 449 (1976).
[19] A. D. Johnson-Winegar and L. Spero, *Curr. Microbiol.* **8,** 311 (1983).
[20] P. M. Elias, P. Fritsch, and G. Mittermayer, *J. Invest. Dermatol.* **66,** 80 (1976).
[21] P. O. Fritsch, G. Kaaserer, and P. M. Elias, *Arch. Dermatol. Res.* **264,** 287 (1979).
[22] C. B. Lillibridge, M. E. Melish, and L. A. Glasgow, *Pediatrics* **50,** 728 (1972).

TABLE I
COMPARATIVE PROPERTIES OF TWO SEROTYPES OF EPIDERMOLYTIC TOXIN

Property	ETA	ETB
Specific extinction coefficient	8.4	8.2
Isoelectric point	7.0	6.95
Molar mass	30,000	29,500
Secondary structure: percentage helix; β sheet; turns	8; 46; 15	9; 45; 14
Heat sensitivity at neutral pH	20 min at 100°	30 min at 60°
Genetic control	Chromosomal	Plasmid

A number of the physicochemical properties of the toxins are also very similar (Table I) as are the amino acid compositions.[7,14,17] The limited amount of amino acid sequence data[14] also indicates that the proteins are related. The major differences found to date are in the location of the structural gene, ETA having a chromosomal origin,[23] ETB being located on a 42-kb plasmid,[2,3,24] and in the sensitivity to heat denaturation. ETB has the typical sensitivity of a soluble globular protein, but ETA is remarkably insensitive to high temperature.[3]

There is as yet no generally accepted basis for the understanding of the molecular action of the epidermolytic toxins. Reports[10,25] that ETA requires certain transition metal ions for activity have not been confirmed. It has been suggested that the toxins are inhibited by a subcellular fraction from epidermal cells containing mitochondria and lysosomes[26] and also by ganglioside extracts.[27] Attention has been drawn toward the mitogenic effect of toxin on cells.[28] Most recently it has been asserted that there is a strong interaction between the epidermolytic toxins and the tissue-specific filaggrin group of proteins, and that this binding ability is important for the action of the toxin.[29]

[23] B. B. Wiley and M. Rogolsky, *Infect. Immun.* **18,** 487 (1977).

[24] M. O'Reilly, G. Dougan, T. J. Foster, and J. P. Arbuthnott, *J. Gen. Microbiol.* **124,** 99 (1981).

[25] S. Sakurai and I. Kondo, *Jpn. J. Med. Sci. Biol.* **31,** 208 (1978).

[26] K. Nishioka, I. Katayama, and S. Sano, *J. Dermatol.* **8,** 7 (1981).

[27] I. Kondo and S. Sakurai, *Zentralbl. Bakteriol., Parasitenk., Infektionskr. Hyg., Abt. 1, Suppl.* **10,** 311 (1981).

[28] B. A. Morlock, L. Spero, and A. D. Johnson, *Infect. Immun.* **30,** 381 (1980).

[29] T. Smith and C. J. Bailey, *FEBS Lett.* **194,** 309 (1985).

[6] Preparation of Toxic Shock Syndrome Toxin-1

By DEBRA A. BLOMSTER-HAUTAMAA and PATRICK M. SCHLIEVERT

Introduction

Toxic shock syndrome (TSS) is a multisystem illness characterized by acute onset of high fever, hypotension or dizziness, rash, desquamation of the skin upon recovery, and variable multisystem involvement.[1-4] Toxic shock syndrome toxin-1 (TSST-1) has been cited by several investigators as a major toxin most likely responsible for the symptoms of TSS.[5-8] In 1981, TSST-1 was identified and characterized independently by Schlievert *et al.*[5] as pyrogenic exotoxin C (PEC) and by Bergdoll and colleagues as enterotoxin F (SEF).[6] Later, PEC and SEF were shown to be immunologically and biochemically identical.[9-11] The name TSST-1 was then adopted for use in reference to this toxin.[10] Other laboratories have subsequently reported on the isolation of this TSS-associated toxin.[11-13] Cumulatively, various physicochemical properties of TSST-1 have been reported with some variation in the molecular weight (20,000–24,000), isoelectric point (6.8–7.2), and amino acid composition. These discrepancies have been addressed in previous reports where it was sug-

[1] J. Todd, M. Fishaut, F. Kapral, and T. Welch, *Lancet* **1,** 1116 (1978).

[2] J. P. Davis, P. J. Chesney, P. J. Wand, and M. LaVenture, *N. Engl. J. Med.* **303,** 1429 (1980).

[3] K. N. Shands, G. P. Schmid, B. B. Dan, D. Blum, R. J. Guidotti, N. T. Hargrett, R. L. Anderson, D. L. Hill, C. V. Broome, J. D. Band, and D. W. Frazer, *N. Engl. J. Med.* **303,** 1436 (1980).

[4] R. W. Tofte and D. M. Williams, *Ann. Intern. Med.* **94,** 149 (1981).

[5] P. M. Schlievert, K. N. Shands, B. B. Dan, G. P. Schmid, and R. D. Nishimura, *J. Infect. Dis.* **143,** 509 (1981).

[6] M. S. Bergdoll, B. A. Crass, R. F. Reiser, R. N. Robbins, and J. P. Davis, *Lancet* **1,** 1917 (1981).

[7] J. P. de Azavedo and J. P. Arbuthnott, *Infect. Immun.* **46,** 314 (1984).

[8] J. K. Rasheed, R. J. Arko, J. C. Feeley, R. W. Chandler, C. Thornsberry, R. J. Gibson, M. L. Cohen, C. D. Jeffries, and C. V. Broome, *Infect. Immun.* **47,** 598 (1985).

[9] P. F. Bonventre, L. Weckbach, J. Staneck, P. M. Schlievert, and M. Thompson, *Infect. Immun.* **40,** 1023 (1983).

[10] M. S. Bergdoll and P. M. Schlievert, *Lancet* **2,** 691 (1984).

[11] H. Igarashi, H. Fujikawa, H. Usami, S. Kawabata, and T. Morita, *Infect. Immun.* **44,** 175 (1984).

[12] S. Notermans and J. B. Dufrenne, *Antonie van Leeuwenhoek* **48,** 447 (1982).

[13] R. F. Reiser, R. N. Robbins, G. P. Khoe, and M. S. Bergdoll, *Biochemistry* **22,** 3907 (1983).

gested that the diversity may be the result of differences in assay methods employed by each laboratory or due to the presence of low levels of contaminants.[14,15]

In this chapter, we describe two methods for the purification of TSST-1: A modification of the method of Schlievert *et al.*[5] for isolation of PEC as reported by Blomster-Hautamaa and colleagues,[14] and a modification of the method of Bergdoll *et al.*[6] as described by Reiser *et al.*[13] These two procedures differ primarily in their means of isolation, isoelectric focusing versus ion-exchange chromatography followed by gel filtration, rather than in culture methods. The isolation of TSST-1 by isoelectric focusing, described by Blomster-Hautamaa and colleagues,[14] yields a highly purified product, as assessed by highly sensitive silver-staining techniques. Furthermore, by successive electrofocusing steps, TSST-1 is resolved into two separate bands, designated TSST-1a and TSST-1b. These two proteins have different isoelectric points (7.08 and 7.22, respectively), but are immunologically and functionally identical. No other purification paper reports the resolution of TSST-1 into two highly purified proteins. Therefore, this chapter concentrates on the details of the isoelectric focusing purification procedure.

Isoelectric Focusing Method

Materials. All reagents and glassware used in the purification of TSST-1 are maintained pyrogen free to prevent endotoxin contamination. Pyrogen-free water was prepared by glass-distilling deionized water. All reagents were prepared with pyrogen-free water. All glassware was heated to 190° for at least 3 hr prior to use. Equipment that could not be heated to remove endotoxin (for example, the isoelectric focusing plate) was soaked in a 10% sodium dodecyl sulfate solution and rinsed with pyrogen-free water.

Bacterial Strains. Staphylococcus aureus strain MN8 is used primarily as the source of TSST-1, TSST-1a, and TSST-1b. Strain MN8 was isolated from a patient who met the established criteria for TSS and is a high TSST-1 producer. The strain is maintained in the lyophilized state in the presence of 20% nonfat dry milk.

Media Preparation. Beef heart medium is prepared from 5 lb of ground heart tissue suspended in 2 liters of pyrogen-free water. The tissue suspension is solubilized by adding 8 g of trypsin (1 : 250, Difco Laboratories, Detroit, MI) and maintaining the pH at 8.0 by frequent additions of

[14] D. A. Blomster-Hautamaa, B. N. Kreiswirth, J. S. Kornblum, R. P. Novick, and P. M. Schlievert, *J. Biol. Chem.* **261,** 15783 (1986).

[15] N. J. Poindexter and P. M. Schlievert, *J. Toxicol. Toxin Rev.* **4,** 1 (1985).

2.5 *M* NaOH. The digestion is complete when the pH remains at 8.0, usually about 30 min. The solubilized tissue is poured into dialysis tubing, 45-mm diameter and 3 ft in length (molecular weight cutoff 12,000–14,000). Filled tubing is rinsed with pyrogen-free water to remove any tissue from the tied ends and is then dialyzed against 10 liters of pyrogen-free water for 4 days at 4° without stirring. The resultant dialysate is sterilized by autoclaving. To complete the medium 12 ml of the following solution is added per liter: 0.33 *M* glucose, 0.5 *M* $NaHCO_3$, 0.68 *M* NaCl, 0.12 *M* Na_2HPO_4, and 0.027 *M* L-glutamine. Dialysates of Todd–Hewitt or beef heart infusion broths (Difco) may also be used as growth media.

Methods

Preparation of TSST-1. Strain MN8 is cultured overnight (until stationary phase) in 1200 ml of dialyzable beef heart medium with shaking (200 rpm gyrotary shaker, New Brunswick Scientific Co., New Brunswick, NJ) at 37°. The toxin is highly resistant to proteolysis and denaturation. Therefore, once made it is stable in culture medium for more than 2 days. TSST-1 is precipitated from the culture fluid by adding 4 vol (4800 ml) of 4° absolute ethanol and storing at 4° for at least 2 days. Removal of bacterial cells prior to addition of ethanol is not necessary. The precipitate is collected by centrifugation (650 *g* for 10 min, 4°), and the residual alcohol is removed by brief lyophilization (30 min). The pellet is resuspended in 10 ml pyrogen-free water and vortexed vigorously, then centrifuged at 10,000 *g* for 20 min, 4°. The supernatant is saved and the remaining precipitate again suspended in 10 ml pyrogen-free water. After centrifugation at 10,000 *g*, 20 min, both supernatant fluids are pooled. The pooled fluids are then dialyzed against 4 liters of pyrogen-free water for 24 hr at 4° (23 mm, 6000–8000 molecular weight cut-off) to remove salts which interfere with the electrofocusing procedure. Two such resultant preparations are pooled and then subjected to thin-layer isoelectric focusing in a pH gradient of 3–10 using commercial ampholytes for 15 hr with the use of the LKB Multiphor apparatus operated according to the manufacturer's specifications.[16] The visible band (due to a change in refractive index of the focused toxin in high concentration) containing TSST-1 is harvested and refocused in a pH 6–8 gradient yielding highly purified TSST-1. The inert matrix used in the flatbed apparatus is Sephadex (Pharmacia Fine Chemicals, Uppsala, Sweden) G75 superfine that has been swelled in water, then exhaustively washed with absolute ethanol, and finally dried; 4.5 g is used per run.

[16] A. Winter, H. Perlmutter, and H. Davis, LKB Instruments, Stockholm, Sweden, 1975.

Preparation of TSST-1a and TSST-1b. TSST-1a and 1b are isolated by one additional electrofocusing step. After focusing TSST-1 on the pH 6–8 gradient, approximately one-half of the Sephadex gel is removed from the anode end. The gel remaining on the cathode end, containing the TSST-1 band, is repoured after the addition of two more grams of Sephadex gel and then focused overnight using the remaining pH gradient, approximately pH 6.5–7.5.

After electrofocusing in a pH 6–8 or 6.5–7.5 gradient, protein bands are located by the zymogram print method.[16] Discrete bands are scraped off the plate and eluted with pyrogen-free water from the Sephadex gel, which is loaded into a 12-ml plastic syringe with a glass wool plug. After the toxin is eluted, remaining fluid is expressed by using the plunger of the syringe. This last procedure forces out additional TSST-1, presumably loosely associated with the gel. The ampholytes are removed by dialysis against 1.0 liter of water for 4 days, at 4°, with daily changes (23 mm, molecular weight cut-off 6000–8000). Protein concentration is determined by the Bradford protein dye-binding assay as modified by Bio-Rad Laboratories, Richmond, California. A typical purification of TSST-1a and 1b from strain MN8 yields approximately 2 mg of each toxin per liter of culture fluid. For *S. aureus* strains other than MN8, 200 μg of each toxin is obtained per liter of culture fluid. Sterile solutions of TSST-1 may be lyophilized, sealed, and stored at −20° indefinitely without detectable loss of biological or immunological activity. Also, toxin remains active for several months at 4° or frozen at −20°.

This purification scheme generates two TSST-1 proteins, TSST-1a and TSST-1b. Both proteins migrate as homogeneous bands, in SDS gels, to a molecular weight of 22,000 and have isoelectric points of 7.08 and 7.22, respectively.

Alternative Method

The isolation procedure described by Reiser *et al.*[13] is based on ion-exchange chromatography and gel filtration. This isolation scheme yields a TSST-1 protein with a molecular weight of 24,000 and an isoelectric point of 7.0.

Cultures containing 3% N-Z Amine NAK (Humko-Sheffield Chemical) and 1% yeast extract (Difco) are inoculated with a TSS-associated *S. aureus* strain. The inoculated flasks are cultured overnight at 37° with shaking.

Ten liters of culture supernatant fluid is absorbed to Amberlite CG-50 resin (150 ml, wet volume) by batch adsorption, for at least 1 hr.

The resin is transferred to a column (4 × 25 cm) and washed with water (500 ml). The column is then washed with 2.0 liters of 0.5 *M* sodium phosphate buffer containing 0.5 *M* NaCl at pH 6.2. The entire phosphate buffer eluate is concentrated 4-fold and then dialyzed against 5 m*M* phosphate buffer at pH 5.6, in preparation for the ion-exchange column.

The dialyzed concentrate is applied to a 2.5 × 45 cm column of CM-Sepharose CL-6B, equilibrated with 5 m*M* phosphate buffer, pH 5.6. The TSST-1 fraction is eluted from the Sepharose with 2250 ml of 0.03 *M* phosphate buffer at pH 6.0 followed by 2000 ml of 0.045 *M* phosphate buffer at pH 6.4. Fractions containing eluate volumes 3100–3400 ml are pooled and concentrated to approximately 15 mg protein/ml. The major amount of TSST-1 is found in this fraction.

The concentrate is then applied to a 2.5 × 117 cm column of Sephacryl S-200, which is equilibrated and washed with 0.05 *M* phosphate buffer at pH 6.8 containing 1.0 *M* NaCl. Elution volumes 410–430 ml are combined, dialyzed against 5 m*M* phosphate buffer at pH 6.8, and lyophilized.

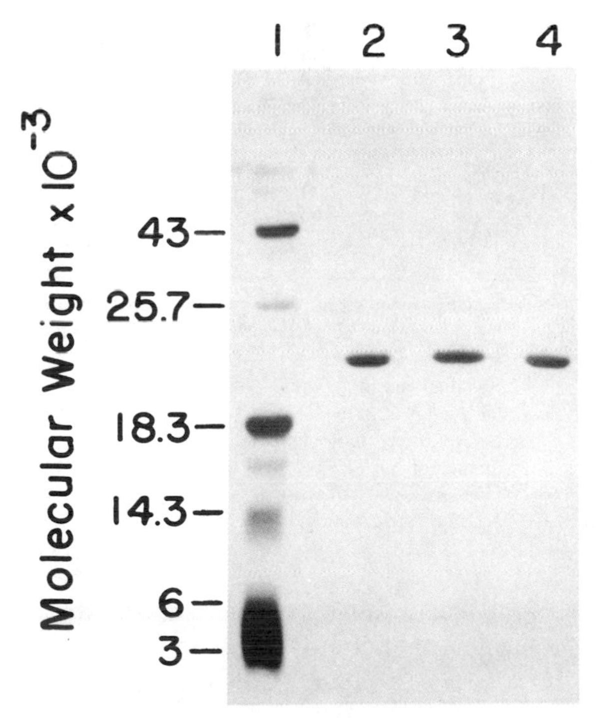

FIG. 1. SDS–polyacrylamide gel electrophoresis. Lane 1, low-molecular-weight standards; lane 2, purified TSST-1 (3.5 μg); lane 3, TSST-1a (3.5 μg); lane 4, TSST-1b (3.5 μg). The protein bands were detected by silver staining. The proteins used as molecular weight standards were ovalbumin (43K), α-chymotrypsinogen (25.7K), β-lactoglobulin (18.4K), lysozyme (14.3K), bovine trypsin inhibitor (6.2K), and insulin (3K).

TABLE I
AMINO ACID COMPOSITIONS OF TSST-1a AND 1b[a]

Amino acid	Amino acid composition		
	TSST-1a residues per mole[b]	TSST-1b residues per mole[b]	TSST-1 clone[b]
Aspartic acid	26	27	25
Threonine	21	20	19
Serine	20	20	21
Glutamic acid	20	20	17
Proline	10	8	10
Glycine	13	14	11
Alanine	4	5	3
Half-cystine	0	0	0
Valine	5	5	5
Methionine	0	0	2
Isoleucine	15	15	17
Leucine	14	16	15
Tyrosine	10	8	9
Phenylalanine	7	7	7
Histidine	5	5	5
Lysine	23	24	21
Tryptophan	ND[d]	ND[d]	3
Arginine	4	5	4
	197	199	194

[a] Isolated from strain MN8, as compared to the inferred amino acid composition of the TSST-1 structural gene.

[b] Residues per mole values are based on a molecular weight of 22,000.

[c] Residues per mole inferred from the DNA sequence of the TSST-1 structural gene. Blomster-Hautamaa and colleagues.[14]

[d] ND, Not determined.

The TSST-1 protein thus isolated is reported to be at least 96% pure by Coomassie blue staining of SDS–PAGE gels and Ouchterlony immunodiffusion.

Physicochemical Properties of TSST-1

The purity of the toxin preparations from isoelectric focusing was evaluated on a 15% SDS–PAGE gel (Fig. 1). The level of purity was determined by silver staining.[17] The gel shows that all three toxins (TSST-1, 1a, and 1b) migrate as single homogeneous bands with molecular weights of 22,000. These results verify that the electrofocusing procedure yields a highly pure preparation of TSST-1.

Previously, Schlievert *et al.*[5] reported an isoelectric point of 7.2 for TSST-1 isolated from the pH 3–10 gradient. Purified TSST-1 focused on a narrow pH gradient (6.5–7.5) resolves the toxin preparation into two separate bands, TSST-1a (p*I* 7.08) and TSST-1b (p*I* 7.22).

The amino acid compositions for TSST-1a and TSST-1b are listed in Table I. The proteins have nearly identical amino acid compositions. In addition, comparisons between the composition of 1a and 1b, with the predicted values derived from the known TSST-1 gene sequence, are also in good agreement.[14] The composition of the two toxins also correlates closely with compositions previously reported for TSST-1,[11,13] excluding the differences in cysteine residues reported to be present by Reiser *et al.*[13] The sequencing studies reveal no cysteine residues. Also, the sequence-predicted amino acid composition yields a molecular weight of 22,049, which is analogous to TSST-1 isolated by isoelectric focusing.

Conclusion

The high purity of TSST-1 generated by successive isoelectric focusing and its close comparisons to the physicochemical data of cloned TSST-1 strongly suggests that the values reported for TSST-1 by Blomster-Hautamaa and colleagues[14] are the most accurate.

[17] B. R. Oakley, D. R. Kirsh, and N. R. Morris, *Anal. Biochem.* **105,** 361 (1980).

[7] Purification of Staphylococcal Enterotoxins

By JOHN J. IANDOLO and RODNEY K. TWETEN

Introduction

At least six distinct enterotoxins (A, B, C_1, C_2, D, and E) are produced by *Staphylococcus aureus* and account for the majority of food poisoning in the United States. Although these toxins are termed enterotoxins, it is a misnomer, since none produces the classic ileal loop response. Nevertheless, all cause emesis and diarrhea in man and higher

TABLE I
CHARACTERISTICS OF ENTEROTOXINS[a]

Enterotoxin	Producing strain	Molecular weight	Isoelectric point	Extinction coefficient
A	13N-2909	27,800	7.3	14.6
B	S6	28,366	8.6	14.4
	10-275			
C_1	137	26,000	8.6	12.1
C_2	361	26,000	7.0	12.1
D	1151 M	27,300	7.4	10.8
E	FRI-326	29,600	7.0	12.5

[a] From Refs. 4 and 14.

primates.[1] The enterotoxins have also been shown to be powerful mitogens and are able to induce DNA synthesis and interferon synthesis in lymphocytes.[2,3] In fact, enterotoxin A is among the most effective T-cell mitogens that stimulate interferon production in mouse spleen cell cultures.[4] However, it seems clear that these mitogenic activities are not associated with either major antigenic regions of the molecules or with their site of action for emesis and diarrhea.[1] It is important to note that no specific mode of action or enzymatic activity has been identified for any of the enterotoxins but continues to be a research area of concern.

Some of the general characteristics of the *S. aureus* enterotoxins and the strains that produce them are presented in Table I. The molecular weights of all the enterotoxins are approximate except that for enterotoxin B (SEB) and enterotoxin C_1 (SEC), which have been completely sequenced.[5,6] All of the toxins are secreted into the extracellular milieu and all are probably synthesized as precursor molecules as in the case of SEA and SEB.[7–10]

[1] M. Bergdoll, in "Staphylococci and Staphylococcal Infections" (C. S. F. Easmon and C. Adlam, eds.), Vol. 2. Academic Press, New York, 1983.

[2] D. L. Peavy, W. H. Adler, and R. T. Smith, *J. Immunol.* **105,** 1453 (1970).

[3] D. L. Archer, J. A. Wess, and H. M. Johnson, *Immunopharmacology* **3,** 71 (1980).

[4] H. M. Johnson, J. E. Blalock, and S. Baron, *Cell. Immunol.* **33,** 170 (1977).

[5] I.-Y. Huang and M. S. Bergdoll, *J. Biol. Chem.* **245,** 3518 (1970).

[6] R. K. Tweten and J. J. Iandolo, *Infect. Immun.* **34,** 900 (1981).

[7] R. K. Tweten and J. J. Iandolo, *J. Bacteriol.* **153,** 297 (1983).

[8] L. Spero, A. Johnson-Winegar, and J. J. Schmidt, *in* "Bacterial Toxins" (M. C. Hardegree and W. Habig, eds.), Vol. 3. Dekker, New York, in press.

[9] K. K. Christianson, R. K. Tweten, and J. J. Iandolo, *Appl. Environ. Microbiol.* **50,** 696 (1985).

[10] D. M. Ranelli, C. L. Jones, M. B. Johns, G. J. Mussey, and S. A. Khan, *Proc. Natl. Acad. Sci. U.S.A.* **82,** 5850 (1985).

The purification of the enterotoxins has been, for the most part, based on classical ion-exchange and gel filtration techniques. However, newer methodologies, such as fast-protein liquid chromatography (FPLC), have been applied to the purification of these toxins.[11] It is not our intent to present the available methods for purification of the enterotoxins in a comprehensive survey. Rather, we present the most widely used methods, both old and new, that are used for the purification of these toxins.

General Methods

The individual purification procedures contain some common steps which will be presented here to avoid needless repetition. For the most part growth of the enterotoxin-producing *S. aureus* strains is similar in all cases. The most widely used general medium for the culture of these organisms contains 3% NZ-amine type A or NAK (Sheffield Chemical Co.), 3% protein hydrolysate powder (Meade Johnson), 0.00005% thiamin, and 0.001% niacin.[1] Protein hydrolysate powder is no longer available and 1% yeast extract has proved to be a suitable substitute for it as well as for the vitamins thiamin and niacin.[8,9] Optimum yields of the enterotoxins are obtained under controlled fermentation where pH, temperature, and oxygen tension are controlled.[6] Any additional growth requirements for the various strains will be discussed under the individual enterotoxin purification schemes. Typically, growth at 37° for 18–24 hr is sufficient for maximum toxin yields. It can be anticipated that yields of enterotoxins B and C_1 and C_2 will be up to several hundred micrograms per milliliter while the yield of the other enterotoxins will be only a few micrograms per milliliter.

All of the enterotoxins are secreted products. Generally, they are produced during the log and stationary stages of cell growth.[1,12] After growth the cells are removed by centrifugation and the toxin-containing supernatant saved. If a large fermentation has been carried out, then cells and supernatant can be quickly separated using a continuous flow centrifuge (Sharples) or Millipore's Pellicon cassette system. One of the initial problems in the purification of these toxins is the concentration of the toxin-containing spent media. Some of the following purification schemes use polyethylene glycol 20M (Carbowax) and dialysis tubing to concentrate the toxin. This procedure is unwieldy and requires substantial

[11] F. Janin, C. Lapeyre, M. L. DeBuyser, F. Dilasser, and M. Borel, *in* "Bacterial Protein Toxins" (J. E. Alouf, F. J. Fehrenbach, J. H. Freer, and J. Jeljaszewicz, eds.). Academic Press, Orlando, Florida, 1984.

[12] J. J. Iandolo and W. M. Shafer, *Infect. Immun.* **16,** 610 (1977).

amounts of dialysis tubing and polyethylene glycol. An attractive substitute method is the use of either Amicon's hollow fiber concentration system or Millipore's tangential flow concentration system. Using the M_r 10,000 cut-off membranes, these systems can concentrate large volumes comparatively rapidly (>11/hr on average) and at the same time eliminate substances of less than M_r 10,000.

To follow toxin purification specific antisera to the enterotoxin of choice should be used in an appropriate immunoassay. There is no enzymatic activity known to be associated with any of the enterotoxins and the use of monkeys[13] or kittens[14] to follow emetic activity is expensive, imprecise, and time consuming. Any one of a number of immunological assays [e.g., immunodiffusion plate assay, radioactive immunoassay (RIA), enzyme-linked immunoassay (ELISA), or paper dot-blot assay] can be used to identify toxin-containing fractions during chromatographic runs. Since there is an abundance of these methods available the specific methodology will not be described here.

Finally, strains used for the production of the various enterotoxins can be obtained from the Food Research Institute (Madison, WI), while some are carried by the American Type Culture Collection (Rockville, MD).

Enterotoxin Purification by Type

Enterotoxin A (SEA)[15]

A high SEA producer such as *Staphylococcus aureus* 13N-2909 is grown in the general medium except that it is made 0.2% in glucose. Initially, 2–3 g/liter of a carboxylic acid-type resin such as Amberlite CG-50 is used for batch isolation. The pH of the resin and culture supernatant (diluted with 4 vol distilled water) is adjusted to pH 5.6. The toxin is then eluted with 0.5 *M* sodium phosphate, pH 6.2. The eluted toxin is dialyzed against 0.008 *M* sodium phosphate, pH 6.0, and applied to a carboxymethyl (CM)-cellulose column preequilibrated with 10 m*M* sodium phosphate buffer, pH 6.0 (4°). The pH of the resin is titrated to pH 6.0 with NaOH prior to chromatography of the toxin. After the toxin has been loaded onto the CM-cellulose column, the resin is washed with a column volume of 8 m*M* sodium phosphate buffer, pH 6.0, containing 8 m*M* NaCl. The toxin is eluted at room temperature with a linear gradient

[13] M. S. Bergdoll and G. Dack, *J. Lab. Clin. Med.* **41,** 782 (1953).

[14] W. G. Clark, G. F. Vanderhooft, and H. L. Borison, *Proc. Soc. Exp. Biol. Med.* **111,** 205 (1962).

[15] E. J. Schantz, W. G. Roessler, M. J. Woodburn, M. J. Lynch, H. M. Jacoby, S. J. Silverman, J. C. Gorman, and L. Spero, *Biochemistry* **11,** 360 (1972).

of 0.01 *M* sodium phosphate, pH 6.0, to 0.05 *M* sodium phosphate, pH 6.8. Gradient size is approximately 2.3 times the size of the column volume.

The fractions containing the toxin are combined and the pH adjusted to 5.7 in preparation for hydroxylapatite chromatography. Hydroxylapatite, 1 g/3.5 mg protein, is resuspended in 8 m*M* sodium phosphate buffer, pH 5.7, allowed to settle, and any fines decanted. The resin is then poured into a column and washed for 24 hr with 30 m*M* sodium phosphate, pH 5.7, at a flow rate of 0.5 ml/min. The toxin is loaded and the column washed with one void volume of 0.2 *M* sodium phosphate, pH 7.5. The toxin is eluted using a linear gradient of phosphate buffer from 0.2 *M*, pH 5.7, to 0.4 *M*, pH 5.7, at a flow rate of 1 ml/min. The gradient volume is four times the column volume.

The final step involves chromatography of the toxin-containing fractions from the hydroxylapatite column. The fractions are lyophilized and resuspended in a volume of distilled water that does not exceed 5% of the column volume. This solution is applied to a column of Sephadex G-75 fine or superfine equilibrated in 0.50 *M* sodium phosphate, pH 6.8, and containing 1 *M* sodium chloride. At this point the toxin should be greater than 99% pure with a yield of 20%.

Enterotoxin B[16]

A strain such as *S. aureus* S6 or 10-275 is used for the production of SEB. The spent medium containing the toxin is diluted two times with water, adjusted to pH 6.4, and approximately 2 g/liter of Amberlite CG-50 (200 mesh) cation ion-exchange resin is added to the toxin mixture. The CG-50 is preequilibrated with 3–4 vol of 0.05 *M* sodium phosphate buffer and titrated to pH 6.4 with sodium hydroxide. The resin is then washed with water to remove buffer salts and stirred with the spent medium for 30 min at room temperature. The resin is then placed into a suitable column and washed with one column volume of water. The toxin is eluted with 0.5 *M* sodium phosphate, pH 6.8, containing 0.25 *M* sodium chloride.

The eluted toxin solution is dialyzed to lower the buffer salts to less than 0.01 *M*. This dialyzed preparation is applied to a column of CG-50 preequilibrated as above, but at pH 6.8. Twenty grams of resin is used per gram of protein. After absorption of the toxin the column is washed with water until the A_{280} decreases to baseline and the toxin is eluted with 0.15 *M* disodium phosphate.

The eluted toxin is dialyzed once more to lower the salt and buffer

[16] E. J. Schantz, W. G. Roessler, J. Wagman, L. Spero, D. A. Dunnery, and M. S. Bergdoll, *Biochemistry* **4,** 1011 (1965).

concentration to less than 0.01 *M*. The toxin is then applied to a column of carboxymethyl cellulose or CM-Sephadex preequilibrated in 0.01 *M* sodium phosphate, pH 6.2. Unbound proteins are eluted with the same buffer until the A_{280} is close to zero. The toxin is eluted by applying a gradient of sodium phosphate from pH 6.2, 0.02 *M*, to pH 6.8, 0.07 *M*. The toxin elutes between 0.03 and 0.04 *M*.

At this point the toxin is essentially homogeneous. However, up to four isoelectric species are often found in SEB preparations.[17] Our laboratory (Tweten, unpublished data) has found that by using chromatofocusing (Pharmacia), these isoelectric species can be separated rapidly (<30 min) using polybuffer 96. Essentially the procedure is as follows. The toxin preparation is exchanged into 0.025 *M* diethanolamine-HCl, pH 9.5, by dialysis or gel filtration. The sample is then loaded onto a Pharmacia Mono P chromatofocusing column using the fast-protein liquid chromatography (FPLC) system and eluted with polybuffer 96, pH 7.5. Exact conditions are provided in the Mono P manual from Pharmacia.

Enterotoxin C_1 (SEC_1)[18]

Culture supernatant from *S. aureus* 137 is concentrated, dialyzed against distilled water, and lyophilized. The toxin is redissolved in 0.10 *M* sodium phosphate buffer, pH 5.5. Some of the lyophilized material may not redissolve and can be removed by centrifugation. The crude toxin is applied to a carboxymethyl cellulose column preequilibrated in the same buffer. The column is washed until the absorbance at 280 nm returns to baseline. The toxin is eluted with a stepwise gradient starting with 0.020 *M* phosphate buffer, pH 6.0. The bulk of the toxin is eluted with 0.06 *M* sodium phosphate, pH 6.8. The toxin peak consists of a sharp peak and a trailing edge. The trailing edge is discarded since it is heavily contaminated with β-hemolysin. The eluted toxin is concentrated and applied to Sephadex G-75 preequilibrated in 0.02 *M* sodium phosphate, pH 6.8. The toxin elutes as a single peak.

The toxin is then concentrated and run twice through a column of Sephadex G-50. The eluted toxin is dialyzed against water and lyophilized.

Enterotoxin C_2 (SEC_2)[19]

Culture supernatant from *S. aureus* 361 is concentrated as for SEC_1 and dialyzed against 20 m*M* sodium phosphate, pH 5.4. The toxin is then

[17] P.-C. Chang and N. Dickie, *Biochim. Biophys. Acta* **236,** 367 (1971).
[18] C. R. Borga and M. S. Bergdoll, *Biochemistry* **6,** 1467 (1967).
[19] R. M. Avena and M. S. Bergdoll, *Biochemistry* **6,** 1474 (1967).

applied to a carboxymethyl cellulose column that has been equilibrated with the same buffer. The column is washed with the starting buffer until the absorbance at 280 nm returns to baseline. SEC_2 is eluted with 0.06 *M* sodium phosphate, pH 6.7.

The eluted toxin is lyophilized, resuspended in distilled water, and dialyzed against 5 m*M* sodium phosphate, pH 5.8. The toxin is applied to a column of carboxymethyl cellulose equilibrated in the same buffer. After washing the column with this buffer, the column is developed with a gradient from 0.005 *M* sodium phosphate, pH 5.8, to 0.05 *M* sodium phosphate, pH 5.8. Toxin elutes at approximately 0.03 *M* sodium phosphate. The gradient volume is about eight times the column volume.

The partially purified toxin is concentrated and applied to a Sephadex G-75 column equilibrated in 0.02 *M* sodium phosphate, pH 6.8. The eluted toxin is concentrated and finally reapplied to a column of Sephadex G-50. Recovery is about 40% with purity exceeding 99%.

Enterotoxin D (SED)[20]

S. aureus strain 1151 M is used for the production of enterotoxin D. The medium was similar to that used for SEA and SEB. After growth and removal of the cells, the pH of the supernatant is adjusted to 5.6 and diluted 5-fold with distilled water. Approximately 2–3 g of Amberlite CG-50 resin, preequilibrated as for SEA, is added per liter of original culture supernatant. The mixture is stirred for 1 hr at room temperature and then allowed to settle. The bulk of the culture supernatant is decanted and the resin washed with 1 liter water in a chromatography column. The toxin is eluted with 0.5 *M* sodium phosphate, pH 6.2, containing 0.5 *M* sodium chloride. The eluted toxin was originally concentrated using 20% w/v polyethylene glycol, 20 *M*; however, this could be achieved using any of the methods described in the general methods section.

The concentrated toxin is dialyzed against 0.020 *M* sodium phosphate, pH 5.6, and applied to a carboxymethyl cellulose column preequilibrated in the same buffer. Approximately 1.5 ml of swollen gel is used per milligram of protein. The column is first washed with at least one column volume of 0.02 *M* sodium phosphate, pH 5.6, then 0.03 *M* sodium phosphate, pH 6.1. To elute the toxin a linear gradient (three column volumes) from 0.03 *M* sodium phosphate, pH 6.1, to 0.06 *M* sodium phosphate, pH 6.8, is used. The eluted toxin is then rechromatographed a second time on CM cellulose using the same conditions. The toxin solution is concentrated and chromatographed on Sephadex G-75 superfine preequilibrated in 0.02 *M* sodium phosphate, pH 6.8. This step is repeated once.

[20] H.-C. Chang and M. S. Bergdoll, *Biochemistry* **18,** 1937 (1979).

Enterotoxin E (SEE)[21]

Staphylococcus aureus strain FRI-326 culture supernatant is concentrated and dialyzed against 0.020 *M* sodium phosphate, pH 5.6. The toxin is then absorbed to a column of carboxymethyl cellulose. The elution of bound protein is carried out in a stepwise fashion using the following sodium phosphate buffers; 0.020 *M*, pH 5.6; 0.02 *M*, pH 6.7; and 0.06 *M*, pH 6.4. Toxin elutes at the last step. The eluted toxin is concentrated and chromatographed twice on Sephadex G-75 superfine, preequilibrated in 20 m*M* sodium phosphate, pH 6.8.

To obtain highly purified SEE the authors found it necessary to chromatograph the toxin once more on G-75 superfine in the presence of 6 *M* urea. After removal of the urea the toxin retained its emetic activity in rhesus monkeys.

Recent Methods

The original methods described by Bergdoll and co-workers are still in general use today. However, with the changing technology of protein purification newer methods have been employed for the purification of certain enterotoxins from *S. aureus*. Some of these methods are presented below.

Enterotoxins A and C_2[22]

A 10-ml culture of *S. aureus* 11N-165 (SEA) or *S. aureus* 361 (SEC_2) is grown overnight at 37°. The cells are washed in sterile 0.0003 *M* phosphate buffer, pH 7.2, and resuspended in 10 ml of the buffer. These cells are then added to 90 ml of sterile phosphate buffer in a 2-liter flask that contains a 500-ml dialysis bag of sterile medium. The cells are grown at 37° on a rotary shaker for 24 hr. The fluid surrounding the dialysis bag is centrifuged to remove the cells and the supernatant serves as the source of the enterotoxin.

The removal of the enterotoxin from the supernatant is carried out as follows. QAE-Sephadex (Pharmacia) is preequilibrated in 0.033 *M* ethylenediamine-HCl, pH 9.6, and then added batchwise to the crude toxin preparation (2 g QAE-Sephadex/liter). The suspension is stirred for 1 hr at room temperature. The ion exchanger, containing the bound toxin, is recovered by filtration on a sintered glass funnel. The toxin is then eluted batchwise by resuspending the ion exchanger in 0.033 *M* ethylenedi-

[21] C. R. Borga, E. Fanning, I.-Y. Huang, and M. S. Bergdoll, *J. Biol. Chem.* **247,** 2456 (1972).

[22] H. Robern, S. Stavric, and N. Dickie, *Biochim. Biophys. Acta* **393,** 148 (1975).

amine-HCl, pH 6.4, for 30 min at room temperature. The eluted toxin is again recovered by filtration on a sintered glass funnel. This procedure is repeated until no further toxin can be eluted and the pH of the filtrate is between 6.4 and 6.7. The eluates are then concentrated by ultrafiltration using any number of devices available.

The concentrated toxin is then passed through a column of Sephadex G-100 preequilibrated in 0.033 *M* ethylenediamine-HCl, pH 6.4. Two peaks absorbing at 280 nm are eluted from the column with the latter containing the enterotoxin. The eluted toxin is again concentrated by ultrafiltration and applied to a column of QAE-Sephadex. This column is preequilibrated in 0.033 *M* ethylenediamine-HCl, pH 6.4, and run isocratically with this buffer. SEA elutes before SEC_2, which elutes as a double peak. The toxin is concentrated and then rerun on a Sephadex G-100 column. The overall recovery is about 30% for SEC_2 and 40–50% for SEA. Both toxins appear homogeneous by sodium dodecyl sulfate-polyacrylamide gel electrophoresis.

Enterotoxins A, C_1, D[11]

This method utilizes Pharmacia's fast protein liquid chromatography (FPLC) system and the high-resolution chromatofocusing Mono P column. A dialysis sac method as described above is used for the growth of the various enterotoxin-producing strains (see Table I for the appropriate strains for each toxin). The concentrated crude supernatant containing SEA is passed over a Sephadex G-75 and the toxin-containing fractions are pooled. For C_1 and D and supernatants are passed over an Amberlite CG-50 column as described for SED and the active fractions pooled. All three toxins are then exchanged into the starting buffer for chromatofocusing and then are separated using the Mono P column FPLC system. Since all of the toxins have isoelectric points in the range of 7–9, the polybuffer system PBE 96 is used for elution. The chromatofocusing is carried out as follows. The column is washed with 0.025 *M* diethanolamine-HCl, pH 9.5, until the pH of the eluent matches that of the start buffer. The sample, which has been equilibrated in this buffer, is applied to the column. The column is then washed with 9 ml of this buffer to elute any unbound proteins. Elution is begun at 1 ml/min with 34 ml of a 10% solution of PBE 96 in water, pH 6.0 (with acetic acid). The authors estimated the purity of SEA, SEC_1, and SED to be 98, 95, and 80%, respectively. SEA elutes as two peaks at pH 8.8 and 8.6. SEC_1 also elutes as two peaks at pH 8.3 and 7.9. SED elutes as three peaks at pH 8.6, 8.3, and 8.0. All of the isoelectric points determined for these toxins in this study are significantly different from those derived by Bergdoll and his co-workers (see Table I). This may be due, in part, to some microheterogeneity as is seen for SEB[23]; however, the isoelectric point for SEA of 8.6 is 1.3 pH units higher than that determined for SEA by Bergdoll and co-workers.

Acknowledgment

Supported in part by NIH Grant AI17474 and in part by U.S. Army Research Contract DAMD17-86-C-6055. Contribution 86-259-J from the Kansas Agricultural Experiment Station, Manhattan, Kansas 66506.

[23] L. Spero, J. R. Warren, and J. F. Metzger, *Biochim. Biophys. Acta* **336**, 79 (1974).

[8] Production, Purification, and Assay of Streptolysin O

By JOSEPH E. ALOUF and CHRISTIANE GEOFFROY

Streptolysin O (SLO) is a toxic, cytolytic, and immunogenic single-chain protein (M_r ~60K) released into the extracellular medium along with other toxins (streptolysin S and erythrogenic toxin), exoenzymes, and mitogens during the growth of most strains of group A and many strains of groups C and G streptococci.[1,2] The toxin is highly lytic for eukaryotic cells, including erythrocytes from mammals and other species (hemolysis). It is lethal to laboratory animals and has very potent cardiotoxic properties.[1,3] SLO is the prototype of a group of at least 16 bacterial cytolytic protein toxins known as oxygen-labile or thiol (or sulfhydryl)-activated cytolysins produced by different gram-positive bacteria of the species *Streptococcus, Bacillus, Clostridium,* and *Listeria.* These proteins have similar physical and biochemical properties and molecular weights of about 60K. They share the following common properties: (1) they are lethal (cardiotoxic) to animals and lytic for eukaryotic cells, including erythrocytes; (2) their biological, cytolytic, and lethal effects are lost by oxidation or by reaction with sulfhydryl group reagents and restored by reduction with thiols or other reducing agents; (3) they are antigenically related, giving rise to cross-reacting precipitating and neutralizing antibodies; (4) they are inactivated by cholesterol and related 3β-hydroxysterols; (5) they are believed to damage cells by binding in the

[1] J. E. Alouf, *Pharmacol. Ther.* **11,** 661 (1980).
[2] B. Dassy and J. E. Alouf, *J. Gen. Microbiol.* **129,** 643 (1983).
[3] S. P. Halbert, *in* "Microbiol Toxins" (T. C. Montie, S. Kadis, and S. J. Ajl, eds.), Vol. 3, p. 69. Academic Press, New York, 1970.

reduced state to membrane cholesterol; (6) they are excreted in the extracellular medium except for pneumolysin, which is intracytoplasmic.[4-6]

The strategy of SLO purification described in this article allows the purification of the other related toxins with some specific modifications as already reported for alveolysin from *Bacillus alvei*.[7]

Production of Steptolysin O

Culture supernatant fluids are the starting material for SLO production which is released during the exponential and stationary phases of bacterial growth.[2] The toxin can be produced in moderate titers in chemically defined media.[2] However, higher titers are obtained in media consisting of diffusates prepared by dialysis of peptones, casein hydrolysates, and similar products supplemented with yeast extracts. Media such as Todd–Hewitt broth should not be used for purification purposes because of the presence of high-molecular-weight peptides. Glucose is necessary (0.5–1%) for optimal production. As lactic acid is produced during the growth, the pH should be maintained between 6.8 and 7.4 to avoid toxin destruction by acidity and (or) by proteolysis by the potent sulfhydryl-dependent streptococcal proteinase, which is released into the medium when the pH drops below 6.7. Most work on toxin production and purification has been carried out by batchwise cultures. However, fermenters under pH control have been used in certain cases.[3,8] The peak of SLO occurs between 6 and 12 hr. All type A, C, and G strains appear to produce a toxin which is identical biologically and immunochemically.[3]

Strains Used for Toxin Production

Individual strains vary considerably in their quantitative ability to produce SLO in suitable media.

The best yields (500–1000 hemolytic U/ml) in optimal culture conditions have been obtained with the following strains: C203S type 3, Richards type 3, Kalback S84 type 3, and strain 814, type 4 (see Ref. 1 for references).

The production of SLS by these strains may be avoided if culture

[4] A. W. Bernheimer, *in* "Mechanisms of Bacterial Toxinology" (A. W. Bernheimer, ed.), p. 85. Wiley, New York, 1976.

[5] C. J. Smyth and J. L. Duncan, *in* "Bacterial Toxins and Cell Membranes" (J. Jeljaszewicz and T. Wadström, eds.), p. 129. Academic Press, New York, 1978.

[6] J. E. Alouf and C. Geoffroy, *in* "Bacterial Protein Toxins" (J. E. Alouf, F. J. Fehrenbach, J. H. Freer, and J. Jeljaszewicz, eds.), p. 165. Academic Press, London, 1984.

[7] C. Geoffroy and J. E. Alouf, *J. Biol. Chem.* **258,** 9968 (1983).

[8] S. Shany, P. S. Grushoff, and A. W. Bernheimer, *Infect. Immun.* **7,** 731 (1973).

media are free of SLS inducers (serum components, yeast RNA active fraction, detergents). This is the case for the dialyzed medium described below. The SLS^- mutant C203U derived from the C203S strain by Bernheimer has also been successfully employed for SLO production and purification.[9] In our laboratory we use the S84 strain (Pasteur Institute Collection, code number A78), which produces up to 1000 HU/ml (testing with rabbit red blood cells as described in Assay of Streptolysin O) in the medium described below.

Preparation of Culture Medium

The medium devised in our laboratory for routine production of SLO allows easy purification of the toxin and avoids detectable production of streptococcal proteinase which, if present, causes important toxin losses in the crude material and during the purification process, particularly when reducing conditions are used. In order to obtain sufficient starting material for a final yield of about 1 mg of pure toxin we routinely use 25 liters of culture medium from peptone diffusate. The medium is made by dissolving 500 g of meat pepsic peptone and 500 g of yeast pancreatic peptone (both provided by Organotechnie, La Courneuve, France) or similar products in about 1.5 liters of quartz-distilled water with moderate heating (45–50°). The solution is poured into three or four 50-cm lengths of 10-cm dialysis tubings (Union Carbide, Chicago, IL) previously boiled in distilled water for 10 min. The dialysis sacs are immersed in 25 liters of distilled water for 96 hr at 4°. The dialysis sacs and contents are discarded and the diffusate is adjusted in volume to 25 liters with distilled water, and distributed in volumes of 4 liters in 5-liter flasks. The remaining liter is distributed in volumes of 400 ml in 500-ml flasks, which will be used for inoculation. The pH is adjusted to 7.8 in all flasks with 10 *N* NaOH.

The flasks are sterilized at 120° for 30 min. Prior to inoculation, 80 ml of 50% (w/v) glucose in distilled water sterilized at 105° for 40 min and 100 ml of 10% (w/v) sodium bicarbonate in distilled water (similarly sterilized) are added into the 5-liter flasks and proportional quantities into the inoculation flasks.

Production of Crude Culture Supernatants Containing Streptolysin O

The strain (S84) preserved in 1-ml aliquots in liquid nitrogen is inoculated into the 500-ml flasks and incubated overnight at 37°. Approximately 100 ml of culture is added per 4 liters of medium. The flasks prewarmed at 37° are incubated at this temperature, then supplemented after 5 hr of

[9] R. Linder, *FEMS Microbiol. Lett.* **5**, 339 (1979).

culture (when the pH has dropped to about 6.7) with 200 ml of 10% (w/v) sodium bicarbonate. After two or three additional hours of culture, the flasks are cooled overnight at 4° and the culture is centrifuged in the cold at 10,000 *g* to remove bacteria. The clear brown yellow supernatant contains approximately 400–600 HU/ml after reduction as described in the assay section. The supernatant is supplemented with sodium azide to a final concentration of 0.02%.

Fractionation of Culture Supernatant and Isolation of Streptolysin O

Concentration of the Crude Material. The culture supernatant is concentrated to a volume of 4 liters (fraction F_1) by ultrafiltration on Amicon DC2 apparatus (Amicon, Lexington, MA) using Amicon H ×10 hollow fibers, which eliminate in the ultrafiltrate the material of M_r below 10,000. The retentate (F_1) contained about 90% of the initial amount of SLO. At this stage the toxin is in the partially oxidized (hemolytically inactive) state. The extent of toxin oxidation is variable and depends on media and culture conditions. The hemolytic activity of the oxidized fraction could be immediately restored upon addition of 20 m*M* cysteine or dithiothreitol. For the medium described here fraction F_1 is to a large extent (75–90%) in the oxidized state.

Precipitation of the Retentate (F_1). The retentate (4000 ml) is supplemented under magnetic stirring with 400 ml of 0.5 *M* Na_2PO_4 and then 150 ml of 1 *M* $CaCl_2$ added dropwise under very gentle stirring. The flask is allowed to stand for 10 min then shaken again briefly, once or twice, and then allowed to stand unstirred until a clear supernatant appears over the calcium phosphate layer. No appreciable hemolytic activity (<50 HU/ml) should remain in the supernatant, which is decanted and kept apart for purification of erythrogenic toxin (see Chapter [10] in this volume). The calcium phosphate phase is then centrifuged at 10,000 *g* for 10 min and the pellet is triturated with about 400 ml of 0.5 *M* NaH_2PO_4, pH 6.8. The supernatant is recovered and the pellet reextracted again under the same conditions. The two supernatants are combined (800 ml) and diluted with bidistilled water to a final volume of 1600 ml. This solution, which contains about 90% of initial SLO, is concentrated to about 600 ml by ultrafiltration in the DC-2 apparatus (fraction F_2).

Covalent Chromatography on Thiopropyl-Sepharose 6B. This procedure is based on thiol–disulfide exchange chromatography for the selective purification of various thiol proteins (Fig. 1).

As fraction F_2 is often in the oxidized state it is first reduced by passing it through a column of thiopropyl-Sepharose 6B in the free thiol form (10 g of gel) connected to another column of thiopropyl-Sepharose 6B in the

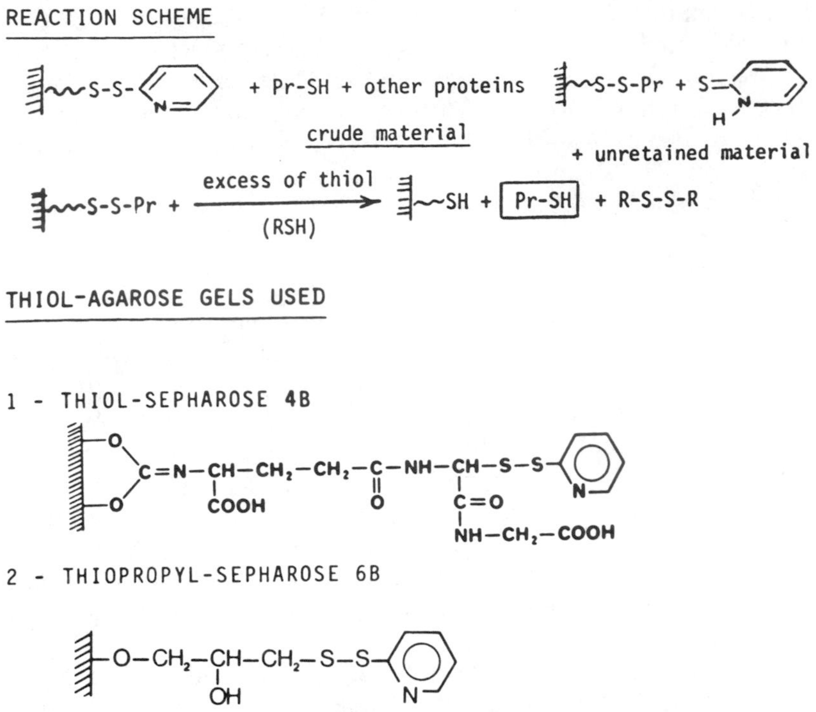

REACTION SCHEME

~S-S-(pyridyl) + Pr-SH + other proteins (crude material) → ~S-S-Pr + S=(pyridine, N-H) + unretained material

~S-S-Pr + excess of thiol (RSH) → ~SH + Pr-SH + R-S-S-R

THIOL-AGAROSE GELS USED

1 - THIOL-SEPHAROSE 4B

$-O\backslash C=N-CH(COOH)-CH_2-CH_2-C(=O)-NH-CH(C(=O)NH-CH_2-COOH)-S-S-$(pyridyl)

2 - THIOPROPYL-SEPHAROSE 6B

$-O-CH_2-CH(OH)-CH_2-S-S-$(pyridyl)

FIG. 1. Purification of sulfhydryl-activated toxins by covalent chromatography on thiol-agarose gels.

disulfide form (5 g of gel). If F_2 (or other SH-activated toxin than SLO) is in the reduced state at this stage the material is directly chromatographed through the gel in the disulfide form. The latter is prepared by allowing freeze-dried thiopropyl-Sepharose 6B (Pharmacia, Uppsala, Sweden) to swell in PBS, pH 6.8, for 15 min at room temperature. The gel is then washed with several aliquots of buffer, using a total volume of 200 ml/g of freeze-dried powder (1 g of freeze-dried powder reswells to give approximately 3 ml of sedimented gel). This gel is converted into the free thiol form by removing the 2-thiopyridyl protecting groups by treatment with 10 m*M* cysteine for 40 min at room temperature with occasional gentle mixing. The gel is deaerated and poured into an appropriate column and then repeatedly washed with a total volume of 600 ml of phosphate-buffered saline (PBS), pH 6.8.

The flow rate of toxin material through either column is about 8–10 ml/hr. The chromatographic process could be run once again through both columns to optimize toxin recovery. The columns are washed with

PBS, pH 6.8, until the effluent no longer absorbs at 280 nm. The two columns are then disconnected and separately eluted (~200 ml) with PBS, pH 7.5, containing 5 m*M* DTT. The material is collected (5-ml fractions) and monitored for optical absorbance at 280 nm and hemolytic activity. The hemolytic fractions are pooled, combined, and concentrated by ultrafiltration to about 20 ml in a stirred cell equipped with Amicon PM10 membrane (fraction F_3).

Gel Filtration on Sephacryl S-200 Column. Fraction F_3 is applied onto a Sephacryl S-200 (Pharmacia) column (2.5 × 100 cm) equilibrated in PBS, pH 6.8, supplemented with 10% (v/v) glycerol. The effluent is monitored as described at the preceding step and the hemolytic fractions are pooled and concentrated by ultrafiltration to 15 ml (fraction 4).

Isolectric Focusing. Fraction 4 is submitted to preparative isoelectric focusing (IEF) performed with a 110-ml column (LKB Produkter, Stockholm, Sweden) as directed in the LKB manual; we use a final concentration of 1% carrier ampholytes (LKB ampholines) which has a pH range of 3.5 to 10 (pH 3.5 to 10 ampholines, 0.5 ml; pH 6 to 8 ampholines, 2 ml) in a sucrose density gradient obtained by mixing high- and low-density sucrose–ampholine solutions prepared as follows. The less dense solution is constituted by the totality of F_4 fraction, 0.8 ml of ampholine mixture adjusted to a final volume of 50 ml with distilled water after dissolution of 0.83 g of sucrose (Merck). The more dense solution contains 1.7 ml of ampholine mixture and 23.3 g of sucrose adjusted to a final volume of 50 ml with distilled water. The central electrode compartment (anode) contains 20 ml of 1% sulfuric acid in 60% sucrose dissolved in distilled water. The cathode, which consistes of 0.2 ml of ethylenediamine in 10 ml of distilled water, is layered on the top of the gradient. A final potential of 1600 V is applied for 16 hr; the gradient is cooled continuously by circulating water at 4°. At the end of the run, column content is collected by 1-ml fractions. The pH (at 4°), optical density at 280 nm, and hemolytic activity of each fraction are determined. The ampholine–sucrose gradient is not itself hemolytic.

The hemolytic material, which focuses around pH 7.8 in a narrow peak, is collected in a volume ranging from 10 to 15 ml (fraction F_5).

Gel Filtration on BioGel P-100. Fraction F_5 is gel filtered on a BioGel P-100 (fine, 100–200 mesh, Bio-Rad Laboratories, Richmond, CA) column (2.5 × 100 cm) equilibrated with PBS, pH 6.8–10% glycerol to eliminate ampholines, sucrose, and some nonhemolytic material absorbing at 280 nm. The hemolytic fractions elute as a sharp peak. In order to obtain the highest specific activity, the fractions around the top of the peak are collected (fraction F_6). The other fractions from both sides of the peak are recovered separately and have slightly lower specific activity.

Characteristics of the Pure Toxin

Fraction F_6 showed generally a single band by SDS–PAGE after staining by Commassie Blue or silver nitrate. The corresponding M_r was calculated as 60K. Another faint band corresponding to an M_r of 55K is sometimes present in certain preparations and very likely represents nicked SLO fragments. Fraction F_6 is also homogeneous when tested by immunodiffusion techniques. It can therefore be considered as the pure toxin preparation with an apparent specific activity of 6×10^5 HU/mg protein when assayed with sheep red blood cells. The overall yield process averaged around 6–10% on the basis of recovered hemolytic activity. This yield could be increased 2-fold if the isoelectric focusing step is omitted. However, in this case, the toxin preparation has a specific activity of about 90% of that obtained by running the whole process. Substantial losses in toxin activity are avoided if the Sephacryl and BioGel columns are previously treated with crude concentrated SLO material and then thoroughly washed and kept in the cold with 0.02% sodium azide.

Two other valuable techniques of purification have been reported.[9,10] The recovery varied from 3.2[9] to 10%.[10] The cloning and expression of SLO gene in *E. coli* has been recently reported.[11]

Assay of Streptolysin O

The quantitative determination of SLO in crude or purified material is usually based on the estimation of the hemolytic activity of toxin expressed as hemolytic units (HU) per milliliter of toxin solution. The HU is arbitrarily defined by most authors as that smallest amount (highest dilution) of toxin previously activated that will liberate half the hemoglobin (50% lysis) from a suspension of washed erythrocytes (usually sheep or rabbit) under fixed conditions of time, temperature, ionic environment, and red blood cell concentration. The HU employed is unfortunately not identical in all instances, since the parameters of the assay system vary in different laboratories. An agreement for a unified titration system by all investigators studying SLO and related toxins is highly desirable and would facilitate the comparison of experimental data. However, in spite of this situation, the figures found in the literature are at least roughly comparable.

A detailed description of our assay system of SLO on rabbit erythrocytes as well as the rationale, limitations, and requirements for accurate

[10] S. Bhakdi, M. Roth, A. Sziegoleit, and J. Tranum-Jensen, *Infect. Immun.* **46,** 394 (1984).
[11] M. Kehoe and K. N. Timmis, *Infect. Immun.* **43,** 804 (1984).

determination of the hemolytic activity have been reported elsewhere.[12] For practical reasons we use at present 2.25% sheep erythrocyte standardized suspensions (6×10^8 cells/ml) instead of 2.25% rabbit erythrocyte suspensions. However, either sheep or rabbit cell suspensions are always standardized so that 0.5 ml lysed by adding 14.5 ml of 0.1% sodium carbonate solution gives an optical absorbance of 0.200 at 541 nm in a Beckman spectrophotometer in a 10-mm light path cuvette. Under these conditions 1 HU of SLO in the sheep red blood cell assay system is equivalent to 2 HU in the rabbit system. The standard sheep red blood cells (SRBC) suspension is kept at 4° and used within 5 days.

SLO is reduced prior to assay by incubation of toxin solution for 10 min at room temperature with 20 m*M* cysteine. Then volumes of appropriately diluted toxin solution decreasing in 0.1-ml amounts from 1 to 0.1 ml are placed in small serology tubes and the volume in all tubes is brought to 1 ml by adding PBS containing 0.1% bovine serum albumin. Then 0.5 ml SRBC is added. The tubes are incubated at 37° for 45 min and then briefly centrifuged. The percentage of hemolysis is estimated by the optical absorbance at 541 nm of the hemoglobin in the supernatant fluid. One HU is the amount of test material which causes 50% lysis of the cells. For improved accuracy the data may be estimated graphically by plotting on a log probit graph the percentage lysis against the toxin dilution.[13]

[12] J. E. Alouf, M. Viette, R. Corvazier, and M. Raynaud, *Ann. Inst. Pasteur, Paris* **108,** 476 (1985).

[13] J. E. Alouf and M. Raynaud, *Ann. Inst. Pasteur, Paris* **114,** 812 (1968).

[9] Production, Purification, and Assay of Streptolysin S

By JOSEPH E. ALOUF and CATHERINE LORIDAN

Two hemolytic and cytolytic toxins, streptolysin O (SLO) and streptolysin S (SLS), are produced by group A streptococci.[1,2] SLO (see Chapter [8] in this volume) is a 60-kDa immunogenic protein released in conventional culture media throughout the exponential phase of growth.[3] In contrast, SLS is a less well-understood, nonimmunogenic toxin which is

[1] J. E. Alouf, *Pharmacol. Ther.* **11,** 661 (1980).

[2] I. Ginsburg, *in* "Microbial Toxins" (T. C. Montie, S. Kadis, and S. J. Ajl, eds.), Vol. 3, p. 99. Academic Press, New York, 1970.

[3] B. Dassy and J. E. Alouf, *J. Gen. Microbiol.* **129,** 643 (1983).

never detected unless a carrier (also called inducer) is added to the culture or to resting cell suspensions.[2,4] Many chemically unrelated inducers have been reported, the most potent being the RNase-resistant fraction of yeast RNA (RNA core) and lipoteichoic acids.[1,5] The hemolytic activity can be transferred from one inducer to another.[1,2] This led to the concept that SLS, as first suggested by Bernheimer,[6] is a complex formed by a peptide associated with an inducer, the inducer acting as a carrier or stabilizer. The peptide was estimated to consist of 32 amino acids.[7] Previous attempts to isolate a carrier-free hemolytic moiety never succeeded, suggesting that this moiety undergoes decay or denaturation when the complex is treated in ways designed to remove or destroy only the carrier.[4] Recently, the successful separation of the hemolytically active peptide of about 1800 Da has been achieved in our laboratory.[8]

The purification, to various extents, of RNA core-induced SLS has been reported by various authors,[1,8] but no biochemical or immunochemical indications of the homogeneity of the purified preparations were provided, due to the lack of sensitive staining methods and immune sera. In a recent work the purification of RNA core SLS was described and shown to be homogeneous by silver staining after electrophoresis in SDS–polyacrylamide gel.[8]

SLS is a very potent membrane-damaging agent. It lyses a wide variety of living cells. Its lytic spectrum is somewhat broader than SLO; it is lytic or cytotoxic not only for eukaryotic cells but also for wall-less forms of some bacteria, notably protoplasts and L forms from various species. All eukaryotic cells tested were found damaged: erythrocytes, lymphocytes, polymorphonuclear leukocytes, platelets, various tissue culture cells, and tumor cells. Intracellular organelles such as mitochondria and lysosomes are also disrupted by SLS.[1,6,9] The lytic effects of SLS are inhibited by phospholipids, suggesting that these ubiquitous biomembrane components are presumably involved in the cytolytic action of the toxin.[10]

[4] A. W. Bernheimer, *J. Exp. Med.* **90,** 373 (1949).

[5] T. Theodore and G. B. Calandra, *in* "Chemistry and Biological Activities of Bacterial Surface Amphiphiles" (G. D. Schockman and A. J. Wicken, eds.), p. 271. Academic Press, New York, 1981.

[6] A. W. Bernheimer, *in* "Streptococcal Infections" (L. W. Wannamaker and J. H. Matsen, eds.), p. 19. Academic Press, New York, 1972.

[7] C. Y. Lai, M. T. Wang, J. B. De Faria, and T. Akao, *Arch. Biochem. Biophys.* **191,** 804 (1978).

[8] C. Loridan and J. E. Alouf, *J. Gen. Microbiol.* **132,** 307 (1986).

[9] J. H. Freer and J. P. Arbuthnott, *in* "Mechanisms in Bacterial Toxinology" (A. W. Bernheimer, ed.), p. 170. Wiley, New York, 1976.

[10] J. L. Duncan and L. Buckingham, *Biochim. Biophys. Acta* **648,** 6 (1981).

Strains Used for Toxin Production

Most strains of group A hemolytic streptococci produce SLS, which is indeed the main lytic agent responsible of the clear zones of β-hemolysis around the colonies in agar media. Naturally, occurring SLS$^-$ strains are seldom seen, therefore SLS is generally considered to be a conserved trait. Over 95% of groups A, B, C, and G strains tested produced serum or RNA-induced SLS.[2] Production of the hemolysin by certain strains from groups belonging to groups E, H, and L was also reported.[2,5]

Individual group A strains vary in their quantitative ability to produce SLS. High SLS yields are obtained with strains C203S (type 3) provided by the American Type Culture Collection and first used by Bernheimer and then in many laboratories,[4,5,8] C203A (ATCC 14289),[7] S23g,[10] and CB112252.[5] These strains produce both SLS and SLO. Blackmore (type 11) strain produces only SLS.[2]

Little information is available about the genetic control of SLS (as well as SLO) production in streptococci. Transposon Tn*916* has been transferred from *Steptococcus faecalis* and inserted into the chromosome of group A streptoccal strain C591, leading to inactivation of gene(s) essential for SLS expression whereas that of SLO was not affected.[11] Upon excision of the transposon SLS was once again expressed.

SLS-Like Toxins

SLS-like toxins have been shown to be produced by *Treponema hyodysenteriae* in the presence of RNA core, Tween 80, and serum albumin, by *Streptococcus mutans* in the presence of Tween 80, and by *Streptococcus agalactiae* in the presence of Tween, starch, or lipoteichoic acid. For the last two organisms RNA core was inactive as an inducer (see Ref. 8 for references). Recently, *Hemophilus pleuropneumonia* has been reported to produce an SLS-like product.[12]

Production of Streptolysin S

Lyophilized *Streptococcus pyogenes* C203S strain is grown overnight at 37° without shaking in brain heart infusion broth (BHI, Difco) supplemented with 1% (w/v) maltose and 2% (w/v) sodium bicarbonate (BHI-BM). A 50-ml volume of an overnight culture is then inoculated in 2.5 liters BHI-BM and grown for 5 hr at 37°. The culture is centrifuged at 12,000 *g* for 25 min at 4° and the cell pellet washed in 100 m*M* potassium

[11] K. Nida and P. Cleary, *J. Bacteriol.* **155,** 1156 (1983).

[12] P. G. Martin, P. Lachance, and D. F. Neven, *Can. J. Microbiol.* **31,** 456 (1985).

phosphate buffer, pH 7.0, before resuspension to a final volume of 40 ml in induction phosphate buffer (IB) medium (100 m*M* KH_2PO_4, 2 m*M* $MgSO_4$; adjusted to pH 7.0 with NaOH) supplemented with 30 m*M* maltose. The cell suspension (about 17 mg dry wt cocci/ml) is incubated for 5 min at 37° and then induced by adding 0.5 mg yeast RNA core (Sigma Chemical Co., St. Louis, MO) per milliliter of suspension (2.9%, w/w, cocci) for 5 min without shaking. The cell suspension is then immediately centrifuged at 4° at 15,000 *g* for 20 min. The supernatant (crude SLS) is collected and supplemented with ammonium acetate buffer, pH 7.0 (100 m*M* final concentration) to stabilize SLS.[7] The pellet is then resuspended in 40 ml of IB supplemented with maltose (15 m*M* final concentration), and induced as described above. Four successive inductions could be made on the same pellet. The combined hemolytic material (about 240 ml) contained 3–5 × 10^5 hemolytic units (HU) of SLS obtained from 2.5 liters of culture (450–750 HU/mg cell dry weight).

Purification of Streptolysin S

All steps were done at 4°. Crude nondialyzed SLS preparation (240 ml) is applied to a column (2.5 × 14 cm) of hydroxylapatite (BioGel-HTP, Bio-Rad Laboratories, Richmond, CA) that has been equilibrated with 100 m*M* potassium phosphate/100 m*M* ammonium acetate buffer (pH 7.0). The column is washed with this buffer and then eluted with 400 m*M* potassium phosphate/100 m*M* ammonium acetate buffer (pH 7.0). The flow rate is 30 ml/hr and 3-ml fractions are collected and analyzed for A_{280} and hemolytic titer. The hemolytic activity of the processed material is recovered as a single sharp peak in fractions 160–190 of the column. The fractions are pooled (ca. 30 ml) and concentrated by ultrafiltration in an Amicon cell fitted with a YM2 membrane (M_r cut-off 1000). After repeated washing in this cell with 20 m*M* potassium phosphate/100 m*M* ammonium acetate buffer (pH 7.0), the retentate (fraction F_1, 4.5 ml) is supplemented with 5 ml of 6 *M* guanidine · HCl as described by Lai *et al.*,[7] allowed to stand at room temperature for 15 min, ice chilled, loaded onto a Sephadex G-100 column (2–6 × 100 cm) equilibrated with 20 m*M* potassium phosphate/100 m*M* ammonium acetate buffer (pH 7.0), and eluted fractions (3.5 ml) are monitored as described above. Two protein peaks are eluted, the first of which coincides with hemolytic activity. Fractions 80–110 have the highest specific activity. They are pooled and concentrated by ultrafiltration (fraction F_2). The apparent M_r of this pool determined by gel filtration on the calibrated column is about 20,000 in the presence of guanidine · HCl and 40,000 in its absence. When submitted to SDS–PAGE and silver staining, fraction F_2 runs as a single band migrating at the same

R_f as bromothymol blue, suggesting an apparent M_r below 4000. Fraction F_2 contains ca. 40% of the hemolytic activity of the initial crude material and is considered to be highly purified SLS.

Characteristics of Purified Streptolysin S

Fraction F_2, submitted to polyacrylamide gel electrophoresis in the absence of detergent, reveals a single hemolytic band on the migration front, as shown either by cutting the gel with subsequent elution and titration or by layering it with sheep red blood cells. Its specific activity ranges from 1.2 to 3.6 × 10^6 HU mg/ml depending on purification batches.

However, at this step the toxin still contains small amounts of free yeast RNA core and associated RNA core protein(s) which are eliminated by submitting F_2 to isoelectric focusing.[8] This is done with a 110-ml column (LKB) as directed in the LKB manual. A pH 3.5–11 gradient is generated with two sucrose solutions containing, respectively, 5 and 15% (v/v) carrier ampholytes (LKB ampholines) and 1.5 and 45% (w/v) sucrose as described for SLO in Chapter [8] in this volume. The less dense solution contains SLS sample. The central electrode solution (anode) consists of 1% (v/v) sulfuric acid and 60% (w/v) sucrose. The cathode solution (0.2 ml ethylenediamine in 10 ml distilled water) is layered on the top of the gradient. A final potential of 1600 V is applied for 16 hr at 2°, then column contents are collected (1-ml fractions), and monitored for pH (4°), A_{280}, and hemolytic titer. This process resolves the F_2 fraction into two sharp hemolytic peaks around pH 3.6 and 9.2 and four nonhemolytic peaks (p*I* 3.9, 4.2, 4.5, 6.0). The p*I* 3.6 fraction is very highly purified SLS–RNA core complex devoid of free RNA core. The p*I* 9.2 fraction is the RNA core-free hemolytic peptide moiety of about 1800 Da.[8]

Streptolysin S Assay

The lytic effect of SLS on erythrocytes is used for toxin assay. Defibrinated sheep blood is centrifuged (5000 *g*, 5 min) and the erythrocyte pellet is washed three times in 150 m*M* sodium-PBS, pH 6.8. Sheep red blood cells (SRBC) are suspended (2.5%, v/v) in PBS such that a 30-fold dilution of this suspension in distilled water gives an A_{541} of 0.200 (see this volume [8]). This standard SRBC suspension (about 6 × 10^8 cells/ml) is kept at 4° and used within 5 days.

Volumes of appropriately diluted toxin solution decreasing in 0.1-ml amounts from 1 to 0.1 ml are placed in serology tubes and the volumes in all tubes are brought to 1 ml by adding PBS; then 0.5 ml SRBC is added. The tubes are incubated at 37° for 45 min and then briefly centrifuged. The percentage of hemolysis is estimated by the A_{541} of the hemoglobin in the supernatant fluid. One hemolytic unit is defined as the amount of test material which causes 50% lysis of the cells. For improved accuracy, the data may be estimated graphically by plotting on a log probit graph the percentage of lysis against the volume of toxin sample.[13] Appropriate controls are set by mixing the material with cholesterol (10 μg/ml), which inhibits the hemolytic activity of SLO but not that of SLS, or with trypan blue (13 μg/ml), which inhibits SLS. The RNA core alone used for induction did not exhibit any hemolytic activity up to 10 mg/ml.

[13] J. E. Alouf and M. Raynaud, *Ann. Inst. Pasteur, Paris* **114,** 812 (1968).

[10] Production, Purification, and Assay of Streptococcal Erythrogenic Toxin

By CHRISTIANE GEOFFROY and JOSEPH E. ALOUF

Erythrogenic toxin (ET), also known as Dick toxin, scarlatinal or scarlet fever toxin and streptococcal pyrogenic exotoxin, is produced by many strains of group A streptococci[1,2] but not by other streptococcal groups. Three antigenically distinct types (A, B, C) have been described.[1,3] These toxins elicit a wide range of biological effects, including the production of erythematous reactions in human, rabbit, and guinea pig skin, pyrogenicity, lethality, enhanced susceptibility to endotoxin shock, alteration of the blood-brain barrier (permitting entry of toxic products and bacteria), depression of reticuloendothelial clearance function, alteration of host antibody response to immunogens cardiotoxicity, and T-cell mitogenicity.[1–6] ET also elicits the production of interferon.[7]

In this chapter we will deal only with type A toxin which has been studied more extensively than the two other types. The toxin is a phage conversion product[8] secreted by the bacteria. The structural gene is car-

[1] J. E. Alouf, *Pharmacol. Ther.* **11,** 661 (1980).
[2] D. W. Watson and Y. B. Kim, *in* "Microbial Toxins" (T. C. Montie, S. Kadis, and S. J. Ajl, eds.), Vol. 3, p. 173. Academic Press, New York, 1970.
[3] P. M. Schlievert, K. M. Beltin, and D. W. Watson, *J. Infect. Dis.* **140,** 676 (1979).
[4] P. M. Schlievert and D. W. Watson, *Infect. Immun.* **21,** 753 (1978).
[5] H. Knöll, D. Gerlach, J. H. Ozegowski, V. Hribalovà, and W. Köhler, *Zentralbl. Bakteriol., Parasitenk., Infektionskr. Hyg., Abt. 1: Orig, Reihe A* **256,** 49 (1983).
[6] J. M. Cavaillon, C. Leclerc, and J. E. Alouf, *Cell. Immunol.* **76,** 200 (1983).
[7] J. M. Cavaillon, Y. Rivière, J. Svab, L. Montagnier, and J. E. Alouf, *Immunol. Lett.* **5,** 323 (1982).
[8] S. K. Nida and J. J. Ferretti, *Infect. Immun.* **36,** 745 (1982).

ried by the bacteriophage[9,10] and its nucleotide sequence has been recently established.[11] The calculated molecular weight of ET is 25,785, in agreement with the values determined on the purified protein which range from 28K to 30K[1,11] although a low-molecular-weight value has been reported for certain preparations.[1,12]

Strains Used for Toxin Production

A list of various strains reported to produce ET toxins A, B, and C (some strains release two or three types) has been published.[1,3,5,8] The NY-5 strain ATCC 12351 has been the most widely used for toxin production and studies.[1]

Production of Erythrogenic Toxin

This toxin is produced in our laboratory by cultivation of Kalback S84 type 3 strain in the medium used for streptolysin O (SLO) production. The culture supernatant obtained for the purification of SLO is concentrated as reported for this toxin by ultrafiltration on Amicon DC2 apparatus (see this volume [8]).

Purification of Erythrogenic Toxin

The retentate (4000 ml corresponding to 25 liters of culture supernatant) is stirred in calcium phosphate gel generated in this fluid as described for SLO (see this volume [8]). The resulting supernatant fraction, devoid of significant amounts of SLO hemolysin, is used as the source for the purification of ET. Fraction S_1 (4500 ml, approximately) is precipitated by adding ammonium sulfate to 80% saturation and allowing to stand overnight at 4°. After centrifugation at 10,000 *g* for 30 min at 4°, the dark brown pellet is dissolved in about 100 ml of isotonic phosphate-buffered saline (PBS), pH 6.8, and dialyzed exhaustively against this buffer until NH_4^+ is no longer detected in the dialysate by Nessler's reagent (Merck, Darmstadt, FRG). The solution obtained (fraction S_2, 150 ml, approximately) is salted out between 50 and 80% of ammonium sulfate. The precipitate is separated by centrifugation, dissolved in about 20 ml of PBS (fraction S_3), and gel filtered on a Bio-Gel P-100 (fine, 100–200 mesh, Bio-Rad Laboratories, Richmond, CA) column (2.5 × 100 cm) equilibrated

[9] C. R. Weeks and J. J. Ferretti, *Infect. Immun.* **46,** 531 (1984).
[10] L. P. Johnson and P. M. Schlievert, *Mol. Gen. Genet.* **194,** 52 (1984).
[11] C. R. Weeks and J. J. Ferretti, *Infect. Immun.* **52,** 144 (1986).
[12] C. W. Houston and J. J. Ferretti, *Infect. Immun.* **33,** 862 (1981).

with PBS, pH 6.8. The fraction corresponding to the volume eluted between 160 and 240 ml is collected and concentrated by ultrafiltration to about 20 ml in a stirred cell equipped with Amicon PM10 membrane (fraction S_4).

Fraction S_4 is then submitted to preparative isoelectric focusing (IEF) performed with a 110-ml column (LKB Produkter, Stockholm, Sweden) as directed in the LKB manual; we used a final concentration of 1% carrier ampholytes (LKB ampholines) which has a pH range of 3.5 to 10 (pH 3.5 to 10 ampholines, 2.5 ml) in a sucrose density gradient prepared as described in this volume [8]. The less dense solution contains S_4 fraction and 0.8 ml of 40% ampholines. The more dense sucrose solution contains 1.7 ml of the ampholines. The central electrode compartment (anode) contains 1% sulfuric acid in 60% sucrose. The cathode, which consists of 0.2 ml of ethylenediamine in 10 ml of distilled water, is layered on top of the gradient. A final potential of 1600 V is applied for 16 hr; the gradient is cooled by circulating water at 4°. At the end of the run, column content is collected in 1-ml fractions. The pH (at 4°) and optical density at 280 nm are determined.

The material, which focuses around pH 4.8 in a narrow peak, is collected and dialyzed in Amicon cell by using PBS to eliminate ampholines and sucrose. This fraction (S_5) constitutes purified ET. Another electrophoretic form of ET of p*I* 4.2 is often separated simultaneously with that of p*I* 4.8 in the IEF process. Both forms show total cross-reactivity against immune sera raised by rabbit immunization with fraction S_3 and therefore can be combined for biological studies.

Characteristics of the Pure Toxin

Fraction S_5 showed a single band by SDS–PAGE after staining by Coomassie blue or silver nitrate. The corresponding M_r value was calculated as ca. 28K, in agreement with other reports[1,11] and with nucleotide sequencing.[11] Purified ET may occasionally contain traces of streptolysin O as evidenced by hemolytic assay. These traces are readily eliminated by adding 1 vol of cholesterol ethanolic solution (10 mg/ml) to 20 vol of ET sample. The micellar cholesterol–SLO precipitate is eliminated by centrifugation at 10,000 *g* for 10 min at 4°.

Biological Assay of Erythrogenic Toxin

Many of the various biological activities reported to be associated with ET (see Introduction) can be used for the assay of the toxin as well as hemagglutination inhibition[4] and enzyme-linked immunosorbent assay.[12] Three different assays could be recommended for practical purposes.

Erythematous Skin Test in Rabbits or Guineas Pigs. The backs of 2.5–3 kg rabbits (chinchilla or New Zealand species are particularly sensitive) or 300–500 g guinea pigs are carefully shaved at least 1 hr before injection and injected intradermally with various doses of ET under a volume of 0.1 ml with 1-ml tuberculin syringes. A positive skin reaction is demonstrated by erythema, which is outlined with a marking pen and its diameter measured. The skins are observed at 24 and 48 hr after injection. One skin test dose is that amount of toxin which elicits an erythema of 1 cm in diameter. In rabbits and guinea pigs, it corresponds to 1 and 0.1 μg of toxin, respectively. A reaction can be elicited by as low as 1 pg in man ($>10^9$ skin test doses/mg toxin).

Lymphocyte Blast Transformation. The lymphocytes are separated from heparinized peripheral human or rabbit blood (10 heparin units/ml blood). They are diluted with an equal volume of 0.9% NaCl, layered on Ficoll-Hypaque (Pharmacia, Fine Chemicals, Uppsala, Sweden) (d = 1.077 g/cm^3), and submitted to gradient centrifugation. The lymphocyte fraction is withdrawn and washed three times with tissue culture medium (RPMI 1640, Gibco Laboratories, Grand Island, NY) buffered with $NaHCO_3$ and supplemented with penicillin (100 U/ml), streptomycin (100 μg/ml), and mycostatin (100 U/ml) and enriched with 10% pooled human (or rabbit) plasma or calf serum. The washed lymphocytes are counted in a hematocytometer and then resuspended at a concentration of 10^6 cells/ml in the medium.

The cells are cultured in 96-well flat-bottomed microtiter plates (Falcon 3072, Falcon Products, Oxnard, CA). Samples of 100 μl of cell suspension are introduced into each well (10^5 cells/well) and then 100 μl of ET dilution in tissue culture medium (0.1–25 μg/well). The plates are incubated at 37° in a humidified air–CO_2 (5%) atmosphere. After a 3- or 6-day incubation, the cultures are pulse labeled with [^{3}H]thymidine (1 μCi/well, sp. act. 5 Ci/mmol) and after a further incubation of 18 hr the cultured cells are collected with a multiple sample harvester on glass fiber filters (Whatman). The filters are dried, placed in 2 ml Lipofluor (J. T. Baker Chemicals, Phillipsburg, NJ), and counted in a liquid scintillation counter at 0.4 min/sample. The median value of triplicate cultures is used for data analysis. Control tubes are run with sterile tissue culture medium in place of the test products. The results are expressed as the stimulation index, which is the ratio of the counts per minute (cpm) of the stimulated culture to the cpm of the nonstimulated control culture.

Immunological Assay of Erythrogenic Toxin. ET can also be detected and quantitated by an enzyme-linked immunosorbent assay (ELISA) as described by Houston and Ferretti.[12] This technique is sensitive and specific. It detects microgram amounts of type A erythrogenic toxin.

[11] Diphtheria Toxin: Purification and Properties

By STEPHEN F. CARROLL, JOSEPH T. BARBIERI, and R. JOHN COLLIER

Monomeric diphtheria toxin (DT) is a complex enzyme that performs at least three functions during the course of intoxicating sensitive mammalian cells.[1-3] It binds to receptors on the cell surface; undergoes an internalization and translocation process whereby at least the enzymatically active portion of the toxin penetrates to the cytosolic compartment; and finally catalyzes transfer of the ADP-ribosyl moiety of NAD into covalent linkage with elongation factor 2 (EF-2), blocking protein synthesis and causing cell death.

DT is secreted by *Corynebacterium diphtheriae* as a single (M_r 58,342) polypeptide proenzyme, but purified preparations of the toxin frequently contain a heterogeneous array of molecular species[4]:

1. In most preparations a fraction of the toxin has been cleaved by contaminating trypsin-like proteases, yielding a "nicked" form containing two large peptide fragments (A and B) linked by a disulfide bridge. It is the nicked form which is believed to be responsible for the biological effects of DT.[5] The A fragment (DTA; M_r 21,167) exhibits full enzymatic activity *in vitro,* but is essentially nontoxic to intact cells. The B fragment (DTB; M_r 37,195), which mediates receptor recognition and membrane translocation, is also nontoxic. Other proteolytic fragments of the toxin are occasionally detected.[6]

2. A fraction of the toxin in most preparations is incapable of binding NAD (or ATP), due to the presence of tightly but noncovalently bound dinucleotides (predominantly ApUp) which block the NAD site.[7,8] Occupancy of the nucleotide-binding site blocks the NAD-glycohydrolase (NADase) activity of whole DT, but has no influence on subsequent toxic-

[1] R. J. Collier, *in* "ADP-Ribosylation Reactions" (O. Hayaishi and K. Ueda, eds.), p. 575. Academic Press, New York, 1982.

[2] A. M. Pappenheimer, Jr., *Harvey Lect.* **76,** 45 (1982).

[3] T. Uchida, *in* "Molecular Action of Toxins and Viruses" (P. Cohen and S. van Heyningen, eds.), p. 1. Elsevier, Press, New York, 1982.

[4] S. F. Carroll, J. T. Barbieri, and R. J. Collier, *Biochemistry* **25,** 2425 (1986).

[5] K. Sandvig and S. Olsnes, *J. Biol. Chem.* **256,** 9068 (1981).

[6] D. M. Gill and L. L. Dinius, *J. Biol. Chem.* **246,** 1485 (1971).

[7] S. Lory, S. F. Carroll, P. D. Bernard, and R. J. Collier, *J. Biol. Chem.* **255,** 12011 (1980).

[8] J. T. Barbieri, S. F. Carroll, and R. J. Collier, *J. Biol. Chem.* **256,** 12247 (1981).

ity when assayed at 37°.[9,10] Neither the source of the endogenous dinucleotides nor their function has been determined with certainty.

3. DT is also found in various states of aggregation. A high-molecular-weight 6.8 S form (relative to the 4.2 S monomer) was first reported by Goor,[11] and other authors subsequently described similar forms in various toxin preparations (see Ref. 4). These aggregates represent noncovalent multimers of DT, which apparently associate by hydrophobic interactions between the B fragments of two toxin monomers. Recent studies have shown[4,12] that the dimeric toxin can be crystallized, and is significantly less toxic to animals or intact cells than is monomer. The reduced toxicity of dimer may result from the loss of receptor recognition.

The present article describes the purification of DT from toxin preparations obtained from commercial sources, and its subsequent fractionation into the various forms. Inasmuch as monomeric and multimeric DT have been found to interconvert during incubation at temperatures below −10°,[13] suitable conditions for freezing and storage are also considered. Methods for assaying the toxic, cytotoxic, and enzymatic activities of DT are described in the companion chapter ([13] in this volume).

The production and purification of toxin from laboratory cultures of *C. diphtheriae* have been discussed elsewhere,[14] and involve sequential chromatography of culture supernatants on phenyl-Sepharose and DEAE-Sephacel. For reasons outlined below, we recommend that these columns be equilibrated and eluted with buffer B containing the appropriate salt. The toxin so obtained is predominantly unnicked, nucleotide-free monomer.

Purification of DT

Commercial Sources of DT. Crude preparations of DT are available from Connaught Laboratories (Ontario, Canada); import licenses are required for shipment into the United States (obtained from the Center for Disease Control, Atlanta, GA). Each vial contains 100,000 LF units (ca. 250 mg) as a concentrated cell-free supernatant derived from fermentor cultures by precipitation with ammonium sulfate,[15] and should be stored

[9] R. L. Proia, L. Eidels, and D. A. Hart, *J. Biol. Chem.* **256,** 4991 (1981).

[10] J. T. Barbieri and R. J. Collier, unpublished observations.

[11] R. S. Goor, *Nature (London)* **217,** 1051 (1968).

[12] R. J. Collier, E. M. Westbrook, D. B. McKay, and D. Eisenberg, *J. Biol. Chem.* **257,** 5283 (1982).

[13] S. F. Carroll and R. J. Collier, unpublished observations.

[14] R. Rappuoli, M. Perugini, I. Marsili, and S. Fabbiani, *J. Chromatogr.* **268,** 543 (1983).

[15] L. A. Robb, D. W. Stainer, and M. J. Scholte, *Can. J. Microbiol.* **16,** 639 (1970).

at −70°. Although the relative percentages of the various toxin forms vary from lot to lot, their proportions within any given lot remain fairly constant. Other sources for DT include List Biochemicals (Campbell, CA) and the Swiss Serum Institute (Bern, Switzerland). We have not analyzed toxins from the latter sources in detail.

Principle. DT, the most abundant protein in culture supernatants of *Corynebacterium diphtheriae,* has an isoelectric point (p*I*) of 4.1 and is readily chromatographed on anion-exchange resins such as DEAE-Sephacel. Appropriate choice of ionic strength and elution conditions resolves the culture supernatant into monomeric, dimeric, and other oligomeric forms of DT. A variable percentage of the toxin in these supernatants is already nicked, and the remainder is generally converted to the nicked form by treatment with low concentrations of trypsin. Each form of the toxin is then subjected to size-exclusion chromatography to remove inactivated trypsin and other trace contaminants, producing homogeneous monomer (99% pure) and dimer (98% pure). Starting with 1.25 g, recovery of DT ranges between 700 and 900 mg (ca. 70%); the relative percentages of the various forms reflect their concentrations in the initial crude culture filtrate. The nucleotide-free and nucleotide-bound toxins can then be fractionated by affinity chromatography on supports to which have been attached native ligands (such as NAD or ATP) or various dyes (see below). Concentrations of the purified toxins are determined optically, using $E^{1\%}_{1\,cm}$ (280 nm) of 13.0 and 13.4 for nucleotide-free and nucleotide-bound DT, respectively. The composition and purity of each fraction are monitored by high-performance size-exclusion chromatography (HPSEC, Fig. 1).

Reagents. Buffers used are as follow:

Buffer A: 10 m*M* Tris–HCl, 1 m*M* EDTA, pH 7.5, containing 1 μg/ml PMSF and 0.02% sodium azide
Buffer B: 25 m*M* Tris–HCl, 1 m*M* EDTA, pH 7.5
Buffer C: 50 m*M* Tris–HCl, 1 m*M* EDTA, pH 7.5
Buffer D: Buffer C containing 4 *M* urea and 50 m*M* mercaptoethanolamine (Sigma)

DEAE-Sephacel. In a typical purification we process one to five bottles of Connaught toxin (0.25 to 1.25 g); outlined below is the protocol for five bottles. Scale-down for lesser quantities is easily accomplished by appropriate reductions in volumes, flow rates, and column cross-sectional areas. Each bottle is allowed to thaw in a tray of cool water, and then diluted with 4 vol of ice-cold buffer A. All subsequent operations are performed at 4°. The diluted culture supernatant is applied by gravity feed to a 2.5 × 16 cm column of DEAE-Sephacel equilibrated in buffer B. This

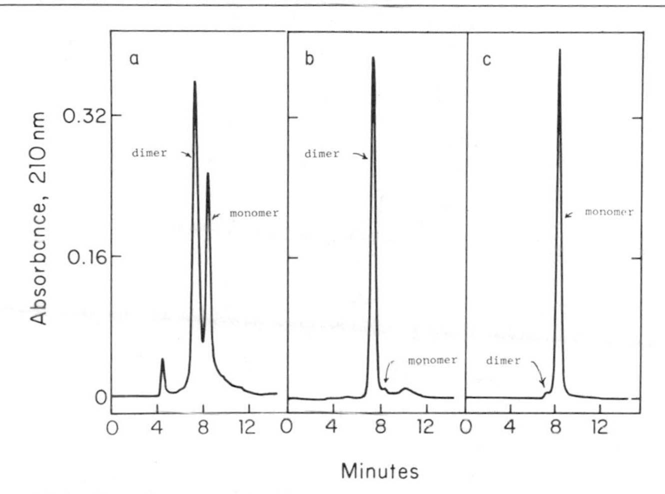

FIG. 1. HPSEC elution profiles of DT samples taken at various stages of purification. At $t = 0$, aliquots were injected onto a Bio-Rad TSK-250 column equilibrated in 20 mM NaH_2PO_4, 100 mM Na_2SO_4, pH 6.8, at 1.0 ml/min, and protein was monitored at 210 nm. (a) commercial preparation of DT; (b) purified DT dimer; (c) purified DT monomer. Percentages of the various forms were determined from the relative peak heights and their position of elution.

step is most conveniently performed overnight using a safety loop attached to the column inlet. The column is then washed with buffer B until the absorbance at 280 nm approaches 0 (two to four column volumes), and eluted with a linear 400-ml gradient of 0 to 0.225 M NaCl in buffer B at a flow rate of 32 ml/hr. Little if any toxin passes through the column during the loading and washing steps, so these are collected in batch. During gradient elution, 2.5-ml fractions are collected with monitoring at 280 nm. Monomeric DT elutes first, followed by dimeric DT (Fig. 2). Each peak is pooled into monomer-rich and dimer-rich preparations and, if necessary, concentrated to >5 mg/ml by ultrafiltration on Amicon YM10 membranes. Higher multimeric forms are found in trailing fractions.

Nicking. Because of variations among toxin batches and lots of trypsin, it is advisable to perform an analytical incubation to optimize digestion conditions. Addition of 1 mM NAD to reaction mixtures containing monomeric DT prevents subsequent degradation of DTA; similar additions to dimer are unnecessary. Digestion mixtures (10 to 20 μl) are prepared on ice and contain pooled monomeric or dimeric DT fractions (ca. 5 mg/ml) with or without NAD. Digestion is initiated by adding trypsin to 1.0 μg/ml and incubating the mixtures at 25°. After 0, 15, 30, 45, and 60 min, 2-μl aliquots are removed, mixed with 10 μl of soybean trypsin inhibitor (10 μg/ml), and electrophoresed in a 12.5% SDS–polyacryl-

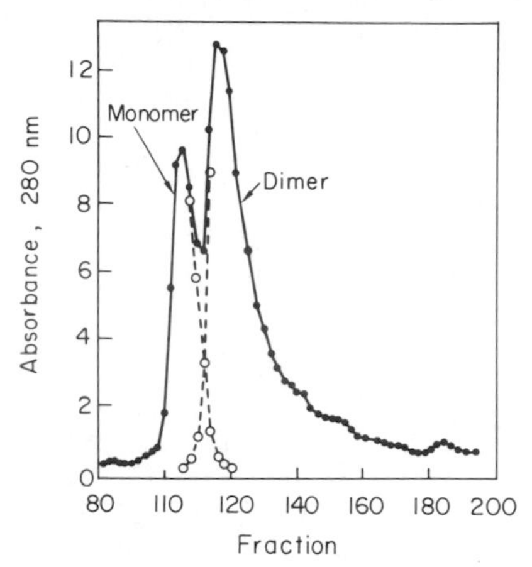

FIG. 2. Fractionation of DT monomer and dimer on DEAE-Sephacel. A commercial DT preparation (2 g) was applied to a column of DEAE-Sephacel equilibrated in 25 m*M* Tris–HCl, 1 m*M* EDTA, pH 7.5, and eluted as described in the text with monitoring at 280 nm. The relative percentages of monomer and dimer in column fractions were determined by HPSEC as described in Fig. 1.

amide gel under reducing conditions.[16] That incubation period which yields complete nicking without further degradation is selected, and preparative digestion is performed by appropriate increases in reaction volumes. Following the addition of trypsin inhibitor and incubation for 5 min at 25°, the samples may be stored at −70°.

Final Purification of Monomer and Dimer. Monomer-rich and dimer-rich fractions, before or after digestion with trypsin, are concentrated by ultrafiltration on YM10 membranes to ca. 20 ml, and individually chromatographed on a 2.5 × 100 cm column of Sephacryl S-200 equilibrated in buffer C. The column is eluted at a flow rate of 25 ml/hr, and 2.5-ml fractions are collected with monitoring at 280 nm. The resultant peaks are individually pooled, concentrated by ultrafiltration (if necessary), and stored in small vials at −70°. Electrophoretic profiles of purified monomeric and dimeric DT in polyacrylamide gels are shown in Fig. 3.

Fractionation of Nucleotide-Free and Nucleotide-Bound DT. Matrex Gel Green A (Amicon), an agarose gel containing a modified triazine dye, can effectively substitute for ATP-Sepharose in the separation of nucleotide-free and nucleotide-bound toxins, and offers greater stability and adsorptive capacity.[4] Monomeric or dimeric DT (20 mg at 2 mg/ml) is applied by gravity feed to a 1.3 × 4.5 cm column of Matrex Gel Green A equilibrated at 4° in buffer C, and 1-ml fractions are collected. The nucleo-

[16] U. K. Laemmli, *Nature (London)* **227,** 680 (1970).

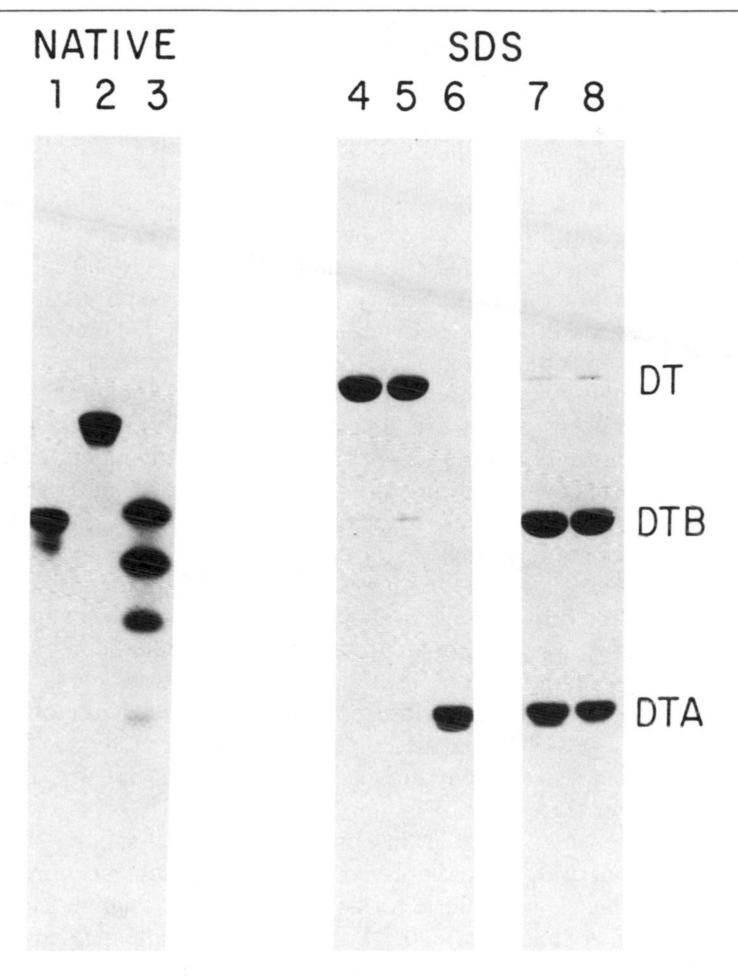

FIG. 3. Polyacrylamide gel electrophoretic profiles of purified DT fractions. Samples (2–4 μg) were electrophoresed in 7.5% native gels (no detergent, lanes 1–3) or in 11.25% SDS gels in the absence (lanes 4–6) or presence (lanes 7 and 8) of 2-mercaptoethanol. Lanes 1, 4, and 7, purified DT monomer; lanes 2, 5, and 8, purified DT dimer; lanes 3 and 6, DT fragment A. Reprinted from Ref. 4.

tide-bound form of DT passes through unretarded. After several column volumes of buffer C are passed through the column such that the absorbance at 280 nm approaches 0, nucleotide-free DT is eluted batchwise with 0.75 *M* NaCl in the same buffer. Each peak is pooled, concentrated by ultrafiltration, dialyzed against 1000 vol of buffer C at 4° for 16 hr, and stored in small aliquots at −70°. The 280/260 nm absorbance ratios for nucleotide-bound and nucleotide-free DT are ca. 1.2 and 1.7, respectively.

Preparation of DTA. To a solution of nicked DT (monomer, dimer, or a mixture of the two; 20 to 120 mg) in a total volume of 6.4 ml of buffer C is added solid urea (4.8 g) to a final concentration of 8 *M*. Solid dithiothreitol (0.15 g) is then added to 0.1 *M*, and the mixture is incubated at 25° for 30 min. The reduced and denatured fragments are chromatographed at 4° on a 1.8 × 115 cm column of Sephacryl S-200 equilibrated in buffer D at a flow rate of 12.5 ml/hr. Two-milliliter fractions are collected. Under these conditions, baseline resolution is achieved between DTB and DTA. Fractions containing DTA are pooled, dialyzed against three 2-liter changes of buffer C containing 10 m*M* mercaptoethanolamine, and stored at −70°. Concentrations are determined optically, versus appropriate blanks, using the extinction $E^{1\%}_{1\,cm}$ (280) = 1.5. If degradation of DTA to lower M_r forms occurs during storage, the fragment can be stabilized by adding adenine to 1.0 m*M*. Prior to using such preparations, the ligand should be removed by dialysis or desalting.

Stability and Storage of Different Forms. Monomeric DT (1–30 mg/ml) is unaffected by storage at 0–4° for several months, or incubation at 37° for several days. Multimeric forms are less stable. Dimer shows slight conversion to monomer (5–10%) over a period of weeks at 4°, and higher order multimers dissociate into complex mixtures of oligomers within several days. At 37° dimer dissociates to monomer at about 5%/day. The addition of carrier protein (such as BSA to 0.2 mg/ml) to DT multimers does not alter their stability.

When frozen at ca. −20° in phosphate-buffered saline (PBS), monomeric or dimeric DT rapidly converts into a mixture of monomeric, dimeric, and higher multimeric forms.[13] This effect is significantly reduced if the toxin is stored at −70°. Various factors influence the extent of interconversion induced by freezing, the most significant of which is the type of buffer used. Conversion is maximal in buffers containing sodium phosphate or borate, but essentially undetectable in Tris, MOPS, HEPES, or potassium phosphate. An excellent correlation exists between buffers that undergo dramatic reductions in pH during freezing (the liquid phase of initially neutral sodium phosphate solutions can approach pH 2–3)[17,18] and those that promote interconversion of monomers and multimers. For storage below −10° sodium phosphate or borate buffers should therefore be avoided. Routinely, samples are stored at or below −70° in Tris–HCl or HEPES.

[17] L. van den Berg and D. Rose, *Arch. Biochem. Biophys.* **81,** 319 (1959).
[18] Y. Orii and M. Morita, *J. Biochem.* (*Tokyo*) **81,** 163 (1977).

Interconversion of Molecular Forms

Conversion of Monomer to Dimer. The effects of freezing DT in certain buffers suggested a simple method for producing dimer from monomer. DT monomer (2–5 mg/ml) is dialyzed or desalted into PBS, then frozen at −22° for 48 hr. The sample is then thawed and chromatographed on Sephacryl S-200 for isolation of dimer and residual monomer.

Solvent-Induced Dissociation of Dimer to Monomer. Urea at concentrations up to 2 *M* causes little dissociation of dimer, whereas higher concentrations convert the toxin into fully denatured polypeptides. The dimer is similarly affected by guanidine hydrochloride, and NaCl or NaSCN (up to 2 *M*) induce only slight changes in structure. In contrast, incubation with dimethyl sulfoxide quantitatively dissociates the dimer into fully functional monomers.[4] Endogenous dinucleotides remain associated with the toxin following such treatments.

Five milliliters of DT dimer (ca. 10 mg/ml) in buffer C is placed in a 10-ml beaker containing a small stir bar. At room temperature, an equivalent volume of 80% dimethyl sulfoxide is added dropwise with constant mixing, then the solution is incubated at 37° for 10 min. The sample is desalted by dialysis or by chromatography on a 2 × 20 cm column of Sephadex G-50F equilibrated in buffer C, collecting 1-ml fractions. If a precipitate forms, the sample should be clarified by centrifugation and rechromatographed on Sephacryl S-200 as described for the initial purification.

Removal of Endogenous Dinucleotides. The predominant dinucleotide associated with DT (ApUp) binds with sufficient avidity (K_d = 9 p*M* at 5.5°)[19] such that it copurifies with the toxin in routine procedures. At 37°, the binding constant decreases to 1.8 n*M*. The following procedure takes advantage of this decreased affinity at 37° to remove endogenous dinucleotides by competitive exchange with another toxin ligand.

Samples of nucleotide-bound toxin (monomer or dimer) and inositol hexaphosphate (IHP, recrystallized from 20% methanol and neutralized with HCl) are mixed to final concentrations of 5 mg/ml and 50 m*M*, respectively, in 8 ml of buffer C and incubated at 37° for 15 min. During this incubation, an equivalent volume of 100 m*M* IHP in the same buffer is warmed to 37° and loaded onto a water-jacketed 3.2 × 24 cm column of Sephadex G-50F thermostatted to 37° in buffer C. This "preload" of IHP substantially reduces the amount of dinucleotides which remain associated with the toxin as it passes through the column. Following the incubation, the sample is chromatographed at 80 ml/hr, collecting 3-ml fractions.

[19] C. M. Collins, J. T. Barbieri, and R. J. Collier, *J. Biol. Chem.* **259,** 15154 (1984).

The toxin-containing peak is pooled and fractionated on Matrex Gel Green A as described above. By this procedure at least 90% of the dinucleotides are removed from monomeric toxin, and ca. 60% from the dimer.

[12] Isolation and Characterization of the *Botulinum* Neurotoxins

By LANCE L. SIMPSON, JAMES J. SCHMIDT, and JOHN L. MIDDLEBROOK

Structure in Relation to Biological Activity[1a]

Clostridium botulinum is an anaerobic, spore-forming organism that is relatively ubiquitous in its distribution.[1] The bacterium is known for its ability to produce two remarkably potent toxins. The first of these is a neurotoxin that acts preferentially on cholinergic nerve endings to block the release of acetylcholine.[2,3] The second is a more diffusely acting toxin that, among other things, appears to promote the movement of fluid across membranes.[4–6] Only the neurotoxin has been associated with human illness. The present chapter describes the isolation, purification, and assay of this toxin.

Botulinum neurotoxin has been isolated in seven immunologically distinct forms designated A, B, C, D, E, F, and G. Although the various neurotoxins are immunologically distinct, i.e., there is little cross-neutralization, they do share certain structural and functional properties. For example, work with monoclonal antibodies has shown that various serotypes may have similar epitopes. In addition, pharmacological experiments suggest that the various serotypes may proceed through the same general sequence of events in producing blockade of transmitter release. However, the neurotoxins do not share a common receptor[7] and there

[1a] The views of the authors do not purport to reflect the positions of the Department or the Army of the Department of Defense (para. 4-3, AR 360-5).

[1] L. D. Smith, "Botulism: The Organism, Its Toxins, the Disease." Thomas, Springfield, Illinois, 1977.

[2] C. B. Gundersen, *Prog. Neurobiol.* **14,** 99 (1980).

[3] L. L. Simpson, *Pharmacol. Rev.* **33,** 155 (1981).

[4] W. I. Jensen and R. M. Duncan, *Jpn. J. Med. Sci. Biol.* **33,** 81 (1980).

[5] I. Ohishi, M. Iwasaki, and G. Sakaguchi, *Infect. Immun.* **31,** 890 (1981).

[6] L. L. Simpson, *J. Pharmacol. Exp. Ther.* **223,** 695 (1982).

[7] S. Kozaki, *Naunyn-Schmiedebergs Arch. Pharmakol.* **308,** 67 (1979).

may be both qualitative and quantitative differences in their subcellular actions.[8–11]

The individual neurotoxins are synthesized as single-chain polypeptides having molecular weights of approximately 150,000.[12] In this form, the molecule is characterized by the presence of at least one intrachain disulfide bond and by relatively low toxicity. When this precursor molecule is exposed to trypsin or trypsin-like enzymes, it is cleaved to yield a dichain molecule. The heavy chain (M_r ~100,000) and the light chain (M_r ~50,000) remain linked by an interchain disulfide bond.

Conversion to the dichain molecule, commonly referred to as "nicking," is usually accompanied by an increase in specific toxicity.[3,12] However, the relationship between nicking and full activation remains unclear.[12] There are three sites at which proteolytic cleavage could lead to an increase in toxicity: (1) proteolysis at the amino terminus, (2) proteolysis within the molecule to generate the dichain structure, and (3) proteolysis at the carboxy terminus. Recent work on sequencing of the neurotoxin has shown that activation does not involve proteolysis at the amino terminus.[13] Additional work is needed to clarify the role of nicking and proteolysis at the carboxy terminus.

Exposing the neurotoxin to disulfide bond-reducing agents produces loss of toxicity.[14] Dialysis of the mixture to remove the reducing agent allows reconstitution of the molecule and restoration of biological activity.[15] Conversely, the reduced molecule can be submitted to electrophoresis in polyacrylamide gels with sodium dodecyl sulfate (SDS). This technique allows resolution of the heavy and light chains, and in the presence of the appropriate markers it permits a determination of apparent molecular weight.

The basic structure of the various *botulinum* toxins is notably similar to that of tetanus toxin.[16] It is interesting that the heavy chain of tetanus toxin can undergo proteolytic cleavage when exposed to papain, and the products are two nontoxic polypeptides having molecular weights of ap-

[8] L. C. Sellin, S. Thesleff, and B. R. DasGupta, *Acta Physiol. Scand.* **119,** 127 (1983).

[9] J. A. Kauffman, J. F. Way, Jr., L. S. Siegel, and L. C. Sellin, *Toxicol. Appl. Pharmacol.* **79,** 211 (1985).

[10] L. S. Siegel, *Toxicol. Appl. Pharmacol.* **84,** 255 (1986).

[11] L. L. Simpson, *Infect. Immun.,* **52,** 858 (1986).

[12] G. Sakaguchi, *Pharmacol. Ther.* **19,** 165 (1983).

[13] J. J. Schmidt, V. Sathyamoorthy, and B. R. DasGupta, *Arch. Biochem. Biophys.* **238,** 544 (1985).

[14] B. R. DasGupta and H. Sugiyama, *Biochem. Biophys. Res. Commun.* **48,** 108 (1972).

[15] B. Syuto and S. Kubo, *J. Biol. Chem.* **256,** 3712 (1981).

[16] B. R. DasGupta and H. Sugiyama, *in* "Perspectives in Toxinology" (A. W. Bernheimer, ed.), p. 87. Wiley, New York, 1977.

proximately 50,000.[17] The same pattern of papain-induced cleavage has been obtained with *botulinum* neurotoxin (B. R. DasGupta, personal communication).

The general features of toxin-induced blockade of transmitter release have been determined, but the precise mechanism of action remains to be established.[3,18,19] The neurotoxin can block acetylcholine release from numerous sites, but it shows greatest affinity for nerve endings that impinge on striate muscle. A variety of experiments suggest, but do not prove, that the toxin proceeds through a sequence of three steps in producing its effects at the neuromuscular junction. There is an initial binding step that involves a receptor, as yet uncharacterized, on the surface of the plasma membrane. The binding step does not alter neuromuscular transmission. The next step is believed to be internalization, which may in fact represent two sequential mechanisms. There is a receptor-mediated endocytosis event that causes the toxin to enter the nerve terminal, and there is a channel-forming event that allows the internalized toxin to reach the cytoplasm. The third and final step is the one that accounts for neurotoxicity. *Botulinum* neurotoxin exerts an effect on the nerve terminal, the end result of which is blockade of nerve stimulus-induced release of acetylcholine.

Work is now being done to establish the structure–function relationships of the *botulinum* neurotoxin molecule. There is convincing evidence that the heavy chain mediates binding. By analogy with tetanus toxin, one might propose that the carboxy terminus of the heavy chain possesses ligand-binding properties.[20,21] There is suggestive evidence that the amino terminus of the heavy chain may play a role in penetration of the endosome membrane. This portion of the molecule has been shown to form pH-dependent channels in artificial membranes.[22]

The precise mechanism for blockade of acetycholine release has not been established, but a host of clues suggests that the toxin is an enzyme.[3,19] It is worth noting that other potent toxins of microbial origin (e.g., diphtheria toxin) have been shown to proceed through a sequence of steps identical to that discussed above, and many of these toxins are

[17] T. Helting and O. Zwisler, *J. Biol. Chem.* **252,** 187 (1977).

[18] L. L. Simpson, *J. Pharmacol. Exp. Ther.* **212,** 16 (1980).

[19] L. L. Simpson, *Annu. Rev. Pharmacol. Toxicol.* **26,** 427 (1986).

[20] N. P. Morris, E. Consiglio, L. D. Kohn, W. H. Habig, M. C. Hardegree, and T. B. Helting, *J. Biol. Chem.* **255,** 6071 (1980).

[21] R. L. Goldberg, T. Costa, W. H. Habig, L. D. Kohn, and M. C. Hardegree, *Mol. Pharmacol.* **20,** 565 (1981).

[22] D. H. Hoch, M. Romero-Mira, B. E. Ehrlich, A. Finkelstein, B. R. DasGupta, and L. L. Simpson, *Proc. Natl. Acad. Sci. U.S.A.* **82,** 1692 (1985).

thought to be enzymes (e.g., to possess ADP-ribosylating activity).[23,24] To the extent that it is appropriate to extrapolate from findings on these other toxins, one might speculate that the light chain of *botulinum* neurotoxin possesses catalytic activity.

Data on the structure–function relationships of the *botulinum* neurotoxin molecule indicate that there are three types of isolation and characterization experiments that should be done. First, the seven immunological forms of the toxin need to be isolated and purified to homogeneity. Second, the heavy and light chains of each toxin must be isolated, and ideally the amino terminus and carboxy terminus of the respective heavy chains should also be isolated. Finally, the primary structure of each neurotoxin should be determined, thus allowing for identification of possible sequence homologies and perhaps pointing to possible active sites within the molecule.

None of these three has been carried entirely to completion. Techniques for isolating types A, B, C, D, E, and F have been reported and are summarized below; a method for isolating and purifying type G has not been published. The technique for reducing the interchain disulfide bond and for separating the heavy and light chains of several toxins has been established. The method for separating the papain products of the heavy chain has not been reported, but the approach is fundamentally the same as that used with tetanus toxin. No one has yet reported the primary structure of an entire neurotoxin molecule.

Isolation of the Neurotoxins

Toxigenic organisms as well as the toxin itself should be handled only by trained personnel who have been immunized with toxoid. A multivalent toxoid can be obtained by qualified investigators; it is supplied by the Centers for Disease Control, Atlanta, Georgia. Isolation of the toxin should be carried out in a facility equipped to handle highly pathogenic organisms and toxic substances. A P-III containment facility is ideal.

There are two forms in which the neurotoxin is customarily isolated from organisms. Most clostridia are proteolytic, and thus they are capable of converting the single chain precursor into the active dichain molecule. For example, type A strains are proteolytic, and the toxin obtained from the culture fluids of these organisms is fully activated. By contrast, type E

[23] D. M. Gill, *in* "Bacterial Toxins and Cell Membranes" (J. Jeljaszewicz and T. Wadström, eds.), p. 291. Academic Press, London, 1978.

[24] O. Hayaishi and K. Ueda (eds.), *in* "ADP-Ribosylation Reactions." Academic Press, New York, 1982.

strains are typically nonproteolytic. The toxin obtained from these organisms is predominantly in the single chain form. An exogenous source of trypsin must be used to promote proteolytic cleavage and activation.

The methods for isolation of the *botulinum* neurotoxins are currently undergoing a substantial change. Traditionally, isolation of the toxins has involved a series of precipitation and/or centrifugation steps, followed by several fractionations that employed gravity flow chromatography. A concise description of this methodology as it pertains to type A neurotoxin is given below. More recently, a high-performance chromatographic procedure has been devised.[25] This technique will be described in greater detail.

Tse *et al.* have published a method for purification of type A toxin.[26] The method begins with precipitation of the toxin from the growth medium by adjustment of the pH to 3.5, followed by centrifugation. The precipitate is extracted with 0.2 *M* sodium phosphate buffer, pH 6.0, and the extract is incubated with ribonuclease, 100 μg/ml, at 34° for 3 hr. Next, ammonium sulfate is added to 60% saturation and the precipitate is collected by centrifugation. It is then redissolved in 0.05 *M* sodium citrate, pH 5.5, and adsorbed batchwise with DEAE-Sephadex. After removal of the resin by centrifugation, the supernatant is loaded onto a column of DEAE-Sephacel, equilibrated with the citrate buffer. Toxin emerges in the flow-through; it is precipitated with ammonium sulfate and redissolved in 0.05 *M* sodium phosphate, pH 6.8.

The preparation is then adsorbed batchwise with the affinity resin, which is *p*-aminophenyl-β-D-thiogalactopyranoside covalently coupled to CH-Sepharose 4B (easily synthesized, or purchased ready-made from Pharmacia P-L Biochemicals, Piscataway, NJ). At this stage, the toxin is in the form of a noncovalently bound complex with a hemagglutinating protein, which in turn binds to the affinity resin with considerable tenacity.[27] The resin–protein slurry is gently stirred at 25° for 2 hr, then transferred to a small chromatography column. After unbound proteins are removed from the column by washing with buffer, toxin is dissociated from hemagglutinin and eluted from the column with 0.1 *M* sodium phosphate, pH 7.9, containing 1.0 *M* NaCl. Hemagglutinin remains firmly bound to the resin. Toxin from the affinity column is next equilibrated with 0.15 *M* Tris hydrochloride, pH 7.9, and applied to a column of DEAE-Sephacel in the same buffer. Most of the toxicity remains in the

[25] J. J. Schmidt and L. S. Siegel, *Anal. Biochem.* **156,** 213 (1986).

[26] C. K. Tse, J. O. Dolly, P. Hambleton, D. Wray, and J. Melling, *Eur. J. Biochem.* **122,** 493 (1982).

[27] L. J. Moberg and H. Sugiyama, *Appl. Environ. Microbiol.* **35,** 878 (1978).

flow-through. This material appears to be pure type A neurotoxin, based on polyacrylamide gel electrophoresis, electrofocusing, and immunodiffusion tests.

Similar types of methods have been used to purify other toxins. For example, Kitamura *et al.* have described a method for isolating type E neurotoxin that is complexed with at least one other protein.[28,29] This complex could be dissociated by chromatography on DEAE-Sephadex at alkaline pH to yield a toxic and a nontoxic component. The former was judged to be pure neurotoxin according to the criteria employed at the time. Later refinements have produced a neurotoxin that appears pure on SDS–polyacrylamide gel electrophoresis (SDS–PAGE).[30]

The methods noted above, like most traditional methods for isolating the neurotoxin, are quite lengthy, and they include ammonium sulfate precipitations and size-exclusion chromatography. Consequently, these methods are not directly compatible with current high-performance chromatography systems. In contrast, a method for the purification of type E neurotoxin has recently been developed which exploits the tendency of the neurotoxin to form a stable, noncovalently bound complex with at least one other intrinsic protein.[25] The procedure is based solely on ion-exchange chromatography; preconcentration steps, ammonium sulfate precipitations, and size-exclusion columns are unnecessary. With a high-performance ion-exchange system and a series of columns obtained from Pharmacia Fine Chemicals (Piscataway, NJ), pure neurotoxin can be obtained after five steps: (1) extraction of neurotoxin from the bacteria; (2) protamine sulfate treatment; (3) cation-exchange chromatography at pH 5.7; (4) anion-exchange chromatography at pH 7.6; (5) cation-exchange chromatography at pH 5.7. These steps are described in detail below.

Extraction of Neurotoxin from the Bacteria. Unlike other serotypes, most of the type E neurotoxin is found inside the bacteria at the end of the growth period. Therefore the cells can be collected by centrifugation and later extracted with phosphate buffer as described.[28,30] However, the ammonium sulfate precipitation step and the initial chromatography step that are characteristic of earlier methods are not needed. Instead, the crude extract is dialyzed against 25 m*M* sodium succinate, pH 5.7 (succinate buffer).

Protamine Sulfate Treatment. The neurotoxin does not bind to a negatively charged resin unless the concentration of nucleic acid in the crude extract is substantially decreased. This can be achieved by incubation

[28] M. Kitamura, S. Sakaguchi, and G. Sakaguchi, *Biochim. Biophys. Acta* **168,** 207 (1968).
[29] M. Kitamura, S. Sakaguchi, and G. Sakaguchi, *J. Bacteriol.* **98,** 1173 (1969).
[30] B. R. DasGupta and R. Rasmussen, *Toxicon* **21,** 535 (1983).

with ribonuclease[28,30] or by addition of protamine sulfate.[25] The latter requires significantly less time than the former and does not include incubation at elevated temperature, which may lead to proteolysis of the neurotoxin. Sufficient protamine sulfate is added to the dialyzed crude extract such that, following centrifugation, the supernatant has an A_{260}/A_{280} ratio of 0.8 or less.

Cation-Exchange Chromatography. Supernatant from the preceding step is applied to a Mono-S (sulfopropyl) column of an HPLC system, equilibrated with succinate buffer. The neurotoxin binds to the resin and is eluted with a linear gradient (10 m*M*/ml) of increasing NaCl concentration. Neurotoxin, at this stage in the form of a noncovalently bound complex with at least one other protein, is found in the first major peak to elute. It is equilibrated with 25 m*M* Tris chloride, pH 7.6 (Tris buffer) by gel filtration on Sephadex G-25 (superfine).

Anion-Exchange Chromatography. The neurotoxin fraction from the Mono-S column is applied to a Mono-Q (QAE) column, equilibrated with Tris buffer. Neurotoxin binds to the column and is eluted, as described above, in the first major peak to emerge. Chromatography at alkaline pH on an anion-exchange resin dissociates the complex and separates the neurotoxin from nontoxic binding component(s).[12] Appropriate fractions are pooled and equilibrated with succinate buffer by gel filtration on Sephadex G-25 (superfine).

Second Cation-Exchange Chromatography. Purification of the neurotoxin is completed by a second application to the Mono-S column, again equilibrated with succinate buffer. A linear gradient (21 m*M*/ml) elutes the neurotoxin at an apparent NaCl concentration of approximately 0.19 *M*, compared to a concentration of about 0.09 *M* in the first cation-exchange step. The different elution profiles suggest that free neurotoxin is more electropositive than the neurotoxin–protein complex. The techniques of SDS–PAGE and protein sequencing can be used to demonstrate the high purity of the product, which is at least equal to that obtained with considerably longer methods.

Beginning with dialyzed crude extract, one can obtain at least 4 mg of pure neurotoxin within two working days with the high-performance ion-exchange system. But note that this purification scheme does not completely depend on high-resolution chromatography; rather, it depends primarily on the substantially different chromatographic behavior of free neurotoxin relative to that of the neurotoxin–protein complex. Therefore, the procedure is readily adapted to user-prepared columns, with the advantage that an entire 50-liter fermentor run can be processed in one batch.

In large-scale runs, chromatography is done with the aforementioned succinate and Tris buffers. One additional step is inserted, equivalent to the first CM-Sephadex step of Kitamura *et al.*[28] Dialyzed crude extract is percolated through a 2.6 × 30 cm column of CM-Sepharose CL-6B (Pharmacia Fine Chemicals, Piscataway, NJ). The neurotoxin-containing flow-through is then treated with protamine sulfate. The same CM-Sepharose column (regenerated) is used in place of the first Mono-S step. Then a 1.6 × 35 cm column of DEAE-Trisacryl (LKB Instruments, Gaithersburg, MD) is substituted for the Mono-Q column. Final chromatography is done on a 1.6 × 35 cm column of CM-Trisacryl, in place of the last Mono-S step. After samples are applied to columns, linear gradients are formed with 300 ml of buffer and 300 ml of buffer/0.3 *M* NaCl. Neurotoxin prepared in this way is equal in purity to that from HPLC system.

Purified neurotoxin can be stored at −70°, and it can be thawed at least once with no loss of activity. Although type E strains are considered to be nonproteolytic, storage of the neurotoxin at 4°, with or without ammonium sulfate precipitation, sometimes leads to the appearance of modified forms with concomitant activation.

The tendency to form complexes with certain other proteins is a characteristic of most of the *botulinum* neurotoxins. This suggests that the isolation procedure developed for type E neurotoxin may prove successful for other neurotoxins as well. Preliminary data indicate that this is true for type B neurotoxin.

Unlike type E, most of type B neurotoxin is found in the culture fluid at the end of the growth period. Therefore, the culture fluid is acidified to pH 3.5 and the precipitate is isolated by centrifugation. It is resuspended in and dialyzed against 0.2 *M* sodium citrate, pH 5.5, and incubated at 37° for 3 hr with 50 μg/ml ribonuclease. The preparation is then percolated through a 2.5 × 50 cm column of DE-52 anion exchanger (Whatman Inc., Clifton, NJ) equilibrated with citrate buffer. Neurotoxin emerges in the flow-through, but the column retains most of the brown pigment and some of the nucleic acid and contaminating proteins.

From this point on, the procedure for type B is analogous to that for type E, with the exception that different pH values must be used for the cation-exchange steps: first cation exchange, pH 4.9; second cation exchange (i.e., the last chromatographic step), pH 5.2. Prior to anion-exchange chromatography at alkaline pH, the neurotoxin, complexed with one or more other proteins, does not bind to cation-exchange resins at pH values above 5.0. But after removal of the binding protein(s) by chromatography at pH 7.6 on Mono-Q or DEAE-Trisacryl, the free neurotoxin binds to cation exchangers at pH 5.2.

One may purify small-to-medium amounts of type B neurotoxin in a short time with the 0.5 × 5 cm columns of the HPLC system. Alternatively, one may utilize the high capacities of user-packed columns to process a 50-liter fermentor run in one batch. Neurotoxin of equal purity is obtained either way.

Separating the Heavy and Light Chains

A method for obtaining heavy and light chains from type A neurotoxin has been published by Kozaki *et al.*[31] It is based in part on methods previously developed for separation of the chains in types B and C neurotoxin.[15,32]

The neurotoxin is dialyzed against sodium borate–sodium phosphate buffer at pH 8.5. This buffer is obtained by mixing 11.0 g $Na_2B_4O_7 \cdot 10H_2O$ and 6.5 g $NaH_2PO_4 \cdot 2H_2O$ in 1 liter of water. The dialyzed protein is then added to a QAE-Sephadex A-50 column (0.9 × 5.0 cm) equilibrated with the borate–phosphate buffer. The column is washed with 20 ml of buffer, followed by 10 ml of the same buffer containing 10 m*M* dithiothreitol (DTT). The column is allowed to stand overnight in buffer containing 100 m*M* DTT and 2 *M* urea.

Elution of protein is initiated by adding borate–phosphate buffer with 10 m*M* DTT and 2 *M* urea. The light chain (M_r ~58,000) elutes as a sharp peak within the first 40 ml. The buffer is then changed to contain 10 m*M* DTT, 2 *M* urea, and 0.2 *M* NaCl. The heavy chain (M_r ~105,000) emerges in the next 40 ml.

Assays for Protein and for Toxicity

The most sensitive method for detecting and quantifying botulinum neurotoxin activity is to conduct an *in vivo* assay. Serial dilutions of a standard are administered to mice, and the time to death for each administered dose is measured. Although the technique is often done with intraperitoneal injections, the intravenous route generates data that are more consistent. The activity of freshly prepared toxin is quantified by comparing its potency to that of the standard. The technique of Boroff and Fleck[33] illustrates the principles of the method and provides representative data. One should note that this method causes pain and suffering to animals, and it should not be used without justification.

[31] S. Kozaki, S. Togashi, and G. Sakaguchi, *Jpn. J. Med. Sci. Biol.* **33,** 61 (1981).
[32] S. Kozaki, S. Miyazaki, and G. Sakaguchi, *Infect. Immun.* **18,** 761 (1977).
[33] D. A. Boroff and U. Fleck, *J. Bacteriol.* **92,** 1580 (1966).

Another form of bioassay involves using the isolated neuromuscular preparation, such as the phrenic nerve–hemidiaphragm preparation.[34] As with the *in vivo* assay, the *in vitro* assay compares the paralysis times of tissues exposed to freshly prepared toxin and to known standards of toxin. The disadvantage of this method is that it is not as sensitive as the live animal assay. The advantages of the method are 2-fold: (1) when used appropriately, the *in vitro* assay can quantify the rates of the three steps involved in the onset of paralysis; and (2) animals can be sacrificed for the method in a way that does not cause pain or suffering.

Immunological techniques have been described for detecting protein that has antigenic determinants characteristic of those in botulinum neurotoxin. An ELISA method has been developed for several of the toxins (e.g., Ref. 35), and more recently a radioimmunoassay has been reported.[36] The latter rivals the sensitivities of the bioassays discussed above. Immunological techniques offer obvious advantages, but nevertheless there is one acknowledged drawback. Immunoassays detect protein, but not biological activity. There is abundant evidence that antigenicity and neurotoxicity can vary independently.

Note added in proof: Shortly after this chapter was finished, the complete primary structure of tetanus toxin was published.[37,38]

[34] E. Bulbring, *Br. J. Pharmacol. Chemother.* **1,** 38 (1946).
[35] S. Notermans, S. Kozaki, Y. Kamata, and G. Sakaguchi, *Jpn. J. Med. Sci. Biol.* **37,** 137 (1984).
[36] A. C. Ashton, J. S. Crowther, and J. O. Dolly, *Toxicon* **23,** 235 (1985).
[37] U. Eisel, *EMBO Journal* **5,** 2495 (1986).
[38] N. F. Fairweather and V. A. Lyness, *Nuc. Acids. Res.* **14,** 7809 (1986)

[13] Purification of Tetanus Toxin and Its Major Peptides

By JOHN P. ROBINSON

Tetanus toxin is a neurotoxic protein which contains no lipid or carbohydrate. It is produced by the anaerobic bacterium *Clostridium tetani* which can infect wounds under favorable conditions. The toxin acts by binding to gangliosides at the presynaptic membrane of certain synapses and blocking the release of transmitter substance.[1] Tetanus toxin is produced as a single peptide which is cleaved in the growth medium by

[1] S. I. Zacks and M. F. Sheff, *in* "Neuropoisons: Their Pathological Actions" (L. L. Simpson, ed.), Vol. 1, pp. 225–262. Plenum, New York, 1971.

proteolytic enzymes that are produced by the organism. The molecular weight of this toxin has been reported to be between 140,000 and 150,000.[2–4] This value is probably high due to aggregation and more recent values of 128,000 ± 3000 for the cleaved toxin in the absence of detectable aggregation and 140,000 ± 5000 for the uncleaved toxin with some aggregation are probably more accurate.[5] Once the single peptide is cleaved by proteases that are endogenous to the culture or artificially by trypsin the toxin consists of two peptides held together by disulfide linkage. These peptides are designated as H (heavy) and L (light) chains of 87,000 and 48,000 Da, respectively.[2,3,6,7] Following disulfide reduction, the H and L chains can be separated and each peptide is atoxic and immunogenic,[2] although it is difficult to completely remove the toxin from the H chain. The fully active toxin can be reconstituted by removing the reducing agent from equal molar mixtures of the H and L chain peptides in the presence of 2 *M* urea.[2]

The usual assay for tetanus toxin preparations is done in mice. About 0.1-ml vol of appropriate dilutions of toxin are injected into the muscle of the hind leg of 18- to 20-g mice. The animals are observed for 4 days for symptoms of paralysis or death. In purified toxin preparations there are generally around 10^8 mouse lethal doses (MLD) per milligram of toxin.

Purification

The most widely used growth medium for tetanus toxin production was developed by Latham *et al.*[8] The medium contains no proteins and this simplifies the problem of separating tetanus toxin from other proteins in the culture filtrate. The uncleaved (cell extract) toxin can be prepared by hypertonic extraction from cells near the end of exponential growth phase.[9] The protease-cleaved toxin (filtrate toxin) is prepared from older culture filtrates. The toxin is concentrated in cell extracts or culture filtrates by salt precipitation at slightly acid[10] or neutral[11] pH or, more conveniently, by pressure dialysis. Precipitates are dissolved and de-

[2] M. Matsuda and M. Yoneda, *Biochem. Biophys. Res. Commun.* **57,** 1257 (1974).
[3] C. J. Craven and D. J. Dawson, *Biochim. Biophys. Acta* **317,** 277 (1973).
[4] J. P. Robinson, J. B. Picklesimer, and D. Pue II, *J. Biol. Chem.* **250,** 7435 (1975).
[5] J. P. Robinson, L. A. Holladay, J. H. Hash, and D. Pue II, *J. Biol. Chem.* **257,** 407 (1982).
[6] T. B. Helting, S. Parschat, and H. Engelhardt, *J. Biol. Chem.* **254,** 10728 (1979).
[7] J. P. Robinson and J. H. Hash, *Mol. Cell. Biochem.* **48,** 33 (1982).
[8] W. C. Latham, D. F. Bent, and L. Levine, *Appl. Microbiol.* **10,** 146 (1962).
[9] M. Raynaud, *C. R. Compt. Rend. Acad. Sci.* **225,** 543 (1947).
[10] B. Bizzini, J. Blass, A. Turpin, and M. Raynaud, *Eur. J. Biochem.* **17,** 100 (1970).
[11] S. G. Murphy and K. D. Miller, *J. Bacteriol.* **94,** 580 (1967).

salted on Sephadex G-50. The toxin is then separated from most other components by elution from a DEAE-cellulose column with a linear NaCl gradient (0–0.1 *M*) in 0.5 *M* Tris-HCl buffer at pH 7.5.[11] This procedure provides sufficient purification for most purposes.

Proteolytic activity, which is a problem in these preparations, can be partially removed by gel filtration on Sephadex G-150.[12] The problem can also be controlled by treatment of purified toxin with phenylsulfonyl fluoride (PMSF).[12,13] Purified toxin prepared in this manner is suitable for most purposes and can be stored at 4° in 0.2 *M* phosphate buffer at pH 6.0 for 4–6 weeks without significant alteration, as judged by polyacrylamide gel electrophoresis.

To prepare toxin or H and L chains that are suitable for characterization by chemical or physical procedures, we use preparative gel electrophoresis as an additional purification step. We find this to be a useful method for consistently preparing large amounts of material of suitable purity.[12,14] The design of the electrophoresis system was based on observations which were made during repeated attempts to separate the H and L chain peptides by column chromatography following reductive cleavage of the interchain disulfide bond.[2,3,7] Isolation of the two peptides was dependent on both the urea concentration used in the reduction mixture and the time which elapsed between the final addition of urea to the reduction mixture before its application to the column.[12,15,16] The urea concentration was found to be optimal at 4 *M* and the time of exposure of the reducing mixture to the urea must be kept as short as possible. Aggregation of the reduction products was evident even after 15–20 min.[12] A gel electrophoresis system was then designed so that the time of exposure of toxin to both urea and reducing agents would be minimized. The stacking gel contained 4 *M* urea, the resolving gel contained no urea, and the upper buffer contained 2 m*M* glutathione or thioglycolic acid as reducing agent. The resolving gel which gave the best separation was 60–90 ml of 4% acrylamide–0.104% bisacrylamide. The stacking gel consisted of a 32-ml volume containing 2.5% acrylamide–0.625% bisacrylamide. The buffer system consists of the following: upper electrode buffer (6.32 g Tris and 3.94 g glycine/1000 ml containing 2 m*M* reducing agent), lower electrode buffer (48.4 g Tris and 200 ml 1 *N* HCl/1000 ml); and elution buffer (12.1 g Tris and 50 ml 1 *N* HCl/1000 ml). The elution buffer was made 1 m*M* in

[12] S. J. DiMari, M. R. Cumming, J. H. Hash, and J. P. Robinson, *Arch Biochem. Biophys.* **214,** 342 (1982).

[13] T. B. Helting and O. Zwisler, *Biochem. Biophys. Res. Commun.* **57,** 1263 (1974).

[14] S. J. DiMari, J. H. Hash, and J. P. Robinson, *Arch. Biochem. Biophys.* **214,** 354 (1982).

[15] M. Matsuda and M. Yoneda, *Biochem. Biophys. Res. Commun.* **86,** 635 (1979).

[16] S. van Heyningen, *FEBS Lett.* **68,** 5 (1976).

thioglycolic acid immediately before use. All buffers were deaerated by sonication before use.

Under these conditions, the toxin and charged reducing agent simultaneously come into contact with 4 *M* urea only while they are moving into and through the stacking gel. The separation of the H and L chains from one another and from any intact toxin is extended in the resolving gel where the various components migrate at different rates based on charge differences. Using this procedure we are able to prepare the H and L chains of tetanus toxin consistently in sufficient quantity and quality to allow physical and chemical characterization.[5,7,12,14] During these experiments it was observed that when filtrate toxin is applied to this gel electrophoresis system, a small amount of H and L chain peptides could be separated from a large toxin peak in the absence of a reducing agent. A comparison of peptides prepared in the presence of reducing agents with those prepared in the absence of a reducing agent revealed no detectable differences in amino acid composition, molecular weight, or immunogenic determinants. The amino-terminal residues were identical and it was concluded that cleavage by the proteolytic enzymes of the culture occurred close to and on either side of the interchain disulfide bond linking the H and L chain peptides in the native toxin. Consequently, there is a microheterogeneity produced at the cleavage site.[12] Figure 1 illustrates the elution profile of the L chain, toxin, and H chain during preparative electrophoresis and shows that separation is due to electrophoretic mobility. The order of elution is L chain, 48,000 Da (most highly charged), toxin, 128,000 Da, and H chain, 87,000 Da (least charged). Figure 2 shows the currently accepted molecular model for the tetanus toxin molecule and the cleavage sites of proteolytic enzymes.

This preparative electrophoresis system is not only useful for preparing the major toxin peptides, it can also be used to prepare the protease-cleaved and -uncleaved toxins in a state of high purity. By applying filtrate toxin preparations to the gel in the absence of reducing agents, one can separate the H and L chain peptides, which are formed by cleavage on the N-terminal side of the interchain disulfide bond, providing a more homogeneous filtrate toxin preparation consisting of toxin which was cleaved on the C-terminal side of the disulfide bond. This toxin is represented by the central peak in Fig. 1 and can be rerun through the apparatus if desired.

It has been particularly difficult to prepare uncleaved tetanus toxin which does not contain considerable amounts of the cleaved form. By

[17] L. Simpson, *Brain Res.* **305,** 177 (1984).

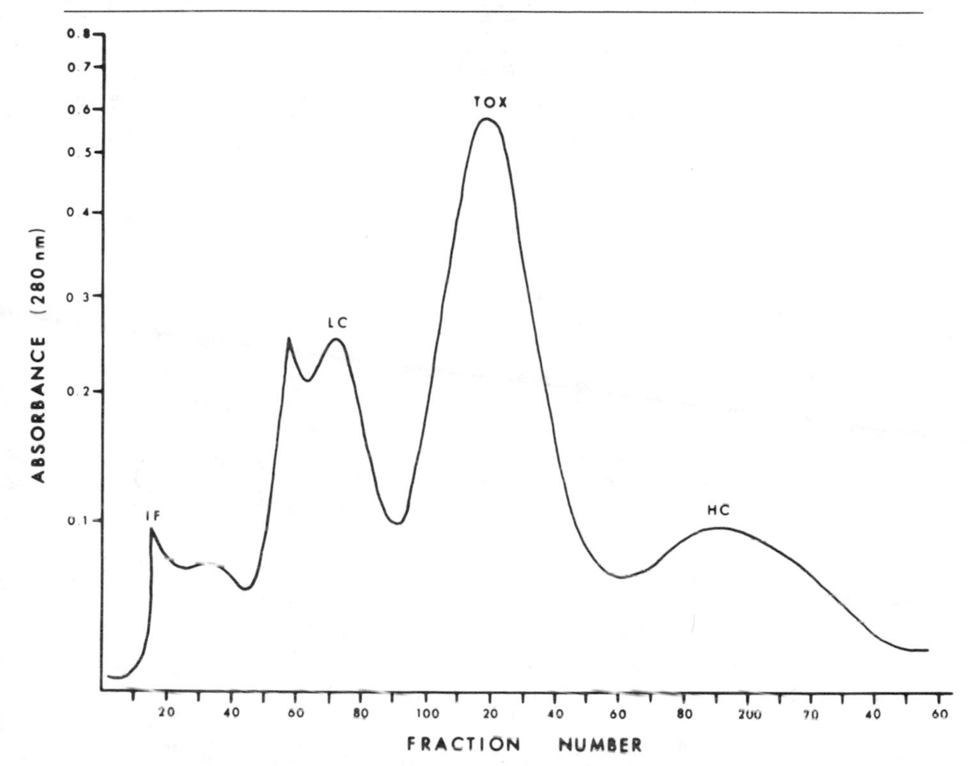

FIG. 1. Elution profile of toxin and the major peptides from preparative gel electrophoresis. Electrophoresis was conducted at 45 mA for 12 to 16 hr or until all components had been eluted from the gel. The flow rate at which elution buffer was pumped through was 12 to 18 ml/hr and 1-ml fractions were collected. The apparatus used was a Büchler Poly-Prep 200 cooled by tap water at 15–25°. The buffer and gel systems are detailed in the text. Tox, Toxin; LC, L chain; HC, H chain; IF, ion front.

applying extract toxin preparations to this gel electrophoresis system with the reducing agent in the upper electrode buffer, the cleaved toxin can be essentially removed and a highly purified cell extract toxin can be prepared.[12,14]

Preparation of Peptide C

Helting and Zwisler reported the preparation of a peptide by gel filtration following controlled papain digestion of filtrate toxin.[13] This peptide, often referred to as papain peptide C, is of particular interest in that it will bind ganglioside, show retrograde transport in axons, is atoxic, and a good immunogen, producing antibodies which neutralize tetanus

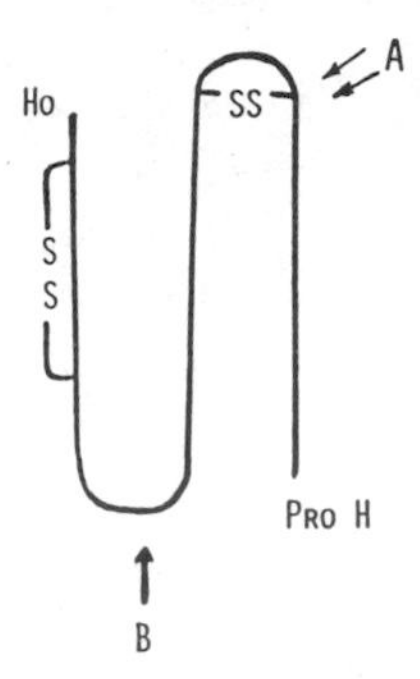

FIG. 2. Molecular model of tetanus toxin. The arrows at A indicate the cleavage sites for the proteolytic enzymes that are endogenous to the culture medium. The L chain occupies the amino-terminal end of the toxin molecule and proline is the amino-terminal residue. The arrow at B indicates the cleavage site during controlled papain digestion. It is located at about the midpoint of the H chain. Papain peptide C occupies the C-terminal end of the H chain.

toxin.[13,18] The peptide is now known to occupy approximately one-half of the H chain at the C-terminal end as illustrated in Fig. 2.[19–21]

Preparation of this interesting peptide is straightforward and reproducible in our hands, although scaling down from the large amounts of toxin used by these authors may require some modification. The method they reported is as follows: To 0.7 g of filtrate toxin in a 45-ml volume of 0.1 *M* phosphate buffer, pH 6.5, containing 1.0 m*M* Na_2EDTA, add 6 mg of cysteine-HCl and about 300 U of papain in about one-tenth of the volume of the toxin solution. Incubate the mixture at 55° for 4 hr, cool, and apply to a column of Sephadex G-100 (5 × 100 cm). Elute with Tris-HCl, pH 8.0, containing 1.0 *M* NaCl. Peptide C elutes as the third peak with the first and second peaks being native toxin, peak A, and the second peak a slightly toxic dipeptide referred to as fragment B.

Papain peptide C has been particularly useful in binding studies,[18,22,23] as well as studies of the pharmacological action of tetanus and botulinum neurotoxins.[17,24,25]

[18] G. Lee, E. F. Grollman, S. Dyer, F. Beguinot, L. D. Kohn, W. H. Habig, and M. C. Hardegree, *J. Biol. Chem.* **254,** 3826 (1979).

[19] T. B. Helting and O. Zwisler, *J. Biol. Chem.* **252,** 187 (1977).

[20] M. Matsuda and M. Yoneda, *Biochem. Biophys. Res. Commun.* **77,** 268 (1977).

[21] V. Neubauer and T. Helting, *Biochim. Biophys. Acta* **668,** 141 (1981).

[22] N. P. Morris, E. Consiglio, L. D. Kohn, W. H. Habig, M. C. Hardegree, and T. B. Helting, *J. Biol. Chem.* **255,** 6071 (1980).

[23] R. L. Goldberg, T. Cosza, W. H. Habig, L. D. Kohn, and M. C. Hardegree, *Mol. Pharmacol.* **20,** 565 (1981).

[24] L. Simpson, *J. Pharmacol. Exp. Ther.* **228,** 600 (1984).

[25] P. Boquet and E. Duflot, *Proc. Natl. Acad. Sci. U.S.A.* **79,** 7614 (1982).

[14] Purification of Alpha Toxin from *Clostridium perfringens:* Phospholipase C

By COLETTE JOLIVET-REYNAUD, HERVÉ MOREAU, and JOSEPH E. ALOUF

Alpha toxin was the first bacterial toxin to be identified as an enzyme.[1] It is a phospholipase C (phosphatidylcholine choline phosphohydrolase, EC 3.1.4.3) which hydrolyzes lecithin to phosphorylcholine and 1,2-diglyceride. The toxin has lethal, necrotizing, and cytolytic activities and is considered to play the most important role in the pathogenesis of gas gangrene.[2]

Clostridium perfringens phospholipase C is a potential tool for studying membrane structure and function because of its high specificity for phosphatidylcholine as compared to sphingomyelin. This enzyme was found to be a zinc metalloenzyme.[3] Calcium plays an essential role in the interaction of phospholipase with the substrate and provides the appropriate charge density.[4,5]

Many attempts at toxin purification have been reported.[3,5–14] However, only a few have obtained purified protein of high specific activity and restricted enzymatic activity specificity suitable for use in membrane studies. *Clostridium perfringens* produces, in parallel with phospholipase C, large amounts of toxins and enzymes which contaminate most preparations of alpha toxin. The two methods described here permit the preparation of *Clostridium perfringens* phospholipase C that is free of other tox-

[1] M. G. MacFarlane and B. C. J. Knight, *Biochem. J.* **35,** 884 (1941).
[2] T. A. Willis, "Clostridia of Wound Infection." Butterworth, London, 1969.
[3] E. L. Krug and C. Kent, *Arch. Biochem. Biophys.* **231,** 400 (1984).
[4] A. D. Bangham and R. M. C. Dawson, *Biochim. Biophys. Acta* **59,** 103 (1962).
[5] H. Moreau, C. Jolivet-Reynaud, and J. E. Alouf, *in* "Bacterial Protein Toxins" II (P. Falmagne, J. E. Alouf, F. J. Fehrenbach, and J. Jeljaszewicz, eds.), Springer–Verlag, p. 65. 1986.
[6] J. Stephen, *Biochem. J.* **80,** 578 (1961).
[7] A. Ito, *Jpn. J. Med. Sci. Biol.* **21,** 379 (1968).
[8] C. J. Smyth and J. P. Arbuthnott, *J. Med. Microbiol.* **7,** 41 (1974).
[9] K. Mitsui, N. Mitsui, and J. Hase, *Jpn. J. Exp. Med.* **49,** 65 (1973).
[10] R. Möllby and T. Wadström, *Biochim. Biophys. Acta* **321,** 569 (1973).
[11] R. A. Bird, M. G. Low, and J. Stephen, *FEBS Lett.* **44,** 279 (1974).
[12] T. Takahashi, T. Sugahara, and A. Ohsaka, *Biochim. Biophys. Acta* **351,** 155 (1974).
[13] Y. Yamakawa and A. Ohsaka, *J. Biochem.* (*Tokyo*) **81,** 115 (1977).
[14] C. Ottolenghi, this series, Vol. 14, p. 188.

ins, particularly theta toxin (perfringolysin O) and delta toxin (in type C strains), which are both hemolytic and cytolytic.

Bacterial Strains

Strains of all five types of *C. perfringens* produce alpha toxin but type A strains usually have much higher amounts of phospholipase C as compared to the other types. However, toxin production also depends on culture conditions and growth medium. The medium and growth conditions described below were developed for optimal production of alpha toxin by strain BP6K N5.

Culture Medium

One liter of medium consists of a dialysate of 25 g of protease peptone 3 (Difco Laboratories, Detroit, MI) and 6 g of yeast extract (Difco), which are dissolved in 150 ml of distilled water and dialyzed for 3 days in the cold against 1 liter of water. The dialysate is supplemented with 2 g of cysteine hydrochloride, 10 g of $NaHCO_3$, 20 g of dextrin (Difco), 10 g of acid-hydrolyzed casein (casamino acids, Difco), and 2 g of KH_2PO_4. The pH is adjusted to 6.7 with 3 *N* NaOH. It is sterilized at 115° for 20 min in a 1.5-liter fermentor (Biolafitte, Poissy, France). The fermentor is allowed to cool and then maintained until inoculation under nitrogen.

Growth and Toxin Production

The lyophilized bacterial strain is inoculated anaerobically into 20 ml of TGY medium (trypticase, 3 g, glucose, 1 g, yeast, 2 g, in 100 ml of medium, pH 7.5) and then incubated for 16 hr at 37°. This culture is used to seed the fermentor medium.

The culture is allowed to grow at 37° for 5 hr under a continuous nitrogen flow and moderate stirring. The pH is kept constant at 6.7 with an automatic titrator by addition of 3 *N* NaOH.

Thirty minutes before stopping the culture growth, 6 ml of an ethanolic solution of cholesterol (20 mg/ml) is slowly added in order to precipitate theta toxin, which is also produced in large amounts and usually contaminates most alpha toxin preparations. Benzamidine (12 m*M* : 2 g/liter) is added at the same time to prevent alpha toxin degradation by proteases.

The bacteria and aggregates of cholesterol with theta toxin are removed by two centrifugations at 10,000 *g* at 4°, for 20 min. The supernatant is used as the starting material for purification of alpha toxin.

Purification

Reagents

Buffer A: Tris–HCl, 20 mM, pH 7.2; NaCl, 150 mM; $ZnCl_2$, 1 mM
Ammonium sulfate [$(NH_4)_2SO_4$]: Reagent grade

Procedure 1

One liter of the cell-free culture medium is precipitated with ammonium sulfate to 60% saturation and allowed to stand overnight, at 4°, before centrifugation at 16,000 *g* for 15 min. The precipitate is then dissolved in 20 ml of buffer A and the protein solution is centrifuged again at 16,000 *g* for 15 min in order to eliminate insoluble residue. Ten milliliters of the resulting supernatant is applied to a 2.6 × 100 cm BioGel P-100 column (Bio-Rad Laboratories, Richmond, CA) equilibrated in buffer A. Alpha toxin is eluted in a single peak as evidenced by phospholipase C activity (see Chapter [41] in this volume). Fractions are pooled and then concentrated to about 5 ml by ultrafiltration with a PM10 membrane (Amicon). Pooled toxin-containing fractions from two gel filtrations are submitted to a second gel filtration on the same column. Fractions which contain phospholipase C activity are concentrated by ultrafiltration with a PM10 membrane and kept at −20° in aliquot samples containing 5% glycerol. Alpha toxin activity is stable for several months.

Comment. This procedure is the simplest technique to obtain alpha toxin without biologically active contaminants such as theta toxin, protease, deoxyribonuclease, or hyaluronidase, which are usually present in commercial preparations. However, alpha toxin prepared by this procedure is not homogeneous. Alpha toxin purity is about 90%; SDS–PAGE analysis reveals fine bands of high-molecular-weight contaminants which might be polymers of alpha toxin.[3,13]

Procedure 2

The toxin is purified as in procedure 1 to the first gel filtration. The pooled active fractions obtained after gel filtration are submitted to preparative isoelectric focusing (IEF) with a 110-ml column (LKB produkter, Stockholm, Sweden) as described previously for delta toxin purification.[15] A final concentration of 1% carrier ampholytes in a pH range of 4 to 8 (pH 3.5 to 5, ampholines, 0.5 ml; pH 4 to 6, ampholines, 1.5 ml; pH 5 to 8, ampholines, 0.5 ml) is used in a sucrose density gradient. A final potential

[15] J. E. Alouf and C. Jolivet-Reynaud, *Infect. Immun.* **31,** 536 (1981).

of 1600 V is applied for 16 hr. At the end of the run, 1-ml fractions are collected and assayed without prior removal of the ampholines or the sucrose. Phospholipase C focuses around pH 5.4. Active fractions are subjected to SDS–PAGE to test homogeneity. Fractions which prove to be homogeneous are pooled and concentrated by ultrafiltration with a PM10 membrane.

Ampholytes are removed by salting out on a small BioGel P-4 column (1.5×25 cm) equilibrated in buffer A. Purified toxin is stocked as in procedure 1.

Comment. Enzymatic activity is not stable in the presence of ampholytes and thus exposure time to ampholytes must be very short. The addition of $ZnCl_2$ or gel filtration in buffer A restores the activity.

Properties

Alpha toxin presents the following characteristics: the molecular weight of the enzyme determined by SDS–PAGE is 43 K and its isoelectric point is pH 5.4. Enzymatic activity requires calcium ions and is activated by zinc ions. Phosphatidylcholine is the main substrate, sphingomyelin is hydrolyzed to a lesser extent (10 times less); all other phospholipids tested are not degraded.

Specific activity of *Clostridium perfringens* phospholipase C usually varies in a range of 100 to 200 U/mg of protein in the titrimetric assay (1 U = 1 μmol H^+ released/min) and 200 U/mg of protein in the turbidimetric assay (see *Clostridium perfringens* phospholipase C assays in Chapter [41] this volume). The hemolytic activity of the purified toxin determined on rabbit erythrocytes is 65,000 HU/mg of protein. In this case one hemolytic unit (HU) corresponds to about 15 ng of protein.

[15] Production, Purification, and Assay of *Clostridium perfringens* Enterotoxin

By JAMES L. MCDONEL and BRUCE A. MCCLANE

The enterotoxin produced by *Clostridium perfringens* type A under conditions inducing sporulation has been studied and characterized in considerable depth according to its physiochemical, biological, and serological properties.[1] The enterotoxin has been shown[1–5] to bind specifically to a wide range of cell membranes, including the small intestinal brush border membrane which is the natural target[6] of enterotoxin. The receptor for enterotoxin in small intestinal epithelial cells may be a 50,000 M_r protein.[7] After attachment, enterotoxin causes a series of poorly understood events which result in membrane damage characterized by increased permeability and a loss of normal membrane structure and integrity, causing extensive blebbing.[6,8–10] Because of these activities, *C. perfringens* enterotoxin can be an interesting and useful tool in probing membrane components as they relate to structure and function, especially in systems such as the tightly configured microvilli found in the brush border membrane of the intestinal epithelial cell.

Since its identification,[11–13] numerous methodologies for purification of the enterotoxin have been described. Table I summarizes four general purification methods with which we have had experience. Table II summarizes some advantages and disadvantages of each method given in Table I. We used the column chromatography method of Stark and Duncan[14] for many years with excellent results. We now routinely use the ammonium sulfate precipitation method[15] because it is faster and less complicated than the former method while generating equally good results.

The details of production, purification, and assay of *C. perfringens* enterotoxin given below are based on our own experience as well as various published reports given in the references.

[1] J. L. McDonel, *Pharmacol. Ther.* **10,** 617 (1980).
[2] J. L. McDonel and B. A. McClane, *Biochem. Biophys. Res. Commun.* **87,** 497 (1979).
[3] J. L. McDonel, *Biochemistry* **21,** 4801 (1980).
[4] T. Jarmund and W. Telle, *Acta Pathol. Microbiol. Immunol. Scand., Sect. B* **90,** 377 (1982).
[5] A. P. Wnek and B. A. McClane, *Microb. Pathogen* **1,** 89 (1986).
[6] J. L. McDonel, L. W. Chang, J. G. Pounds, and C. L. Duncan, *Lab. Invest.* **39,** 210 (1978).
[7] A. P. Wnek and B. A. McClane, *Biochem. Biophys. Res. Commun.* **112,** 1099 (1983).
[8] B. A. McClane and J. L. McDonel, *Biochim. Biophys. Acta* **600,** 974 (1980).
[9] B. A. McClane and J. L. McDonel, *J. Cell. Physiol.* **99,** 191 (1979).
[10] B. A. McClane, *Biochim. Biophys. Acta* **777,** 99 (1984).
[11] C. L. Duncan and D. H. Strong, *J. Bacteriol.* **100,** 86 (1969).
[12] A. H. W. Hauschild, L. Niilo, and W. J. Dorward, *Can. J. Microbiol.* **16,** 331 (1970).
[13] A. H. W. Hauschild, L. Niilo, and W. J. Dorward, *Can. J. Microbiol.* **16,** 339 (1970).
[14] R. L. Stark and C. L. Duncan, *Infect. Immun.* **6,** 662 (1972).
[15] P. E. Granum and J. L. Whitaker, *Appl. Environ. Microbiol.* **39,** 1120 (1980).

TABLE I
Clostridium perfringens ENTEROTOXIN PURIFICATION PROCEDURES

Procedure	Techniques	Reported purification (-fold)
Ammonium sulfate[a,b] precipitation	$(NH_4)_2SO_4$ precipitation gel filtration	12.3
Column chromatography[c,d]	Gel filtration; ion-exchange chromatography hydroxyapatite	9.5
Affinity chromatography[e–g]	Affinity chromatography with or without gel filtration	7.2–8.4
Preparative gel electrophoresis[h]	Gel filtration and preparative gel electrophoresis	4.4

[a] P. E. Granum and J. L. Whitaker, *Appl. Environ. Microbiol.* **39,** 1120 (1980).
[b] G. Sakaguchi, T. Uemura, and H. P. Riemann, *Appl. Microbiol.* **26,** 762 (1973).
[c] A. H. W. Hauschild and R. Hilsheimer, *Can. J. Microbiol.* **17,** 1425 (1971).
[d] R. L. Stark and C. L. Duncan, *Infect. Immun.* **6,** 662 (1972).
[e] V. N. Scott and C. L. Duncan, *Infect. Immun.* **12,** 536 (1975).
[f] T. Uemura and R. Skjelkvale, *Acta Pathol. Microbiol. Scand., Sect. B* **84,** 414 (1976).
[g] H. M. Barnhart, L. B. Bullerman, E. M. Ball, and F. W. Wagner, *J. Food Sci.* **41,** 903 (1976).
[h] G. L. Enders and C. L. Duncan, *Infect. Immun.* **17,** 425 (1977).

Production of Enterotoxin

Clostridium perfringens enterotoxin is produced primarily by type A strains and then only during sporulation. Therefore, obtaining good sporulation of the organism during culture is of paramount importance. We have always cultured our organisms in Duncan–Strong (D–S)[16] medium and have routinely enjoyed 90% or more sporulation over the past 10 years. However, we are also aware of numerous laboratories around the world having serious problems with sporulation levels. We cannot overemphasize that even the slightest modification in the procedures, ingredients, or conditions listed below could result in unsuccessful preparation, isolation, purification, and assay of purified, biologically active *C. perfringens* enterotoxin.

Organisms. Clostridium perfringens type A, strain NCTC 8239, H-3 (Hobbs type 3) is used.

Culture Stocks. Grow cells in screw-cap tubes containing 10 ml each of fluid thioglycolate (FTG) medium at 37° overnight. Centrifuge, decant

TABLE II
COMPARISON OF DIFFERENT PROCEDURES FOR *C. perfringens* TYPE A ENTEROTOXIN PURIFICATION

Method	Advantages	Disadvantages
$(NH_4)_2SO_4$ precipitation	Easy, fast, good purity, retains biological and serological activities; will work for large-scale enterotoxin purification	Few
Column chromatography methods	Good purity, enterotoxin retains biological and serological activities	Very time consuming, particularly for large-scale purification
Affinity chromatography	Easy, fast, good purity	Not for large-scale purification, may lose biological activity of enterotoxin, requires highly specific anti-enterotoxin antibodies
Preparative gel electrophoresis	Easy, fast, enterotoxin retains biological and serological activities	Not practical for large-scale purification, purity of end product questionable

supernatant, resuspend pellet in each tube with 1 ml sterile skim milk, distribute to lyophiles, freeze dry, seal, and store at −70°.

Working Stocks. Grow cells overnight at 37° in screw-cap tubes containing 10 ml FTG. Innoculate 1 ml of this culture into each screw-cap tube containing 10 ml of cooked meat medium, incubate at 37° overnight, then seal tightly and store frozen at −20°. These working stocks should be stable for at least 1 year.

Purification of Enterotoxin

The 6-day procedure outlined below is designed for optimal conformity to the normal work day.

Day 1, 9 P.M. Thaw a working stock, transfer 1 ml into each of ten 10-ml FTG tubes. Immediately place the tubes into a 70° water bath. Tubes are heated at 70° for 20 min. This "heat shock" step will kill vegetative cells and select for spore-producing cells. Transfer the tubes from the 70° bath to a 37° incubator and incubate for 18 hr.

Day 2, 9 A.M. Prepare eight 100-ml FTG flasks (150-ml Erlenmeyer flasks with cotton plugs should suffice). Bring to 37° before use.

Prepare 20 liters of D–S medium,[16] following the recipe exactly. For each 5-liter flask add the following: 75 g proteose peptone; 20 g yeast extract; 5 g sodium thioglycolate; 50 g $Na_2PO_4 \cdot 7H_2O$, and 20 g soluble starch. Add all but the starch to 3.5 liters distilled water and stir until all is dissolved. Put the starch in a 1-liter glass beaker with 150 ml distilled water and swirl, then add boiling distilled water to bring the volume to about 800 ml. Bring to a boil for 2–3 min while stirring to dissolve all the starch. Add to the rest of the medium and bring to a final volume of 5 liters. Do *not* prepare D–S medium until the day it is to be used. Distribute the medium such that there are 4 liters in each of five 6-liter flasks. Autoclave, and while still hot take one flask and aseptically transfer 1 liter to each of the remaining four flasks so that each finally contains 5 liters. This decreases the surface area of the medium in each flask, which will enhance growth and sporulation. In the absence of equipment or handling skills to make the aseptic transfer, prepare the medium in each flask as full as the available autoclave will allow without boiling over during autoclaving or depressurization.

Place the hot flasks in an 85° water bath for 20 min. Do not shake. Then let them sit at room temperature for several hours before placing in a 37° room. The object is to have the flasks cooled down to 37° (but not less than 37°) by the time they are to be inoculated. These large flasks are significant heat sinks and incorrect temperature at inoculation time is a prime cause of poor sporulation.

Day 2, 3 P.M. Inoculate each of the eight 100-ml FTG flasks with the full contents of one 10-ml FTG tube. Incubate at 37° for 9 hr.

Also, begin rinsing the Sephacryl-200 (S-200, Pharmacia) column, and allow it to rinse overnight (see column and buffer description below).

Day 2, Midnight. Inoculate each 5-liter D–S flask with the entire contents of a 100-ml FTG 9-hr culture. Incubate for 8 hr.

Day 3, 8 A.M. After an 8-hr incubation take samples from several flasks and view under a phase-contrast microscope. Spores, if present, will be readily evident, primarily as refractile terminal or subterminal bodies. The conditions described above should routinely produce very high levels of sporulation. If the percentage of cells containing spores is under 80%, autoclave your cultures, discard, and begin again.

If sporulation is sufficiently high, at 8 hr distribute the culture medium to smaller flasks, place in an ice bath, and begin pelleting. The best method depends upon the centrifuge available. We have performed this

[16] C. L. Duncan and D. H. Strong, *Appl. Microbiol.* **16,** 82 (1968).

step with good results using fixed angle and swinging bucket rotors, and a continuous flow apparatus.[17] The important point is to pellet the cells as quickly and conveniently as possible, all at 4°.

Meanwhile, prepare 80% saturated ammonium sulfate by weighing 86 g $(NH_4)_2SO_4$ and bringing to 200 ml with double-distilled water (ddw). Filter through Whatman #4 filter paper into a 500-ml Erlenmeyer flask with a stir bar. Autoclave (along with a separate vial of 1 *M* NaOH) for 15 min, cool to 4°, and adjust the pH to 7.0 with 1 *M* NaOH.

Cell Pellet. Combine all pellets and suspend in 20 ml buffer A. (Buffer A: put 1 liter of 75 m*M* K_2HPO_4 into a beaker and begin adding 75 m*M* KH_2PO_4 until the pH is stable at 6.8. Autoclave for 15 min and store at 4°.) All procedures described below are to be performed at 4° and with autoclaved buffers and distilled H_2O. Autoclaving is necessary to assure protease-free buffer. Otherwise, artifact forms of enterotoxin may appear.[18]

Sonication of the cell paste suspended in 20 ml buffer A is most easily done with a sonication rosette immersed in an ice bath, which allows for good circulation and cooling of the suspension during energization. A small beaker may be used, but it is very important to use short bursts of the sonicator to prevent heating which will destroy the enterotoxin.

Sonicate at about 60% of maximum power in 2-min intervals alternated with 2-min pauses for cooling.[19] After four such cycles, check a small sample of the sonicate under the phase scope to determine the degree of cell rupture and spore release. Continue sonication until approximately 90% of spores are free. One can usually tell when this stage is being reached in that the paste begins to assume a more liquid, less viscous consistency.

Centrifuge the entire contents from the sonicating vessel at 10,000 *g* for 20 min at 4°. Withdraw the supernatant, being very careful not to disturb or take any of the pellet. Transfer the supernatant to another centrifuge tube and spin a second time. Again, remove the supernatant very carefully. This is the crude toxin preparation. A rich green sheen is usually a sign of a good preparation while too much gray is a sign of a poor one.

Determine the total volume of the supernatant, transfer to a sterile 250-ml flask with a stir bar, and slowly add an equal volume of ice-cold sterile ammonium sulfate (80%, brought to pH 7.0 with 1 *M* NaOH), with gentle stirring. After all the ammonium sulfate has been added, allow

[17] We currently use a Sorvall GS-3 rotor, centrifuging at 10,000 *g* for 20 min at 4°.

[18] B. R. DasGupta and M. W. Pariza, *Infect. Immun.* **38,** 592 (1982).

[19] Sonication settings are given for a Heat Systems–Ultrasonics W-225R sonicator, but settings should be similar for other sonicators.

stirring to continue at 4° for 30 min. Centrifuge in an SS-34 rotor at 10,000 *g* for 20 min at 4°. Discard the supernatant and resuspend the pellet in 25 ml buffer A. Determine the total volume and adjust the solution to 15% ammonium sulfate by addition of 80% saturated ammonium sulfate as follows: milliliters of 80% ammonium sulfate to be added = (0.23) (milliliters of suspended pellet). Add the ammonium sulfate slowly and let it stir for 30 min at 4°. Centrifuge as before and discard the supernatant, followed by resuspension of the pellet in 7 ml of buffer A. Add 0.25 g sucrose and vortex gently until the sucrose dissolves. Load the entire sample onto an S-200 column [2.5 × 70 cm; equilibrated with buffer A, pH 6.8; a Sephadex G-100 (Pharmacia) column can also be used but will not allow as fast a flow rate as can be achieved with the S-200 column bed]. Run overnight. We use a flow rate of 40 ml/hr, collect 160-drop fractions, and use 120 fraction tubes.

Day 4, A.M. Normally the second peak to come off contains the enterotoxin. Take a sample from all tubes within this peak as well as several on either side to use in 8-well immunodiffusion slides in which the inner well contains 8 μl of rabbit antitoxin,[20] and each outer well, 8 μl from each fraction. At the same time perform polyacrylamide gel electrophoresis[21] on each fraction as follows: 60 μl sample, 40 μl running buffer, 10 μl tracking dye, and 2 drops of concentrated sucrose solution for a total volume to be added to each gel of 125 μl. Run at 3 mA/gel, and about 80 V. Stain the gels for 30 min to 1 hr, and destain until a band is visible (usually 24 hr).

In the afternoon, based on immunodiffusion results, pool all fractions showing a positive result, and dialyze overnight against 4 liters ddw (as close to pH 7.0 as possible; adjust pH with 1 *M* NaOH if necessary) using 6000–8000 M_r exclusion range dialysis tubing (Spectrapor).

Day 5, A.M. Place a dialysis sac in 4 liters of fresh ddw. At the end of the day take the contents of the sac, read the absorbance at 276 nm [milligrams protein = (OD/1.33) (total vol of the sample)], and lyophilize overnight in a suitable vessel.

Day 6, A.M. Suspend a small portion of the lyophilized enterotoxin (fluffy white crystals) at a concentration of about 1 mg/ml (determine by absorbance at 276 nm or by Lowry protein assay[22]) to be used for checking purity (single band) with polyacrylamide gel electrophoresis and to perform the bioassay outlined below.

[20] B. A. McClane and R. J. Strouse, *J. Clin. Microbiol.* **19,** 112 (1984).

[21] R. L. Stark and C. L. Duncan, *Infect. Immun.* **4,** 89 (1971).

[22] O. H. Lowry, N. J. Rosebrough, A. L. Farr, and R. J. Randall, *J. Biol. Chem.* **193,** 265 (1951).

Bioassay of Enterotoxin

It is important to mention that serological and biological activities of this toxin are not synonymous.[1,23] It is possible to have serological activity while having no biological activity. Therefore serological assays are effectively useless in quantitating enterotoxin for cell system studies dealing with biological effects. Bioassay of the enterotoxin is best performed in two steps. The first step is the preliminary estimation of the activity in erythemal units (EU) with the guinea pig skin test. This is a relatively fast and simple procedure but lacks sensitivity (detection limit of approximately 250 ng[24]). The second step is to more precisely determine activity with a highly sensitive cell culture assay[23] (detection limit of approximately 0.1 ng[23]). This assay determines activity in plating efficiency units (PEU).

Guinea Pig Skin Test. The hair on the back and sides of two guinea pigs is removed first with clippers and then with depilatory cream, which is allowed to stand for about 20 min, followed by removal with paper towels (water will upset the animals and make them shiver, which will hinder the assay). An indelible black felt-tip marker is used to mark the back and sides into four rows (head–tail), each with six squares of equal size. Into the skin (intradermal) in the center of each square is injected with a tuberculin syringe exactly 0.05 ml of enterotoxin solution. If done properly a nice spherical bubble will appear in the skin as the injection is made, and there will be little or no leakage. If the needle accidentally goes under the skin, do not inject, but instead try again as far from the first site as possible, still within the same square.

It is best to assume about 2400 EU/mg enterotoxin when making dilutions for the first time. Using this assumption, prepare enough to make four injections of 0.05 ml each of 5, 2.5, 1.25, 0.612, 0.31, and 0.15 EU (i.e., a serial dilution starting with 5 EU/0.05 ml, which should be about 42 μg/ml). This serial dilution is best done in plastic microcentrifuge tubes with attached snap caps. Avoid glass tubes of any kind. The enterotoxin has a strong affinity for glass, which can lead to adsorption and gross underestimation of potency.

The next day, measure the diameter of all 24 zones of erythema at the injection sites and average the 4 values for each dilution. Then, use the dilution that gave an average spot size as close to, but under, 0.8 cm. In the event that one dilution gives 1.0 cm or larger and the next dilution gives 0.5 cm or smaller, calculate erythemal unit values as described

[23] J. L. McDonel and B. A. McClane, *J. Clin. Microbiol.* **13,** 940 (1981).

[24] M. D. Stringer, *in* "Clostridia in Gastrointestinal Disease" (S. P. Borriello, ed.), p. 118. CRC Press, Boca Raton, Florida, 1985.

below for both dilutions and average them together for the final value. Do *not* use only one dilution to calculate if that dilution gives an average zone of erythema ≤ 0.5 cm or ≥ 0.9 cm. If all spots are either above or below this range, alter the starting concentration accordingly and repeat the assay.

Calculate the activity as follows:

$$[\text{Average diameter (cm)}/0.8](\text{dilution factor}) = \text{EU/ml test solution}$$

From this can be calculated the number of EU/mg toxin. The EU is defined as that amount of enterotoxin causing a zone of erythema 0.8 cm in diameter.[21]

Tissue Culture Assay. Further details of this procedure are found in previously published reports.[9,23] Vero (African green monkey kidney) cells have been shown to be very sensitive to the action of *C. perfringens* enterotoxin.[2,5,7–10] The enterotoxin attaches to specific binding receptors[2–5,7] followed by structural and functional membrane damage which causes leakage of essential cells components and eventual cell death.

The basis of the assay is mixing serial dilutions of the enterotoxin with a small, fixed number (200) of Vero cells which are then seeded into the wells of Microtest II (Falcon) tissue culture plates. After 18 hr of incubation about 80% of control cells have settled to the bottom of the plates, attached, and assumed the normal appearance of healthy Vero cells. Cells treated with various dilutions of the enterotoxin fail to attach with increasing frequency as the toxin concentration increases, in a proportional relationship. This assay is the most sensitive biological assay available for *C. perfringens* enterotoxin. The PEU (plating efficiency inhibiting unit) is defined as that amount of enterotoxin, under the conditions specified in this assay, that causes a 25% inhibition of plating efficiency.[23] There are normally at least about 400,000 PEU/mg enterotoxin.

Vero Cells. Vero cells are cultured in plastic disposable 75-cm^2 tissue culture flasks containing medium 199 supplemented with 5% newborn calf serum and 0.075% sodium bicarbonate. When monolayers reach confluency, they are removed by trypsinization with 0.25% trypsin in calcium and magnesium-free Hanks' balanced salts solution and reseeded into 75-cm^2 flasks, or counted for inoculation into test wells in the assay described below.

Determination of Plating Efficiency. An aliquot of trypsinized cells is removed to determine cell concentration with a hemocytometer. An appropriate dilution is made such that there are 4000 cells/ml. Then, 0.5 ml of this suspension is added to a plastic microcentrifuge tube and the appropriate amount of enterotoxin is added and the volume made up to 1.0 ml. This suspension of cells and enterotoxin contains 2000 cells/ml.

To each of at least six wells is added 100 μl of this suspension (200 cells/well). A similar preparation is done for each enterotoxin dilution and control (medium containing enterotoxin heat inactivated at 60° for 20 min, or medium alone) being run. The recommended enterotoxin dilution is a serial dilution from 10 ng/100 μl down to 0.312 ng/100 μl medium.

After inoculation, the plates are incubated for 18 hr in a humidified 5% CO_2 atmosphere. Then, the supernatant medium containing nonadherent cells is discarded and the remaining cells are washed very gently with 100 μl of 0.85% saline solution. Adherent cells are stained for 1 min with 100 μl of a 1% methylene blue solution to aid in counting. Individual wells are divided into quadrants by etching with a razor to aid in counting. The number of cells per well is determined by direct counting under the light microscope. Plating efficiency of enterotoxin-treated cells is expressed as a percentage of control plating efficiency with control values being taken as 100%.

Finally, a graph is prepared with nanograms of toxin on the x axis and percentage inhibition of plating on the y axis. The slope should be linear in the range of about 0.5 to 10 ng enterotoxin. The PEU is defined as that amount of enterotoxin that will cause a 25% inhibition of plating (compared to control values) of 200 cells under the culture and volume conditions described. It is highly recommended that a standard calibrated solution of enterotoxin be freeze dried in numerous aliquots to be used, at three different concentrations, to prepare a standard curve each time the assay is repeated with new enterotoxin unknowns.

[16] Production and Purification of Anthrax Toxin

By STEPHEN H. LEPPLA

Bacillus anthracis secretes three proteins which are collectively known as anthrax toxin.[1-5] The protective antigen (PA, 85 kDa), lethal factor (LF, 83 kDa), and edema factor (EF, 89 kDa) proteins individually have no known toxic activities. Simultaneous injection of PA and LF causes death of rats, while PA and EF together produce edema in skin.[5]

[1] H. Smith and J. Keppie, *Nature (London)* **173,** 689 (1954).

[2] C. B. Thorne, D. M. Molnar, and R. E. Strange, *J. Bacteriol.* **79,** 450 (1960).

[3] P. Hambleton, J. A. Carman, and J. Melling, *Vaccine* **2,** 125 (1984).

[4] S. H. Leppla, *in* "Microbiology—1985" (L. Lieve, ed.), p. 63. Am. Soc. Microbiol., Washington, D.C., 1985.

[5] J. L. Stanley and H. Smith, *J. Gen. Microbiol.* **26,** 49 (1961).

Thus, "anthrax toxin" is actually two toxins, each of which is like staphylococcal leukocidin[6,7] and botulinum C_2 toxin[8] in having receptor-binding and effector domains on separate proteins. The PA protein appears to play a dual role as the B moiety for two different A proteins. Binding studies have shown that PA must be present in order for EF to bind to cells.[9] EF is a calcium- and calmodulin-dependent adenylate cyclase which causes large increases in intracellular cAMP concentrations.[10] The mechanism of action of LF is unknown.

For *B. anthracis* to be fully virulent, it must produce two materials, the anthrax toxin and a polyglutamic acid capsule.[11] Recent work has shown that virulent strains possess two large plasmids. Plasmid pXO1 (114 megadaltons, MDa) codes for all three toxin components,[12-16] while pXO2 (60 MDa) codes for the polyglutamic acid capsule.[17,18] Methods are available to eliminate either or both plasmids. The genes for PA[13] and for LF[16] have been cloned, and the former has been sequenced.[19]

Of particular value in toxin research are strains possessing only pXO1 since these strains are at least 10^5-fold less virulent than strains that produce both toxin and capsule. The most widely used toxinogenic, noncapsulated strain is that designated Sterne, after its originator.[20] Suspensions of Sterne spores are employed as a vaccine for livestock and have also been used as a vaccine in man.[3]

The protective antigen protein was originally recognized by its ability

[6] A. M. Woodin, *in* "Microbial Toxins" (T. C. Montie, S. Kadis, and S. J. Ajl, eds.), Vol. 3, p. 327. Academic Press, New York, 1970.

[7] M. Noda, T. Hirayama, F. Matsuda, and I. Kato, *Infect. Immun.* **50,** 142 (1985).

[8] M. Iwasaki, I. Ohishi, and G. Sakaguchi, *Infect. Immun.* **29,** 390 (1980).

[9] S. H. Leppla, *Adv. Cyclic Nucleotide Protein Phosphorylation Res.* **17,** 189 (1984).

[10] S. H. Leppla, *Proc. Natl. Acad. Sci. U.S.A.* **79,** 3162 (1982).

[11] C. B. Thorne, *Ann. N.Y. Acad. Sci.* **88,** 1024 (1960).

[12] P. Mikesell, B. E. Ivins, J. D. Ristroph, and T. M. Dreier, *Infect. Immun.* **39,** 371 (1983).

[13] M. H. Vodkin and S. H. Leppla, *Cell* **34,** 693 (1983).

[14] C. B. Thorne, *in* "Microbiology—1985" (L. Lieve, ed.), p. 56. Am. Soc. Microbiol., Washington, D.C., 1985.

[15] EF gene was shown to be on pXO1 by demonstrating adenylate cyclase activity in *B. cereus* $pXO1^+$ transcipients; C. B. Thorne and S. H. Leppla, unpublished observations (1985).

[16] D. L. Robertson and S. H. Leppla, *Gene* **44,** 71 (1986).

[17] I. Uchida, T. Sekizaki, K. Hashimoto, and N. Terakado, *J. Gen. Microbiol.* **131,** 363 (1985).

[18] B. D. Green, L. Battisti, T. M. Koehler, C. B. Thorne, and B. E. Ivins, *Infect. Immun.* **49,** 291 (1985).

[19] S. L. Welkos, J. R. Lowe, F. Eden-McCutchan, M. Vodkin, S. H. Leppla, and J. J. Schmidt. Submitted to *Gene,* 1988.

[20] M. Sterne, *Onderstepoort J. Vet. Sci. Anim. Ind.* **8,** 279 (1937).

to induce immunity to infection.[21] Extensive work led to the partial purification and eventual licensing of an aluminum hydroxide-adsorbed PA preparation as a human vaccine.[3,22] The work described below was initiated to find improved methods for vaccine production, and was later extended to include purification and characterization of the other two toxin components.

Safety Considerations

The anthrax toxins are not of extremely high potency,[23,24] and are not known to be absorbed from the digestive tract. Therefore, the normal precautions used in handling toxic chemicals provide adequate protection. An added degree of safety applies when the toxin components have been separated, since they have no known action individually.

The principal hazard is possible infection by the organism. If recognized early, infections are treated successfully with antibiotics, but symptoms are often nonspecific. In our laboratory, where both virulent and avirulent strains are under study, all bacteriological work is done in containment suites operated at the BL3 level. Personnel are immunized with the licensed human vaccine, purchased from the Michigan Department of Public Health. Virulent strains are grown in flasks in a shaking incubator, but not in fermentors, and swabs are taken at weekly intervals to detect any contamination of laboratory surfaces. Suspected accidental inoculation of organisms is sufficient basis for prophylactic administration of penicillin. Although the virulent strains present a much greater risk than Sterne-type strains, possible infection by the latter cannot be entirely discounted. Thus, it has recently been shown that relatively low doses of Sterne (10^3 spores) are lethal for certain inbred mouse lines.[25] Therefore, the possibility that some persons may be unusually susceptible to infection by Sterne strains should be kept in mind, and large-volume cultures, in particular, should be treated with appropriate care. For laboratories planning to grow only Sterne-type strains in volumes of less than 5 liters, the rigorous measures described above would be excessive. Some commercial producers of vaccine are believed to grow substantial volumes of Sterne with the usual concern for bacteriological safety, but do not immunize personnel. In those cases where the intent is to produce toxin rather

[21] G. P. Gladstone, *Br. J. Exp. Pathol.* **27,** 394 (1946).

[22] M. Puziss, L. C. Manning, J. W. Lynch, E. Barclay, I. Abelow, and G. G. Wright, *Appl. Microbiol.* **11,** 330 (1963).

[23] D. M. Gill, *Microbiol. Rev.* **46,** 86 (1982).

[24] J. W. Ezzell, B. E. Ivins, and S. H. Leppla, *Infect. Immun.* **45,** 761 (1984).

[25] S. L. Welkos, T. J. Keener, and P. H. Gibbs, *Infect. Immun.* **51,** 795 (1986).

than spores, cultural conditions should be chosen which prevent sporulation, thereby simplifying the problem of decontamination.

Bacterial Strain Selection

All natural isolates of *B. anthracis* appear to produce approximately the same amount of toxin (within 50%), and to produce all three toxin components. Thus, any $pXO1^+$, $pXO2^-$ strain could be selected for toxin production. Investigators should use either well-characterized Sterne-type strains or verify that locally derived, nonencapsulated strains lack pXO2. It is not correct to assume that every noncapsulated strain lacks pXO2, since some virulent strains can generate capsule-negative variants while retaining the pXO2 plasmid; these variants can revert to full virulence.[26] Previous concern that true Sterne-type strains ($pXO1^+$, $pXO2^-$) might revert to virulence can now be dismissed, since the pXO2 plasmid carries essential capsule genes.

A readily available strain developed for protective antigen production[22] and used by the Michigan Department of Public Health for human vaccine preparation is V770-NP1-R (American Type Culture Collection, Accession No. 14185). This strain was selected as a nonproteolytic, noncapsulated mutant. While theoretically an advantage, the nonproteolytic characteristic does not seem to improve toxin yields when the medium described below is used. Another readily available, attenuated strain, Pasteur vaccine No. 1, ATCC 4229, is not suitable for toxin production since it is $pXO1^-$, $pXO2^+$.[18]

The procedures described in this chapter were optimized for the Sterne strain, use of which is recommended. The Sterne strain is available from this and other laboratories, but is not currently deposited with the ATCC. Another suitable strain is a spontaneous, rifampicin-resistant Sterne mutant designated SRI-1, which has been found to produce 50–75% more toxin than Sterne. This strain appears defective in septum formation and grows in long filaments. This strain is not a good choice if a method generating high shear forces, such as tangential flow filtration, is used for removal of bacteria.

Strains are stored as either spore suspensions or frozen vegetative cells, and are revived on blood agar or other appropriate media. In this laboratory, aliquots of vegetative cells grown in RM medium (described below) and stored at −70° are thawed and spread on solid RM medium lacking bicarbonate and grown 24 hr at 32° to prepare a fermentor inocu-

[26] T. M. Koehler, R. E. Ruhfel, B. D. Green, and C. B. Thorne, *Am. Soc. Microbiol.* Abstr. H-178 (1986).

lum. Growth at temperatures above 37° should be avoided since plasmid curing may occur.[12,14] If virulent *B. anthracis* are in use in the laboratory, aliquots of the fermentor inoculum should be grown in parallel on serum- or bicarbonate-containing medium and incubated in a CO_2 atmosphere to detect any capsulated contaminants.

Culture Medium and Growth Conditions

A number of investigators developed media that promote production of PA[22,27] or toxin.[2,28,29] Recent work at this laboratory developed "R" medium,[29] which was derived from one of the more successful semi-synthetic formulations[28] by replacement of casamino acids with an equivalent L-amino acid mixture (except that alanine was omitted to limit sporulation). Further media development work[30] was done to optimize yields of LF and EF in addition to PA and to facilitate product recovery. In these trials, yields of all three toxin components increased or decreased in parallel. The modified medium developed through these trials was designated RM, and is described below.

The features of a medium which appear important for toxin production are as follows: (1) inclusion of $NaHCO_3$, which aids pH control, enhances gene transcription,[29–30a] and may also have a chemical effect, perhaps by permeabilizing the bacteria[31]; (2) maintenance of pH > 7 (achieved here by $NaHCO_3$ and Tris), which serves to limit the action of proteolytic enzymes[32]; and (3) growth under essentially anaerobic conditions. Besides enhancing yields, anaerobic growth is advantageous because it places fewer demands on the fermentation equipment and decreases the potential for contamination of the laboratory.

RM medium contains the following ingredients at the indicated final concentrations (mg/liter), with all amino acids being of the L configuration: tryptophan (35), glycine (65), tyrosine (144), lysine hydrochloride (230), valine (173), leucine (230), isoleucine (170), threonine (120), methionine (73), aspartic acid (184), sodium glutamate (612), proline (43), histidine hydrochloride (55), arginine hydrochloride (125), phenylalanine (125), serine (235), NaCl (2920), KCl (3700) adenine sulfate (2.1), uracil (1.4), thiamin hydrochloride (1.0), cysteine (25), KH_2PO_4 (460), 2-amino-

[27] C. B. Thorne and F. C. Belton, *J. Gen. Microbiol.* **17,** 505 (1957).
[28] B. W. Haines, F. Klein, and R. E. Lincoln, *J. Bacteriol.* **89,** 74 (1965).
[29] J. D. Ristroph and B. E. Ivins, *Infect. Immun.* **39,** 483 (1983).
[30] S. H. Leppla, unpublished observations (1984).
[30a] J. M. Bartkus and S. H. Leppla, Am. Soc. Microbiol. Abstr. H-180 (1988).
[31] M. Puziss and M. B. Howard, *J. Bacteriol.* **85,** 237 (1963).
[32] R. E. Strange and C. B. Thorne, *J. Bacteriol.* **76,** 192 (1958).

2-(hydroxymethyl)-1,3-propanediol (Tris) (9060), glucose (5000), $CaCl_2 \cdot 2H_2O$ (7.4), $MgSO_4 \cdot 7H_2O$ (9.8), $MnCl_2 \cdot 4H_2O$ (1.0), and $NaHCO_3$ (8000). The RM medium differs from R in containing NaCl, KCl, Tris, increased glucose (0.5 vs 0.25%), decreased potassium phosphate (3.4 vs 17.2 m*M*), and in substitution of cysteine for cystine. The latter substitution was done for convenience and is not known to enhance toxin yields. The medium is prepared with good quality distilled or deionized water. If extremely high-quality water is used, addition of iron and trace metals may be required to support growth.

To prepare medium for a 50-liter fermentor culture, the first 21 ingredients in the above list (ending at thiamin) are added as solids to 40 liters of distilled water and sterilized in the fermentor vessel. The remaining eight ingredients are individually dissolved in sterile water and sequentially pumped through a disposable capsule filter into the vessel, followed by water as needed to reach 50 liters. The medium in the vessel is then titrated to pH 8.0. For RM medium to be used in flasks, it is more convenient to group the ingredients into several stock solutions (for examples, see Haines *et al.*[28]); these are separately filter sterilized and added to the flasks. Stocks of $NaHCO_3$ should be made fresh and filtered by pressure to avoid loss of CO_2, and media to which this stock has been added should be kept in tightly closed flasks. Failure of *B. anthracis* to grow in aged medium is usually due to alkaline conditions resulting from CO_2 loss.

One medium modification that consistently increases yields of all three toxin components by 50–100% is addition of horse serum to 3–5%.[24] Adding serum precludes purification (except by immunoadsorption—see below), but the increased yields are useful when the goal is small-scale production of immunochemically or enzymatically active toxin. In cases where growth conditions are not optimal, adenylate cyclase activity may be detectable only in serum-supplemented cultures. Other proteins or putative protective agents and dialyzed horse serum seem to be less effective than horse serum. The basis of this effect is not known.

To grow a 50-liter fermentor culture, the bacteria on five RM agar plates are suspended in 25 ml RM medium, giving an $A_{540\ nm} = 8$; this suspension is added to the vessel. The culture is stirred at 150 rpm, regulated at 35°, controlled at pH 8.0 by addition of 1 *M* NaOH, and no aeration is used. If dissolved oxygen is measured, it is found that this falls to 0 once significant growth of the culture has occurred. The culture grows to $A_{540\ nm} = 2–2.5$ by 18–24 hr, with logarithmic growth evident from the rate of NaOH consumption. Approximately 1–3 mol of NaOH is consumed. Fermentor cultures are harvested promptly after growth has ceased.

For growth in flasks, the containers are half-filled with medium, inoculated, tightly capped, and shaken at a speed just sufficient to maintain the bacteria in suspension. Cultures grown in flasks or in fermentors without pH control will show a fall in pH to 7.2–7.4. Toxin yields in such cultures are usually comparable to those in pH-controlled fermentor cultures.

Recovery of Toxin from Culture Supernatants

In cultures grown as described above, PA, LF, and EF are present at approximately 20, 5, and 1 μg/ml respectively, and collectively constitute more than half of the extracellular protein. Since separating the components from each other and from impurities is not difficult (see below), the principal challenge lies in recovering the dilute proteins from the culture supernatant. Immunoadsorbent chromatography has been used to recover and purify LF,[33] and could also be applied to PA and EF, but may not be economical if applied to unconcentrated supernatants. Ultrafiltration or batch adsorption to ion-exchange resins can be used if appropriate equipment is available. Two alternate methods for toxin recovery will be described here, one that is currently in use with 50-liter fermentor cultures, and a second which may be more convenient for small-scale cultures.

Fermentor Cultures (>10 Liters). When growth has ceased, 1,10-phenanthroline hydrochloride is added to give a final concentration of 0.05 m*M*; EDTA is added to 2 m*M*. These chelators inhibit a *B. anthracis* metalloprotease,[30] which is similar to that produced by *Bacillus cereus*.[34] Phenylmethylsulfonyl fluoride is added to 0.1 m*M*, although it has not been proved that this enhances yields. Mercaptoethanol is added to 2 m*M*. The culture is chilled if the fermentor has this capability, and is pumped via a stainless steel coil immersed in an ice bath into a continuous-flow centrifuge (Sorvall TZ-28 rotor, or a comparable system that does not generate an aerosol). The TZ-28 rotor is operated at 12,000 rpm, and a flow rate of 400 ml/min is maintained with a peristaltic pump. The supernatant is moved to a cold room, and all subsequent steps are performed at 4°. The supernatant is sterilized by filtration using a Pellicon tangential flow ultrafiltration unit containing 10 ft^2 of Durapore 0.45-μm membrane (Millipore Corp.). Filtration rates exceeding 500 ml/min can be obtained in this unit. The membrane is treated with bovine serum albumin before its first use in order to prevent adsorptive losses. The membranes may be used to process a number of 50-liter cultures, provided that each use is followed by washing with 0.1 *M* NaOH. Tangential flow filtration

[33] E. J. Machuga, G. J. Calton, and J. W. Burnett, *Toxicon* **24,** 187 (1986).

[34] B. Holmquist, *Biochemistry* **16,** 4591 (1977); see also this series, Vol. 19, p. 569.

systems having screens on the retentate side have not been useful as a substitute for the initial centrifugation, because the bacterial chains (average 10 cells/chain for Sterne strain) plug the screens.

In the next step, the proteins are concentrated from the sterile supernatant by a "salting out" adsorption process.[35] Approximately 1 liter of cross-linked agarose beads (Sepharose CL-4B, Pharmacia, or a similar resin) is added to the supernatant, followed by the slow addition of 25 kg $NH_4(SO_4)_2$. The suspension is stirred gently until the salt dissolves (2–3 hr) and then the agarose beads are allowed to settle. Every effort is made to reach this point on the same day the culture is harvested. If successful, it is convenient to let the resin settle overnight. The supernatant is pumped off, passing it through a porous plastic funnel (Bel-Art Plastics) if necessary to collect any resin that has not settled. The agarose is then placed in a 14-cm diameter column and eluted at 25 ml/min with 2 liters of 50 m*M* Tris, 1 m*M* EDTA, 2 m*M* 2-mercaptoethanol, pH 8.0. The fractions containing >95% of the protein are pooled and precipitated by slow addition of solid $NH_4(SO_4)_2$ to 75% saturation. After 2–24 hr the precipitated toxin is collected by centrifugation, redissolved in 100 ml 10 m*M* Tris, 0.05 m*M* 1,10-phenanthroline, 2 m*M* 2-mercaptoethanol, pH 8.0, and dialyzed against the same buffer. At the author's facility, it was found convenient and feasible at this stage to filter sterilize the toxin and remove it from the BL3 laboratory. Subsequent purifications are done at the BL1 containment level. Successful preparations contain 2–4 g protein, as determined by UV absorption, assuming $E_{1\%}$ (280 nm, 1 cm) = 10. The phenanthroline in the sample absorbs at 280 nm, but this does not interfere if dialysate is used to blank the spectrophotometer.

Small-Scale Cultures (0–10 Liters). For these cultures, it may be more convenient to use a protocol employing adsorption to hydroxyapatite. LF and EF can be adsorbed from undiluted culture supernatants due to the low phosphate concentration in RM medium (3.4 m*M*, 20% of that in R medium). PA will not be completely adsorbed under the same conditions, but recovery of PA can be made nearly complete by addition of polyethylene glycol. Cultures grown in flasks or small fermentors are centrifuged to remove bacteria, chilled, adjusted to pH 7.0 with acetic acid, and protease inhibitors are added as described for large cultures, with the exception that EDTA is omitted[36] and 1,10-phenanthroline is increased to

[35] F. von der Haar, *Biochem. Biophys. Res. Commun.* **70,** 1009 (1976); see also R. K. Scopes, "Protein Purification: Principals and Practice," p. 141. Springer-Verlag, New York, 1982.

[36] EDTA is omitted because it would extract the calcium from hydroxyapatite ($CaSO_4$), and might dissolve the adsorbant and prevent protein binding. Phenanthroline does not dissolve hydroxyapatite, even when present at 20 m*M*.

0.2 m*M*. To each liter of supernatant, 5 g of hydroxyapatite (Fast Flow type, Calbiochem) is added and, if high recovery of PA is desired, 100 g of polyethylene glycol 8000 (PEG) is also added. The mixture is gently agitated at 5° until the PEG is dissolved and then an additional 1–3 hr, avoiding use of a magnetic stir bar. The hydroxyapatite is transferred to a small column or filter funnel and washed with 10 m*M* Tris, pH 7.5. If PEG is not used, the lower viscosity makes it feasible to pass the culture supernatant directly through a column of hydroxyapatite. The toxin is eluted with 0.66 *M* potassium phosphate, pH 7.0. The eluate is supplemented with EDTA to 0.01 *M*, and is dialyzed against 10 m*M* Tris, 2 m*M* EDTA, pH 8.0, or other suitable buffer, depending on the intended use of the preparation. Toxin prepared in this way has not been as fully characterized as that made in fermentors. The reported presence of aldehydes and peroxides in PEG and related polyether detergents[37] raises concerns that subtle chemical damage may exist in proteins made using PEG.

Chromatographic Separation of Toxin Components

A number of chromatography methods are available to separate the toxin components. Details are given here for processing the amount of toxin obtained from a 50-liter fermentor, by using sequential chromatography on hydroxyapatite and DEAE-Sepharose. This order is preferred over the inverse, since EF was found to elute in several distinct peaks when crude toxin was run on DEAE, a result that may reflect a weak interaction of EF with PA. In the protocols described below, all operations are performed at 4°, and dialysis times should not exceed 16 hr, except for the final product.

The dialyzed crude toxin (2–4 g) is pumped onto a 2.6 × 38 cm (200 ml) column of hydroxyapatite (Fast Flow type, Calbiochem) previously equilibrated to 0.005 *M* potassium phosphate, 0.1 *M* NaCl, 0.05 m*M* 1,10-phenanthroline, 2 m*M* 2-mercaptoethanol, pH 7.0 (buffer A). The column is washed with at least 150 ml buffer A at 50 ml/hr and then eluted with a gradient of 500 ml each of 0 and 0.5 *M* potassium phosphate, pH 7.0, both in buffer A. Fractions of 10 ml are collected in tubes containing 0.1 ml 100 m*M* EDTA. The components elute in the order PA > LF > EF, and each is evident as a peak of UV-absorbing material. The EF peak is small compared to the others, but is easily identified because it is the last significant peak. PA and LF are concentrated from pooled fractions by

[37] W. J. Ray, Jr., and J. M. Puvathingal, *Anal. Biochem.* **146,** 307 (1985); see also this series, Vol. 52, p. 145, and Vol. 56, p. 742.

ammonium sulfate precipitation at 75% saturation and dialysis against 0.01 *M* Tris, 0.025 *M* NaCl, 0.05 m*M* 1,10-phenanthroline, 2 m*M* 2-mercaptoethanol, 1% glycerol, pH 8.0 (buffer B). The EF is concentrated by ultrafiltration to about 25 ml and is dialyzed against buffer B.

Final purification of each component is achieved by chromatography on DEAE-Sepharose CL-4B (Pharmacia), with a column containing about 1 ml resin/5–10 mg input protein. As an example, 700 mg PA is purified on a 1.6 × 50 cm (100 ml) column, with a gradient of 750 ml each of 0 and 0.25 *M* NaCl in buffer B. For LF and EF, the high-salt buffers should contain 0.40 and 0.25 *M* NaCl, respectively. The protein in the pooled fractions is dialyzed against 5 m*M* HEPES, 50 m*M* NaCl, pH 7.5, filter sterilized with low protein-binding filters (Millex-GV, Millipore), quick frozen in small aliquots, and stored at −70°.

Successful purifications yield 400 mg PA, 75 mg LF, and 20 mg EF. The LF and EF proteins appear as single species on SDS gels and analytical ion-exchange HPLC (Mono-Q resin, Pharmacia); these components appear homogeneous. In contrast, two types of heterogeneity in PA have been observed. A variable fraction (usually <10%) of the PA contains a cryptic polypeptide cleavage approximately 330 residues from the N terminus. Electrophoresis under denaturing conditions reveals fragments of 37 and 47 kDa. The site which is cleaved by the endogenous bacillus protease is also highly susceptible to specific cleavage by other proteases, including chymotrypsin.[38] The two large peptide fragments produced by intentional cleavage at this site will be useful in structure–function analyses of PA. The other type of heterogeneity in PA consists of differences in net charge. During the final chromatography on DEAE, most preparations show one or two partially separated, trailing species. These species are indistinguishable on SDS gels, but migrate as two to four evenly spaced bands after isoelectric focusing or nondenaturing, continuous slab gel electrophoresis. Careful selection of fractions, or chromatography on Mono-Q with shallow gradients (Tris buffer, pH 8.0, 0.25–0.35 *M* NaCl) yields distinct, homogeneous preparations. The species with the least negative charge (eluting first from DEAE and Mono-Q) has been shown to have significantly greater potency (with LF) in the macrophage lysis assay described below. The nature of the alteration that introduces additional negative charge and decreases potency is not known. A similar type of heterogeneity has been observed in *Pseudomonas* exotoxin A.[39]

[38] S. H. Leppla, manuscript in preparation (1988).

[39] L. T. Callahan, III, D. Martinez, S. Marburg, R. L. Tolman, and D. R. Galloway, *Infect. Immun.* **43,** 1019 (1984).

Toxin Assay

The anthrax toxins can be detected and quantitated according to their immunochemical, enzymatic, or toxic activity, or by direct chemical measurement.

Chemical Assays. For laboratories beginning purification of the toxin and not yet having component-specific antisera, it may be most convenient to perform direct chemical measurement of the toxin components by rapid electrophoresis or chromatography methods. The large size (83–89 kDa) and the high concentrations of the proteins in RM medium cultures make it relatively simple to locate the bands or peaks of PA and LF. While the three toxin components are similar in size, they can be distinguished on 8 or 10% polyacrylamide SDS slab gels[40] if small amounts of protein (0.1 μg/band) are loaded. The three proteins are well separated during ion-exchange high-performance liquid chromatography (Mono-Q), eluting in the order PA > EF > LF when the column is developed with an NaCl gradient at pH 8.

Immunochemical Assays. Antisera to the toxin components can be elicited using standard techniques.[41] Purified PA and LF give rise to high-titer sera, while EF seems less immunogenic and has not yielded sera useful in gel diffusion systems. Antisera to PA may be obtained by immunization with the licensed human vaccine (Michigan Department of Public Health), which contains principally this toxin component. Immunization of guinea pigs and rats with Sterne-strain spores induces antibodies to PA,[41a] and small amounts of antibodies to LF and EF. Goat and rabbit sera have been used successfully in gel diffusion,[27] rocket immunoelectrophoresis,[24] immunoblots,[13] and ELISA.[13,16] For routine detection of PA and LF in column eluates, gel diffusion in agar is preferred because antigens can be detected over a wide range of concentrations. For quantitative measurement of PA, radial immunodiffusion employing specific goat antiserum has been most useful. Monoclonal antibodies to all three toxin components have been developed[42] and can be expected to replace the polyclonal sera in some assays. Due to the low potency of the available sera, EF is usually assayed enzymatically (see below).

Adenylate Cyclase Assay. Of the three toxin proteins, only EF is known at this time to have enzymatic activity. The EF adenylate cyclase has enzymatic properties resembling those of the *Bordetella pertussis*

[40] U. K. Laemmli, *Nature (London)* **227,** 680 (1970).

[41] B. A. L. Hurn and S. M. Chantler, this series, Vol. 70, p. 104.

[41a] B. E. Ivins and S. L. Welkos, *Infec. Immun.* **54,** 537 (1986).

[42] S. F. Little, E. M. Cora, and S. H. Leppla, submitted to *Infect. Immun.* (1988).

adenylate cyclase.[43–45] An exception is that in all preparations of EF, the enzyme activity has been totally dependent on addition of calmodulin; no calmodulin-independent forms like those of the *B. pertussis* enzyme have been detected. Because EF has high enzymatic activity (V_{max} = 1.2 mmol cAMP/min/mg[9]), any of the methods developed for the more difficult task of measuring eukaryotic adenylate cyclases may be used. The assay described below is modified from that of Salomon[46] by dilution of [^{32}P]ATP to lower specific activity, omission of the phosphodiester inhibitor and of the [^{3}H]cAMP added to determine chromatographic recoveries, and optional use of [^{3}H]- or [^{14}C]ATP as a substitute for [^{32}P]ATP. This method has a higher sensitivity than is needed, but is preferred because the Dowex and alumina columns are reusable. Manganese ion is used in order to make the enzyme activity independent of calcium ion concentration. Samples and buffers should be chosen so as to exclude phosphate, which is a strong inhibitor (K_i about 0.01 m*M*).

Reagents

Assay buffer (5×): pH 7.5
 100 m*M* HEPES, 25 m*M* $MnCl_2$, 2.5 m*M* $CaCl_2$, 2.5 m*M* EDTA, 0.25 m*M* cAMP, 2.5 m*M* dithiothreitol, 0.5 mg/ml bovine serum albumin (calmodulin free[44])
[^{32}P]ATP (10×): 0.02 mCi/ml, 5 m*M* ATP
 If assays are done infrequently, it may be preferred to use [^{3}H]- or [^{14}C]ATP. These provide adequate sensitivity and long half-life, but require counting in scintillation fluid
Calmodulin, bovine (10×): 0.05 mg/ml

Protocol. The reaction mixtures contain 0.020 ml assay buffer (5×), 0.010 ml calmodulin (10×), EF sample (2–10 ng) and water totalling 0.060 ml, and 0.010 ml [^{32}P]ATP (10×). Mixtures are preincubated 20 min before addition of ATP, to allow association of EF, calmodulin, and calcium. Controls should include reactions (1) with excess EGTA to remove calcium, (2) omitting calmodulin, and (3) with excess EF (0.01 mg/ml) to cause complete ATP conversion. The latter allows calculation of chromatographic recovery. After 60 min at 23°, 0.10 ml of stopping solution (2% sodium lauryl sulfate, 45 m*M* ATP, 1.3 m*M* cAMP) is added, the samples are heated 5 min at 95–100°, and 1.0 ml of water is added. The samples are poured into disposable plastic chromatography columns hav-

[43] E. Hewlett and J. Wolff, *J. Bacteriol.* **127,** 890 (1976).

[44] J. Wolff, G. H. Cook, A. R. Goldhammer, and S. A. Berkowitz, *Proc. Natl. Acad. Sci. U.S.A.* **77,** 3841 (1980).

[45] D. L. Confer and J. W. Eaton, *Science* **217,** 948 (1982).

[46] Y. Salomon, *Adv. Cyclic Nucleotide Res.* **10,** 35 (1979).

ing integral 10-ml reservoirs (Bio-Rad) and packed with 1.0 ml Dowex AG 50W-X4 (Bio-Rad), followed by two portions of 3.0 ml water, taking care to let the fluid run completely into the resin after each addition. The Dowex columns are then placed over columns of the same design containing 1.0 ml neutral alumina WN-3 (Sigma), and 7 ml water is added to elute the cAMP from the Dowex and transfer it to the alumina. After the 7.0 ml has drained through the alumina columns, these are placed above scintillation vials, and the cAMP is eluted with 7.0 ml 0.1 *M* imidazole hydrochloride, pH 7.0. The [^{32}P]cAMP is measured by Cerenkov counting. If [^{3}H]- or [^{14}C]ATP is used as substrate, an aliquot is transferred to another vial containing an aqueous scintillation fluid. When the Dowex and alumina columns are first set up, ATP and cAMP standards should be run and fractions collected to verify elution positions. The results obtained may show that the [^{3}H]- or [^{14}C]cAMP can be collected in a smaller volume so as to facilitate scintillation counting.

Toxicity Assays. Anthrax toxin was originally defined as an agent causing edema in skin (now known to reflect the action of PA with EF), and subsequently was found to contain a material lethal for guinea pigs and rats (LF, when combined with PA). The skin edema assay is laborious and rather variable, and has been replaced in this laboratory by the Chinese hamster ovary (CHO) cell elongation assay.[9] The rat lethality assay remains useful as a quick and accurate measure of potency for PA and LF.[24] These toxicity assays can be performed directly on *B. anthracis* culture supernatants, since these do not contain other substances having measurable toxicity. All bioassays of unfractionated anthrax toxin must take into account the competitive action of LF and EF[5,10,24]; these components will inhibit the toxicity of the heterologous component unless diluted to concentrations below 0.1 μg/ml.

Rat lethality assays require use of male Fischer 344 rats weighing 250–300 g. Toxin samples or mixtures of components are diluted in buffer containing 0.1% bovine serum albumin. Measured volumes of 1.0–2.0 ml are injected into the dorsal penile vein. Times to death (40–200 min) are recorded and related to a standard curve.[24,28] If good injections are made, duplicate animals will have times to death differing by 2–5 min. Either PA (5–25 μg) or LF (2–10 μg) may be accurately measured if the complementary protein is injected in excess (>60 μg). With PA and LF at 5 : 1 (wt/wt), the minimum lethal dose is 3 μg PA and 0.6 μg LF.[24] In this assay, injection of 2.0 ml of a successful RM culture results in a time to death of 50–60 min.

Although most cultured cells do not show acute effects when exposed to anthrax toxin, several cell systems do provide useful assays for the toxin. Most fibroblast cell lines treated with PA and EF show a rise in concentration of intracellular cAMP,[10] which can be measured by radioimmunoassay[47] after extraction into dilute acid. CHO cells also undergo a morphological change, which forms the basis of a convenient and sensitive assay for PA and EF, but one which is difficult to quantitate. The combination of PA and LF does not kill most cells. The exception is mouse and rat macrophages, which lyse after a 2-hr treatment with PA and LF at 0.1 μg/ml each.[48] This combination also slows the growth of certain cells, such as BHK-21, but this effect can be demonstrated only if cells are seeded at very low density, exposed to toxin, and cultured for several days.

Summary and Future Developments

Study of anthrax toxin has been essential in improving our understanding of the virulence of *B. anthracis* and in design of improved vaccines. In addition, study of the anthrax toxins may show them to be useful tools in cell biology. The adenylate cyclase toxin (PA + EF) has already been employed to study the effect of increased intracellular concentrations of cAMP. For this purpose it may be preferred over cholera toxin because the effect is rapidly reversed after toxin removal.[10] Expanded use of the anthrax toxins as pharmacological tools may occur in the future when the molecular basis of LF action is determined and when methods for toxin production are developed which do not require growth of *B. anthracis*. In the latter regard, recent work at this Institute has succeeded in cloning the PA gene into *B. subtilis* on plasmid pUB110.[41a] Even without medium optimization, PA production and secretion by this strain equalled that by Sterne. It can be expected that use of protease-deficient *B. subtilis* hosts,[49] optimized medium, and alteration of the DNA sequences controlling expression will combine to further increase yields.

Acknowledgment

In conducting the research described in this report, the investigators adhered to the "Guide for the Care and Use of Laboratory Animals," as promulgated by the Committee on Care and Use of Laboratory Animals of the Institute of Laboratory Animal Resources, National Research Council. The facilities are fully accredited by the American Association for Accreditation of Laboratory Animal Care. The views of the authors do not purport to reflect the positions of the Department of the Army or the Department of Defense.

[47] G. Brooker, J. F. Harper, W. L. Terasaki, and R. D. Moylan, *Adv. Cyclic Nucleotide Res.* **10,** 1 (1979).

[48] A. M. Friedlander, *J. Biol. Chem.* **261,** 7123 (1986).

[49] F. Kawamura and R. H. Doi, *J. Bacteriol.* **160,** 442 (1986).

[17] Production and Isolation of Neissereal IgA Proteases

By ANDREW G. PLAUT

IgA proteases are a group of bacterial enzymes that are synthesized and secreted into the external environment by microorganisms pathogenic for human beings.[1,1a] These bacteria colonize and/or infect human mucosal surfaces protected by secreted IgA antibodies of both IgA_1 and IgA_2 isotypes. Human IgA_1 is the sole substrate known for all IgA proteases with the exception of the *Clostridium ramosum* enzyme[1b] that cleaves both isotypes. Each enzyme cleaves either a Pro-Thr or Pro-Ser peptide bond among several such bonds grouped in a duplicated, proline-rich octapeptide at the center of the heavy (alpha) polypeptide chain of IgA_1. The type 2 protease of *Neisseria* species cleaves peptide bond number 235–236, and type 1 enzyme bond 237–238 in the human IgA_1 alpha chain.[1] Hydrolysis divides the Fc_α and the Fab_α regions of the molecule and identification of these intact fragments form the basis of the protease assay. Human IgA_2 proteins lack the susceptible hinge-region octapeptides and are therefore protease resistant. IgA proteases have been described in numerous bacterial genera, including *Neisseria, Hemophilus, Streptococcus, Bacteroides, Clostridium,* and *Capnocytophaga.*[2] The protease gene (*iga*) cloned from *Neisseria* and *Hemophilus* into *Escherichia coli* encodes an active enzyme consisting of a single polypeptide chain of approximately 105,000 relative molecular weight.

The various proteases have many biochemical and antigenic differences, and therefore isolation procedures for each is different. No IgA protease has been purified to homogeneity in high yield, presumably because the enzyme protein is a minor constituent of culture supernatants. An exception may be the protease of *Bacteroides melaninogenicus,* a thiol-activated enzyme[3] which is smaller than other IgA proteases and has been purified by high-performance liquid chromatography. Although difficulties in purification have impeded biochemical characterization of the

[1] A. G. Plaut, *Annu. Rev. Microbiol.* **37,** 603 (1983).

[1a] M. H. Mulks, *in* "Bacterial Enzymes and Virulence" (I. A. Holder, ed.) pp. 81–104, CRC Press, Boca Raton, FL, 1985.

[1b] Y. Fujiyama, K. Kobayashi, S. Senda, *et al., J. Immunol.* **134,** 573 (1985).

[2] M. Kilian, B. Thomsen, T. E. Petersen, and S. H. Bleeg, *Ann. N.Y. Acad. Sci.* **409,** 612 (1983).

[3] S. B. Mortensen and M. Kilian, *Infect. Immun.* **45,** 550 (1984).

enzymes, the available preparations are highly active and useful for immunochemical studies involving both secretory and serum IgA_1.

Isolation and Partial Purification of the IgA_1 Protease of *Neisseria gonorrhoeae*

Preparation of Inoculum. A fresh clinical isolate of *Neisseria gonorrhoeae* or a strain preserved at −70° is heavily streaked on chocolate agar plates and incubated at 37° for 16–24 hr in an atmosphere of 5% CO_2 in air. The confluent growth on each plate is removed by sterile swab and suspended in 10 ml prewarmed medium (see below). The entire content of one plate is used to inoculate each liter of broth culture.

Broth Culture. When a fully defined medium is desired that of Morse and Bartenstein is used,[4] dispensed in 1-liter volumes, preferably into flasks having their largest diameter at the base, e.g., Erlenmeyer or Low-Form culture flasks (Corning 4422). To improve enzyme yield three constituents of the medium are not added until just prior to inoculation; these are $Fe(NO_3)_3$, $CaCl_2$, and $NaHCO_3$. The medium is prewarmed to 37°, inoculated, and incubated in room air at 37° for 16 hr with gentle swirling rotation (about 50 rpm). Bubbles or foaming should be avoided. After confirming by gram stain that the complete culture consists of gram-negative diplococci the bacteria are removed by centrifugation and the clear supernatant fluid harvested for enzyme isolation. It is usually possible to detect IgA protease activity in this unconcentrated fluid.

Neisseria strains elaborating type 1 protease typically grow much more slowly and yield less enzyme activity than type 2 strains. To increase growth we modify our culture in three ways. First, each liter of medium is inoculated with the bacteria from six chocolate agar plates. Second, each liter is supplemented with 10 ml of soluble starch. This is prepared by suspending 8 g soluble starch (Difco) into 80 ml deionized H_2O and autoclaving for 20 min to bring into solution. Third, the amount of hypoxanthine and uracil used for culturing type 1 is one-fifth that used for type 2-producing bacteria, i.e., 0.01 g of each per liter.

Neisseria species also yield type 1 or type 2 protease when cultured in standard nondefined culture medium, e.g., Difco GC broth. The medium is supplemented with IsoVitaleX (BBL Division, Becton-Dickinson) as specified by the manufacturer.

Enzyme Assay. Detection of IgA protease activity requires the use of human IgA_1 protein substrate. Detailed methods for qualitative[5] and quantitative[6,7] assay have been published.

[4] S. A. Morse and L. Bartenstein, *Can. J. Microbiol.* **26,** 13 (1980).

[5] A. G. Plaut, R. Wistar, Jr., and J. D. Capra, *J. Clin. Invest.* **54,** 1295 (1974).

During enzyme isolation the location of activity can be monitored qualitatively by using as substrate the IgA in the unfractionated serum of a patient with multiple myeloma. The paraprotein should be present in sufficient amounts to show a tall peak when the serum is electrophoresed on agarose gels or nitrocellulose. When incubated with enzyme-containing fractions the peak is progressively reduced in size and new bands (Fab and Fc fragments) can usually be seen.[5] Approximately 10% of myeloma IgA proteins are of the IgA_2 isotype and will be unaffected by IgA proteases.

For quantitative assay purified IgA_1 substrate is used and Fab and Fc fragments are separated by more precise techniques, e.g., polyacrylamide gel electrophoresis (PAGE) in sodium dodecyl sulfate (SDS) and 2-mercaptoethanol.

Purification of Type 2 Neissereal Protease. Spent, bacteria-free culture supernatants are treated as follows:

1. Ammonium sulfate (Ultra pure, Schwartz-Mann Corp.), 390.0 g/liter of fluid, is added slowly with stirring and the mixture held overnight at 4°. Following centrifugation the brown pellet containing the enzyme is redissolved in several milliliters of Tris–HCl buffer, 0.05 *M*, pH 7.5. When defined medium has been used the pellet is small.
2. Concentrated protease is applied to a 2.5 × 70 cm gel filtration column containing Bio-Gel P-100 (Bio-Rad Corporation) and eluted at 4° with Tris–HCl buffer containing 0.1% sodium azide. Enzyme monitored by qualitative assay elutes in the void volume. Following this step, or in its place, active fractions are filtered on Sephadex G-150 (Pharmacia Corp.), from which enzyme elutes just following the void volume.
3. Gel filtrates containing activity are pooled and concentrated by positive pressure dialysis using a PM10 or YM10 membrane (Amicon Corp.). The concentrate is applied to a column (approximately 2 × 12 cm) containing the anion-exchanger diethylaminoethyl(DEAE)-cellulose (DE-52; Whatman Chemical Separation, Ltd.) which has first been extensively washed with normal saline and then Tris–HCl buffer, 0.05 *M*, pH 7.5. Type 2 neissereal protease does not bind under these conditions while several contaminating proteins are retained. The fall-through solution contains active IgA protease but no other detectable proteolytic enzymes. When examined in 9% polyacrylamide gels (PAGE) the product has three or four weakly staining protein bands using Coomassie Blue stains. Activity is stable for months at −70°.

[6] M. H. Mulks, S. J. Kornfeld, B. Frangione, and A. G. Plaut, *J. Infect. Dis.* **146,** 266 (1982).

[7] A. G. Plaut, J. V. Gilbert, G. Leger, and M. Blumenstein, *Mol. Immunol.* **22,** 821 (1985).

Preliminary studies show that type 1 neissereal protease may be more acidic than type 2 and therefore may bind to anion-exchange material using conditions outlined above. The method provided should at this time be regarded as suitable only for the type 2 enzyme. The other microbial IgA proteases have major differences in charge and structure and purification methods should be adapted for each independently. Halter *et al.*[8] and Blake and Swanson[9] have reported similar purification schemes for neissereal proteases using different gel filtration and ion-exchange materials and incorporation of isoelectric focusing and affinity chromatography methods. Purification schemes for the enzymes of *Streptococcus sanguis*[10] and *Bacteroides melaninogenicus*[3] have been reported; the latter requires high-performance liquid chromatography as a final step.

Acknowledgment

Supported by Grant #DE06048 HD #20962 and Core Center Grant #PO 30AM34928 from the National Institutes of Health.

[8] R. Halter, J. Pohlner, and T. F. Meyer, *EMBO J.* **3,** 1595 (1984).
[9] M. S. Blake and J. Swanson, *Infect. Immun.* **22,** 350 (1978).
[10] R. S. Labib, N. J. Calvanico, and T. B. Tomasi, Jr., *Biochim. Biophys. Acta* **526,** 547 (1978).

[18] Purification of Pertussis Toxin

By V. Y. PERERA

Pathogenic strains of *Bordetella pertussis,* the causative organism of whooping cough, produces at least four toxins: heat-labile toxin, endotoxin, tracheal cytotoxin, and pertussis toxin (PT).[1] Pertussis toxin (also referred to as histamine-sensitizing factor, lymphocytosis-promoting factor, islet-activating protein, and pertussigen) has many biological activities. The molecular basis of some of these activities (e.g., islet activation) is better understood than others (e.g., histamine sensitization). The toxin is found as both cell-free and cell-associated protein in growing cultures.[2] Current data indicate that PT is a multimeric aggregate consisting of five

[1] A. C. Wardlaw and R. Parton, *Pharmacol. Ther.* **19,** 1 (1983).
[2] V. Y. Perera, A. C. Wardlaw, and J. H. Freer, *in* "Zentralblatt für Baktenologie Mikrobiologie und Hygiene I. Abteilung." Supplement 15. Bacterial Protein Toxins (P. Falmagne *et al.,* eds.), p. 383. Grustav Fischer, Stuttgart, New York. (1986).

dissimilar subunits.[3] The toxin appears to consist of two functional moieties; the A protomer (subunit S_1) with ADP-ribosyltransferase activity and a B oligomer (subunits S_2, S_3, S_4, and S_5) which mediates binding of the toxin to target cells. Estimates of the apparent M_r of the holotoxin range from 90K to 107K.[1]

Pertussis toxin is proving to be a useful tool for studying cellular events, particularly in the fields of mammalian cell physiology, enzymology, and pharmacology. The toxin has been shown to increase cAMP levels by activation of the adenylate cyclase pathway.[4] This is achieved by inactivation (by ADP ribosylation) of the GTP-binding receptor protein (the α receptor, N_i) in mammalian cells such as rat C6 glioma cells, human red blood cells and platelets, rat adipocytes, and cyc^- S49 cells.[3,5] This protein which has an apparent M_r of 40K to 41K, and is distinct from the chlorea toxin substrate.[6] In pancreatic B cells, PT blocks the α-adrenoreceptor by ADP ribosylation and thus causes a slow influx of Ca^{2+}.[7,8] This in turn activates adenylate cyclase and the increased cAMP (and Ca^{2+}) levels stimulate secretion of insulin (hyperinsulinemia).

Other biological activities of PT include sensitization of mice to histamine[9] and lymphocytosis and leukocytosis.[10,11] The toxin also induces agglutination of erythrocytes.[12] Important questions still unanswered are whether or not transfer of ADP-ribose is the sole biologically significant activity associated with the S_1 subunit of PT, and whether the remaining subunits contribute solely to binding specificities[3,13] or have other more significant biological functions associated with pathogenesis.

Over the last decade, considerable progress has been made on the purification of PT. The most widely used starting material is culture supernatants of *B. pertussis,* phase 1, although processing of cytoplasmic extracts has also been reported.[12] Some of the earlier methods for purification of PT include cesium chloride density gradient centrifugation[11] and

[3] M. Tamura, K. Nogimori, S. Murai, M. Yajima, K. Ito, T. Katada, and M. Ui, *Biochemistry* **21,** 5516 (1982).

[4] D. M. Gill, *in* "Bacterial Toxins and Cell Membranes" (J. Jelaszewicz and T. Wadstrom, eds.), p. 271. Academic Press, London, 1978.

[5] R. D. Sekura, F. Fish, C. R. Manclark, B. Meade, and Y.-L. Zhang, *J. Biol. Chem.* **258,** 14647 (1983).

[6] T. Katada and M. Ui, *J. Biol. Chem.* **257,** 7210 (1982).

[7] T. Katada and M. Ui, *J. Biol. Chem.* **254,** 469 (1979).

[8] T. Katada and M. Ui, *J. Biochem.* (*Tokyo*) **89,** 979 (1981).

[9] J. J. Munoz and R. K. Bergman, *Bacteriol. Rev.* **32,** 103 (1968).

[10] H. Arai and Y. Sato, *Biochim. Biophys. Acta* **444,** 765 (1976).

[11] S. I. Morse and J. H. Morse, *J. Exp. Med.* **143,** 1483 (1976).

[12] L. I. Irons and A. P. MacLennan, *Biochim. Biophys. Acta* **580,** 175 (1979).

[13] M. Tamura, K. Nogimori, M. Yajima, K. Ase, and M. Ui, *J. Biol. Chem.* **258,** 6756 (1983).

adsorption to hydroxylapatite followed by gel filtration and affinity chromatography on concanavalin A–Sepharose.[14] More recent reports have exploited the binding of PT to sialic acid-containing glycoproteins such as human haptoglobin or fetuin, in affinity chromatography for isolation of PT.[5,12] The yield of purified PT varies with the strain used and is generally minute, ranging from 1.8 to 10.8 mg from approximately 20 liters of culture supernatant. Recently, Imaizumi *et al.*[15] have reported that addition of heptakis(2,6-*O*-dimethyl)-β-cyclodextrin to the culture medium enhanced toxin production although the high value of 50 mg PT/liter of culture broth has not been confirmed by others.

Materials

Organism. Bordetella pertussis strain 18334, phase 1 (Connaught Laboratories) is used. Other phase 1 strains would probably be suitable but it should be noted that strain 165 is a better toxin producer than strain Tohama.[5]

Growth Media and Buffers. Bordet-Gengou (BG) blood agar [500 ml BG base (Gibco), is autoclaved at 121° for 15 min, cooled to 45°, and mixed with 100 ml defibrinated horse blood (Gibco)]; cyclodextrin liquid (CL) medium[15] (see Table I); phosphate-buffered saline (PBS) (0.01 *M* sodium phosphate, pH 7.5, containing 0.15 *M* NaCl); coupling buffer (0.1 *M* $NaHCO_3$, pH 8.0, containing 0.5 *M* NaCl); 1 *M* ethanolamine hydrochloride, pH 9.0; 0.1 *M* sodium acetate buffer, pH 4.0; buffer A (0.1 *M* Tris–HCl, pH 7.5, containing 0.5 *M* NaCl); buffer B (0.1 *M* Tris–HCl, pH 7.5, containing 2 *M* NaCl), and buffer C (0.1 *M* Tris, pH 10.0, containing 0.5 *M* NaCl and 3 *M* KSCN).

Purification Procedure

Step 1: Growth of B. pertussis. Strain 18334 is initially grown on BG agar plates for 72 hr at 37° and the cells are resuspended in 10 ml sterile 1% casamino acids. One liter of CL medium (contained in a 2-liter Erlenmeyer flask) is inoculated with the above cell suspension to an initial optical density (at 600 nm) of 0.05. Flasks are shaken (150 cpm) for 48 hr at 37°.

Step 2: Culture Supernatants. Bacteria from 4 liters of culture are harvested by centrifugation at 10,000 *g* for 30 min and the pH of the

[14] M. Yajima, K. Hosoda, Y. Kanbayashi, T. Nakamura, K. Nogimori, Y. Mizushima, Y. Nakase, and M. Ui, *J. Biochem.* (*Tokyo*) **83,** 295 (1978).

[15] A. Imaizumi, Y. Suzuki, S. Ono, H. Sato, and Y. Sato, *Infect. Immun.* **41,** 1138 (1983).

TABLE I
COMPOSITION OF CL MEDIUM[a]

Component	Weight (g/liter)
Sodium L-glutamate	10.7
L-Proline	0.24
NaCl	2.5
KH_2PO_4	0.5
KCl	0.2
$MgCl_2 \cdot 6H_2O$	0.1
$CaCl_2$	0.02
Tris	6.1
L-Cysteine[b]	0.04
$FeSO_4 \cdot 7H_2O$[b]	0.01
Niacin[b]	0.004
Glutathione (reduced)[b]	0.15
Ascorbic acid[b]	0.40
Casamino acids (Difco)	10.0
Methylated β-cyclodextrin[c]	0.5

[a] Based on that of Imaizumi *et al.*[15] Dissolve in 1 liter distilled water, adjust pH to 7.6 with 5.0 *N* HCl, and sterilize by autoclaving at 121° for 15 min.

[b] Sterilize by membrane (0.45-μm pore size) filtration and add aseptically to the sterile medium just before inoculation.

[c] From Teijin, Ltd., Medical and Pharmaceutical Division, 1-1, Uchisaiwai-Cho, 2-Chome, Chiyoda-Ku, Tokyo, Japan.

resulting culture supernatant is adjusted to 7.5 with 5 *N* HCl. The culture supernatant is concentrated 20-fold by ultrafiltration (PM10 hollow fiber cartridge, Amicon) and may be stored at −20°.

Step 3: Fetuin-Sepharose Column. Fetuin (Sigma; 150 mg) is dissolved in 75 ml of coupling buffer and added to 30 ml (settled volume) of cyanogen bromide-activated Sepharose 4B (Pharmacia). After mixing the contents for 18 hr at 4°, the gel is washed sequentially (under negative pressure on a sintered funnel, No. 3) with 200 ml each of coupling buffer, sodium acetate buffer, buffer C, and finally coupling buffer. Unreacted groups on Sepharose are blocked with ethanolamine for 2 hr at 25° and the gel is poured into a column (2.5 × 10 cm). The fetuin–Sepharose column is finally washed with buffer A containing 0.02% NaN_3 as a preservative and stored at 4°.

Step 4: Affinity Chromatography. Concentrated culture supernatant is passed through the fetuin–Sepharose column at a flow rate of approximately 10 ml/hr at 25° and the column is washed with 200 ml of buffer A and 1 liter of buffer B at 4° using a flow rate of approximately 40 ml/hr. After washing the column with a further 200 ml of buffer A, bound PT is eluted with 60 ml of buffer C using a flow rate of approximately 10 ml/hr. Fractions (3 ml) containing protein are pooled, neutralized with 1 *N* HCl, and concentrated by ultrafiltration (YM30 membrane, Amicon). The re-

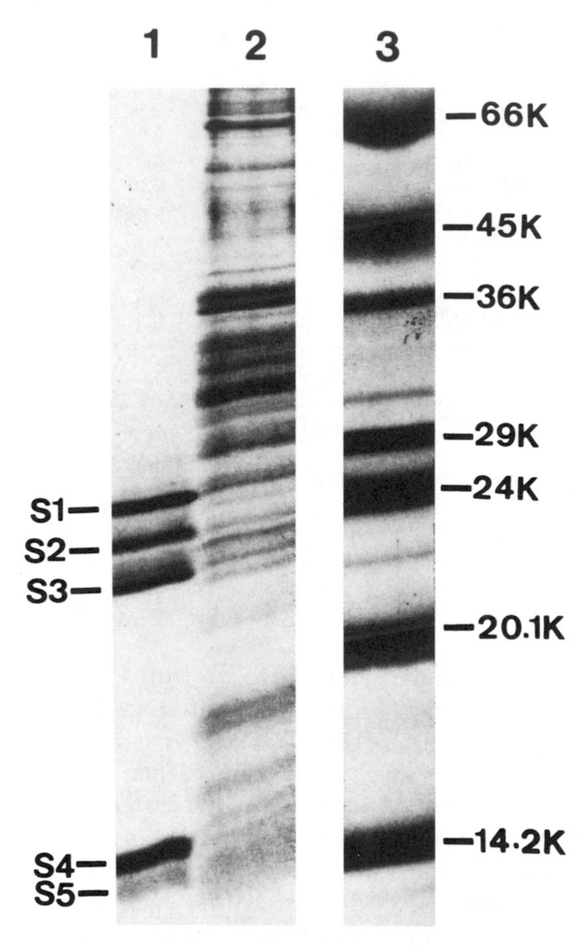

FIG. 1. The SDS–PAGE pattern of (1) purified PT (20 μg), (2) culture supernatant (40 μg), and (3) protein standards (from top to bottom: bovine serum albumin, ovalbumin, glyceraldehyde-3-phosphate dehydrogenase, carbonate dehydratase, trypsinogen, trypsin inhibitor, and α-lactalbumin). Samples are run in a 13 to 20% (w/v) linear acrylamide gradient. Purified PT and culture supernatant are solubilized in SDS and urea.[3] Protein standards are reduced in the presence of 2-mercaptoethanol. Polypeptides are stained with Coomassie Brilliant Blue. Subunits S_1, S_2, S_3, S_4, and S_5 of PT are shown.

TABLE II
In Vivo AND *In Vitro* ASSAYS FOR QUANTITATION OF PT

Assay method	Limit of detection of PT	Reference
Release of glycerol from rat adipocytes	300 pg/tube	18
Chinese hamster ovary (CHO) cell assay	120 pg/ml	19
Hemagglutination[a]	100 ng/well	20
Enzyme-linked immunosorbent assay	1 ng/well	21
Lymphocytosis-promoting activity	20 ng/mouse	12,18
Histamine-sensitizing activity[b]	3 to 10 ng/mouse	12,18

[a] With chicken erythrocytes; more sensitive with goose erythrocytes.
[b] Varies with the strain of mouse used.

tentate is dialyzed for 16 hr against 100 vol of PBS containing 10 μM phenylmethylsulfonyl fluoride (Sigma). The retentate containing purified PT is stored frozen at $-20°$. (Approximate yield of PT is 8 mg from 4 liters of culture supernatant. This corresponds to about 60% recovery from the culture supernatant.)

The fetuin–Sepharose column is regenerated by washing with a further 100 ml of buffer C followed by 100 ml of buffer A containing 0.02% NaN_3 and stored at 4°.

Purity and Biological Activities of Purified PT

The sodium dodecyl sulfate-polyacrylamide gel electrophoresis (SDS–PAGE) pattern of PT purified from a culture supernatant is shown in Fig. 1, lane 1. The toxin is pure as judged by SDS–PAGE and subunits S_1, S_2, S_3, S_4, and S_5 (with apparent M_r value of 26,500, 24,000, 22,000, 14,500 and 12,000, respectively) are detectable. The polypeptides present in the culture supernatant (lane 2) are also shown for comparison. Purified PT sensitizes mice to histamine[16] (histamine-sensitizing dose giving 50% death in mice is approximately 17 ng), reacts in an enzyme-linked immunosorbent assay which utilizes both polyclonal and monoclonal antibodies to PT[16] (lowest concentration detectable is 10 ng/ml), displays ADP-ribosyltransferase activity (M. Houslay, personal communication), and at a dose of 150 ng/mouse causes hypoglycemia in mice.[17] The purified PT has no detectable *B. pertussis* adenylate cyclase activity (less than 0.05 pmol cAMP/μg protein generated in the presence of calmodulin).

[16] V. Y. Perera, A. C. Wardlaw, and J. H. Freer, *J. Gen. Microbiol.* **132,** 553 (1986).
[17] B. L. Furman, F. M. Sidey, and A. C. Wardlaw, *Proc. Br. Pharmacol. Soc.* C67 (1985).

The endotoxin content is about 3% as measured by the *Limulus* lysate assay but this may be further reduced by passing through Detoxi-gel (Pierce) after fetuin affinity chromatography. The purity of the final product is greater than 95% and the method described achieves approximately 50-fold purification with respect to concentration of total protein in the supernatant.

Summary of Assay Procedures

Several *in vitro* and *in vivo* assays are employed for quantitation of PT (Table II).[12,18-21] The CHO cell assay appears to be the most sensitive of all the assays for PT. One of the drawbacks of using an *in vivo* assay such as measuring lymphocytosis-promoting activity is the interference of endotoxin. Endotoxin may also interfere with the histamine-sensitizing assay but the responses due to endotoxin and PT can be distinguished from each other by following a time course of sensitization in mice.[9] The hemagglutinating activity of the fimbrial hemagglutinin (FHA) is five to seven times greater than that of PT. Thus FHA (if present as a contaminant) may interfere if this type of assay is used to quantify PT.

Acknowledgment

This work was supported by a Medical Research Council (UK) Grant Number G8306837SB. I thank Dr. R. Parton for his advice.

[18] M. Endoh, M. Soga, and Y. Nakase, *Microbiol. Immunol.* **24,** 887 (1980).
[19] E. L. Hewlett, K. T. Sauer, G. A. Myers, J. L. Cowell, and R. L. Guerrant, *Infect. Immun.* **40,** 1198 (1983).
[20] H. Sato, A. Ito, J. Chiba, and Y. Sato, *Infect. Immun.* **46,** 422 (1984).
[21] H. Sato, Y. Sato, and A. Ito, *J. Microbiol. Methods* **1,** 99 (1983).

[19] Purification of *Escherichia coli* Heat-Stable Enterotoxin

By Richard N. Greenberg and Abdul M. K. Saeed

Enterotoxigenic *Escherichia coli* (ETEC) may induce clinically significant intestinal fluid secretion (i.e., diarrhea) in humans and animals by production of enterotoxins. Several types of ETEC enterotoxins are recognized, each with a unique structure and mechanism of action.[1-17] This volume as well as recent review articles describe these toxins in de-

[1] H. W. Smith and C. L. Gyles, *J. Med. Microbiol.* **3,** 387 (1970).

tail.[14–17] Our chapter is restricted to the purification of ETEC enterotoxin ST_a, which is the heat-stable mouse-active enterotoxin.[18,19] ST_a will activate particulate guanylate cyclase in intestinal epithelial cells.[7,8]

Early attempts to purify ST_a found that the enterotoxin diffused through dialysis tubing and by ultrafiltration had a molecular weight less than 10,000.[1–4] Partially purified preparations resulted from utilization of precipitation, molecular sieve filtration, and ion-exchange chromatography.[4,9,20–24] Improvement in purification has resulted in part due to a defined growth medium,[9,10,25–28] use of nonionic polymeric adsorbant res-

---

[2] H. W. Smith and C. L. Gyles, *J. Med. Microbiol.* **4,** 403 (1970).

[3] E. M. Kohler, *Ann. N.Y. Acad. Sci.* **176,** 212 (1971).

[4] R. J. Bywater, *J. Med. Microbiol.* **5,** 337 (1971).

[5] H. W. Moon and S. C. Whipp, *Ann. N.Y. Acad. Sci.* **176,** 197 (1971).

[6] D. G. Evans, D. J. Evans, Jr., and N. F. Pierce, *Infect. Immun.* **7,** 873 (1973).

[7] J. M. Hughes, F. Murad, B. Change, and R. L. Guerrant, *Nature (London)* **271,** 755 (1978).

[8] M. Field, L. H. Graf, Jr., W. J. Laird, and P. L. Smith, *Proc. Natl. Acad. Sci. U.S.A.* **75,** 2800 (1978).

[9] J. F. Alderete and D. C. Robertson, *Infect. Immun.* **19,** 1021 (1978).

[10] S. J. Staples, S. E. Asher, and R. A. Giannella, *J. Biol. Chem.* **10,** 4716 (1980).

[11] W. S. Dallas and S. Falkow, *Nature (London)* **288,** 499 (1980).

[12] D. M. Gill, J. D. Clements, D. C. Robertson, and R. A. Finkelstein, *Infect. Immun.* **33,** 677 (1981).

[13] E. K. Spicer, W. M. Kavanaugh, W. S. Dallas, S. Falkow, W. H. Konigsberg, and D. E. Schafer, *Proc. Natl. Acad. Sci. U.S.A.* **7,** 50 (1981).

[14] R. L. Guerrant, R. K. Holmes, D. C. Robertson, and R. N. Greenberg, *in* "Microbiology—1985," pp. 68–73. Am. Soc. Microbiol., Washington, D.C., 1985.

[15] C. H. Lee, S. L. Moseley, H. W. Moon, S. C. Whipp, C. L. Gyles, and M. So. *Infect. Immun.* **42,** 264 (1983).

[16] D. J. Kennedy, R. N. Greenberg, J. A. Dunn, R. Abernathy, J. S. Pyerse, and R. L. Guerrant, *Infect. Immun.* **46,** 639 (1984).

[17] R. N. Greenberg and R. L. Guerrant, *Pharmacol. Ther.* **13,** 507 (1981).

[18] A. G. Dean, Y. Ching, R. G. Williams, and L. B. Harden, *J. Infect. Dis.* **125,** 407 (1972).

[19] R. A. Giannella, *Infect. Immun.* **14,** 95 (1976).

[20] T. M. Jacks and B. J. Wu, *Infect. Immun.* **9,** 342 (1974).

[21] N. A. Mullan, M. N. Burgess, and P. N. Newsom, *Infect. Immun.* **19,** 779 (1978).

[22] R. A. Kapitany, A. Scoot, G. W. Forsyth, S. L. McKenzie, and R. W. Worthington, *Infect. Immun.* **24,** 965 (1979).

[23] R. A. Kapitany, G. W. Forsyth, A. Scott, S. F. McKenzie, and R. W. Worthington, *Infect. Immun.* **26,** 173 (1979).

[24] S. Stavric and T. M. Gleeson, *Toxicon Suppl.* **3,** 413 (1982).

[25] W. M. Johnson, H. Lior, and K. G. Johnson, *Infect. Immun.* **20,** 352 (1978).

[26] E. Olsson, *Vet. Microbiol.* **7,** 253 (1982).

[27] A. M. K. Saeed, N. Sriranganathan, W. Cosand, and D. Burger, *Infect. Immun.* **40,** 701 (1983).

[28] A. M. K. Saeed, N. S. Magnuson, N. Sriranganathan, D. Burger, and W. Cosand, *Infect. Immun.* **45,** 242 (1984).

ins,[8–10] isoelectric focusing,[27–29] reversed-phase silica batch adsorption chromatography,[30] and reversed-phase high-performance liquid chromatography (RP-HPLC).[27–32]

Materials and Methods

Bacterial Strains

We have purified ST_a from bovine and human *Escherichia coli* strains including B41 (0101 : K^-;K99), B44 (09 : K30, K99), M490 (0101 : K30, K99), M524 (08 : K85, K99) kindly furnished by Dr. C. Gyles, Ontario Veterinary College, Guelph, Canada, and $WSUH_1$ (020 : KX106 : H4). One additional strain, Cl-4, was kindly supplied by Dr. Richard Guerrant, University of Virginia, Charlottesville, Virginia.

Protein Staining and Determination

ST_a in gels has been reported to be resistant to staining with conventional protein stains.[9,29,32] In our work the determination of protein concentration in various fluids and ST_a samples has been performed by the modified method of Lowry *et al.* using BSA as a standard.[33] The method gives results consistent with those obtained by the more cumbersome ninhydrin test and by those derived from the amino acid compositional analysis.

ST_a Bioassay

A modification suckling mouse assay (SMA) is used for detection and quantitation of ST_a throughout the purification steps.[18,19] A group of three Swiss-Webster or DUR/ICR strain 1- to 3-day-old mice are used per test sample. Suckling mice are starved for 2 hr prior to inoculation with the test materials. Each mouse is inoculated orally through polyethylene tubing connected to a 0.5-ml syringe. The size of each inoculum is 0.1 ml. After 3 hr of incubation at room temperature, the mice are killed by chloroform anesthesia or neck dislocation and dissected. The ratio of the entire gut weight (excluding the stomach) to remaining body weight (including stomach) is calculated. One mouse unit (MU) of ST_a is the minimal amount of ST_a protein which induces a gut-to-remaining body weight

[29] L. A. Dreyfus, J. C. Frantz, and D. C. Robertson, *Infect. Immun.* **42,** 539 (1983).

[30] G. L. Madsen and F. C. Knoop, *Infect. Immun.* **28,** 1051 (1980).

[31] A. M. K. Saeed and R. N. Greenberg, *Anal. Biochem.* **151,** 431 (1985).

[32] S. Aimoto, T. Takao, Y. Shimonishi, T. Takeda, Y. Takeda, and T. Miwatani, *Eur. J. Biochem.* **129,** 257 (1982).

[33] G. L. Peterson, *Anal. Biochem.* **83,** 346 (1977).

ratio of 0.087. In our laboratory the gut weight-to-remaining body weight ratio in negative control mice is 0.058 ± 0.006.[28]

Growth Medium and Chemical Reagents

The growth medium we use for ST_a production is that described by Staples *et al.*[10] It is composed from the following ingredients (in g/liter of medium): L-asparagine (crystalline, anhydrous, Sigma Chemical Company, Saint Louis, MO), 5.00; NaCl, 2.52; sodium acetate, 10.00; K_2HPO_4 (anhydrous), 8.71; sodium sulfate (anhydrous), 0.14; magnesium sulfate (nhydrous), 0.05; $MnCl_2 \cdot 4H_2O$, 0.005; and $FeCl_3 \cdot 6H_2O$, 0.005. The latter two salts are prepared as 1% solutions and added as 0.5 ml/liter of the medium after dissolving the rest of ingredients.

The bacteria is grown in 15 liters of culture medium in 20-liter fermentation carboys. Bacterial strains are maintained at −70° in 15% glycerol in 5 ml trypticase soy broth (BBL) in glass vials. A thawed vial is used to inoculate 300 ml of the medium in a 1-liter flask. The bacteria is grown at 39° in a shaker incubator at 200 rpm for 24 hr. The 300-ml culture is used to inoculate the 15-liter medium carboy. Carboys are purged with filter-sterilized oxygen and air at rates of 300 ml and 3000 ml/min, respectively. This fermentation is at 39° for 18 hr. Following fermentation, bacterial cell-free filtrate is obtained by tangential flow filtration through a 0.2-μm cassette in a Millipore Pellicon system (Millipore Corp., Bedford, MA). The ST_a is then absorbed from the filtrate by Amberlite XAD-2 batch adsorption.

Amberlite XAD-2 Batch Adsorption Chromatography

The Amberlite XAD-2 resin is extensively washed with purified water to remove contaminants. Five hundred grams of the washed resin is suspended into the culture filtrate. The suspension is maintained overnight under gentle stirring at 4°. The resin is collected by pouring the suspension into an 8-cm i.d. × 40-cm long glass column and washed with 5 liters of pure water. The loosely bound contaminants are eluted with 1 liter of 1% acetic acid in 20% methanol. ST_a is eluted with 1 liter of 1% acetic acid in 99% methanol followed by 1 liter of 1% acetic acid in 80% methanol. A final wash to elute ST_a is 1 liter of 0.1% trifluoroacetic acid (TFA) in acetone. The first three fractions are pooled and concentrated by flash evaporation of the solvents and then lyophilized.

Acetone Fractionation

The lyophilized crude ST_a is dissolved in distilled water acidified with acetic acid or trifluoroacetic acid to pH 4.0. Then up to 5 vol of acetone is gradually added to completely solubilize the crude ST_a. The ST_a is left for

1 hr at room temperature. The subsequent suspension is centrifuged at 20,000 *g* for 15 min. The supernatant is recovered and the acetone is flash evaporated. The acetone-fractionated ST_a is lyophilized and is ready for isoelectric focusing or preparative HPLC.

Preparative Isoelectric Focusing in Flatbed Granulated Gel (PIFGG)

Our technique is similar to that described by Radola.[34] For the flat-bed gel the anticonvective surface is made from prewashed and dried Sephadex G-200 superfine (Pharmacia Fine Chemicals, Inc., Uppsala, Sweden). A slurry of the gel is made by suspending 3.5 g of the gel powder in 130 ml of purified water. To this slurry is added ampholines [2.75 ml (pH 2.5 to 4.0) and 2.25 ml (pH 3.5 to 5.0); LKB Produkter, Bromma, Sweden]. The crude ST_a is mixed with the gel slurry and left for 2 hr at 4° before degassing and pouring onto a glass tray (10.0 × 25 cm) to form a flat bed. Anodic electrofocusing strips are soaked in 1 *M* phosphoric acid, cathodic strips are soaked in 1 *M* sodium hydroxide, and the strips are applied at appropriate ends of the gel tray. As previously described,[28] the isoelectric focusing is not initiated until approximately one-third of the water has been evaporated from the poured flat bed. Evaporation is achieved under a gentle stream of air. The focusing requires a power setting of 8 W and a maximum voltage setting of 1400. The run is 15 hr in an LKB 2117 multiphore chamber (LKB Produkter, Bromma, Sweden). During the run the gel is maintained at 4°.

At completion, the flat bed is viewed under UV illumination (Chromato-Vue transilluminator model 0-63, Ultraviolet Products, Inc., San Gabriel, CA). Bright fluorescent bands on the anodic end of the flat bed will be evident. This area (fractions 5 through 8) correspond to the highest UV-absorbing fractions at 280 nm (Fig. 1). At the cathodic end of the flatbed, a sharply focused band is observed at pH 4.3 with minor bands between pH 3.8 and 4.1. Under tungsten light illumination, only the focused ST_a band is evident. Under initial running conditions, however, it is recommended that the pH be measured on the gel at 0.75-cm intervals and that the gel be sliced into 30 fractions. Protein is eluted from each fraction and tested for ST_a biological activity by the suckling mouse assay. The majority of ST_a activity is found in the fraction with a pH of 4.3. Table I shows the loads and recovery of ST_a from several ETEC strains. It should be noted that more than 60% of the loaded ST_a is focused in a sharp band at pH range of 4.2–4.6. ST_a from the minor bands can be recovered and recycled.

[34] B. J. Radola, *Biochim. Biophys. Acta* **386,** 181 (1974).

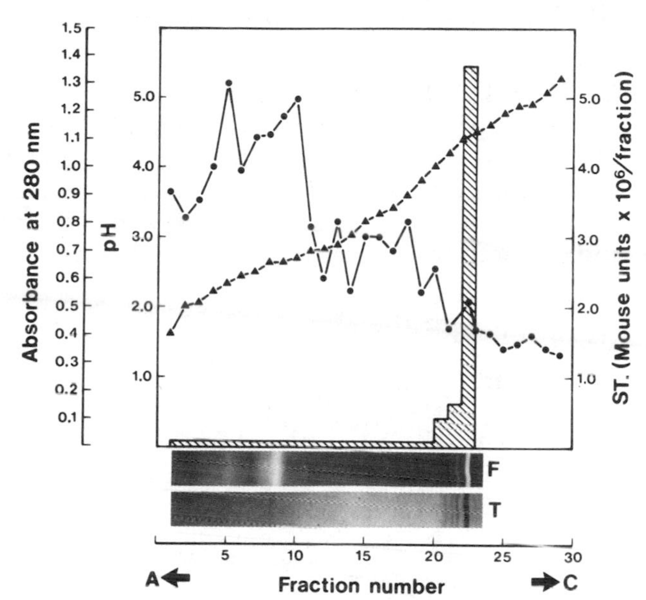

FIG. 1. PIFGG of acetone-fractionated ST_a. Anticonvective stabilizing gel, Sephadex G-200 superfine; ▲—▲, pH gradient (1.8 to 5.2); load, 570 mg of crude ST_a; voltage, 1400 V for 15 hr; power, 8 W, A, Anode; C, cathode; ●—●, absorbance at 280 nm; hatched area, ST_a concentration on mouse units; F, gel viewed under UV illumination; T, gel viewed under tungsten light.

TABLE I
RECOVERY OF ST_a PURIFIED FROM FOUR STRAINS OF BOVINE ETEC BY PIFGG

Strain	ST_a load[a] (MU × 10^6)[b]	Recovery from the major band (MU × 10^6) (%)	pH	Number of experiments
B41	32	22 (68)	4.3–4.6	4
B44	18	11 (61)	4.2–4.4	3
M490	50	30 (60)	4.2	2
M524	16	10 (60)	4.3	2

[a] Loads represent the biologically active ST_a's from 10-liter batch cultures. Minor bands are harvested, pooled, and recycled.

[b] MU, Mouse units.

Reversed-Phase High-Performance Liquid Chromatography (RP-HPLC)

Samples for RP-HPLC (from isoelectric focusing) are taken in 2 ml of 0.1% TFA and injected into a semipreparative RP-HPLC column Aquapor RP-300 (7.1 mm i.d. × 25 cm long; Brownlee Laboratories, Santa Clara, CA). HPLC is performed on a Waters Associates liquid chromatograph system equipped with two model 510 solvent-delivery pumps, an automated gradient programmer, model U6K injector, and model 441

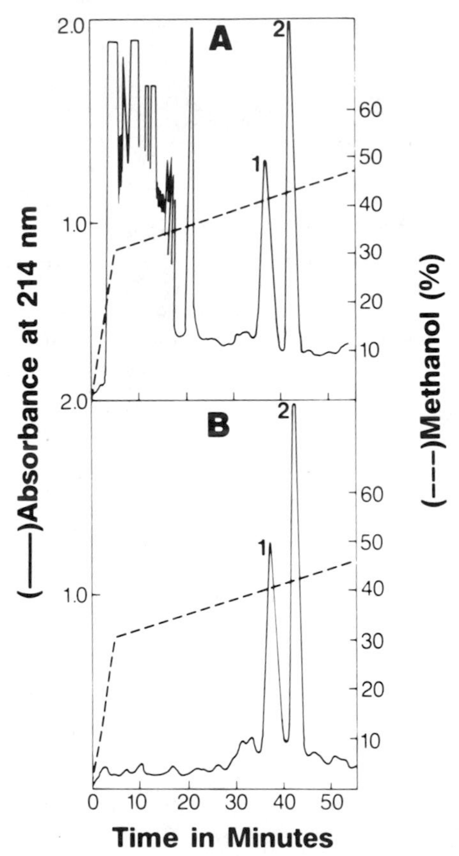

FIG. 2. Purification of ST_a by preparative RP-HPLC. Column: Aquapore RP-300 run at room temperature at 120 ml/hr. Solvent A: 0.1% TFA. Solvent B: 0.1% TFA in methanol. Gradient: 30% B in 5 min, 30–60% B in 80 min. (A) Crude ST_a. (B) ST_a cleaned by Sepralyte chromatography. Peak 2 contained more than 98% of the ST_a biological activity as determined by SMA. Peak 1 had little ST_a biological activity.

TABLE II
PURIFICATION SUMMARY OF ST_a FROM BOVINE ETEC STRAIN ($WSUH_1$)

Procedure	Protein (mg)	ST_a, total (MU × 10^6)	Minimal effective dose (ng)	Purification (-fold)	Recovery (%)
Culture supernatant filtrate (10) liters	12,380	1,000	12.5	1	100
Amberlite XAD-2 chromatography	2,500	840	3.0	4	85
Acetone fractionation	850	980	0.9	14	98
Preparative isoelectric focusing	162	1,000	0.2	63	100
Reversed-phase HPLC	12	1,000	0.012	1,000	100

absorbance detector equipped with appropriate light bulbs, filters, and aperture kits to allow detection at various wave lengths (range 214–436 nm). Elution of ST_a is performed by a two-step linear gradient with methanol in 0.1% TFA as solvent B (0–30% in 5 min and 30–60% in 80 min). The UV-absorbing peaks are detected at 214 nm. Peaks are collected and the methanol is flash evaporated before reconstitution in physiological saline and testing for ST_a biological activity in suckling mice. Most of the biologically active ST_a is eluted in the form of one major peak (peak 2) at 42% methanol. A minor peak (peak 1) is also evident at 40% methanol and has little ST_a biological activity (Fig. 2). ST_a purified by the tabulated scheme (Table II) results in a highly purified preparation.

ST_a from peak 2 is found homogeneous by the following criteria: (1) a single symmetrical peak upon analytical RP-HPLC with several different elution gradients (Fig. 3), (2) a single symmetrical peak by size-exclusion HPLC, and (3), a single band by analytical isoelectric focusing (p*I* 4.1). Amino acid composition (analysis performed as previously reported[27,28]) indicates a stoichiometric relationship among the residues constituting the ST_a peptide (Table III). The analysis shows single amino-terminal and single carboxy-terminal residues. The purified ST_a has a high specific activity in the suckling mouse assay.

Carboxymethylation of ST_a

Approximately 50 nmol of the purified ST_a is carboxymethylated by the method of Crestfield *et al.*[35] The reduced and carboxymethylated peptide is purified by RP-HPLC to remove excessive salts.

[35] A. M. Crestfield, S. Moore, and W. H. Stein, *J. Biol. Chem.* **238,** 622 (1963).

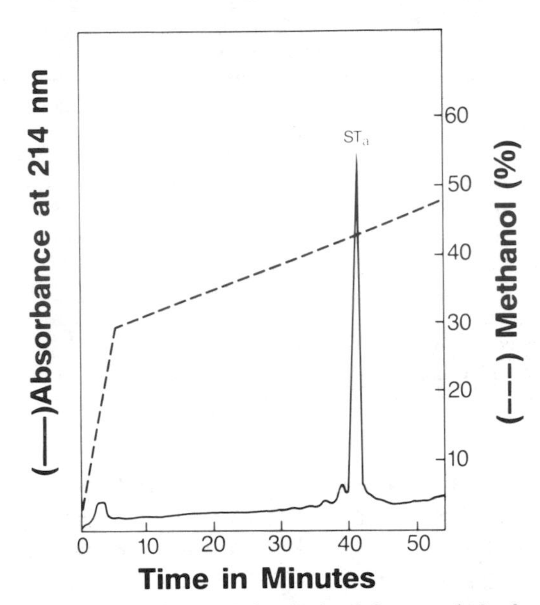

FIG. 3. Chromatography of an ST_a sample obtained from peak 2 of preparative RP-HPLC. Conditions: Same as in the preparative run except that the approximately 1-mg sample was loaded on an analytical RP-HPLC column at 60 ml/hr flow rate.

TABLE III

AMINO ACID ANALYSIS OF ST_a PURIFIED FROM FIVE STRAINS OF BOVINE ETEC

Amino acid residue	HPLC peaks for the following strains[a]						
	B41		B44	M490		M524	
	Major	Minor	(major)	Major	Minor	(major)	$WSUH_1$
Alanine	1.96 (2)	1.93 (2)	1.96 (2)	1.97 (2)	1.97 (2)	1.96 (2)	2.02 (2)
Asparagine[b]	2.00 (2)	2.04 (2)	1.99 (2)	1.99 (2)	1.99 (2)	1.99 (2)	1.99 (2)
Cysteine	6.02 (6)	4.89 (6)	5.10 (6)	5.63 (6)	4.97 (6)	4.75 (6)	4.73 (6)
Glutamic acid	1.07 (1)	1.04 (1)	1.01 (1)	1.08 (1)	1.04 (1)	1.01 (1)	1.08 (1)
Glycine	1.13 (1)	1.11 (1)	1.03 (1)	1.28 (1)	1.14 (1)	1.03 (1)	1.02 (1)
Leucine	1.01 (1)	1.00 (1)	1.02 (1)	1.08 (1)	1.04 (1)	0.98 (1)	1.00 (1)
Phenylalanine	0.97 (1)	0.95 (1)	0.98 (1)	0.95 (1)	0.99 (1)	0.99 (1)	0.95 (1)
Proline	1.00 (1)	0.93 (1)	0.98 (1)	1.05 (1)	1.00 (1)	1.02 (1)	1.02 (1)
Threonine	0.98 (1)	0.99 (1)	0.99 (1)	0.95 (1)	0.97 (1)	0.96 (1)	0.98 (1)
Tyrosine[c]	1.66 (2)	1.65 (2)	1.97 (2)	1.64 (2)	1.64 (2)	1.97 (2)	1.78 (2)

[a] Values in parentheses are rounded to the nearest integer (residues per molecule).

[b] Identified as aspartic acid from the peptide lysate. However, sequence analysis revealed that it is asparagine and the amino-terminal residue in the sequenced ST_a peptides of the four strains.

[c] Carboxy-terminal residue of the ST_a peptide.

Sequence Analysis

Sequence analysis of ST_a peptides is performed by the manual microsequencing technique of Chang using the double-coupling method with dimethylaminoazobenze isothiocyanate/phenyl isothiocyanate.[36] After each cycle of derivatization, the labeled amino-terminal amino acid is cleaved. The dimethylaminoazobenzenethiohydantoin amino acids are identified by RP-HPLC on a Zorbax-ODS column maintained at 52°. Ten-milliliter samples are injected. Peaks are detected at 436 nm. The derivatized amino acids are identified by their retention times in comparison with amino acid standards derivatized and chromatographed under the same conditions.

To confirm our results of the manual microsequencing techniques, a ^{14}C-carboxymethylated ST_a has been sequenced on a model 890B Beckman sequencer (Beckman Instruments, Inc., Fullerton, CA) with 5 mg of Polybrene according to the method of Edman and Begg[37] and the 0.1 *M* Quadrol program of Brauer *et al.*[38] The phenylthiohydantoin amino acids were analyzed by RP-HPLC. The results of the manual microsequencing of the reduced and alkylated ST_a and the native ST_a peptide from peaks 1 and 2 suggest the following sequence: Asn-Thr-Phe-Tyr-Cys-Cys-Glu-Leu-Cys-Cys-Asn-Pro-Ala-Cys-Ala-Gly-Cys-Tyr. By performing automated sequence analysis of the ST_a, the positions of Cys residues in the ST_a peptide were confirmed by using iodo[^{14}C]acetic acid in the alkylation of the reduced peptide and measurement of counts of the sequentially cleaved amino acid residues in a scintillation counter (Fig. 4).

ST_a's purified from most ETEC strains of human, porcine, and bovine origins have the same composition and primary sequence.[28,39–42] However, ST_a's purified from certain strains of human ETEC are reported to have 19 residues and have minor differences in their primary sequence from that of ST_a's of other ETEC strains.[32,43] Nevertheless, ST_a's purified from all strains of ETEC to date have the same number and positions of cysteine residues in their primary sequences (Fig. 5). This may explain the assumed identical bioactive configuration and biological relatedness among ST_a's from ETEC strains of different host origin.

[36] J. Y. Chang, *Biochem. J.* **199,** 557 (1981).

[37] P. Edman and G. Begg, *Eur. J. Biochem.* **1,** 80 (1977).

[38] A. W. Brauer, M. N. Mangolies, and E. Haber, *Biochemistry* **14,** 3029 (1975).

[39] M. R. Thompson and R. A. Giannella, *Infect. Immun.* **47,** 834 (1985).

[40] B. Rönnberg and T. Wadström, *FEBS Lett.* **155,** 183 (1983).

[41] C. Gerady, M. Herman, J. Olivy, N. G. Otthiers, D. Art, E. Jaquemin, A. Kaeckenbeeck, and J. V. Beeumen, *Vet. Microbiol.* **9,** 399 (1984).

[42] R. Lallier, F. Bernard, M. Gendreau, C. Lazure, N. G. Seidah, M. Chretein, and S. A. St-Pierre, *Anal. Biochem.* **127,** 267 (1982).

[43] S. Stavric, T. M. Gleenson, and N. Dickie, *Period. Biol.* **86,** 251 (1984).

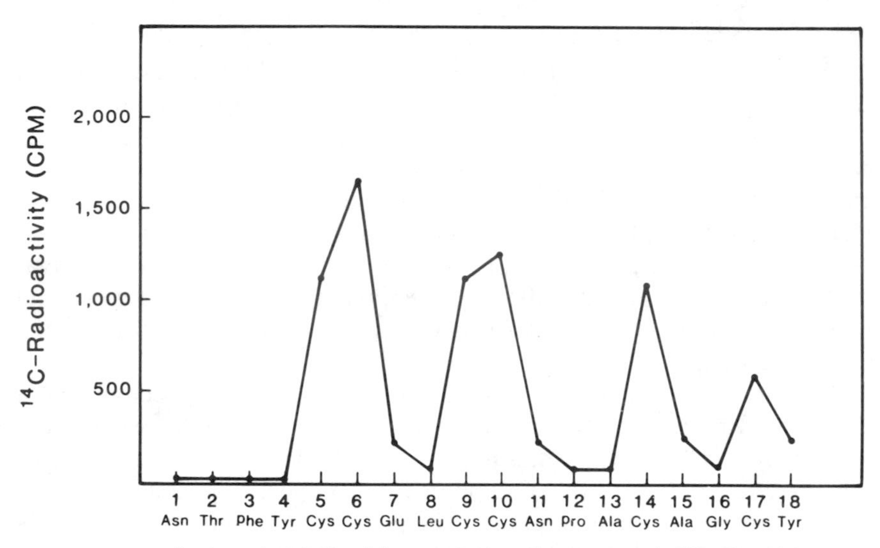

FIG. 4. Amino acid residues of the sequenced ST_a peptide showing the residues which were carboxymethylated with iodo[^{14}C]acetic acid and the counts of each residue.

Biochemical Characteristics of ST_a

Biochemical characterization of our purified ST_a has been performed as previously reported.[10,20,24,27,29] The native toxin is biologically stable after heat treatment at 30 and 100° for 15 min, at pH range of 1–8 for 2 hr, and after treatment with the proteolytic enzymes (pepsin, trypsin, pronase, protease V8, and subtilisin). However, the treatment of the toxin with the reducing agents 2-mercaptoethanol (0.1 *M*) and dithiothreitol (4×10^{-5} *M*) caused rapid loss of the ST_a biological activity, presumably

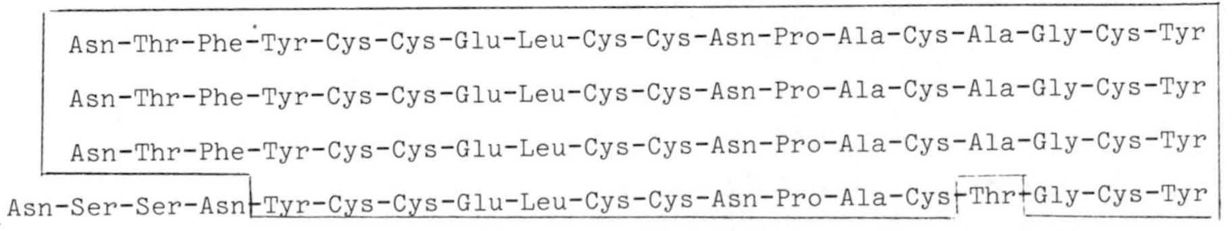

FIG. 5. Amino sequences of ST_as of ETEC from various host origins. (a) Amino acid sequence of ST_a purified from bovine ETEC strains B44, M490, and $WSUH_1$ (this study and see Chapter [43] in this volume). (b) Amino acid sequence of ST_a purified from porcine ETEC strain.[42] (c) Amino acid sequence of ST_a purified from human ETEC strain 18D.[39] (d) Amino acid sequence of ST_a purified from human ETEC strain SK-1,[32] which is also consistent with composition of ST_as purified from other human ETEC strains 73081.[42] Boxed region indicates amino acid sequence homology among ST_a peptides.

due to reduction of the disulfide bonds which are essential for retaining its biological activity. In these aspects ST_a purified by our scheme exhibited biochemical homology to purified ST_a reported by others.[10,29]

Preparative Purification of ST_a

For preparative purposes the described scheme can be scaled upward to produce larger amounts of ST_a. The flat-bed gel isoelectric focusing can be replaced with preparative reversed-phase silica batch adsorption chromatography[31] as follows. The lyophilized crude ST_a is solubilized in 500 ml of 0.1% TFA at 4°. To this solution, 100 g of Sepralyte (trimethylsilyl controlled pore silica, preparative grade, 40 μm, Analyticum International, Harbor City, CA) is slowly added and the slurry is kept under gentle stirring at 4° for 2 hr. The slurry is poured into a sintered glass funnel and washed with 300 ml of 0.1% TFA. Stepwise elution of the adsorbed protein is performed with (1) 300 ml of 0.1% TFA in 20% methanol (HPLC grade), (2) 300 ml of 0.1% TFA in 30% methanol, and (3) 500 ml of 0.1% TFA in 50% methanol. The majority of ST_a (> 80%) will be recovered from the 50% methanol fraction while the 20 and 30% methanol fractions will eliminate contaminants. The methanol is flash evaporated and the residues are dissolved in 0.1% TFA and purified further by RP-HPLC on a semipreparative column using gradients as previously described.

Acknowledgments

The work was supported in part by research funds from the Institute for Medical Education and Research, Saint Louis, Missouri. We thank Dr. Donald Kennedy for encouragement and support, Dr. Nancy S. Magnuson for helpful suggestions, and Mrs. Sue Stevens for excellent secretarial assistance.

[20] Purification of *Escherichia coli* α-*Hemolysin*

By G. A. BOHACH, S. J. CAVALIERI, and I. S. SNYDER

Introduction

α-Hemolysin (AH) is a virulence factor for many clinical isolates of *Escherichia coli*. Its role in the pathogenesis of nonintestinal infections is evaluated in a recent review.[1] Investigations into the toxicity of *E. coli*

[1] S. J. Cavalieri, G. A. Bohach, and I. S. Snyder, *Microbiol. Rev.* **48,** 326 (1985).

hemolysin have been difficult because of inability to purify the protein in an active form. Thus the toxicity of other bacterial products, such as lipopolysaccharide (LPS), in partially purified AH preparations must also be considered in evaluating the importance of the hemolysin in disease.

The hemolysin structural gene codes for a 106,000- to 110,000-Da protein.[2,3] The hemolytic protein is transiently located on the outer membrane of the bacterial cell wall,[4] but is subsequently exported and can easily be obtained from culture supernatants in a crude cell-free form. Purification of AH is difficult because the hemolytic protein forms large heterogeneous aggregates with lipids, carbohydrates and other proteins, probably originating from the cell wall.[5] As a result, physical measurements of the size of active AH are usually in excess of 150,000 to 500,000 Da.[6–8] The denaturing conditions required to dissociate these aggregates result in loss of hemolytic and other biological activities. Thus, studies of AH biological properties use preparations consisting of purified AH complexes rather than homogeneous protein preparations. In this section we describe four methods for purification of these AH complexes and point out the advantages and limitations of each.

Measurement of Hemolytic Activity

AH lyses erythrocytes from a variety of mammalian species by a mechanism that requires divalent cations. The hemolytic reaction is time and temperature dependent. The following modification of the assay developed by Snyder and Koch[9] is based on the ability of a test preparation to lyse 50% of the erythrocytes in a standard 1% suspension of sheep red blood cells.

Standardization of Indicator Erythrocyte Suspensions

1. Sheep blood is obtained by venipuncture of the jugular vein and mixed with an equal volume of modified Alsever's solution (2.5% glucose,

[2] T. Felmlee, S. Pellett, and R. A. Welch, *J. Bacteriol.* **163,** 94 (1985).
[3] W. Goebel and J. Hedgpeth, *J. Bacteriol.* **151,** 1290 (1982).
[4] W. Wagner, M. Vogel, and W. Goebel, *J. Bacteriol.* **154,** 200 (1983).
[5] G. A. Bohach and I. S. Snyder, *J. Bacteriol.* **164,** 1071 (1985).
[6] S. J. Cavalieri and I. S. Snyder, *J. Med. Microbiol.* **15,** 11 (1982).
[7] S. E. Jorgensen, E. C. Short, Jr., H. J. Kurtz, H. K. Mussen, and G. K. Wu, *J. Med. Microbiol.* **9,** 173 (1976).
[8] R. P. Rennie and J. P. Arbuthnott, *J. Med. Microbiol.* **7,** 179 (1974).
[9] I. S. Snyder and N. A. Koch, *J. Bacteriol.* **91,** 763 (1966).

0.80% sodium citrate, 0.055% citric acid, and 0.42% NaCl).[10] The blood is stored at 4° and discarded after 4 weeks.

2. For use in hemolytic assays the erythrocytes are washed three times in saline and adjusted to a concentration of approximately 1% (v/v).

3. One-half milliliter of the suspension is added to 3.5 ml of water to cause erythrocyte lysis. The cell debris is pelleted by centrifugation.

4. The absorbance at 545 nm (A_{545}) of the supernatant is measured. If necessary, adjustments are made in the volume of the suspension so that lysis of erythrocytes in this manner gives an A_{545} of 0.30 ± 0.02.

Hemolytic Assay Procedure

1. Samples are initially diluted at least 2-fold in saline containing 10 m*M* $CaCl_2$ (Ca–S). Additional doubling dilutions are made in Ca–S, keeping the volume in all tubes at 1.0 ml.

2. One milliliter of the standard erythrocyte suspension is added to each tube. The contents are mixed and incubated in a 37° water bath.

3. After 60 min, 2 ml of cold saline is added to each tube (total volume = 4 ml). Unlysed cells are removed by centrifugation.

4. The A_{545} of each lysate is measured against a control for background lysis. The control tube contains erythrocytes without AH.

5. The $\log_{10}$ of each dilution and resulting A_{545} are plotted on semilogarithmic graph paper. The dilution that produces 50% lysis (A_{545} = 0.30 ± 0.02) is determined from the graph and designated as the end point of the titration. The reciprocal of this value expressed in HU_{50}/ml gives a relative measure of the hemolytic activity of the preparation.

Production of AH

Culture Medium Preparation

Several types of common complex bacteriological media can be used to produce AH. However, maximum production and stability of hemolytic activity is obtained when *E. coli* is grown in the chemically defined medium of Snyder and Koch.[9] Use of this medium facilitates purification by eliminating the need to remove medium proteins. The basal medium contains 0.23% K_2HPO_4, 0.078%, KH_2PO_4, 0.10% $(NH_4)_2SO_4$, 0.01% $MgSO_4$, and 0.06% sodium citrate, pH 7.4. To eliminate batch variation, the ingredients are prepared as 10× or 100× stock concentrations and

[10] S. C. Bukantz, C. R. Rein, and J. F. Kent, *J. Lab. Clin. Med.* **31**, 394 (1946).

stored frozen at −20°. The 1× basal medium is aliquoted and sterilized by autoclaving prior to use. Filter-sterilized glucose is added to a final concentration of 0.2%.

Culture Conditions

Maximum production of AH is achieved by rapidly growing, well-aerated cultures. High yields are obtained only for a short time during the late-exponential growth phase. Afterward, the hemolytic titer rapidly declines. Therefore, the use of a standard inoculum and controlled conditions in order to achieve known growth kinetics is necessary. A standard inoculum is prepared by mixing a mid-exponential phase *E. coli* culture with an equal volume of sterile glycerol. Aliquots (17 ml) are stored frozen at −70°. As needed, the contents of a vial are rapidly thawed in a 37° water bath and 15 ml is used to inoculate 750 ml of prewarmed medium in a 2.8-liter Fernbach flask. The culture is incubated at 37° with shaking (200 rpm). Kinetics of growth and AH production are recorded to determine the incubation time that results in peak hemolytic activity. Cultures for large-scale production of AH, using the same standard inoculum, should be harvested at the predetermined incubation time. We routinely inoculate six flasks (750 ml each) with 15 ml of standard inoculum and incubate as described above. Frozen inocula can be stored up to 12 months without any noticeable change in growth or AH production patterns.

Preparative Purification Techniques

For large-scale purification of AH we use either equilibrium centrifugation in glycerol gradients[6] or a column chromatography procedure[5] involving both gel filtration and ion-exchange chromatography. Slightly higher yields are obtained by equilibrium centrifugation. However, AH purified by the chromatographic technique has a higher specific activity and is separated into various fractions based on charge heterogeneity of AH complexes (see Table I).

Preparation of Crude AH Concentrates

The first step in any AH purification technique involves the removal of bacterial cells and the concentration of AH in culture supernatants.

All procedures are performed at 0–4°.

1. Bacterial cells are pelleted by centrifugation (13,700 *g*).
2. The supernatant is recovered and filtered through a 0.45-μm Millipore Durapore membrane filter.

3. The filtrate is concentrated 40-fold by ultrafiltration (Amicon XM300 membrane) under 20 psi of N_2. This step takes advantage of the large size of AH complexes. Molecules with sizes smaller than 300,000 Da pass through the membrane whereas AH does not and is concentrated.

4. The concentrate is clarified by ultracentrifugation at 50,000 rpm (235,000 *g*) in an SW 50.1 rotor. This results in pelleting of nonhemolytic cell debris.

The concentrate can be analyzed in its crude state or subjected to additional purification. For short-term storage (up to 2 days) crude AH can be kept at 4° without significant loss of activity. The concentrate should be frozen (−80°) if prolonged storage is desired.

Column Chromatography

AH in crude concentrates can be further purified by a two-step chromatographic procedure. Most of the contaminants are removed by gel filtration but additional purification is achieved by ion-exchange chromatography.

Gel Filtration. The large pore size (particle exclusion limit = 300–400 nm) of Sephacryl S-1000 (Pharmacia Fine Chemicals) makes this gel especially useful for purification of AH. This is the only gel filtration medium we have found that retards the flow of AH, thereby allowing its separation from large (>300 nm) particles.

The gel is washed with 0.05 *M* acetic acid/sodium acetate buffer, pH 4.0, containing 0.2 *M* NaCl (ACB–0.2 *M* NaCl) and packed to a height of approximately 90 cm in a 2.6-cm-diameter column. A crude AH concentrate (7–10 ml) is applied to the top of the gel and eluted with ACB–0.2 *M* NaCl (flow rate = 88.8 ml/hr). An ultraviolet monitor capable of detecting A_{280} readings of less than 0.025 can be used to measure protein elution. Fractions (7 ml) are collected, dialyzed against saline to reduce acidity, and analyzed for hemolytic activity.

Crude AH from most strains of *E. coli* is resolved into three protein peaks by the procedure. Hemolytic activity elutes from the column simultaneously with a single broad protein peak (k_{av} = 0.35) that is well separated from voided contaminants and an additional peak of smaller proteins. Fractions containing peak hemolytic activity are pooled and can be further purified by ion-exchange chromatography.

Ion-Exchange Chromatography. AH complexes are more acidic than other proteins in crude AH concentrates and bind to anion-exchange resins at low pH. After unbound contaminants are washed from the column, hemolytic activity can be eluted by increasing ionic strength.

DEAE-Sephadex (Pharmacia) is hydrated, equilibrated in ACB–0.2 *M*

NaCl, and packed to a height of approximately 10 cm in a 2.6-cm-diameter column. Up to 50 ml of partially purified AH, eluted from gel filtration columns, is loaded onto the top of the column and allowed to slowly enter the gel bed. The gel is flushed with two bed volumes of ACB–0.2 *M* NaCl to elute nonhemolytic contaminants. The elution buffer is then changed to ACB–0.3 *M* NaCl. Two bed volumes of ACB–0.3 *M* NaCl are passed through the column to elute bound AH (fraction 1). A more highly charged AH fraction is eluted from the gel with ACB–0.4 *M* NaCl (fraction 2).

This elution procedure is used for recovery of AH prepared from cultures grown in chemically defined medium. AH prepared from cultures in complex medium, such as heart infusion broth, is more acidic and requires higher salt concentrations (up to 0.7 *M* NaCl) for complete elution of hemolytic activity. Table I shows that the more acidic fraction 2 has a much higher specific activity than fraction 1. This is likely the result of aggregation of AH complexes in different proportions.

Equilibrium Centrifugation

This procedure is presented as an alternative to purification of AH by column chromatography.

1. Stock glycerol solutions [5% (v/v) and 30% (v/v)] are prepared in a solution containing 0.1% $(NH_4)_2SO_4$, 0.01 *M* $MgSO_4$, and 0.06% sodium citrate, pH 5.4. These stock solutions are sterilized by autoclaving and stored at 4°.

TABLE I
COMPARISON OF PREPARATIVE TECHNIQUES FOR PURIFICATION OF AH

Procedure	Yield (%)[a]	Specific[b] activity
Equilibrium centrifugation	60	228
Column chromatography		
Step 1. Gel filtration	82	319
Step 2. Ion exchange	54	
Fraction 1		88
Fraction 2		990

[a] Percentage of hemolytic activity in crude AH concentrate recovered.
[b] $HU_{50}/\mu g$ of protein.

2. A 5–30% gradient (32 ml) is poured into polyallomer ultracentrifuge tubes (1 in. × 3.5 in.) using a gradient mixer.

3. Crude AH concentrate (5 ml) is layered over the gradient and the tubes are filled with mineral oil.

4. The gradients are centrifuged at 50,000 rpm (89,000 *g*) in an SW 27 rotor for 20–21 hr.

5. Fractions (1 ml) are collected by siphoning from the bottom of the gradient and screened for hemolytic activity.

6. Hemolytic fractions are pooled and dialyzed for removal of glycerol.

Purification of AH Using Immunoadsorbents

Immunization of animals with AH results in production of antibodies reactive with contaminating *E. coli* products as well as the hemolytic protein. Therefore, polyclonal antibodies are not useful for immunoaffinity purification of AH. In contrast, monoclonal antibodies (MAbs) with specificity for the active hemolysin protein can be selected by their ability to neutralize hemolytic activity. Thus, immunoadsorbents prepared with AH-neutralizing MAb ligands are used to obtain biologically active AH preparations with maximum purity. This technique is useful when small amounts of highly purified AH are required. However, the large amount of MAb required for preparing immunoadsorbents makes it impractical for large-scale purification.

Preparation of Anti-AH MAbs

In this section, we summarize the methodology we have used for successful production of anti-AH MAbs. Details are provided only when applicable specifically to AH. For a more detailed description of basic MAb techniques the interested reader should consult any of several pertinent publications.[11-13]

Immunization. Male BALB/c mice are immunized with crude AH concentrates. For 1 week prior to immunization, small quantities of diluted antigen (10–100 HU_{50}) are administered daily to each of four 8-week-old mice by the intraperitoneal route to induce tolerance to endo-

[11] G. Galfre and C. Milstein, this series, Vol. 73, p. 3.

[12] B. Caterson, J. E. Christner, and J. R. Baker, *J. Biol. Chem.* **258,** 8848 (1983).

[13] J. F. Kearney, A. Radbruch, B. Liesegang, and K. Rajewsky, *J. Immunol.* **123,** 1548 (1979).

toxin in the preparation. The animals are then immunized for 2 months according to the following schedule.

Day 1: 500 HU_{50} in Freund's complete adjuvant
Day 3: 500 HU_{50} in Freund's incomplete adjuvant
Day 5–56: 500 HU_{50} in saline every two days

Each dose of antigen (volume 0.5 ml) is administered bilaterally as subcutaneous injections into the axillary and inguinal areas and in each hind footpad as described by Caterson *et al.*[12]

Hybridoma Production. Two days after the last immunization dose, the regional lymph nodes (axillary, inguinal, brachial, and popliteal) draining the sites of immunization are removed and teased apart in RPMI 1640 tissue culture medium. Lymph node cells are harvested and fused with X63–Ag8.653 mouse myeloma cells[13] in the presence of polyethylene glycol. The mixture is plated in 96-well culture plates and grown in the presence of hypoxanthine, aminopterin, and thymidine to select for hybridomas.

Hybridoma Screening. When growth in the majority of the wells reaches 50% confluency (approximately 21 days) the culture supernatants are tested for anti-AH antibody in a hemolysis neutralization microassay as follows:

1. One hundred microliters of culture supernatant from each well is transferred to 96-well round-bottom polystyrene microtiter plates.
2. One hundred microliters of crude AH concentrate diluted in CA–S (25 HU_{50}/ml) is mixed with each culture supernatant.
3. The plates are incubated at 37° for 30 min to allow antigen–antibody binding.
4. Fifty microliters of a washed sheep erythrocyte suspension (2.5%, v/v, in saline) is added to each well.
5. The plates are incubated for 1 hr at 37° and observed visually for inhibition of hemolysis.

Antibody Quantitation. AH-neutralizing MAbs are obtained from ascitic fluid after injection of BALB/c mice with hybridoma cells. The antibody should be quantitated to ensure that sufficient amounts are present for immunoadsorbent preparation. The following assay can be used to quantitate AH-neutralizing titers.

1. A stock AH solution (approximately 120 HU_{50}/ml) is prepared by dilution of a crude AH concentrate with saline.
2. Varying volumes (2 to 20 μl) of stock AH are added to a series of tubes containing 1 ml of Ca–S.

3. One milliliter of a standardized 1% (v/v) sheep erythrocyte suspension is added and the tubes are incubated for 1 hr in a 37° water bath.

4. Saline is added to adjust the volume in each tube to 4 ml and the unlysed cells pelleted by centrifugation.

5. The A_{545} of each lysate is measured. The volume of AH stock containing 1.0 HU_{50} (lyses 50% of the erythrocytes) is determined from the lysate with an A_{545} of 0.30 ± 0.02.

6. Serial 2-fold dilutions of ascitic fluid are made in tubes containing 1 ml of Ca–S.

7. Three HU_{50} (determined from step 5) of stock AH is added to each tube. The contents are mixed and incubated for 15 min in a 37° water bath to allow antigen–antibody binding.

8. Residual hemolytic activity is detected by addition of 1 ml of a standardized 1% (v/v) sheep erythrocyte suspension and incubation for an additional hour at 37°.

9. The volume in each tube is adjusted to 4 ml by adding saline and the cells and unlysed cells are pelleted The A_{545} of each lysate is measured. The reciprocal of the antibody dilution that neutralizes 2 HU_{50} and results in an A_{545} of 0.30 ± 0.02 (1 HU_{50} remaining) is designated the antibody titer.

All ascitic fluids possess low levels of nonspecific AH-neutralizing activity. Therefore, it is important to include a control titration of ascitic fluid induced by unfused myeloma cells or hybridomas producing antibodies against an irrelevant antigen.

Immunoadsorbent Preparation

The major consideration in choosing a solid phase for immunoadsorbent preparation is steric interference of binding among AH complexes due to their large size. Using gels with reactive groups at the end of long spacer arms for coupling of proteins reduces the problem. The following method has been effective for production of efficient immunoadsorbents.

1. The MAbs are precipitated from ascitic fluid with 33% $(NH_4)_2SO_4$ and redissolved in coupling solution (0.1 *M* $NaHCO_3$ containing 0.5 *M* NaCl) to give an AH-neutralizing titer of 1600–3200. The antibody solution is dialyzed against several changes of coupling solution to remove residual $(NH_4)_2SO_4$.

2. A slurry (5–10 ml) of Affi-Gel 15 (Bio-Rad Laboratories, Richmond, CA) is washed with cold (4°) 2-propanol and distilled water on a sintered glass filter.

3. The washed gel and 0.5–2.0 vol of MAb solution are transferred to a 15-ml tube and the total volume is adjusted to 12.5 ml with coupling solution.

4. The contents of the tube are mixed by rocking overnight at 4° for maximum antibody coupling.

5. Residual reactive sites are blocked by adding 0.5 ml of 1 *M* ethanolamine, pH 8.0, and incubating for an additional hour.

6. The gel is washed exclusively with wash buffer (0.1 *M* acetic acid–sodium acetate buffer, pH 4.0, containing 0.5 *M* NaCl) and then with running buffer (0.01 *M* Tris/HCl, pH 8.5, containing 0.5 *M* NaCl).

Affinity Chromatography Procedure

Consideration should be given to the lability of hemolytic activity, especially during the elution procedures. We find that AH is relatively resistant to elution by low pH treatment but acidity should be neutralized immediately. The following procedure results in maximum antigen binding and subsequent elution of hemolytic activity from MAb ligands.

1. The washed immunoadsorbent is resuspended in 7.5 ml of running buffer and mixed with 1–3 ml of crude AH concentrate (900–2000 HU_{50}) in a 15-ml tube. The contents are mixed for 3 hr at room temperature to allow the antigen–antibody reaction to proceed.

2. The gel is washed exhaustively, first with wash buffer and then with running buffer, to remove unbound AH and contaminants.

3. The gel is resuspended in 10 ml of running buffer, poured into a column (1.6 × 20 cm), and rinsed with several bed volumes of running buffer.

4. AH is eluted by flushing the column with at least five bed volumes of 0.1 *M* glycine–HCl buffer, pH 2.5. Fractions (2 ml) are collected in an equal volume of 0.1 *M* Tris/HCl buffer, pH 7.5, containing 0.85% NaCl. The fractions are kept in an ice bath while screening for hemolytic activity is performed. Hemolytic fractions are pooled and stored at −80°.[14]

Acknowledgments

This research was supported in part by Public Health Service Biomedical Research Support Grant SO7RR-05433 from the National Institutes of Health and by funding from the West Virginia University School of Medicine, Department of Urology.

[14] See addenum (page 399) for *Polyethylene Glycol-Precipitated Hemolysin* procedure.

[21] Preparation of *Pseudomonas* Exotoxin A

By KENNETH J. KOZAK and CATHARINE B. SAELINGER

Pseudomonas exotoxin A (PE) is one of the major virulence factors of *Pseudomonas aeruginosa*. Purified toxin is cytotoxic for a variety of cell lines,[1] and specifically inhibits protein synthesis by the ADP ribosylation of elongation factor 2.[2] A low protease-producing strain PA103 is usually used for exotoxin production.[3] The medium routinely used for the culture of *P. aeruginosa* and the production of PE is the dialyzable portion of trypticase soy broth (DTSB) originally described by Liu.[3] A chemically defined medium also has been described.[4] Toxin purification has been accomplished by both electrophoretic[5] and NAD affinity chromatographic[6] methods; however, the activity and purity of the different preparations vary. We routinely use a modification of the method of Leppla[7] to purify PE.

Required Materials

Trypticase soy broth—500 g (BBL Microbiology Systems, Cockeysville, MD)
Agar (BBL Microbiology Systems, Cockeysville, MD)
L-Glutamic acid monosodium salt (2.0 *M*; filter sterilized)
Glycerol (50%, v/v; filter sterilized)
Chelex 100, 50–100 mesh sodium form—100 g (Bio-Rad, Rockville Centre, NY)
Preswollen anion-exchanger DE-52—1000 g (Whatman, Inc., Clifton, NJ)
Hydroxyapatite—100 g (Boehringer-Mannheim Biochemicals, Indianapolis, IN)
Hemoglobin–agarose (Sigma Chemical Co., St. Louis, MO)
Antifoam B emulsion—100 ml (Sigma Chemical Co., St. Louis, MO)
Fermentor—15 liter capacity
Pseudomonas aeruginosa strain PA103 (ATCC 29260)

[1] J. L. Middlebrook and R. B. Dorland, *Can. J. Microbiol.* **23,** 183 (1977).
[2] B. H. Iglewski and D. Kabat, *Proc. Natl. Acad. Sci. U.S.A.* **72,** 2284 (1975).
[3] P. V. Liu, *J. Infect. Dis.* **128,** 506 (1973).
[4] R. M. DeBell, *Infect. Immun.* **24,** 132 (1979).
[5] L. T. Callahan III, *Infect. Immun.* **14,** 55 (1976).
[6] G. Cukor and N. A. Nowak, *J. Bacteriol.* **149,** 1162 (1982).
[7] S. H. Leppla, *Infect. Immun.* **14,** 1077 (1976).

Büchner filter funnel, table style–8-liter capacity
Plastic garbage can—32-gal capacity
Dialysis tubing (27 or 10 mm, M_r cut-off 12,000–14,000) (Spectra/POR; Fisher Scientific)
Sigma iron detection kit
Variable-speed electric motor and paddle
5 × 30 cm chromatography column
1.6 × 30 cm chromatography column
Peristaltic pump
Sheep anti-*Pseudomonas* exotoxin A (Swiss Serum and Vaccine Institute, Berne, Switzerland)
Phosphate-buffered saline (8.0 g NaCl, 0.2 g KCl, 1.15 g Na_2HPO_4, 0.2 g KH_2PO_4 in 1000 ml; pH 7.2)

Growth Media Preparation

The undefined growth medium for PE production is the dialysate of trypticase soy broth (TSB). TSB is dissolved in double-distilled water (ddH_2O) to one-tenth the final required volume (i.e., to prepare 1 liter dialysate, 30 g TSB is dissolved in 100 ml ddH_2O). Chelex 100 resin is added to the TSB at a concentration of 5 g Chelex to 30 g TSB for the removal of free iron, and the mixture is stirred for 2.5 hr at room temperature. The TSB–Chelex mixture is then dialyzed against ddH_2O for 18 hr at 4° at a ratio of 100 ml TSB–Chelex to 1000 ml ddH_2O. Double-knotted dialysis bags are required because of increase in pressure during dialysis. The dialysate (DTSB) is used for PE production. The concentration of iron in the dialysate must be monitored carefully since excess iron depresses toxin production; the optimum iron concentration is 0.05 μg/ml.[8] The DTSB is autoclaved and then supplemented by the addition of 50% (v/v) sterile glycerol and 2.0 *M* monosodium glutamate (filter sterilized) to final concentrations of 1.0% and 0.05 *M*, respectively.

Dialyzed trypticase soy agar (DTSA) for initial colony growth is made by the addition of agar to a final concentration of 1.5% (w/v) to the DTSB, sterilized, then supplemented with glycerol and monosodium glutamate to the concentrations described above.

Growth Conditions

Pseudomonas aeruginosa strain PA103 (ATCC 29260) is subcultured daily on DTSA at 32° for 5 days prior to starter culture inoculations.

[8] M. J. Bjorn, B. H. Iglewski, S. K. Ives, J. C. Sadoff, and M. L. Vasil, *Infect. Immun.* **19,** 785 (1978).

Twenty milliliters of DTSB in a 250-ml Erlenmeyer flask is inoculated with an isolated colony from a DTSA plate, and incubated for 18 hr in an environmental shaker incubator at 32° rotating at 200 rpm. One milliliter of this culture is added to 100 ml of DTSB in a 1.0-liter flask and incubated as above for 5.5 hr. An aliquot is taken, bacteria removed, and supernatant tested for PE production by the precipitin ring test (see below). A 100-ml starter culture giving a positive ring test is used to inoculate 10 liters of DTSB in a 15-liter fermentor. The culture is stirred at 340 rpm, aerated at 20 liters/min, and maintained at 32°. Four hours after the start of the incubation foaming is controlled by the slow addition of antifoam B emulsion by a peristaltic pump. The amount of antifoam added throughout the entire fermentation should not exceed 10 ml. Approximately 10 hr into the incubation, the ADP-ribosylation activity of cell-free culture supernatant is determined (for method, see chapter [32] in this volume). Enzyme activity is measured every 0.5 hr to identify peak toxin production; peak toxin production should be seen 12–15 hr after primary inoculation. Once a peak is reached, the culture is cooled to 4°, bacteria are removed by centrifugation at 10,000 *g* for 30 min, and the bacteria-free culture supernatant is stored at 4°.

Batch Diethylaminoethyl (DEAE)-Cellulose Chromatography

All purification steps are done at 4°. The 10 liters of culture supernatant is diluted with 60 liters of ddH_2O in a 32-gal plastic garbage can to decrease the ionic strength. Unwashed DEAE-cellulose (DE-52) (750 g) is added and the suspension stirred gently with a variable speed electric motor and paddle for 2.5 hr. The beads are collected by vacuum filtration onto two pieces of Whatman No. 1 filter paper using a large Büchner filtering funnel. The beads should not be allowed to dry during this process. The beads are transferred to a large (4.0 liter) beaker and resuspended in 1500 ml of 50 m*M* NaCl–10 m*M* tris(hydroxymethyl)aminomethane hydrochloride (Tris), pH 8.1. The slurry is gently mixed for 1.0 hr, the beads collected, and the washing process repeated. Exotoxin is eluted from the beads by resuspending the beads in 1.0 liter of 200 m*M* NaCl–10 m*M* Tris, pH 8.1, for 1.0 hr. The beads are harvested while saving the 200 m*M* NaCl–10 m*M* Tris wash. The process is repeated for a total of three washes. The three washes are pooled, and the protein precipitated by the slow addition of solid $(NH_4)_2SO_4$ to 70% saturation [480g $(NH_4)_2SO_4$/liter]. The solution is stirred for 1.0 hr and the precipitate is collected by centrifugation for 20 min at 10,000 *g*. The precipitate is redissolved in approximately 200 ml of 50 m*M* NaCl–10 m*M* Tris–3.0 m*M* dithiothreitol (DTT), pH 8.1, and dialyzed against 8.0 liters of the same buffer overnight, changing the buffer once.

Hemoglobin-Agarose and DEAE-Cellulose Chromatography

The dialyzed protein from the batch elution is collected and the insoluble material is removed by centrifugation for 20 min at 15,000 *g*. Contaminating proteases are removed from the dialyzed PE by passage through a 40-ml hemoglobin–agarose column.[9] The sample is then applied to a 1.6 × 30 cm column of DEAE-cellulose (DE-52; 70 ml) which is equilibrated in 50 m*M* NaCl–3.0 m*M* DTT–10 m*M* Tris, pH 8.1. The column is washed until no protein can be detected in the effluent as determined by absorbance at 280 nm. The exotoxin is eluted from the column by a 1.0-liter linear gradient of NaCl, 50 to 200 m*M* NaCl in 3.0 m*M* DTT–10 m*M* Tris, pH 8.1. Ten-milliliter fractions are collected, the protein profile is determined by absorbance, and the exotoxin is located by the precipitin ring test. The exotoxin-containing fractions, emerging at an approximate NaCl concentration of 125 m*M*, are pooled and diluted to 50 m*M* NaCl with ddH_2O. Sodium phosphate is added to final concentration of 5.0 m*M* and the pH is adjusted to 7.0.

Hydroxyapatite Column Chromatography

A 5.0 × 30.0 cm column is packed with 50 g of hydroxyapatite and equilibrated in 5.0 m*M* sodium phosphate–50 m*M* NaCl–3.0 m*M* DTT, pH 7.0. The pooled and adjusted fractions from the DEAE-cellulose chromatography step are applied to the column and the column is washed with 5.0 m*M* sodium phosphate–50 m*M* NaCl–3.0 m*M* DTT, pH 7.0, until no protein can be detected in the effluent. PE is eluted by a linear gradient of sodium phosphate, 5.0 to 100 m*M*, in 50 m*M* NaCl–3.0 m*M* Tris, pH 7.0, using a total volume of 1.0 liter. Ten-milliliter fractions are collected, protein profiles determined, and exotoxin located by the precipitin ring test. The exotoxin-containing fractions are pooled and the exotoxin precipitated by the addition of solid $(NH_4)_2SO_4$ to 70% saturation. The solution is stirred for 1.0 hr and the solid is collected by centrifugation. The resultant pellet is redissolved in a small volume of 0.01 *M* Tris, pH 8.1, and dialyzed against 4.0 liters of the same buffer for 18 hr, changing the buffer three times. The dialyzed exotoxin is collected and the insoluble material removed by centrifugation. The final product is diluted to 1.0 mg/ml (determined by the Lowry protein assay[10]) in 0.1 *M* Tris, pH 8.1, aliquoted into small volumes, and stored at −70°. Analysis of a typical purification is shown in Table I.

[9] S. Lory and R. J. Collier, *Infect. Immun.* **28,** 494 (1980).

[10] O. H. Lowry, N. J. Rosebrough, A. L. Farr, and R. J. Randall, *J. Biol. Chem.* **193,** 265 (1951).

TABLE I
PURIFICATION OF *Pseudomonas* EXOTOXIN A

Purification stage	Total volume (ml)	Total protein (mg)[a]	Total toxin (mg)[b]	Specific activity (%)[c]
Undiluted culture supernatant	11,100	4,750	320	6.7
Post DE-52 batch[d]	275	1,340	155	11.6
DE-52 peak[e]	278	125	105	84.0
HAP peak[f]	290	60	60	100
Final product[g]	10	56	56	100

[a] Determined by optical density where $E_{280}^{1\%} = 11.9$.
[b] Determined by measuring ADP ribosylation activity using activated toxin.
[c] [Toxin (mg)/total protein (mg)]100.
[d] DE-52 pooled batch elutions after precipitation and dialysis.
[e] Pooled toxin containing fractions after hemoglobin and DE-52 column chromatography.
[f] Pooled toxin containing fractions from hydroxyapatite column chromatography.
[g] Hydroxyapatite peak after precipitation and dialysis.

Precipitin Ring Test for Exotoxin Detection

Antitoxin is diluted 1 : 4 in phosphate-buffered saline containing 0.5 *M* sucrose. Diluted antiserum (100 μl) is pipetted into a 6 × 500 mm test tube, and 50 μl of the solution to be tested (culture supernatant, column fraction) is gently overlaid on this pad. After standing 5–10 min at room temperature a precipitin ring develops at the interface if exotoxin is present. This method detects as little as 0.5 μg PE/ml.

Properties of Purified Exotoxin

Electrophoresis on 8% sodium dodecyl sulfate–polyacrylamide gels under reducing conditions shows a homogeneous preparation with a molecular weight of 66,000 when stained with Coomassie Brilliant Blue. Isoelectric focusing studies give an apparent p*I* of 5.1, which is consistent with amino acid analyses showing the presence of a large number of acidic amino acids. The N-terminal amino acid is arginine and the arginine-to-lysine ratio is unusually high.[7] The toxin molecule contains four disulfide bonds but no free sulfhydryl groups.[7] The detection of carbohydrates by gas chromatography is negative. PE has a maximum absorbance at 280 nm, an $E_{280}^{1\%} = 11.9$ (diluted in 0.5 *M* Tris–HCl buffer, pH 8.1) and a 280 to 260-nm absorbance ratio of 1.92.[7]

PE was first identified by its mouse lethality. When tested by intraperitoneal injection into 20-g mice, purified PE routinely has a median lethal dose of approximately 0.1 μg. Cytotoxicity for mouse fibroblasts and ADP-ribosyl transferase activity can be assayed as described in this volume [32]. Under the conditions used in our laboratory, the concentration of purified PE required to inhibit protein synthesis in mouse LM fibroblasts by 50% is typically 30–40 ng/ml. Similarly, 500 ng PE result in incorporation of 18,000 cpm in a standard ADP-ribosylation assay; the level of incorporation may vary with wheat germ preparation.

[22] Shiga Toxin: Production and Purification

By Gerald T. Keusch, Arthur Donohue-Rolfe, Mary Jacewicz, and Anne V. Kane

Shiga toxin is among the oldest known protein toxins derived from gram-negative bacilli. It was first clearly described in the prototypic species, *Shigella dysenteriae* 1 (or Shiga's *Bacillus*), in 1903.[1] Because parenteral injection of Shiga toxin into susceptible animals resulted in a delayed limb paralysis followed by death, it was called Shiga neurotoxin (or, simply Shiga toxin).[2] It has long held interest for microbiologists as one of the most deadly microbial toxins, ranking near to tetanus and botulinum toxins in LD_{50} dose.[3] In contrast, until the 1970s there was little continuing interest in its potential role in the pathogenesis of shigellosis, because *S. dysenteriae* 1 alone of the various species of *Shigella* produced the toxin and because a reasonable and relevant biological effect had not been described.[2] However, in 1972, it was reported that Shiga toxin reproduced the two hallmarks of the disease in an animal model, intestinal fluid production and inflammatory enteritis.[4,5] Within a few years, it was also shown that other *Shigella* species produced the same toxin.[6,7] Most re-

[1] H. Conradi, *Dtsch. Med. Wochenschr.* **20,** 26 (1903).
[2] G. T. Keusch, A. Donohue-Rolfe, and M. Jacewicz, *Pharmacol. Ther.* **15,** 403 (1982).
[3] D. M. Gill, *Microbiol. Rev.* **46,** 86 (1982).
[4] G. T. Keusch, G. F. Grady, L. J. Mata, and J. McIver, *J. Clin. Invest.* **51,** 1212 (1972).
[5] G. T. Keusch, G. F. Grady, A. Takeuchi, and H. Sprinz, *J. Infect. Dis.* **126,** 92 (1972).
[6] G. T. Keusch and M. Jacewicz, *J. Infect. Dis.* **135,** 552 (1977).
[7] A. D. O'Brien, M. R. Thompson, P. Gemski, B. P. Doctor, and S. B. Formal, *Infect. Immun.* **15,** 796 (1977).

cently, evidence has been presented that the majority of *flexneri* and *sonnei* strains make a biologically similar, but immunologically distinct cytotoxin.[8] In addition, certain types of *E. coli* capable of causing intestinal disease, including classical enteropathogenic *E. coli* (EPEC) and the newly described agents of hemorrhagic colitis, *E. coli* 0157 : H7 and 026 : H11 produce a cytotoxin either identical to or highly related to Shiga toxin.[9]

In addition to its potential involvement in pathogenesis of enteric infections, shiga toxin has two other properties of current biological interest, including the capacity to inhibit mammalian protein synthesis at the ribosomal level,[10] and the ability to destroy sensory but not motor nerve projections of the vagus nerve.[11] These properties are described in other chapters in this volume. Techniques for the production, purification, and assay of Shiga toxin are described below.

Growth of *Shigella* Species

Safety. Organisms of the genus *Shigella* are unique among the enteric pathogens because an exceedingly small inoculum is capable of causing disease. The oral ID_{50}, determined in human volunteers, is only 200 *S. dysenteriae* 1 or 5000 *S. flexneri* 2a.[12,13] It is obvious that laboratory infections will occur unless standards of sterile microbiological techniques are maintained. Unfortunately, the most virulent species, *S. dysenteriae* 1, is the ideal one to use for the production of toxin, since the yield is around 1000-fold greater than that obtained from other *Shigella* species.[7] However, there are two ways to control this hazard. For example, for 15 years now we have successfully used a rough mutant strain of *S. dysenteriae* 1, originally described by Dubos and Geiger in 1946 as strain 60R,[14] without a single instance of laboratory-acquired infection because this strain is incapable of colonization of the human. An alternative safety measure is based on the essential role of mucosal cell invasion by the organism for the pathogenesis of shigella infection.[2] Because invasiveness is dependent on the presence of a large (120–140 MDa) plasmid

[8] A. V. Bartlett, III, D. Prado, T. G. Cleary, L. K. Pickering, *J. Infect. Dis.* **154,** 996 (1986).

[9] A. D. O'Brien, R. K. Holmes, *Microbiol. Rev.* **51,** 206 (1987).

[10] R. Reisbig, S. Olsnes, and K. Eiklid, *J. Biol. Chem.* **256,** 8739 (1981).

[11] R. G. Wiley, A. Donohue-Rolfe, and G. T. Keusch, *J. Neuropathol. Exp. Neurol.* **44,** 496 (1985).

[12] M. M. Levine, H. L. DuPont, S. B. Formal, R. B. Hornick, A. Takeuchi, E. J. Gangarosa, M. J. Snyder, and J. P. Libonati, *J. Infect. Dis.* **127,** 261 (1973).

[13] H. L. DuPont, R. B. Hornick, A. T. Dawkins, M. J. Snyder, and S. B. Formal, *J. Infect. Dis.* **119,** 296 (1969).

[14] R. Dubos and J. W. Geiger, *J. Exp. Med.* **84,** 143 (1946).

in different *Shigella* species,[15,16] use of a plasmid-cured strain would also be safe.

Media. An essential principle for selection of media is to control its iron content, for it has been amply demonstrated that toxin production is in some way regulated by iron.[2,14] Optimal iron concentration is between 0.1 and 0.15 μg/ml of Fe^{3+}; below this level the organism does not grow well, and above this level toxin yield decreases. Indeed, an increase in iron from only 0.1 to 0.2 μg/ml reduces the yield of toxin by over 60%.[2] Some media also require addition of nicotinic acid and tryptophan to support growth of the organism. A variety of media have been successfully used by different investigators, including peptone broth, modified syncase broth, NZ-amine, and CCY medium.[2] When iron is adjusted to an optimal level, however, we have not found any major differences in toxin yield when organisms are grown in any of these media.[2]

We use modified syncase broth (MSB), because this contains an optimal iron content for toxin production without the need for adjustment of iron.[17] MSB contains (in g/liter): casamino acids, 10; Na_2HPO_4, 5; K_2HPO_4, 5; NH_4Cl, 1.18; Na_2SO_4, 0.089; $MgCl_2 \cdot 6H_2O$, 0.042; and $MnCl_2 \cdot 4H_2O$, 0.004. Five hundred milliliters of this stock is added to a 2-liter Erlenmeyer flask and autoclaved. The final medium is made by addition of 5 ml of a filter-sterilized solution of 20% (w/v) glucose and 0.4% tryptophan to each flask.

Culture. Toxin production is reduced under anaerobic conditions.[2] Therefore, cultures are aerated by vigorous agitation on a rotatory shaker at 300 rpm.[17] Aerated fermentor chambers can also be used.[18] In this case, it is possible to remove a portion of the growth each 24 hr and add fresh medium to increase the final yield of organisms and toxin.

Cultures are always initiated from lyophilized stock, which is grown overnight on a nutrient agar plate. A single colony is then inoculated into 50 ml of MSB in a 250-ml Erlenmeyer flask and incubated with shaking overnight at 37°. Five milliliters of the starter culture is then used to inoculate 500 ml of MSB in 2-liter Erlenmeyer flasks capped with aluminum foil-covered cotton plugs. Incubation proceeds at 37° with shaking at 300 rpm.

Harvest. Either the bacterial cell pellet or the spent growth medium can be used for isolation of the toxin.[2] However, because Shiga toxin is

[15] D. J. Kopecko, O. Washington, and S. B. Formal, *Infect. Immun.* **29,** 207 (1980).

[16] P. J. Sansonetti, H. d'Houteville, C. Ecobichon, and C. Pourcel, *Ann. Microbiol. (Paris)* **134A,** 295 (1983).

[17] A. Donohue-Rolfe, G. T. Keusch, C. Edson, D. Thorley-Lawson, and M. Jacewicz, *J. Exp. Med.* **160,** 1767 (1984).

[18] J. McIver, G. F. Grady, and G. T. Keusch, *J. Infect. Dis.* **131,** 559 (1975).

a periplasmic protein which is released into the medium primarily upon death and lysis of the organism,[19,20] the greatest yield of toxin is from the bacterial cells themselves. Toxin is produced during logarithmic phase growth,[17] and maximum amounts are present by the time organisms reach the stationary phase. We monitor optical density and harvest the bacteria when cultures reach an A_{600} value of 3–3.5, or more simply we inoculate cultures one afternoon and harvest the next morning.

Bacteria are chilled to 4° and pelleted by centrifugation at 10,000 *g* for 10 min. All subsequent procedures are conducted at 4°. The pellet is washed twice in 10 m*M* Tris–HCl, pH 7.4, by resuspension and centrifugation, and suspended in one-fiftieth of the original volume in wash buffer containing 2 m*M* phenylmethylsulfonyl fluoride (PMSF) to inhibit protease activity. The washed bacteria are then broken up to release toxin. This may be accomplished by osmotic shock lysis,[21] providing periplasmic proteins relatively free of contamination by cytoplasmic proteins.[19] Periplasmic proteins can also be easily extracted with the detergent-like antibiotic, polymyxin B. Incubation of the bacterial cell pellet for 2 min at 4° in a 2 mg/ml solution of polymyxin B in 25 m*M* phosphate buffer, pH 7.3, containing 0.14 *M* NaCl, releases over 90% of the total cytotoxic activity contained in a French pressure cell lysate with less than 5% of the cytoplasmic proteins.[19] The purification process described below also works well if the pellet is sonicated at 4° until greater than 95% lysis occurs, which may be estimated by determining A_{600}. The lysate is then spun at 5000 *g* to remove unbroken cells, followed by centrifugation at 37,000 *g* for 45 min. The clarified crude supernatant is the starting material for purification.

Purification of Shiga Toxin

Several procedures are described for purification of Shiga toxin, utilizing combinations of molecular sieve and/or ion-exchange chromatography, affinity chromatography on acid-washed chitin or over antibody-affinity columns, isoelectric focusing in either sucrose gradients or in polyacrylamide gels, and sucrose density gradient centrifugation.[22–24] The major drawback of these schemes is the tiny yields of toxin, usually much

[19] A. Donohue-Rolfe and G. T. Keusch, *Infect. Immun.* **39,** 270 (1983).
[20] D. E. Griffin and P. Gemski, *Infect. Immun.* **40,** 425 (1983).
[21] H. C. Neu and J. Chou, *J. Bacteriol.* **94,** 1934 (1967).
[22] S. Olsnes and K. Eiklid, *J. Biol. Chem.* **256,** 284 (1980).
[23] J. E. Brown, S. W. Rothman, and B. P. Doctor, *Infect. Immun.* **29,** 98 (1980).
[24] A. D. O'Brien, G. D. LaVeck, D. E. Griffin, and M. R. Thompson, *Invect. Immun.* **30,** 170 (1980).

less than 5% of the starting biological activity. The method described below uses only three steps and increases the yield to almost 50% of initial cytotoxicity.

Chromatography on Blue Sepharose. The crude cell lysate from 3 liters of growth is applied at room temperature to a 2.5 × 50 cm column containing the dye Cibacron Blue F3G-A coupled to Sepharose CL-6B (Blue Sepharose, Pharmacia Fine Chemicals, Piscataway, NJ), equilibrated with 20 m*M* Tris–HCl, pH 7.4[17] To increase toxin recovery, the flow-through can be continuously recycled overnight. The column is then washed with 10 column volumes of 10 m*M* Tris–HCl, pH 7.4, and the bound toxin is eluted in fractions of 10 ml in the same buffer containing 0.5 *M* NaCl, as described by Olsnes and Eiklid.[22] The protein-containing fractions are identified by absorbance at 280 nm, pooled, and dialyzed against 25 m*M* Tris–acetate, pH 8.3.

Chromatofocusing. The dialyzed eluate is then subjected to chromatofocusing.[21] The crude toxin is applied to a 0.9 × 20 cm column of Polybuffer exchanger 94 (Pharmacia Fine Chemicals, Piscataway, NJ) equilibrated with 25 m*M* Tris–acetate, pH 8.3. Elution of the bound toxin is then initiated with a degassed 1 : 13 dilution in water of Polybuffer 96 (Pharmacia Fine Chemicals), adjusted to pH 6.0 with acetic acid. Fractions of 1.5 ml are then continuously collected and the pH and absorbance at 280 nm are determined. Initially, it may be desirable to assay these fractions for cytotoxin activity, which may be accomplished by one of the assays described below. However, with experience it is possible to simply pool the A_{280} peak at pH 7.0–7.1 for further processing.

BioGel P-60 Chromatography. The final step is to remove the ampholytes from the chromatofocusing step.[17] The pooled pH 7.0–7.1 fraction is transferred to a dialysis bag, which is then incubated under dry polyethylene glycol (M_r = 20,000) at room temperature until the volume is reduced to around 2 ml. This concentrated toxin is then applied to a 1.5 × 100 cm column of BioGel P-60 (Bio-Rad, Richmond, CA) equilibrated with 20 m*M* ammonium bicarbonate. The cytotoxin-containing fractions are eluted in the same buffer, identified, pooled, and lyophilized. During this process the salts are volatilized and removed, leaving pure toxin in the vial.

In Fig. 1, results of SDS–polyacrylamide gel electrophoresis of the toxin at various stages of the purification scheme in the presence of 2-mercaptoethanol are shown. The purified toxin contains two peptide bands, which represent the M_r 32,000 A subunit and the M_R 7691 B subunit monomer. Table I shows the yield and specific activity of toxin derived from 3 liters of growth in six 2-liter Erlenmeyer flasks. We obtain close to

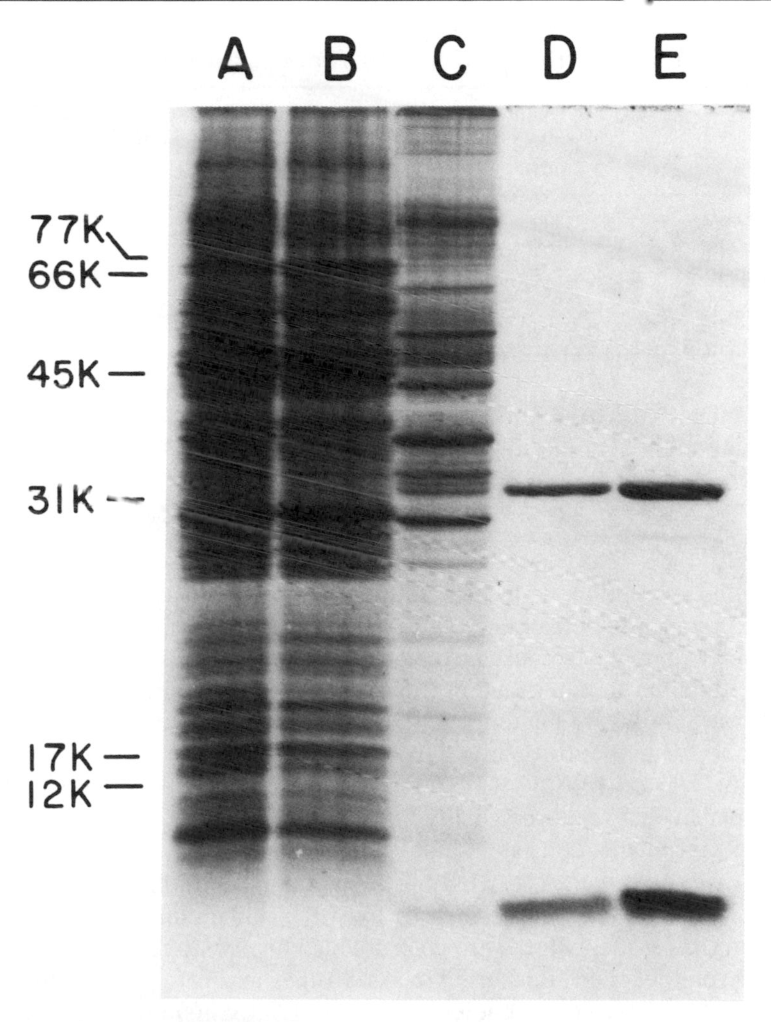

FIG. 1. SDS–polyacrylamide gel electrophoresis of Shiga toxin during purification by the method described in the text. Samples were dissolved in SDS sample buffer containing 2-mercaptoethanol, heated in boiling water for 10 min, and applied to a 15% polyacrylamide gel. Lane A, crude lysate of *S. dysenteriae* 1; lane B, Blue Sepharose flow-through; lane C, Blue Sepharose salt eluate; lane D, pH 7.1 fraction from chromatofocusing step; lane E, BioGel P-60 purified toxin. From Donohue-Rolfe *et al.*[17]

TABLE I
YIELD AND SPECIFIC ACTIVITY OF TOXIN DURING PURIFICATION

Procedure	Total protein (mg)	Cytotoxin specific activity (TC_{50}/mg)	Increase in specific activity (-fold)	Final yield (%)
Cell lysate	2300	2.5×10^4	—	100
Blue Sepharose	165	3.0×10^5	12	86
Chromatofocusing (pH 7.1 peak)	1	3.1×10^7	1240	54
BioGel P-60	0.8	3.4×10^7	1360	47

1 mg of pure toxin from this procedure. The lyophilized pure toxin is stable indefinitely at −70°, and for at least 4 weeks (and probably longer) in solution at 4°.[25a]

Assay of Shiga Toxin

With the exception of tissue culture methods, other bioassays for Shiga toxin are time consuming, subject to considerable day-to-day variability, and expensive because of the costs of purchasing and maintaining animals. In the latter category are the assays for the enterotoxin activity of the toxin (employing ligated small bowel loops in White New Zealand rabbits) and for its neurotoxin activity (using Swiss-Webster mice to determine the LD_{50}).[25] Assay of the cytotoxin activity is relatively simple and inexpensive, it is more sensitive than either of the other two bioassays, and it produces consistent and reliable results.[25,26]

Cytotoxicity Assay. A susceptible mammalian epithelial cell line is selected. We use HeLa cells (CCL-2) from the American Type Culture Collection (Rockville, MD). Vero cells are also highly susceptible,[27] however many epithelial cell lines used to assay other toxins (e.g. Y-1 adrenal cells or CHO cells used for cholera or *E. coli* LT toxins) are resistant to Shiga toxin (see Ref. 28 and our unpublished observations). In general, nonepithelial cell lines tested to date have also been resistant.[29,30] Some HeLa lines are highly resistant as well, and it is necessary to be certain that a responsive cell line is used.[10]

[25a] For scale-up procedure, see addendum on page 399.
[25] G. T. Keusch and M. Jacewicz, *J. Infect. Dis.* **131,** S33 (1975).
[26] G. T. Keusch, M. Jacewicz, and S. Z. Hirschman, *J. Infect. Dis.* **125,** 539 (1972).
[27] K. Eiklid and S. Olsnes, *J. Receptor. Res.* **1,** 199 (1981).
[28] G. T. Keusch and S. T. Donta, *J. Infect. Dis.* **131,** 58 (1975).
[29] G. T. Keusch, *Trans. N.Y. Acad. Sci.* **35,** 51 (1973).
[30] S. Olsnes, R. Reisbig, and K. Eiklid, *J. Biol. Chem.* **256,** 8732 (1981).

Sensitive assays have been developed using HeLa monolayers in 96-well flat-bottom tissue culture microtiter plates. In the assay we routinely use, wells are seeded with approximately 2×10^4 cells in 0.2 ml of tissue culture medium. We use McCoy's 5a modified medium containing 10% fetal bovine serum, but MEM can be used instead. Unfortunately, newborn calf serum cannot be substituted for FBS because it contains a toxin inhibitor.[30] Following overnight incubation in a moist incubator in the presence of 5% CO_2, a near-confluent monolayer is obtained. Cells are washed and duplicate wells are inoculated with 10-fold serial dilutions of toxin in 0.2 ml of medium, and incubated at 37°. The duration of incubation depends upon the assay system selected. One method measures the detachment of dead cells from the monolayer by cell counting.[26] After an initial period during which nothing appears to happen (varying from 30 min to several hours, depending upon the concentration of toxin), cells begin to round up and detach from the plastic surface, reaching a maximum after approximately 12–16 hr of incubation. Therefore, we inoculate microtiter plates one afternoon and complete the assay the following morning, which is very convenient for laboratory personnel. Monolayers are then washed vigorously with PBS, pH 7.4, by aspiration with a Pasteur pipet or a mechanical pipet to remove dead and loosely attached cells. Viable, firmly attached cells are then removed by incubation in 0.2 ml of 0.25% trypsin in calcium-free PBS, mixed and dispersed by vigorous trituration, and then counted in duplicate chambers of a Neubauer brightline hemocytometer. These results are compared with control cells incubated in medium alone, and the cell mortality is calculated. The tissue culture 50% lethal dose (TC_{50}) can then be calculated by the standard method of Reed and Muench.[31] With an experienced cytologist to do the counting, this method is simple, sensitive, and reproducible.

Spectrophotometric methods can also be used to assess the proportion of surviving cells. For example, Gentry and Dalrymple[32] set up their assay in a similar fashion, except for the volume of toxin used (0.1 ml) and the number of cells present in the wells (16,000). Toxin-exposed, washed HeLa cells are then fixed in a 2% solution of formalin in 0.067 *M* PBS, pH 7.2. After 1 min, the fixative is removed and the plates are stained with 0.13% crystal violet (w/v) in 5% ethanol–2% formalin–PBS for 20 min. The excess stain is removed by rinsing in water and the plates are air dried. For quantitation, stain is removed by elution with four successive 50-μl samples of 50% ethanol and diluted in 0.9 ml of PBS. The concentration of dye is determined by measuring absorbance at 595 nm. If available,

[31] L. J. Reed and H. Muench, *Am. J. Hyg.* **27,** 493 (1938).

[32] M. K. Gentry and J. M. Dalrymple, *J. Clin. Microbiol.* **12,** 361 (1980).

an automated ELISA reader with the correct filter can be used. The dilution of toxin producing 50% cell detachment, that is 50% of maximum dye uptake, is calculated from the plot of the dose–response curve. While this method eliminates the direct counting which we use, it is less accurate and no more convenient if an ELISA reader is not available.

Measurement of Amino Acid Incorporation. A more rapid tissue culture assay involves measuring the effect of toxin on amino acid incorporation into protein,[10,22,23,27] an effect which precedes observable cytotoxicity by many hours. Depending on the toxin concentration used, inhibition of amino acid incorporation can be detected as early as 30 min after exposure of cells. Although Olsnes *et al.* allow overnight exposure to toxin before pulsing HeLa cells with [^{14}C]leucine, by which time toxin-affected cells are dead and the assay actually samples the surviving cells,[10,22,27] reasonable titrations can be obtained when monolayers are incubated with toxin for only 4–6 hr. Labeled amino acid (for example, 0.5 μCi [^{14}C]leucine, specific activity = ~350 Ci/mmol) is then added for 1 hr. The monolayers are washed and the reaction is terminated by addition of an equal volume of 10% TCA. The precipitated proteins are harvested and washed with 5% TCA using a MASH cell harvester. Filters are removed, dried, and prepared for counting in a scintillation counter.

Specificity of Tissue Culture Assays. In all of these assays, it must be confirmed that the observed effect is due to shiga toxin by neutralization with high-titer specific antitoxin antibody. We and others have prepared and characterized such reagents, including polyclonal rabbit antitoxin and monoclonal subunit-specific mouse hybridoma antibodies[17,24,33] Because of the lethal neurotoxin activity of shiga toxin in both of these animal species, it is necessary to convert purified shiga toxin to a toxoid for immunization. This is accomplished by incubating 100 μg of toxin in 0.5 ml of 0.1 *M* PBS, pH 8, containing 1% formalin, for 3 days at 37°. The inactivated protein is then dialyzed against PBS and used for immunization. We inject rabbits subcutaneously in four sites with a total of 100 μg of toxoid in 0.5 ml PBS mixed with 0.5 ml of Freund's complete adjuvant.[17] Booster doses of 50 μg of toxoid in Fruend's incomplete adjuvant are given at 5-week intervals until high titer antibody is obtained. Bleeding is performed 8–10 days after the last booster. We have obtained subunit-specific mouse monoclonal antibodies by intraperitoneal injection of female 4- to 8-week old BALB/c mice with 20 μg of toxoid in Freund's complete adjuvant.[17] After 1 month, animals are given an intraperitoneal booster dose of 20 μg of toxoid in incomplete Freund's adjuvant. Three weeks later, an intravenous booster dose of 20 μg of toxoid in PBS is

[33] D. E. Griffin, M. D. Gentry, and J. E. Brown, *Infect. Immun.* **41,** 430 (1983).

given. Finally, spleen cells are harvested 4 days later and fused to myeloma cells by standard hybridoma methods. Antitoxin antibody is screened in the supernatant of all resulting colonies by immunoprecipitation of ^{125}I-labeled toxin, as described below. Positive colonies are cloned and expanded twice by limiting dilution methods. Antibody-containing ascites fluid is obtained by injecting stable antibody-producing hybridoma lines into pristane-primed mice by standard methods.

Immunoprecipitation. Toxin or toxin-containing fractions and all antibodies are serially diluted in PBS, pH 7.4, containing 100 μg/ml ovalbumin as carrier protein. Consistent results are obtained when 100 μl of PBS containing 1 ng of ^{125}I-labeled toxin (using the modified chloramine T procedure described in Ref. 17, we obtain ~30,000 cpm/ng) is incubated overnight at 4° with 10 μl of an appropriately diluted antibody. Antibody–toxin complexes are immunoprecipitated using fixed protein A-positive *Staphylococcus aureus* by the method of Kessler[34] and radioactivity in the precipitate is measured. Using an end point of >80% immunoprecipitation of 1 ng of labeled toxin, we have obtained immunoprecipitation titers varying from 1 : 40,000 for rabbit polyclonal antitoxin down to 1 : 800 for 5B2, an A subunit-specific mouse monoclonal antibody.[17]

ELISA Assay. We have also developed a sensitive indirect ELISA method[35] to measure toxin antigen by a modification of the procedure of Voller *et al.*[36] For this purpose, a B subunit-specific mouse monoclonal ascites fluid antibody, 4D3,[17] diluted to a protein concentration of 10 μg/ml in 50 m*M* sodium carbonate buffer, pH 9.6, is used to capture antigen by addition of 0.2 ml to each well of a microtiter plate (Nunc-Immuno Plate I, Nunc, Kamstrup, Denmark). The plates are incubated overnight at 4°, wells are emptied, and 0.2 ml of 1% bovine serum albumin in carbonate buffer is added for 1 hr at room temperature. After washing five times with 0.2 ml of phosphate-buffered saline–0.05% Tween 20 (PBS-T), 0.2 ml of antigen samples diluted in PBS-T is placed into duplicate wells. Plates are again incubated overnight at 4°, washed five times with PBS-T, and incubated with 0.2 ml of an experimentally determined optimal dilution of rabbit polyclonal antitoxin[17] for 2 hr at room temperature. After washing five times with PBS-T, 0.2 ml of goat anti-rabbit IgG–alkaline phosphatase conjugate (Sigma Chemical Co., St. Louis, MO), diluted

[34] S. W. Kessler, *J. Immunol.* **115,** 1617 (1975).

[35] A. Donohue-Rolfe, M. Kelley, M. Bennish, and G. T. Keusch, *J. Clin. Microbiol.* **24,** 65 (1986).

[36] A. Voller, D. Bidwell, and A. Bartlett, *in* "Manual of Clinical Immunology" (N. R. Rose and H. Friedman, eds.), 2nd Ed., p. 359. Am. Soc. Microbiol., Washington, D.C., 1980.

1 : 1000 in PBS-T, is added for 1 hr at room temperature. Wells are again washed and the reaction developed by addition of enzyme substrate [0.2 ml of *p*-nitrophenyl phosphate, 1 mg/ml in diethanolamine buffer, pH 9.8 (Sigma Chemical Co.)] for 1 hr at room temperature. The absorbance in each well is measured at 405 nm with an automated ELISA reader. Net absorbance is determined by subtracting the absorbance in wells treated with PBS-T in place of test sample from the absorbance in well with toxin-containing test samples. A standard curve is constructed using pure toxin, and unknown samples can be assayed by plotting results from serial dilutions of the unknown on the standard curve. Using the antitoxin antibodies we have described,[17] this assay gives reproducible results with as little as 10 pg toxin/well.

Acknowledgment

The work in our laboratory was funded by Grants AI-16242, AI-20325, and AM-39428, from the National Institutes of Health, Bethesda, Maryland, Grant 82008 from the Programme for Control of Diarrhoeal Diseases, World Health Organization, Geneva, Switzerland, and a grant in geographic medicine from the Rockefeller Foundation, New York, NY.

[23] Preparation of *Yersinia pestis* Plague Murine Toxin

By THOMAS C. MONTIE

Introduction

Plague murine toxin appears to be an envelope protein component of the bacterium *Yersinia pestis*, formerly *Pasteurella pestis*. As isolated, the toxin may contain one or usually two related protein species of different molecular weights (120,000 and 240,000). At the end of the growth cycle toxin is released into the medium following autolysis of the cells. The method described below utilizes this observation to facilitate toxin isolation, although toxin also can be released following sonication of the bacterial envelope.[1] The two proteins isolated are toxic for mice and rats

[1] T. C. Montie and S. J. Ajl, *J. Gen. Microbiol.* **34,** 249 (1964).

at low levels (0.1–1.0 μg/20 g mouse). Mouse lethality then must be an important assay for ultimately determining the biological activity of the protein product. Immunodiffusion and gel electrophoresis assays also serve to facilitate routine assays of the toxin.

Although both toxin A and B appeared routinely in the *Y. pestis* strain Tijwidej (TJW) preparations, toxin A seemed to be the most biologically active of the two proteins. Also, a survey of different avirulent strains indicated it to be the most predominant if not the only toxin present in some *Y. pestis* extracts.

It is difficult to explain the toxicity in terms of any unusual feature of the primary sequences of the protein. However, the sulfhydryl and tryptophan residues are important for toxicity.[2] Evidence to date indicates that the toxin acts as a beta blocking agent (see review by Montie[3]).

Isolation and Preparation

Growth of Cells

A number of different strains of avirulent *Y. pestis* have been used to prepare toxin. We have generally used strain TJW maintained on agar (1.6%) slants containing casein hydrolysate or casamino acids plus mineral salts as modified from Englesberg and Levy.[4] This same medium (CHMG) is used to grow up to 7 liters of batch culture to be used for isolation. The CHMG medium contains the following, in grams per liter: casein hydrolysate or casamino acids (30.0), NH_4Cl (1.0), $MgSO_4 \cdot 7H_2O$ (0.5), $CaCl_2$ finely ground (0.01), and $FeCl_3 \cdot 6H_2O$ (0.0025) or (1.0 ml from a stock solution containing 0.25 g/100 ml H_2O). Stock solutions of potassium phosphate and glucose are autoclaved and added separately; final concentrations are potassium phosphate (0.05 *M*) and D-glucose (2%).

A 24-hr-old CHMG slant culture is washed into 250 ml of the 7 liters to be used as starter culture. The latter culture is grown at 27–28° for 24 hr on a shaker. The starter is inoculated into a 9-liter carboy containing the remaining 7 liters of medium. This 7-liter culture is aerated and stirred by means of a finely performed, bubbling stone apparatus connected to forced air for 7 days at room temperature (maximum toxin release occurs after approximately 5 to 7 days). Antifoam may be added if necessary. Cells and cell debris are then removed by centrifugation for 30 min at 9000 rpm (13,000 *g*) at 4° and the supernatant is saved for processing.

[2] T. C. Montie and D. B. Montie, *Biochemistry* **12,** 4958 (1973).
[3] T. C. Montie, *Pharmacol. Ther.* **12,** 491 (1981).
[4] E. Englesberg and S. B. Levy, *J. Bacteriol.* **68,** 57 (1954).

Ammonium Sulfate Fractionation—Preparation of Crude Toxin

1. The supernatant is brought to 100% saturation (70 g/100 ml) with ammonium sulfate and stirred slowly in the cold overnight. The precipitated toxin is centrifuged or removed with a beaker from the surface of the culture if floating. The precipitate is dissolved in a small amount of distilled water and dialyzed against water with a number of changes until no $BaSO_4$, precipitate is detected with $Ba(OH)_2$ [two drops saturated $Ba(OH)_2$ solution/ml crude toxin solution].

2. Ammonium sulfate is again added slowly with stirring at 4° until 35% saturation is reached (24.5 g/100 ml). The precipitate is removed by centrifugation and discarded.

3. The supernatant containing the toxin is brought slowly to approximately 70% saturation by addition of ammonium sulfate at 4°. The solution can be stored overnight at 4° to obtain the maximum amount of precipitated toxin. The toxin precipitate is removed by centrifugation (4°), redissolved in a small amount of water, and dialyzed exhaustively against water and lyophilized. The crude toxin preparation is best stored desiccated at −70° to avoid denaturation.

Purification and Resolution of Toxins A and B by Molecular Seiving

The crude toxin, containing two toxic proteins as major components, can be further purified by molecular seiving procedures using BioGel or Sephadex columns.[5] Typically, 4.0 mg of crude toxin is dissolved in 2 ml 0.01 *M* potassium phosphate buffer (pH 8.0) and centrifuged to remove any undissolved or denatured toxin materials. The toxin sample is passed through a Sephadex G-100 or G-200 column (2.5 × 40 cm) which has been previously equilibrated with the pH 8.0 phosphate buffer. Elution of toxin is detected at 280 nm. Toxin A (240 kDa) appears at the void volume (eluate volume of approximately 58 ml in Sephadex G-100) and continues completely resolved from toxin B (120 kDa) for approximately 6 to 10 ml of eluate. Both toxins elute next in the trailing position of the first peak, and toxin B emerges as a distinct shoulder trailing the first major ultraviolet-positive peak. The ultraviolet pattern does not clearly distinguish the resolution of the two toxins so that assay by disc or slab gel electrophoresis (PAGE) or by immunodiffusion using immune rabbit serum[5] is required. Toxin obtained exhibited activity in 16- to 18-g female mice Swiss albino mice of an approximate LD_{50} of 1 μg following intraperitoneal injection. Fractions are pooled and dialyzed with a final dialysis step against 0.001 *M* phosphate buffer before lyophilization. The latter step

[5] T. C. Montie, D. B. Montie, and S. J. Ajl, *Biochim. Biophys. Acta* **130,** 406 (1966).

aids in preventing denaturation. Storage of the dried material should be at −70° desiccated.

Isolation by Polyacrylamide Gel Electrophoresis

Elution of Toxin from Acrylamide Gels[6]

Some of the earliest isolation techniques involved elution of toxin protein from disc gels. This technique could be readily applied to slab gel preparation to obtain greater amounts in continuous or discontinuous systems.[7,8] The disadvantage of the disc gel technique is that only microgram amounts of toxin can be obtained. In this technique crude toxin between 1 and 2 mg is loaded onto 0.8 × 5 cm acrylamide disc gel-columns placed in cut sections of 10-ml pipets following the general method of Davis.[9] The run is carried out in an alkaline Tris–HCl buffer (pH 8.9). With the toxins both sample and spacer gels can be eliminated and 20% sucrose or Sephadex are acceptable substitutes for loading the sample. Toxin is premixed with these components and layered onto the small-pore running gel. Parallel disc gel or slab gel lanes are used to locate by the location of toxin following 0.5 hr of rapid staining with Amido black or Coomassie blue. For elution unstained parallel discs or slab sections are cut into 2- to 3-mm sections (up to 6 mm for long slab gels) and eluted with either 0.05 sodium maleate buffer (pH 6.6) or 0.01 *M* phosphate buffer (pH 7.6) at 4°. The gel sections are agitated during the elution period lasting 24 to 48 hr. In some cases gels may be eluted directly by encasing the slice in dialysis tubing. An approach somewhat more time consuming, but affording a higher yield, involves homogenization of the gel slice in buffer using a motor-driven Teflon pestle. The homogenates are then centrifuged and the supernatants dialyzed against distilled water. Although the slightly acid buffer resulted in elution of lower amounts of toxin compared to phosphate elution at this pH it seemed to deter elution of nonspecific UV-absorbing material which was more evident in alkaline buffers. Protein eluted was toxic and could be detected by immunodiffusion. For quick assay to detect toxin, entire uneluted gel sections placed in antigen wells were adequate sources of antigen for production of immunoprecipitin bands.

[6] T. C. Montie, D. B. Montie, and S. J. Ajl, *J. Exp. Med.* **120,** 1201 (1964).

[7] "Hoefer Scientific Instruments Catalogue," p. 96. Hoefer Scientific Instruments, San Francisco, California, 1983.

[8] A. Chrambach and D. Rodbard, *Science* **172,** 440 (1971).

[9] B. J. Davis, *Ann. N.Y. Acad. Sci.* **121,** 404 (1964).

Preparative Gel Electrophoresis[10]

A large preparative gel electrophoresis [Poly-Prep 100, Büchler Instruments Div., Nuclear-Chicago Corporation (Searle and Co.), Fort Lee, NJ]. The advantage of this technique is that milligram amounts of highly purified toxin can be obtained directly from the crude ammonium sulfate prepared toxin. For example, from 250 mg of crude toxin (two runs), 11.3 mg of purified toxin A was obtained. No evidence for lipopolysaccharide contamination was found in biological studies.[11] In some preparations a toxic dose was at the level of 0.1 μg Lowry protein.

Toxin (80 to 150 mg) sample is dissolved in up to 15 ml of sample buffer (2.89 μg Tris and 12.8 ml 1 *M* H_3PO_4 brought to 100 ml) and layered onto a preformed concentrating gel using riboflavin-polymerized acrylamide (see components and solutions, Table I) or by directly layering a toxin solution of Tris–glycine or Tris–phosphate sample buffer with 10% sucrose onto a 6-cm resolving gel. One drop of concentrated bromphenol blue is added as a marker dye. The column is continuously cooled at 6–8°. Run time is approximately 36 hr in a Tris–glycine pH 8.9, 7.5% acrylamide running gel system with an elution flow rate of 20 ml/hr. Toxin bands are identified on analytical gels using 0.02 ml of a 2- to 4-ml collected fraction. A constant amperage is required, set at 50 mA. The voltage will rise slowly during the run from 250 to 400–450 V. The individual toxins are resolved into two peaks of approximately 20 consecutive 2-ml fractions, for example, with toxin B in fractions 58 to 70, and toxin A in fractions 135 to 155 (see analytical gel profile in Ref. 10). The toxin samples are pooled, dialyzed against 0.001 *M* KH_2PO_4, pH 7.0, buffer, and lyophilized. Storage of lyophilized material is at −70°, desiccated.

Preparation of Toxin Subunit Components

Isolation of Biologically Active Subunits Using Sodium Dodecyl Sulfate (SDS)[12,13]

Although large-molecular-weight murine toxin is partially denatured by treatment with 0.5–1% SDS, it apparently retains partial activity even though dissociation occurs to a subunit of approximate MW 24,000. These subunits may be obtained by dissolving toxin protein in 1.0% SDS and incubating at 37° for 3 hr. Excess SDS is removed by dialysis against 0.1%

[10] T. C. Montie and D. B. Montie, *J. Bacteriol.* **100,** 535 (1969).

[11] S. D. Brown and T. C. Montie, *Infect. Immun.* **18,** 85 (1977).

[12] T. C. Montie, D. B. Montie, S. A. Leon, C. A. Kennedy, and S. J. Ajl, *Biochem. Biophys. Res. Commun.* **33,** 423 (1968).

[13] T. C. Montie and D. B. Montie, *Biochemistry* **10,** 2094 (1971).

TABLE I
REQUIRED SOLUTIONS FOR BÜCHLER POLY-PREP ACRYLAMIDE GEL ELECTROPHORESIS

Buffers	Chemicals	Amount	Concentration (-fold)
Upper	Tris	12.64 g	—
(per 2 liters)	Glycine	7.88 g	—
Lower	Tris	72.6 ml	3×
(per 2 liters)	1 *N* HCl	300 ml	
Membrane	Tris	48.4 g	4×
(per liter)	1 *N* HCl	200 ml	
Sample	1 *M* H_3PO_4	12.8 ml	—
(per 100 ml)	Tris	2.85 g	—
Elution	Lower buffer	1 part	1×
	H_2O	2 parts	

Stock solutions

Gel	Volume desired	Volume ratio	Component	Concentration /100 ml
Resolving	100 ml	25 ml	Acrylamide	30.0 g
			Bisacrylamide	2.0 g
		25 ml	1 *N* HCl	24.0 ml
			Tris	18.15 g
			TEMED	0.2– 0.4 ml
		50 ml	Ammonium persulfate	0.15 g
Concentrating[a]	50 ml	30ml	Acrylamide	5.0 g
			Bisacrylamide	0.3– 1.25 g
		15 ml	1 *M* H_3PO_4	12.8 ml
			Tris	2.85 g
			TEMED	0.1 ml
		15 ml	Riboflavin	2.0 mg
Additional solutions		100 ml H_2O	Sucrose	40 g
		100 ml H^2O	Bromphenol blue	5.0 mg

[a] This gel is hardened using a bank of fluorescent lights with riboflavin as the catalyst.

SDS for 17 to 24 hr. Under these conditions at least 60% of the toxic activity is retained.[11] To avoid dialysis, passage through a Sephadex G-75 or G-100 column will remove excess SDS. The subunit form is retained and identified as the major retarded peak. In more dilute incubation solutions a polypeptide of 10,000–12,000 Da is obtained; however, the toxic activity of this unit may be eliminated. Specific conditions for subunit isolation and/or molecular weight determination include passing 1 to 2 mg/ml toxin in SDS–phosphate through a Sephadex G-100 column (1.5 × 90 cm) equilibrated with 0.1 *M* sodium phosphate buffer (pH 7.1).

Isolation of Inactivated Subunits with Organic Acids: Citric Acid and Acetic Acid Plus EDTA[12]

Under these conditions lethal activity is lost, although attempts to reverse the acid denaturation have not been explored fully. In this method toxin (1–4 mg) is rapidly dissolved in 0.5 to 2.0 ml of citric acid (0.1 *M*, pH 2.2) and incubated at 37° for 1 hr or 26° for 24 hr (partial depolymerization). Complete depolymerization to a 12,000-Da fragment occurs after exposure at 37° for 24 hr. Alternatively, 0.1 *M* acetic acid with 10^{-3} *M* EDTA can be substituted for citric acid with approximately the same result. Toxin subunits are isolated by passing toxin dissolved in acid through a Sephadex G-50 column equilibrated with 0.001 *M* potassium phosphate (pH 7.0) and 0.1 *M* KCl. The high salt is added if column adsorption is a problem, but usually it can be omitted. Excess acid may be removed by passing the pooled two or three peak fractions through a second G-50 column. Acetic acid is the acid of choice for peptide determinations since radioactive citrate is bound to toxin even after tryptic digestion and separation of peptides by high-voltage electrophoresis.

Isolation of Inactive Polypeptides by Gel Electrophoresis in a Phenol–Acetic Acid–Urea Gel[10]

Lyophilized native toxin fractions from the Büchler Poly–Prep gel electrophoresis are further separated into polypeptides by gel electrophoresis in a denaturing environment. Microgram amounts of toxin A or B (5 to 25 μg) are dissolved in 0.1 to 0.2 ml phenol–acetic acid–water (2 : 1 : 0.5, v/v/v) and subjected to electrophoresis for 2 hr at room temperature through 7.5% (w/v) acrylamide gels containing 35% (v/v) acetic acid and 5 *M* urea. Protein bands are stained in the usual way with Amido black or Coomassie blue. Toxin A gives two bands: the least electropositive is designated No. 1 and the more electropositive band is designated No. 2. Toxin B exhibits two bands: a band corresponding to the No. 2 band of toxin A, and an additional No. 3 band of greater electrophoretic mobility than the No. 2 band. The No. 2 band (common to both toxins) always appears as the heaviest, in greatest proportion, or most densely stained band of the three. This method has been used by Rotten and Razin[14] to identify mycoplasma membrane proteins and by Takayama *et al.*[15] to distinguish mitochondrial membrane proteins.

[14] S. Rottem and S. Razin, *J. Bacteriol.* **94,** 359 (1967).

[15] K. Takayama, D. H. MacLennon, A. Tzagoloft, and C. D. Stoner, *Arch. Biochem. Biophys.* **114,** 223 (1966).

[24] Production and Purification of Cholera Toxin

By JOHN J. MEKALANOS

Cholera toxin (choleragen, permeability factor, cholera enterotoxin) is an extracellularly secreted protein produced by toxinogenic strains of *Vibrio cholerae*.[1-4] Cholera toxin (CT) is highly related to the heat-labile enterotoxin (LT) of *Escherichia coli* in structure and activity. This family of enterotoxins causes diarrhea through the activation of adenylate cyclase and the alteration of ion transport in intestinal epithelial cells. CT is the most well-studied member of this group and is capable of elevating cAMP levels in a variety of eukaryotic cells that include nonmammalian examples.

CT is composed of two types of subunits, A (MW 27,215) and B (MW 11,677), which are present in the native molecule in a ratio of 1 : 5, respectively.[1,5,6] The A subunit is composed of two disulfide-linked peptides A_1 (MW 21,817) and A_2 (MW 5398).[5,6] These two peptides are derived from a common precursor by an internal proteolytic cleavage or nick that occurs after extracellular secretion of the holotoxin.[7] The five B subunits form a ring-shaped pentamer which can be purified as a separate molecular species called choleragenoid.[1] Each B monomer has a high-affinity binding site for the toxin's cell surface receptor, ganglioside GM_1.[2-4,8] After binding to this glycolipid, the A_1 peptide is translocated to the cell interior where it catalyzes the ADP ribosylation of the membrane-associated G_s subunit of adenylate cyclase.[9,10] The ADP-ribosyltransferase activity of CT has been very useful in characterizing the molecular nature of G_s and in understanding the regulation of adenylate cyclase activity. CT is also the pharmacological reagent of choice for producing elevated cAMP levels in mammalian cells for investigative purposes.

[1] R. A. Finkelstein and J. J. LoSpalluto, *J. Infect. Dis.* **121** (Suppl.), S63 (1970).
[2] D. M. Gill, *Adv. Cyclic Nucleotide Res.* **8,** 85 (1977).
[3] J. Holmgren, *Nature (London)* **292,** 413 (1981).
[4] J. J. Mekalanos, R. J. Collier, and W. R. Romig, *Infect. Immun.* **20,** 552 (1978).
[5] D. M. Gill, *Biochemistry* **15,** 1242 (1976).
[6] J. J. Mekalanos, D. J. Swartz, G. D. N. Pearson, N. Harford, F. Groyne, M. de Wilde, *Nature (London)* **306,** 551 (1983).
[7] J. J. Mekalanos, J. R. Collier, and W. R. Romig, *J. Biol. Chem.* **254,** 5855 (1979).
[8] W. E. van Heyningen, C. C. J. Carpenter, N. F. Pierce, and W. B. Greenough, *J. Infect. Dis.* **124,** 415 (1971).
[9] D. Cassel and T. Pfeuffer, *Proc. Natl. Acad. Sci. U.S.A.* **75,** 2669 (1978).
[10] D. M. Gill and R. Meren, *Proc. Natl. Acad. Sci. U.S.A.* **75,** 3050 (1978).

There are several methods for purification of CT that have been reported.[1,11,12] The one described here[13] is a rapid and simple method that gives high yields of CT and choleragenoid.

Reagents and Media

All chemicals used should be of analytical grade. Sodium hexametaphosphate is purchased from Fisher Scientific Co. (Pittsburgh, PA). Phosphocellulose P11 is purchased from Whatman and must be cycled as described below before use. CYE medium is prepared by dissolving 30 g of acid-hydolyzed casein (United States Biochemical, Cleveland, OH) and 5 g of yeast extract (Difco, Detroit, MI) per liter of deionized water. The pH is adjusted with 4 *N* NaOH to 7.0 and the medium is autoclaved for 20 min.

Assay Procedures

CT antigen is quantitated by a ganglioside binding-dependent ELISA.[14] Microtiter plates (Cat. No. 76-301-05, Flow Laboratories, McLean, VA) are coated by addition to each well of 0.2 ml of 60 m*M* sodium carbonate buffer, pH 9.9, containing 10 μg/ml of type III gangliosides (Sigma). After incubation at 37° overnight, the plates are washed with PBS (10 m*M* sodium phosphate, pH 7.4, 0.14 *M* sodium chloride) and then blocked by addition of PBS–BSA (PBS containing 5 mg/ml of albumin, fraction V, Sigma) and incubation for 4 hr at 37°. The plates are washed again with PBS, air dried, and stored at 4°. Solutions containing toxin (culture fluids or column fractions) are diluted in PBS–BSA and 0.2 ml of the diluted samples is added to the microtiter wells. After incubation at 37° for 1 hr, the plates are washed with PBS and the wells are filled with 0.2 ml of PBS–BSA containing a 1/1000 dilution of rabbit antitoxin serum. After incubating at 37° for 1 hr the plates are washed with PBS and the wells are filled with 0.2 ml of PBS–BSA containing a 1/1000 dilution of goat anti-rabbit IgG–alkaline phosphatase conjugate (Miles Laboratories). After incubation at 37° for 1 hr, the plates are washed with PBS and then the wells are filled with 1 *M* Tris, pH 9, containing 2 mg/ml of *p*-nitrophenyl phosphate (Sigma). The yellow color that develops after 1–6 hr can be read visually to determine end points relative to a diluted

[11] R. S. Rappaport, B. A. Rubin, and H. Tint, *Infect. Immun.* **9,** 294 (1974).

[12] J. L. Taylor, J. Holmgren, L. Svennerholm, M. Lindblad, and J. Tardy, *Eur. J. Biochem.* **113,** 249 (1981).

[13] J. J. Mekalanos, R. J. Collier, and W. R. Romig, *Infect. Immun.* **20,** 552 (1978).

[14] J. Holmgren, *Infect. Immun.* **8,** 851 (1973).

purified CT standard solution (made up in PBS–BSA) or read spectrophotometrically at 405 nm.

ADP-ribosylation activity of cholera toxin is measured by incorporation of radioactivity from [U-^{14}C]adenine NAD into trichloroacetic acid-precipitable material in the presence of synthetic poly(L-arginine).[15] The reaction is performed at 30° in 0.15 ml of 25 m*M* sodium phosphate buffer, pH 7.0, containing 20 m*M* dithioerythritol, 2 μ*M* [^{14}C]adenine NAD (Amersham, >225 mCi/mmol), and 1 mg of poly(L-arginine) (type II-B, Sigma) per milliliter. The reaction is allowed to procede for 1–4 hr and then the reaction is spotted on Whatman 3MM paper that had been soaked with a 10% (w/v) solution of trichloroacetic acid (TCA) in ether and air dried. The spotted samples are washed in 5% TCA and counted in a suitable liquid scintillation solution. This method is relatively insensitive, detecting only CT concentrations above 100 μg/ml in the reaction. More sensitive methods for detecting the ADP-ribosyltransferase activity of cholera toxin are available[10,11,15] but are generally more difficult to set up. Any one of several animal and tissue culture assays can be used to access the biological potency of CT preparations.[2] Some of these utilize secondary responses of tissue to elevated cAMP levels while others directly measure adenylate cyclase activation by CT or the accumulation of cAMP.

Purification Steps

1. Toxin Production. The strain of *V. cholerae* universally used for toxin production and purification is 569B[1,11,12] or hypertoxinogenic mutant derivatives of this strain.[13,16] While the mutant strains give higher yields of CT, they are also genetically unstable and revert on occasion to hypotoxinogenic strains.[14] For this reason, it is advisable to use 569B although care must be taken that the particular 569B subculture being used is capable of producing at least 10 μg/ml of crude toxin when grown in CYE at 30° with vigorous aeration. Scaling up the growth of such a strain in large fermentor cultures is usually successful. Any culture vessel that can be modified to allow efficient aeration through the use of spargers and compressed oxygen can be used for production of CT. Alternatively, the strain can be grown in batch cultures if a large-scale culture apparatus is not available. The method described in detail below is a high-density growth method that gives crude CT yields of about 10–20 μg/ml for strain 569B.

The 569B strain being used is recovered from a previous CYE culture

[15] J. J. Mekalanos, R. J. Collier, and W. R. Romig, *J. Biol. Chem.* **254,** 5849 (1979).
[16] J. J. Mekalanos, R. Sublett, and W. R. Romig, *J. Bacteriol.* **139,** 859 (1979).

that was frozen in the presence of 15% glycerol and stored at −70°. The glycerol starter inoculum is added directly into 5 ml of CYE broth and incubated at 30° with shaking. When this culture is saturated it is used to inoculate 4 liters of CYE distributed in eight 2-liter flasks and incubated at 200 rpm at 30°. The cells from these 4 liters of overnight shaker culture are isolated by centrifugation (10,000 *g*, 40 min), resuspended in CYE, and inoculated to an initial cell density of 2×10^9 to 3×10^9 cells/ml in a Kluyver culture flask or fermentor chamber containing CYE broth (5–10 liters). The supernatant fluid from the overnight batch cultures is saved at 4°. The large fermentor culture is incubated at 30° and aerated vigorously with compressed oxygen until stationary phase is reached as judged by the optical density at 600 nm (about 3 to 4 hr). The cells are recovered by centrifugation and used to inoculate fresh CYE broth to the same initial cell density and another cycle of growth is completed. The second culture is centrifuged and the supernatant fluids from the two cycles are pooled and cooled to 4°. The bacterial cell pellets are autoclaved before disposing.

2. *Hexametaphosphate Coprecipitation*. Crude CT is recovered from the culture supernatant by coprecipitation with sodium hexametaphosphate.[11] Sodium hexametaphosphate is added to the culture supernatant to a final concentration of 2.5 g/liter, and after adjustment of the pH to 4.5 with concentrated HCl, the mixture is stirred for 2 hr at room temperature. The precipitate is then collected by centrifugation (10,000 *g*, 15 min) and redissolved in 0.05 vol of 0.1 *M* sodium phosphate, pH 8.0. This concentrate is dialyzed against at least two changes of 10 m*M* sodium phosphate, pH 7.0, at 4° in the presence of 0.02% sodium azide. After dialysis, any insoluble material is removed by centrifugation at 10,000 *g*, 1 hr. The latter step is important because the insoluble material contains lipopolysaccharide that will become entrapped in the phosphocellulose column and contaminate the eluted toxin.

3. *Chromatography on Phosphocellulose*. Phosphocellulose (50 g) is first exposed to four cycles of alternate 2-liter basic (0.1 *N* NaOH) and acidic (0.1 *N* HCl) washes, allowing the resin to settle out in between. It is then resuspended in 10 m*M* sodium phosphate buffer, pH 7.4 or 8.3, and the pH adjusted to that of the buffer with either NaOH or phosphoric acid. After collecting the resin by filtration, the resuspension and titration is repeated until the resin does not alter the pH of the resuspension buffer by more than 0.05 pH unit. The equilibrated resin is then ready for packing into a column but is washed further with at least two column volumes of buffer before application of the sample. The amount of phosphocellulose to use is about 3 ml of packed resin per liter of original culture

supernatant. The geometry of the column should be such that the height of the packed resin is about two times its width.

The dialyzed and clarified hexametaphosphate concentrate is passed through the column at room temperature under a fast flow rate (usually about 90 ml/hr) and the column is washed with at least two column volumes of initial buffer or until the OD_{280} of the flow through is less than 0.02. The column is then eluted with 0.2 *M* sodium phosphate buffer, pH 7.4. A single peak of protein is eluted by this buffer and this peak should contain essentially all of the CT and choleragenoid originally present in the hexametaphosphate concentrate. Although this eluted material contains a mixture of pure CT and choleragenoid, the amount of the latter is variable and usually represents less than one-half of the protein in the preparation. This level of contamination by choleragenoid does not significantly affect the biological potency of CT preparations. Accordingly, the column fractions that contain the highest levels of protein (OD_{280} >1.0) can be stored at 4° without further manipulation and have remained stable for over 2 years in the presence of 0.02% sodium azide as a preservative.

Alternatively, if cholera toxin substantially free of choleragenoid is desired, the fractions making up this eluted peak can be pooled and dialyzed against 10 m*M* sodium phosphate, pH 8.3, at 4°. A second phosphocellulose column equilibrated in the same buffer is built and used to separate CT from choleragenoid. In this case the required column size is small (about 0.2 ml of packed resin per OD_{280} unit present in the dialyzed pooled

TABLE I
PURIFICATION OF CT AND CHOLERAGENOID[a]

Phosphocellulose chromatography	Volume (ml)	A_{280}	Protein (mg/ml)	Toxin antigen (μg/ml)
		pH 7.0		
Sample[b]	570	0.96	1.0	200
Peak 1	580	0.74	0.75	<2
Peak 2	86	1.4	1.5	1800
		pH 8.3		
Sample[c]	90	1.32	1.5	1500
Peak 1	95	0.7	0.78	700
Peak 2	20	1.0	0.93	1850

[a] From Ref. 13.
[b] Hexametaphosphate concentrate of CYE culture supernatant.
[c] Dialyzed peak 2 from phosphocellulose chromatography, pH 7.0.

fractions from the first column). The dialyzed mixture of CT and choleragenoid is passed through the second column at room temperature, washed with starting buffer, and the flow-through fractions monitored for OD_{280} absorbance. In 10 mM sodium phosphate, pH 8.3, CT does not bind to the

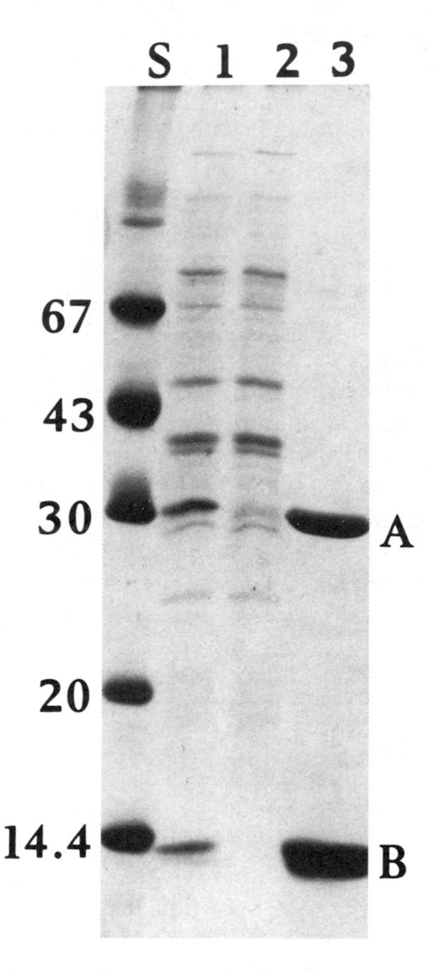

FIG. 1. SDS–polyacryamide gel analysis of cholera toxin purification by phosphocellulose chromatography. Samples were boiled in SDS sample buffer without reducing agents, fractionated by electrophoresis through a 15% polyacryamide slab gel in the presence of SDS, and then stained with Comassie Brilliant Blue. Lane S, molecular weight standards (MW $\times$ 10^{-3}). Lane 1, hexametaphosphate concentrate of a CYE culture supernatant of *V. cholerae* strain 569B (sample). Lane 2, peak 1 from phosphocellulose chromatography, pH 7.0 (the unbound material). Lane 3, peak 2 from phosphocellulose chromatography, pH 7.0, eluted with 0.2 M sodium phosphate buffer, pH 8.0. The A and B subunits of cholera toxin are indicated. Photograph courtesy of Dr. Maja Huber-Lulac.

phosphocellulose and emerges from the column unretarded while choleragenoid is bound. The latter is eluted if desired by once again eluting with 0.2 *M* sodium phosphate, pH 8.0. The first peak containing the purified CT can be stored at 4° after addition of additional sodium phosphate buffer, pH 8, to a final concentration of 0.2 *M* and sodium azide to 0.02%.

Table I shows a typical analysis of fractions obtained by this purification method.

Properties

Although CT has an isoelectric point of 6.6,[1] the chromatographic properties of the toxin on phosphocellulose suggest that a cationic surface is exposed on the CT molecule at pH 7.0. This cationic surface charge distribution apparently resides in the B subunits, which show a high affinity for phosphocellulose even at pH 8.3. This accounts for the chromatographic behavior of these proteins on phosphocellulose.

Electrophoresis on SDS–polyacrylamide gels of toxin purified by this method is shown in Fig. 1. Purified CT appears as two bands, A (M_r 30K) and B (M_r 14K), after electrophoresis when toxin samples are heated at 100° in SDS in the absence of reducing agents (Fig. 1, lane 3). Toxin samples which are not boiled contain the undissociated B protomer, which migrates in SDS gels with an M_r of about 54K.[5] Reduction of the toxin converts the A subunit to the A_1 (M_r 24K) and A_2 (M_r 5–7 K, but does not stain well with Coomassie brilliant blue) peptides while the B subunit shows a reduction in its mobility from an M_r of 12 K to an M_r of about 15K.[5,7,13] Depending on how rapidly the toxin is purified from culture supernatants, the purified CT may contain some fraction of the A subunit in its "unnicked" form which can be nicked by treatment with low levels of trypsin if desired.[7]

CT purified by the method described above is available from several commercial sources. These commercial preparations as well as freshly prepared CT retain high biological activity. Typically, a positive reaction in the rabbit skin test[17] can be obtained with as little as 80 pg of phosphocellulose-purified CT and about 10 pg represents a cytotoxic dose in the S49 tissue culture assay.[18]

[17] J. P. Craig, *J. Bacteriol.* **92,** 793 (1966).

[18] F. E. Ruch, J. R. Murphy, L. H. Graf, and M. Field, *J. Infect. Dis.* **137,** 747 (1978).

[25] Cytolytic Toxins of *Vibrio vulnificus* and *Vibrio damsela*[1]

By ARNOLD S. KREGER, MAHENDRA H. KOTHARY, and LARRY D. GRAY

Introduction

Prior to 1976, *Vibrio cholerae, Vibrio parahaemolyticus,* and *Vibrio alginolyticus* were the only *Vibrio* species clearly demonstrated to be pathogenic for humans. The first two of these species are known to produce cytolytic toxins ("hemolysins"), and highly purified preparations of the toxins have been obtained by Japanese investigators[2-7] in order to characterize the toxins and examine their possible importance in the pathogenesis of diseases caused by *V. parahaemolyticus* and *V. cholerae.* The purification and characterization of the extensively studied thermostable direct hemolysin (Kanagawa phenomenon-associated hemolysin) of *V. parahaemolyticus* is described in another chapter in this volume [see Chapter 26].

Eight new *Vibrio* species (*V. vulnificus, V. fluvialis, V. mimicus, V. damsela, V. metschnikovii, V. hollisae, V. furnissii,* and *V. cincinnatiensis*) have been recognized in the past 12 years as etiologic agents of

[1] The studies performed by the authors were supported by Public Health Service Grant AI-18184 from the National Institutes of Health. We thank Dr. Robert Weaver, Dannie Hollis, and Frances Hickman-Brenner (Centers for Disease Control, Atlanta, GA) for supplying the *V. vulnificus* and *V. damsela* isolates used in our studies. We also thank Dr. Mark Lively (Protein Sequencing Laboratory, Oncology Research Center of the Bowman Gray School of Medicine) and Dr. Lowell Ericsson (AAA Laboratory, Mercer Island, WA) for amino acid analyses, and Dr. Jacqueline Testa (Department of Microbiology and Immunology, Bowman Gray School of Medicine), Dr. Larry Daniel, and Lori Etkin (Department of Biochemistry, Bowman Gray School of Medicine) for examining toxin preparations for phospholipase activity.

[2] T. Honda and R. A. Finkelstein, *Infect. Immun.* **26,** 1020 (1979).

[3] K. Yamamoto, M. Al-Omani, T. Honda, Y. Takeda, and T. Miwatani, *Infect. Immun.* **45,** 192 (1984).

[4] K. Yamamoto, V. Ichinose, N. Nakasone, M. Tanabe, M. Nagahama, J. Sakurai, and M. Iwanaga, *Infect. Immun.* **51,** 927 (1986).

[5] M. Yoh, T. Honda, and T. Miwatani, *Infect. Immun.* **52,** 319 (1986).

[6] T. Honda, S. Taga, T. Takeda, M. A. Hasibuan, Y. Takeda, and T. Miwatani, *Infect. Immun.* **13,** 133 (1976).

[7] Y. Miyamoto, Y. Obara, T. Nikkawa, S. Yamai, T. Kato, Y. Yamada, and M. Ohashi, *Infect. Immun.* **28,** 567 (1980).

infectious diseases in humans.[8–13] Our research interests have focused on biologically active products of *V. vulnificus* and *V. damsela,* two halophilic species which are unique in that they can cause rapidly progressing and life-threatening nonenteric diseases. Mouse-virulent strains of the bacteria produce large amounts of extracellular cytolytic toxins *in vitro,*[14,15] and we have developed purification schemes for the toxins[16,17] in order to characterize their properties and enable us to examine their possible importance in the pathogenesis of diseases caused by *V. vulnificus* and *V. damsela*. The availability of highly purified preparations of the new cytolytic toxins also should prompt research to elucidate the toxins' mechanisms of action and to determine whether the toxins are useful, as are other membrane-damaging toxins, as analytical tools to study the structure of biomembranes.

For simplicity, and in agreement with a long-used nomenclatural convention, we will refer to the cytolytic toxins produced by *V. vulnificus* and *V. damsela* as "vulnificolysin" and "damselysin," respectively. This chapter describes procedures for isolating milligram amounts of vulnificolysin and damselysin free of detectable contamination with medium constituents and other bacterial products, and summarizes the currently available information concerning the physicochemical and biological properties of the purified toxins.

Assay Methods

The ability of vulnificolysin and damselysin to lyse washed mouse erythrocytes suspended in isotonic phosphate-buffered saline (67 m*M* Na_2HPO_4 and 77 m*M* NaCl, pH 7) affords a rapid, sensitive, and reproducible means for quantitating their cytolytic activity. The procedure is essentially that used by Bernheimer and Schwartz[18] to assay staphylococcal alpha toxin, and is described in detail in another chapter in this volume

[8] P. A. Blake, R. E. Weaver, and D. G. Hollis, *Annu. Rev. Microbiol.* **34,** 341 (1980).
[9] L. G. Wickboldt and C. V. Sanders, *J. Am. Acad. Dermatol.* **9,** 243 (1983).
[10] D. L. Tison and M. T. Kelly, *Diagn. Microbiol. Infect. Dis.* **2,** 263 (1984).
[11] J. G. Morris, Jr. and R. E. Black, *N. Engl. J. Med.* **312,** 343 (1985).
[12] J. A. Coffey, Jr., R. L. Harris, M. L. Rutledge, M. W. Bradshaw, and T. W. Williams, Jr., *J. Infect. Dis.* **153,** 800 (1986).
[13] P. R. Brayton, R. B. Bode, R. R. Colwell, M. T. MacDonell, H. L. Hall, D. J. Grimes, P. A. West, and T. N. Bryant, *J. Clin. Microbiol.* **23,** 104 (1986).
[14] A. Kreger and D. Lockwood, *Infect. Immun.* **33,** 583 (1981).
[15] A. S. Kreger, *Infect. Immun.* **44,** 326 (1984).
[16] L. D. Gray and A. S. Kreger, *Infect. Immun.* **48,** 62 (1985).
[17] M. H. Kothary and A. S. Kreger, *Infect. Immun.* **49,** 25 (1985).
[18] A. W. Bernheimer and L. L. Schwartz, *J. Gen. Microbiol.* **30,** 455 (1963).

[see Bernheimer]. One hemolytic unit (HU) is defined as the amount of toxin which causes the release of 50% of the hemoglobin in a standardized (ca. 0.7%, v/v) washed erythrocyte suspension. The stability of purified cytolysin preparations is enhanced by the addition of crystalline bovine albumin (Miles Inc., Kankakee, IL) to the phosphate-buffered saline used to dilute the cytolysin preparations.

The protein content of the toxin preparations is estimated by the method of Bradford,[19] with bovine gamma globulin as the standard. The standard and the assay reagent are obtained from Bio Rad Laboratories, Inc., Richmond, California.

Bacteria and Preparation of Seed Cultures

Vibrio vulnificus strain E4125 and *V. damsela* strain 1421-81 are used to produce the cytolytic toxins. Frozen specimens of the bacteria are prepared by growing the two strains, with shaking at 30°, to the midexponential phase of growth in heart infusion broth (Difco Laboratories, Detroit, MI) and freezing portions (1 ml) of the cultures with sterile fetal calf serum (0.5 ml) at −60°. Seed cultures are prepared by inoculating Columbia agar (BBL Microbiology Systems, Cockeysville, MD) contained in 100- by 15-mm Petri dishes for confluent growth with two or three full loops of rapidly thawed specimens of the bacteria. After incubation at 30° for 16 to 18 hr for the *V. vulnificus* culture and for 24 hr for the *V. damsela* culture, the surface growth on the Columbia agar is collected with sterile 0.85% NaCl (5 ml/plate), the optical densities of the suspensions are determined at 650 nm in 1-cm light path cuvettes, and the seed culture suspensions are used to inoculate the heart infusion diffusate broth used for toxin production.

Preparation of Heart Infusion Diffusate Broth[15]

To facilitate purification of the toxins, the bacteria utilized for their production are grown in heart infusion diffusate broth, a medium whose constituents are dialyzable. The dialysis tubing (ca. 3 in. wide by 17 in. long) and deionized water used to prepare the heart infusion diffusate broth are sterilized by ethylene oxide treatment and autoclaving, respectively. Approximately 1440 ml of 10× (250 g/liter) heart infusion broth (Difco) is prepared by dissolving the powdered medium in boiling deionized water. The solution is divided equally among 16 dialysis bags made of regenerated cellulose tubing (American Scientific Products, Charlotte,

[19] M. M. Bradford, *Anal. Biochem.* **72,** 248 (1976).

NC) and is dialyzed at 4° against 7.2 liters of deionized water in a model C Pope multiple dialyzer (A. H. Thomas Co., Philadelphia, PA). The bottom baffle of the dialyzer is removed to facilitate movement of the dialysis bags. After 3 days, the contents of the bags are discarded, and the diffusate is autoclaved in 2-liter Erlenmeyer flasks (usually 26 flasks containing 200 to 210 ml/flask). The medium has a Na^+ concentration of approximately 210 m*M* (0.5%, w/v) and an $A_{280}:A_{260}$ ratio of approximately 0.7.

Purification of Vulnificolysin[16]

Unless otherwise noted, all steps are done at approximately 4°.

Step 1: Culture Supernatant Fluids. Twelve 2-liter Erlenmeyer flasks containing 200 ml of heart infusion diffusate broth per flask are inoculated with 0.3 to 0.4 ml of seed culture suspension containing 10 optical density units at 650 nm and approximately 10^{10} colony-forming units of bacteria. The culture supernatant fluids are harvested by centrifugation (16,000 *g*, 20 min) after incubation of the cultures for 4 to 6 hr at 37° or for 6 to 8 hours at 30° (midexponential growth phase) in a gyratory shaker incubator (model G-25, New Brunswick Scientific Co., Edison, NJ) operating at 220 cpm. The culture supernatant fluids always should be harvested before the stationary growth phase is reached because extensive loss of extracellular cytolytic activity occurs during that stage of the growth cycle.

Step 2: Ammonium Sulfate Precipitation. Ammonium sulfate (enzyme grade; Schwarz/Mann, Cambridge, MA) is dissolved slowly with gentle stirring in the pooled culture supernatant fluids to a final concentration of approximately 50% saturation (350 g/liter). After 16 to 18 hr, the precipitate is recovered by centrifugation (16,000 *g*, 20 min) and dissolved in 60 ml of deionized water. The solution is centrifuged (20,000 *g*, 20 min) to remove a small amount of insoluble material, and ammonium sulfate is dissolved in the supernatant fluids to a final concentration of approximately 40% saturation (280 g/liter). After 16 to 18 hr, the precipitate is recovered by centrifugation (20,000 *g*, 20 min) and is dissolved in 10 ml of 10 m*M* glycine–NaOH buffer (pH 9.8) supplemented with 20 m*M* NaCl. A small amount of insoluble material is removed from the solution by centrifugation.

Step 3: Gel Filtration with Sephadex G-75. The step 2 preparation is applied to a column (2.6 × 96 cm) of Sephadex G-75 (Pharmacia Fine Chemicals, Piscataway, NJ) equilibrated with the same glycine buffer used in step 2. The column is washed at a flow rate of approximately 20 ml/hr and fractions (5 ml) are assayed for absorbance at 280 nm and for cytolytic activity. Three peaks of absorbance are observed, and the eight or nine fractions containing the bulk of the cytolytic activity in peak 2 (K_{av} = 0.29) are pooled.

Step 4: Hydrophobic Interaction Chromatography with Phenyl-Sepharose CL-4B. The step 3 pool is applied to a column (1.6 × 30 cm; 60-ml bed volume) of phenyl-Sepharose CL-4B (Pharmacia) equilibrated with the same glycine buffer used in step 2, and the column is washed (40 ml/hr) with 25% ethylene glycol in 10 m*M* glycine–NaOH buffer (pH 9.8) until the effluent has an absorbance at 280 nm of 0 (usually three to four bed volumes are required). The bulk of the toxin is eluted as a sharp peak by washing the column (8 ml/hr; 2-ml fractions collected) with 50% ethylene glycol in 2 m*M* glycine–NaOH buffer (pH 9.8). The 10 most active fractions, which usually elute between 60 and 80 ml of effluent, are pooled.

Step 5: Isoelectric Focusing in a Liquid Density Gradient. The step 4 pool is fractionated by high-speed electrofocusing in an ethylene glycol, pH 6 to 8 gradient formed at 15 W for 18 hr with an LKB 8100-1 electrolysis column (LKB Instruments, Inc., Gaithersburg, MD). Five fractions (20 ml, total volume) focusing between pH 7.0 to 7.4 contain the peak of cytolytic activity. They are pooled and stored at −60°.

Typical quantitative results obtained with the five-step purification scheme described above are summarized in Table I. The recovery and specific activity of vulnificolysin in six different preparations ranged from 20 to 25% and from 70,000 to 90,000 HU/mg protein, respectively. If desired, twice the amount of starting culture supernatant fluids (step 1) can be processed by the above procedure, except the preparation obtained by ammonium sulfate precipitation (step 2) is dissolved in 20 ml of glycine buffer, and two separate gel filtration steps (step 3) are performed with 10- to 12-ml portions of the step 2 preparation.

TABLE I
PURIFICATION OF VULNIFICOLYSIN

Step	Volume (ml)	Total protein (mg)	Total activity (HU)	Recovery (%)	Specific activity (HU/mg protein)
1. Culture supernatant fluids	2275	421	25.0 × 10^5	100	5.9 × 10^3
2. Ammonium sulfate precipitation	12.2	122	17.1 × 10^5	68	14.0 × 10^3
3. Sephadex G-75 gel filtration	47	16.2	10.3 × 10^5	41	63.6 × 10^3
4. Hydrophobic interaction chromatography	19.6	10.8	8.6 × 10^5	34	79.6 × 10^3
Isoelectric focusing	20	6.6	5.4 × 10^5	22	81.8 × 10^3

Properties of Vulnificolysin[14–16,20,21]

Purity. Vulnificolysin obtained by the procedure described above is homogeneous by crossed immunoelectrophoresis and analytical thin-layer isoelectric focusing in polyacrylamide gel, and does not contain detectable amounts of protease activity against azocasein (<0.25 protease U/ml) or lipopolysaccharide endotoxin activity against *Limulus* amoebocyte lysate (<0.1 ng/ml). The electrophoretic patterns obtained by sodium dodecyl sulfate–polyacrylamide gradient gel electrophoresis (SDS–PAGE; 4 to 30% gradient) are influenced by the method used to denature the toxin preparations before electrophoresis. Toxin preparations that are denatured but not reduced exhibit a major band (ca. 99% of the Coomassie blue-staining material) with a molecular weight of approximately 56,000 and a minor band consisting of an aggregate with a molecular weight of >200,000. However, preparations that are denatured and reduced do not show the minor, high-molecular-weight component seen in nonreduced preparations, but exhibit a major band (88% of the Coomassie blue-staining material) with a molecular weight of approximately 56,000 and two minor bands with molecular weights of approximately 48,000 and 16,500 (8 and 4% of the Coomassie blue-staining material, respectively). We believe the M_r 48,000 and 16,500 bands are fragments of M_r 56,000-molecular-weight toxin molecules that are proteolytically nicked and held together by intrachain disulfide bonds. The nicked molecules copurify with the intact toxin molecules and comigrate with the intact molecules if both are denatured but not reduced before SDS–PAGE, however, the nicked molecules separate into M_r 48,000 and 16,500 fragments if they are reduced before SDS–PAGE. Proteolytically nicked *Penicillium notatum* phospholipase B,[22] staphylococcal enterotoxin C_3,[23] and *Pseudomonas aeruginosa* exotoxin A[24] also migrate as single bands if they are not reduced but as multiple bands if they are reduced before electrophoresis.

Stability and Inactivation. Highly purified (step 5) vulnificolysin can be stored for at least 3 years at −60° without detectable loss of activity. The toxin is heat labile (56°, 30 min), is partially inactivated by trypan blue (25 to 50 μg/ml) and various commercially available proteases (pronase, trypsin, chymotrypsin, subtilisin BPN, subtilopeptidase A, papain, thermolysin, and proteinase K; 100 μg/ml), and is almost totally inactivated by a concentration of cholesterol (100 μg/ml) that is ca. 100-fold more

[20] L. D. Gray and A. S. Kreger, *Infect. Immun.* **51,** 964 (1986).

[21] L. D. Gray and A. S. Kreger, *J. Infect. Dis.* **155,** 236 (1987).

[22] T. Okumura, S. Kimura, and K. Saito, *Biochim. Biophys. Acta* **617,** 264 (1980).

[23] R. F. Reiser, R. N. Robbins, A. L. Noleto, G. P. Khoe, and M. S. Bergdoll, *Infect. Immun.* **45,** 625 (1984).

[24] Y. Sanai, K. Morihara, H. Tsuzuki, J. Y. Homma, and I. Kato, *FEBS Lett.* **120,** 13 (1980).

than the concentration commonly used to inactivate thiol-activated (oxygen-labile) cytolytic toxins. In addition, activity is lost at pH 4 to 10 (4°, 24 hr) unless crystalline bovine albumin is added (1 mg/ml) to the buffers. The limited sensitivity of vulnificolysin to various proteases suggests that the toxin has a tertiary structure in which the active site(s) is only partially exposed.

The toxin is not affected by various phospholipids (cardiolipin, sphingomyelin, phosphatidylcholine, phosphatidylserine, phosphatidylinositol, and phosphatidylethanolamine; 100 μg/ml), mixed gangliosides (100 μg/ml), glycophorin (100 μg/ml), dithiothreitol (5 mM), chelating agents (EDTA and EGTA, 1 mM), and divalent cations (Ca^{2+}, Mg^{2+}, and Zn^{2+}; 1 mM).

According to Shinoda *et al.*,[25] hemolytic activity in a partially purified preparation of vulnificolysin is heat labile and is inhibited by cholesterol (0.1 μg/ml), divalent cations (25 mM), and dithiothreitol, but is not affected by oxygen, sulfhydryl-blocking agents, and antiserum against streptolysin O.

Hydrophobicity. The ability of vulnificolysin to bind to phenyl-Sepharose CL-4B in the presence of 10 mM glycine–NaOH buffer (pH 9.8) containing 25% ethylene glycol indicates that the toxin is capable of strong interaction with hydrophobic surfaces.

Molecular Weight. The molecular weight of the step 5 toxin is approximately 56,000 determined by SDS–PAGE and approximately 15,000 determined by gel filtration with Sephadex G-75. The most likely explanation for this discrepancy is that the highly purified, hydrophobic toxin interacts with the dextran-based gel and is retarded in its passage through the gel column. Therefore, the M_r 56,000 value obtained by SDS–PAGE is more accurate than the lower value obtained by gel filtration. Similar observations have been made during the estimation of the molecular weight of *Clostridium perfringens* delta toxin.[26]

Amino Acid Composition and Isoelectric Point. The amino acid composition of vulnificolysin is shown in Table II. Basic, acidic, and nonpolar hydrophobic amino acid residues account for approximately 13, 22, and 36% of the total residues, respectively. However, many of the aspartic acid and glutamic acid residues must occur as asparagine and glutamine because the isoelectric point of the toxin is approximately 7.1 as determined by analytical thin-layer isoelectric focusing. The presence of eight half-cystine residues suggests that the toxin could have four disulfide bonds. The first 10 amino-terminal amino acid residues are Gln-Glu-Tyr-

[25] S. Shinoda, S. Miyoshi, H. Yamanaka, and N. Miyoshi-Nakahara, *Microbiol. Immunol.* **29,** 583 (1985).

[26] J. E. Alouf and C. Jolivet-Reynaud, *Infect. Immun.* **31,** 536 (1981).

TABLE II
AMINO ACID COMPOSITIONS OF VULNIFICOLYSIN AND DAMSELYSIN

Amino acid	Vulnificolysin[a] mol%	Vulnificolysin[a] Number of residues[c]	Damselysin[b] mol%	Damselysin[b] Number of residues[d]
Lysine	5.34	26	8.40	51
Histidine	1.78	9	1.90	12
Arginine	5.68	29	3.48	22
Aspartic acid	12.93	63	16.29	98
Threonine	5.92	28	5.07	30
Serine	9.46	44	10.54	60
Glutamic acid	8.92	44	9.36	57
Proline	3.64	17	3.83	22
Glycine	7.76	33	5.27	28
Alanine	5.90	26	6.64	37
Half-cystine	1.67	8	0.64	4
Valine	7.36	35	3.79	22
Methionine	0.42	2	1.12	7
Isoleucine	3.43	17	5.93	35
Leucine	5.90	28	6.69	40
Tyrosine	4.54	23	3.99	25
Phenylalanine	3.72	19	4.70	29
Tryptophan	5.63	29	2.34	15
Total residues:		480		594

[a] From L. D. Gray and A. S. Kreger, *Infect. Immun.* **48,** 62 (1985).
[b] From M. H. Kothary and A. S. Kreger, *Infect. Immun.* **49,** 25 (1985).
[c] Based on a molecular weight of 56,000.
[d] Based on a molecular weight of 69,000.

Val-Pro-Ile-Val-Glu-Lys-Pro, and only a single amino acid sequence is detected in the step 5 toxin preparation.

Antigenicity. Rabbits and mice vaccinated with vulnificolysin produce antibody that can neutralize the toxin, and anti-vulnificolysin antibody has been detected in the sera of mice and a human convalescing from *V. vulnificus* disease. The results of antiserum neutralization studies indicate that vulnificolysin is antigenically distinct from damselysin, the thermostable direct hemolysin of *V. parahaemolyticus,* and the cytolysin produced by the El Tor biotype of *V. cholerae.*

Cytolytic Activity. Vulnificolysin is a broad-spectrum cytolysin that i active against erythrocytes from at least 18 animal species (Table III). T

TABLE III
SENSITIVITY OF ERYTHROCYTES FROM VARIOUS ANIMAL SPECIES TO VULNIFICOLYSIN AND DAMSELYSIN

	Vulnificolysin	Damselysin
Species	Relative sensitivity (%)[a]	Relative sensitivity (%)[b]
Mouse	100	100
Rat	76 to 88	68 to 71
Damselfish	Not determined	7
Rabbit	37 to 41	1 to 6
Guinea pig	41 to 44	<0.01
Hamster	67	<0.01
Gerbil	86	<0.01
Sheep	83 to 300	<0.01
Goat	33	<0.01
Pig	200	<0.01
Horse	69	<0.01
Cow	54	<0.01
Burro	167	<0.01
Cat	167	<0.01
Dog	92	<0.01
Pigeon	83 to 100	<0.01
Chicken	62	<0.01
Cynomolgus monkey	153 to 193	<0.01
Human	33 to 50	<0.01

[a] Compared with sensitivity of mouse erythrocytes (ca. 8×10^4 HU/mg protein). From L. D. Gray and A. S. Kreger, *Infect. Immun.* **48,** 62 (1985), except for gerbil erythrocytes (unpublished observation).

[b] Compared with sensitivity of mouse erythrocytes (ca. 2×10^6 HU/mg protein). From M. H. Kothary and A. S. Kreger, *Infect. Immun.* **49,** 25 (1985), except for gerbil erythrocytes (unpublished observation).

toxin also is cytolytic for Chinese hamster ovary cells (CHO cells) in tissue culture; the minimal toxic dose for CHO cells is approximately 0.1 HU (ca. 1 ng). Data from a multihit survival curve constructed as described by Inoue *et al.*[27] indicate that hemolysis of mouse erythrocytes requires the interaction of two or three molecules of vulnificolysin with

[27] K. Inoue, Y. Akiyama, T. Kinoshita, Y. Higashi, and T. Amano, *Infect. Immun.* **13,** 337 (1976).

the erythrocyte membrane. The rate of mouse erythrocyte lysis by the toxin is temperature dependent and is optimal between 30 and 37°. Hemolysis is not observed at 4° and is at least a two-step process consisting of a temperature-independent, toxin-binding step, followed by a temperature-dependent, membrane-perturbation step(s) that leads to cell lysis. The energy of activation (E_a) for the toxin, estimated from the Arrhenius equation and the slopes of two distinct regions of an Arrhenius plot, is substantially higher at 15 to 20° (28,800 cal/mol) than at 25 to 37° (9580 cal/mol). Similar results have been observed with tetanolysin and streptolysin O.[28] It is possible that the change in the E_a value at a critical temperature is due to a conformational change in the toxin or to an alteration in the fluidity of the erythrocyte membrane.

In Vivo Activities. The LD_{50} of the step 5 toxin preparation for mice is approximately 3 μg/kg (ca. 240 HU/kg) by the intravenous route and 2.2 mg/kg by the intraperitoneal route. Vulnificolysin also produces dermonecrosis and increases vascular permeability in guinea pig, mouse, and rabbit skin. One blueing dose in guinea pig skin is 0.3 to 0.6 μg (25 to 50 HU). Light and electron microscopy of mouse skin damage caused by a single intradermal injection of the toxin reveals acute cellulitis characterized by extensive extracellular edema, disorganization of collagen bundles, large accumulations of cell debris and plasma proteins, damaged or necrotic fat cells, capillary endothelial cells, and muscle cells, and mild inflammatory cell infiltration. These changes are very similar to those produced by *V. vulnificus* skin infections in mice.

Enzyme Activity. The step 1 and step 2 toxin preparations have phospholipase A_2 and lysophospholipase activities; however, both activities are removed by gel filtration with Sephadex G-75[29] (step 3 in the purification scheme). The final purified toxin preparation (step 5) does not have detectable phospholipase A, phospholipase C, phospholipase D, or lysophospholipase activity.

Cloning of Vulnificolysin Gene. The gene(s) coding for production of vulnificolysin has been cloned in *Escherichia coli* by using the lytic cloning vector λ1059; a 3.2-kb DNA fragment containing the toxin gene was isolated by subcloning in plasmid pBR325.[30] The fragment was used as a DNA probe to detect homologous gene sequences in 54 *V. vulnificus* isolates. Homologous sequences were not detected in 96 isolates from 29 other bacterial species.

[28] A. W. Bernheimer, *J. Gen. Physiol.* **30,** 337 (1947).

[29] J. Testa, L. W. Daniel, and A. S. Kreger, *Infect. Immun.* **45,** 458 (1984).

[30] A. C. Wright, J. G. Morris, Jr., D. R. Maneval, Jr., K. Richardson, and J. B. Kaper, *Infect. Immun.* **50,** 922 (1985).

Purification of Damselysin[17]

Unless otherwise noted, all steps are done at approximately 4°.

Step 1: Culture Supernatant Fluids. Six 2-liter Erlenmeyer flasks containing 200 ml of heart infusion diffusate broth per flask are inoculated with 1.0 to 1.4 ml of seed culture suspension containing 10 optical density units at 650 nm and approximately 10^{10} colony-forming units of bacteria. The culture supernatant fluids are harvested by centrifugation (16,000 *g*, 20 min) after incubation of the cultures for 14 to 15 hr at 30° (early stationary growth phase) in a gyratory shaker incubator (model G-25, New Brunswick Scientific Co., Edison, NJ) operating at 220 cpm.

Step 2: Ammonium Sulfate Precipitation. Ammonium sulfate (enzyme grade; Schwarz/Mann, Cambridge, MA) is dissolved slowly with gentle stirring in the pooled culture supernatant fluids to a final concentration of approximately 60% saturation (420 g/liter). After 16 to 18 hr, the precipitate is recovered by centrifugation (16,000 *g*, 20 min) and dissolved in 8 ml of 100 m*M* ammonium bicarbonate (pH 7.8). A small amount of insoluble material is removed by centrifugation (20,000 *g*, 20 min).

Step 3: Gel Filtration with Sephadex G-100. The step 2 preparation is applied to a column (2.6 × 95 cm) of Sephadex G-100 (Pharmacia) equilibrated with 100 m*M* ammonium bicarbonate. The column is washed at a flow rate of approximately 20 ml/hr and fractions (5 ml) are assayed for absorbance at 280 nm and for cytolytic activity. Three peaks of absorbance are observed, and the 12 to 14 fractions containing the bulk of the cytolytic activity in peak 1 ($K_{av} = 0.20$) are pooled.

Step 4: Hydrophobic Interaction Chromatography with Phenyl-Sepharose CL-4B. Ammonium sulfate is dissolved in the step 3 pool to a final concentration of 500 m*M*, and the preparation is applied to a column (1.6 × 30 cm) of phenyl-Sepharose CL-4B (Pharmacia) equilibrated with phosphate-buffered saline (67 m*M* Na_2HPO_4 and 77 m*M* NaCl, pH 7) supplemented with 500 m*M* ammonium sulfate. Some lots of gel recently have been found to require 600 m*M* ammonium sulfate in order to bind the toxin. The column is washed (50 ml/hr) with approximately three bed volumes (180 ml) of equilibrating buffer, and the bulk of the toxin is eluted as a sharp peak by washing the column (16 to 20 ml/hr; 4- to 5-ml fractions collected) with phosphate-buffered saline (pH 7). The four most active fractions, which usually elute between 60 and 80 ml of effluent, are pooled and stored at −60°.

Typical quantitative results obtained with the four-step purification scheme described above are summarized in Table IV. The recovery and specific activity of damselysin in our three most recent preparations ranged from 42 to 54% and from 1.7×10^6 to 2.0×10^6 HU/mg protein, respectively.

TABLE IV
PURIFICATION OF DAMSELYSIN

Step	Volume (ml)	Total protein (mg)	Total activity (HU)	Recovery (%)	Specific activity (HU/mg protein)
1. Culture supernatant fluids	1120	196	112×10^6	100	571×10^3
2. Ammonium sulfate precipitation	11.6	152.3	109×10^6	97	716×10^3
3. Sephadex G-100 gel filtration	62	57	74×10^6	66	1298×10^3
4. Hydrophobic interaction chromatography	18	30.1	61×10^6	54	2027×10^3

Properties of Damselysin[15,17,31,32]

Purity. Damselysin obtained by the procedure described above is homogeneous by SDS–PAGE and crossed immunoelectrophoresis, does not contain detectable amounts of protease activity against azocasein (<0.25 protease U/ml) or lipopolysaccharide endotoxin activity against *Limulus* amoebocyte lysate (<0.1 ng/ml), and is microheterogeneous by analytical thin-layer isoelectric focusing in polyacrylamide gel. The microheterogeneous nature of purified damselysin is not unique. Various other bacterial cytolysins are also known to exhibit microheterogeneity.[33]

Stability and Inactivation. Highly purified (step 4) concentrated preparations of damselysin (1 mg/ml) can be stored for at least 3 years at −60° without detectable loss of activity. The toxin is very heat labile (37°, 30 min) but crystalline bovine albumin (1 mg/ml) prevents inactivation at 37°. The toxin also is unstable at pH 4, 6, and 10 (4°, 24 hr) and is inactivated by pronase and trypsin. Activity is not affected by phospholipids (cardiolipin, sphingomyelin, phosphatidylcholine, phosphatidylserine, phosphatidylinositol, and phosphatidylethanolamine; 100 μg/ml), mixed gangliosides (100 μg/ml), glycophorin (100 μg/ml), cholesterol (100 μg/ml), dithiothreitol (5 m*M*), trypan blue (50 μg/ml), chelating agents (EDTA and EGTA, 1 m*M*), and divalent cations (Ca^{2+}, Mg^{2+}, and Zn^{2+}; 1 m*M*).

[31] A. S. Kreger, A. W. Bernheimer, L. A. Etkin, and L. W. Daniel, *Infect. Immun.* **55,** 3209 (1987).

[32] R. W. Wilcox, R. L. Wykle, J. D. Schmitt, and L. W. Daniel, *Lipids* **22,** 800 (1987).

[33] A. W. Bernheimer, *in* "Microbial Toxins" (S. J. Ajl, S. Kadis, and T. C. Montie, eds.), Vol. I, p. 183. Academic Press, New York, 1970.

Molecular Weight and Isoelectric Point. The molecular weight of damselysin is approximately 69,000 determined by SDS–PAGE and 52,000 determined by gel filtration with Sephadex G-100. The major cytolytic toxin in the stage 4 preparation has an isoelectric point (p*I*) of approximately 5.60, and the minor components have p*I* values of 5.45 and 5.55.

Amino Acid Composition. The amino acid composition of damselysin is shown in Table II. Basic, acidic, and nonpolar hydrophobic amino acid residues account for approximately 14, 26, and 35% of the total residues, respectively. The presence of four half-cystine residues suggests that the toxin could have two disulfide bonds. The first 10 amino-terminal amino acid residues are Phe-Thr-Gln-Trp-Gly-Gly-Ser-Gly-Leu-Thr. The amino-terminal amino acid of the thermostable direct hemolysin of *V. parahaemolyticus* also is phenylalanine [see Chapter 26, this volume].

Antigenicity. Rabbits and mice vaccinated with damselysin produce antibody that can neutralize the toxin. The toxin is antigenically distinct from vulnificolysin, the thermostable direct hemolysin of *V. parahaemolyticus,* and the cytolysin produced by the El Tor biotype of *V. cholerae.*

Cytolytic Activity. Damselysin is a narrow spectrum cytolysin that is very active against erythrocytes from 4 of the 19 animal species examined so far (Table III), and it is most active against mouse erythrocytes. The specific activity of the step 4 toxin (ca. 2×10^6 HU/mg protein; vs mouse erythrocytes) is similar to the values reported for the most potent bacterial cytolytic toxins, streptolysin S[34,35] and thiol-activated cytolysins such as cereolysin.[36] The purified toxin also is cytolytic for Chinese hamster ovary cells in tissue culture [minimal toxic dose ca. 2.5 HU (ca. 1 ng)]. Data from a multihit survival curve constructed as described by Inoue *et al.*[27] indicate that hemolysis of mouse erythrocytes requires the interaction of two molecules of damselysin with the erythrocyte membrane. The rate of erythrocyte lysis by the toxin is temperature and pH dependent and is optimal at 37 to 47° and at pH 7 to 9. Hemolysis is not observed at 4° and is at least a two-step process consisting of a temperature-independent, toxin-binding step, followed by a temperature-dependent, membrane-perturbation step(s) that leads to cell lysis. The energy of activation for the toxin, estimated from the Arrhenius equation and the slope of an Arrhenius plot, is approximately 11,370 cal/mol.

Lethal Activity. The LD_{50} of the step 4 toxin preparation for mice is

[34] C. Y. Lai, M. T. Wang, J. B. de Faria, and T. Akao, *Arch. Biochem. Biophys.* **191,** 804 (1978).

[35] C. Loridan and J. E. Alouf, *J. Gen. Microbiol.* **132,** 307 (1986).

[36] J. L. Cowell, P. S. Grushoff-Kosyk, and A. W. Bernheimer, *Infect. Immun.* **14,** 144 (1976).

approximately 1 μg/kg (ca. 2000 HU/kg) by the intraperitoneal route, 2 μg/kg by the intravenous route, and 18 μg/kg by the subcutaneous route. Mice injected subcutaneously with 1 LD_{50} of toxin are lethargic and have ruffled fur, encrustations around their eyelids, and severe local edema similar to that observed[15] in mice infected with *V. damsela*.

Phospholipase Activity. Damselysin is a phospholipase D which so far is known to be active against phosphatidylcholine and phosphatidylethanolamine [see Daniel *et al.* in this volume], sphingomyelin, L-ET-18-OCH_3 (an ether-linked phospholipid that exhibits selective cytotoxicity toward several types of tumor cells and is relatively inactive toward normal cells), platelet-activating factor, and lyso-platelet-activating factor. Thus, damselysin appears to be unique in that, to our knowledge, phospholipases D from other sources have not been reported to possess hemolytic activity.

Treatment of sheep erythrocytes with minute amounts of phospholipase D (sphingomyelinase D) produced by *Corynebacterium pseudotuberculosis* and by the brown recluse spider (*Loxosceles reclusa*) is known to protect the cells from lysis by staphylococcal sphingomyelinase C (β-lysin) and helianthin, a nonenzymatic but sphingomyelin-binding cytolysin produced by the sea anemone *Stoichactis helianthus*.[37] Sheep erythrocytes also can be protected from lysis by these two sphingomyelin-dependent cytolysins by incubating the cells with subnanogram, sublytic amounts of damselysin.

[37] A. W. Bernheimer, B. J. Campbell, and L. J. Forrester, *Science* **228,** 590 (1985), and references contained therein.

[26] Thermostable Direct Hemolysin of *Vibrio parahaemolyticus*

By YOSHIFUMI TAKEDA

Thermostable direct hemolysin is produced only by Kanagawa phenomenon-positive strains of *Vibrio parahaemolyticus*.[1] Since a close correlation between Kanagawa phenomenon-positive strains and human enteropathogenicity is well recognized epidemiologically,[2] thermostable

[1] J. Sakurai, A. Matsuzaki, Y. Takeda, and T. Miwatani, *Infect. Immun.* **9,** 777 (1974).
[2] Y. Miyamoto, T. Kato, Y. Obara, S. Akiyama, K. Takizawa, and S. Yamai, *J. Bacteriol.* **100,** 1147 (1969).

direct hemolysin is thought to play an important role in *V. parahaemolyticus* infection.[3]

Assay Method

The standard reaction mixture (2.5 ml) for assay of hemolytic activity contains 10 m*M* tris(hydroxymethyl)aminomethane–HCl buffer (pH 7.2), 5 m*M* $CaCl_2$, 1.25 ml of a 1% suspension of either human or rabbit erythrocytes (about 8.3×10^7 cells/ml), 0.9% NaCl, and the hemolysin. The reaction mixture is incubated at 37° for 30 min and then centrifuged at 3000 rpm (4°) for 5 min, and the hemolytic activity is determined by measuring the absorbance of the resulting supernatant fluid at 540 nm. The hemolytic activity determined in this way is proportional to the amount of hemolysin added up to the amount causing complete hemolysis.

Purification Procedure

A Kanagawa phenomenon-positive strain of *V. parahaemolyticus* is cultured in medium containing 30 g of sodium chloride, 10 g of peptone (Difco), 5 g of dibasic sodium phosphate, and 5 g of glucose per liter in distilled water (pH 7.6–7.8). Then 1 ml of culture is inoculated into 2 liters of the same medium in 6.5-liter Erlenmeyer flasks and incubated at 37° for 15 hr with shaking. The culture supernatant is collected by centrifugation. Further operations are carried out at 0–4°.

Step 1: Preparation of Crude Hemolysin from the Culture Supernatant

Solid ammonium sulfate (35.1 g/100 ml) is added to the culture supernatant, and the resulting precipitate is dissolved in a small amount of 0.01 *M* phosphate buffer (Na_2HPO_4–KH_2PO_4, pH 7.0) and dialyzed overnight against the same buffer.

Alternatively, 1 g of alumina C_γ is added per liter of culture supernatant. The alumina is collected by centrifugation, washed once with 0.1 *M* phosphate buffer (pH 7.0), and the adsorbed hemolysin is eluted by washing the alumina C_γ four times with 10-ml vols of 0.5 *M* potassium phosphate buffer (pH 7.0). The hemolysin is precipitated by adding 40 g of solid ammonium sulfate/100 ml of buffer solution. The resulting precipitate is dissolved in a small volume of 0.01 *M* phosphate buffer (pH 7.0) and dialyzed overnight against the same buffer.

[3] Y. Takeda, *in* "Pharmacology of Bacterial Toxins" (F. Dorner and J. Drews, eds.), pp. 183–205. Pergamon, Oxford, England, 1986.

Step 2: DEAE-Cellulose Column Chromatography

Crude hemolysin prepared from 11,5000 ml of culture supernatant is applied to a DEAE-cellulose column (2.2 × 50 cm) equilibrated with 0.01 *M* phosphate buffer (pH 7.0). Material is eluted with 1000 ml of the same buffer containing 0.2 *M* NaCl and then with 1000 ml of a linear gradient of 0.2 to 1.0 *M* NaCl in the same buffer. Fractions with heat-stable hemolytic activity (crude hemolysin contains both heat-labile and heat-stable hemolysins,[1] and so the heat stability of fractions with hemolytic activity should be checked by heating samples at 100° for 10 min) are collected and concentrated by adding solid ammonium sulfate (40.0 g/100 ml). The resulting precipitate is dissolved in a small volume of 0.01 *M* phosphate buffer (pH 7.0) and dialyzed against the same buffer.

Step 3: Hydroxylapatite Column Chromatography

The hemolysin is concentrated to a small volume and applied to a hydroxylapatite column (2.2 × 35 cm) equilibrated with 0.01 *M* phosphate buffer (pH 7.0). Material is eluted with about 500 ml of 0.1 *M* phosphate buffer (pH 7.0) and then with 800 ml of a linear gradient of 0.1 to 0.3 *M* phosphate buffer (pH 7.0). Fractions containing hemolytic activity are collected and concentrated to a small volume on an Amicon PM10 membrane.

Step 4: Sephadex G-200 Gel Filtration

The preparation from the hydroxylapatite column is applied to a Sephadex G-200 column (2.2 × 50 cm) equilibrated with 0.01 *M* phosphate buffer (pH 7.0). Material is eluted with the same buffer, and fractions with hemolytic activity are collected and concentrated to a small volume on an Amicon PM10 membrane.

These procedures yield a preparation that gives a single band staining for protein on polyacrylamide disc gel electrophoresis. Typical data obtained at each purification step are summarized in Table I.[4]

Properties

The thermostable direct hemolysin is composed of two subunit molecules of approximately 21,000 Da. Its p*I* value is about 4.2. The hemolytic activity of the thermostable direct hemolysin is destroyed by pepsin or chymotrypsin, but not by trypsin. The gene encoding the thermostable

[4] T. Honda, K. Goshima, Y. Takeda, Y. Sugino, and T. Miwatani, *Infect. Immun.* **13,** 133 (1976).

TABLE I
PURIFICATION OF THERMOSTABLE DIRECT HEMOLYSIN[4]

Fraction	Total volume (ml)	Total protein (mg)[a]	Total activity (HU)[b]	Specific activity (HU/mg)	Relative activity	Yield (%)
Culture filtrate	46,000	25,760	223,744	8.7	(1.0)	100
Ammonium sulfate fraction	400	14,036	201,142	14.3	1.64	89.9
DEAE-cellulose column eluate	10.8	59.4	68,314	1,150.1	132.2	30.6
Hydroxylapatite column eluate	1.2	11.6	46,966	4,048.8	465.4	21.1
Sephadex G-200 column eluate	1.76	7.4	34,520	4,664.9	536.2	15.5

[a] Protein was determined by the method of Lowry *et al.*[5]
[b] One hemolytic unit (HU) is defined as the amount giving $A_{540} = 0.5$.

direct hemolysin has been cloned[6,7] and its sequence has been reported.[8] The amino acid sequence of the purified thermostable direct hemolysin has also been reported.[9]

The thermostable direct hemolysin shows various biological activities, such as hemolytic activity, cytotoxic activity, enteropathogenic activity, and lethal activity. High hemolytic activity is observed on erythrocytes of various species, the activity decreasing in the order: rat, dog, mouse, monkey, man, rabbit, and guinea pig erythrocytes. No activity is detectable on horse erythrocytes. The hemolytic activity is not inactivated by heating at 100° for 10 min.

The thermostable direct hemolysin has been shown to be cytotoxic to various cultured cells, such as HeLa cells, L cells, FL cells, neuroblastoma cells, primary cultures of fetal mouse heart cells, and CCL-6 cells derived from human intestine.

Injection of 500 μg of thermostable direct hemolysin causes turbid, bloody fluid accumulation in ligated rabbit ileal loops. Histopathological

[5] O. H. Lowry, N. J. Rosebrough, A. L. Farr, and R. J. Randall, *J. Biol. Chem.* **193,** 265 (1951).
[6] J. B. Kaper, R. K. Campen, R. J. Seidler, M. M. Baldini, and S. Falkow, *Infect. Immun.* **45,** 290 (1984).
[7] H. Taniguchi, H. Ohta, M. Ogawa, and Y. Mizuguchi, *J. Bacteriol.* **162,** 510 (1985).
[8] M. Nishibuchi and J. B. Kaper, *J. Bacteriol.* **162,** 558 (1985).
[9] S. Tsunasawa, A. Sugihara, T. Masaki, F. Sakiyama, Y. Takeda, T. Miwatani, and K. Narita, *J. Biochem. (Tokyo)*, **101,** 111 (1987).

examination shows erosive lesions and desquamation of necrotic mucosa accompanied by marked neutrophil infiltration into the intestinal wall.

Low doses of the thermostable direct hemolysin rapidly kill various experimental animals, such as mice, rats, and guinea pigs. For example, intravenous injection of 5 μg of the hemolysin killed mice within 1 min. The rapid death of the animals was shown to be due to the cardiotoxicity of the thermostable direct hemolysin.[4,10]

[10] K. Goshima, T. Honda, M. Hirata, K. Kikuchi, Y. Takeda, and T. Miwatani, *J. Mol. Cell. Cardiol.* **9,** 191 (1977).

[27] Aerolysin from *Aeromonas hydrophila*

By J. THOMAS BUCKLEY and S. PETER HOWARD

Aerolysin is a soluble cytolytic protein produced by the gram-negative bacteria *Aeromonas hydrophila*.[1,2] It is released to the extracellular medium as an inactive protoxin (proaerolysin) which is converted to the active protein by one or more extracellular proteases which are also exported by *A. hydrophila*.[3] The mechanism of action of the toxin is very similar to the mechanism proposed for staphylococcal alpha toxin.[4] The toxin binds to a specific glycoprotein receptor on the surface of eukaryotic cells, inserts into the lipid bilayer, and forms holes approximately 3 nm in diameter. This results in destruction of the membrane permeability barrier and cell death.[5]

Assay Method

Principle. By far the easiest methods of measuring the activity of aerolysin are based on the ability of the toxin to cause the hemolysis of red cells. Plate assays measure the clearing of red cell suspensions in wells and the absence of cell pellets at the bottom of wells. For convenience, human red cells obtained from outdated blood are normally used;

[1] F. H. Castelitz and R. Gunther, *Zentralbl. Bakteriol., Parasitenk. Infektionskr. Hyg. Abt. 1: Orig.* **180,** 30 (1960).
[2] B. Wretlind, R. Möllby, and T. Wadström, *Infect. Immun.* **4,** 503 (1971).
[3] S. P. Howard and J. T. Buckley, *J. Bacteriol.* **163,** 336 (1985).
[4] S. Harshman, *Mol. Cell. Biochem.* **23,** 143 (1979).
[5] S. P. Howard and J. T. Buckley, *Biochemistry* **21,** 1662 (1982).

however, sensitivity can be increased more than 100-fold by using rat erythrocytes.[6]

Reagents

PBS–albumin: 10 m*M* sodium phosphate, 150 m*M* NaCl, pH 7.4, containing 0.1% bovine serum albumin

Washed erythrocytes: Packed human erythrocytes from outdated human blood, washed 3× with PBS and resuspended to 0.8% (v/v) in PBS–albumin

Assay of the Active Toxin. Hemolytic activity determinations are carried out in microtiter plates after 2-fold serial dilutions of samples in 0.1 ml of PBS–albumin. An equal volume of the 0.8% erythrocytes in PBS–albumin is added to each well and the plates are incubated at 37° for 1 hr. Hemolytic activity is expressed as the inverse of the largest dilution at which complete lysis occurs.

Assay of the Inactive Protoxin. The assay procedure is essentially the same as the procedure for the mature protein. The protoxin is activated in the first well of the titer plate, which for this purpose contains PBS but no albumin. Activation is achieved by adding trypsin to a final concentration of 0.1–1 μg/ml and incubating at room temperature for 5 min. Serial dilutions into PBS–albumin are performed and the assay continued as above. The presence of trypsin inhibitor at a final concentration of 1 mg/ml in the PBS–albumin used for serial dilution has no affect on the results of the assay.

Enzyme Production and Isolation

Medium. Aeromonas hydrophila is grown in 2-liter flasks, each containing 200 ml of medium prepared by dissolving 8 g of yeast extract (Difco) and 4 g of casamino acids (Difco) in a total of 156 ml of water containing 0.2 ml of $3.3 \times 10^{-3}\%$ thiamin and 0.2 ml of 0.12% nicotinic acid stock and adjusting the pH to 7.2. Yeast RNA (20 ml of a 10% solution, pH 7.2) is also added to each flask followed by 2 ml of 0.1 *M* $MgSO_4$, 2 ml of 0.01 *M* $CaCl_2$, and 20 ml 10× M9 salts (60.34 g Na_2HPO_4, 10.19 g NH_4Cl, 30.4 g KH_2PO_4, and 5.19 g of NaCl dissolved in water to 1 liter). The hemolytic activity of extracellular culture supernatants is far higher when RNA is present in the culture medium.[7] The reason for this has not been established, but may be due to the ability of the RNA to

[6] A. W. Bernheimer, L. S. Avigad, and G. Avigad, *Infect. Immun.* **11,** 1312 (1975).

[7] A. W. Bernheimer and L. S. Avigad, *Infect. Immun.* **9,** 1016 (1974).

complex divalent cations and thereby inhibit one or more extracellular proteases.

Growth Conditions for the Isolation of Mature Active Toxin.[8] The flasks are inoculated with 1% of an overnight culture containing 2×10^9 to 5×10^9 viable cells per milliliter and shaken at 250 rpm in a New Brunswick gyrorotatory shaker. After 8 hr at 35°, the incubation temperature is reduced to 25° and shaking is continued for an additional 16 hr before the cell-free supernatant is recovered by centrifugation. At this stage the cells should be in stationary phase of growth.

Growth Conditions for the Isolation of Inactive Protoxin. The same medium is used as for mature toxin isolation except that the RNA is omitted because it has no effect on the yield of the protoxin. The flasks are inoculated as above and shaken at 35° until the optical density of the culture at 660 nm is between 8 and 10. This is the point at which the cells begin to leave log phase. It usually takes about 8 hr under these conditions. Timing of bacterial growth is critical. Very little protein is released before an OD_{660} of 8 is attained, and after the cells have passed an OD_{660} of 10, so much protease has been released that it is impossible to prevent conversion of the proaerolysin to the mature toxin during isolation.

Purification of the Mature Toxin. Each stage in the purification is monitored using the titer assay. Under the conditions described here, aerolysin is the only hemolytic protein which is detected. Ammonium sulfate is added to the culture supernatant to 60% of saturation at 0°. After 4 hr the precipitate is collected by centrifugation at 10,000 *g* for 30 min and extracted twice with 0.03 *M* sodium borate, pH 8.2 (10 ml of this buffer is used for every 1 liter of the original culture supernatant). Nearly all of the hemolytic activity is recovered in the supernatant of the second wash. This is desalted by passing it down a 2.5 × 25 cm column of Sephadex G-25 equilibrated in the 0.03 *M* sodium borate buffer, pH 8.2. It is essential that this and all subsequent steps be performed at 0° as concentrated solutions of the active toxin aggregate spontaneously and irreversibly upon warming.

The void volume fraction from the Sephadex G-25 column is applied to a 1.5 × 16 cm column of DEAE-Sepharose CL-6B also equilibrated in the 0.03 *M* borate buffer. The column is eluted at 15 ml/hr with a linear gradient obtained by mixing 250 ml of the 0.03 *M* borate with 0.3 *M* NaCl in same buffer. Four-milliliter fractions are collected. The hemolytic protein, which is eluted with 0.12 *M* NaCl, is precipitated by adding ammo-

[8] J. T. Buckley, L. N. Halasa, K. D. Lund, and S. MacIntyre, *Can. J. Biochem.* **59,** 430 (1981).

nium sulfate to 60% of saturation at 0°. The precipitate is dissolved in a small amount of 0.03 *M* borate, pH 8.2, and applied to a 1.5 × 60 cm Sephadex G-150 column. Active fractions, eluted from this column at 12 ml/hr in the same buffer, are combined and applied to a hydroxyapatite column equilibrated in 0.01 *M* phosphate, pH 7. This column is eluted at 15 ml/hr using a linear gradient of 250 ml 0.01 *M* phosphate and 250 ml of 0.25 *M* phosphate, pH 7. The hydroxyapatite is obtained from Bio-Rad and prepared according to the company's instructions. Pure aerolysin is recovered from the column in fractions which are approximately 0.05 *M* phosphate. The results of a representative purification are presented in Table I.

Purification of the Protoxin.[3] Ammonium sulfate is added to culture supernatants to 85% of saturation of 0° and the mixture is allowed to stand for at least 4 hr. The precipitate is collected by centrifugation at 10,000 *g* for 30 min and resuspended in 20 m*M* phosphate–0.3 *M* NaCl, pH 6 (10 ml/liter of culture supernatant). At this stage the protoxin is quite stable as long as the suspension is not warmed. During purification, stability is higher at pH values below 7 and in solutions of high ionic strength, especially solutions containing divalent anions such as sulfate and phosphate. All subsequent steps should be carried out without delay. The suspension is centrifuged (15,000 *g*, 30 min) and the clear supernatant is desalted on a 2.5 × 30 cm column of Sephadex G-25 equilibrated in the same buffer. Void volume fractions are applied directly to a 1.6 × 15 cm column of hydroxyapatite (see above) also equilibrated in the NaCl–phosphate buffer. This is eluted at 20 ml/hr with a linear gradient formed by mixing 250 ml of the starting buffer and 250 ml of 0.15 *M* phosphate–0.3 *M* NaCl (pH 6). Fractions in a broad, sometimes double peak, eluting at ~0.05 *M* phosphate, contain protoxin and varying amounts of the mature

TABLE I
PURIFICATION OF AEROLYSIN[a]

Step	Total protein (mg)	Total HU	Activity (%)	Specific activity ($HU \cdot mg\ protein^{-1}$)
Culture supernatant[b]	657	5.66×10^6	100	0.86×10^4
Second wash	189	3.09×10^6	55	1.66×10^4
DEAE	17	3.63×10^5	6.4	2.14×10^4
Sephadex G-150	7.5	1.76×10^5	3.1	2.46×10^4
Hydroxyapatite	4.5	1.22×10^5	2.2	2.68×10^4

[a] From Buckley *et al.*[8]
[b] An aliquot of 1200 ml of culture supernatant was used.

toxin (Fig. 1A). At this and all other stages, the relative amounts of toxin and protoxin can be estimated by determining the degree of activation obtained with trypsin, or by SDS-PAGE.[9] Fractions from the first half of the double peak obtained from the hydroxyapatite column are enriched in the protoxin (Fig. 1B). These are combined and the protein precipitated by adding ammonium sulfate to 85% of saturation. The precipitate is recovered by centrifugation at 15,000 *g* for 30 min and dissolved in 20 m*M* Tris–5 m*M* EDTA–0.5 m*M* phenylmethylsulfonyl fluoride, pH 8. The protein is then applied to a DEAE-Sepharose CL-6B column equilibrated in the same buffer and eluted at 20 ml/hr with a linear gradient of 250 ml of 20 m*M* Tris and 250 ml of 20 m*M* Tris–0.5 *M* NaCl, pH 8. The protoxin elutes immediately after the mature protein.

Properties

Purity and Molecular Weight. Aerolysin is isolated free of measurable proteases by these procedures and migrates as a single band on SDS–PAGE, accounting for at least 95% of the applied protein. In nondenaturing acrylamide gels,[10] the toxin separates into two bands of approximately equal intensities, both of which are hemolytic. This is consistent with the observation that two aerolysin isomers are isolated together, with isoelectric points of 5.39 and 5.46.[8] Our preliminary evidence suggests that the existence of isomers is due to posttranslational modification of the protein (unpublished observations).

Proaerolysin samples often appear free of contaminating aerolysin as well as other proteins by SDS–PAGE, but they always have some residual hemolytic activity. This may be due to traces of contaminating toxin, although it is possible that the protoxin is weakly active. The apparent molecular weight of the mature toxin, determined by SDS–PAGE, is about 51,500. A similar value is obtained by gel filtration on Sephadex G-150. The molecular weight of the protoxin, also determined by SDS–PAGE, is 54,000.

Activation of the Protoxin.[3] Proaerolysin is completely converted to the mature active protein by treatment with 0.1–1.0 μg/ml of trypsin (ratio of protease to protoxin 1 : 100) for as little as 5 min. Conversion is due to the removal of ~20 amino acids from the C-terminus of the protoxin. The mature toxin is not further degraded by trypsin even when the concentration of the protease is increased 10-fold and the incubation is continued for 1 hr. The protoxin can also be activated by treatment wi[th]

[9] D. M. Neville, *J. Biol. Chem.* **246,** 6328 (1971).
[10] B. J. Davis, *Ann. N.Y. Acad. Sci.* **121,** 404 (1964).

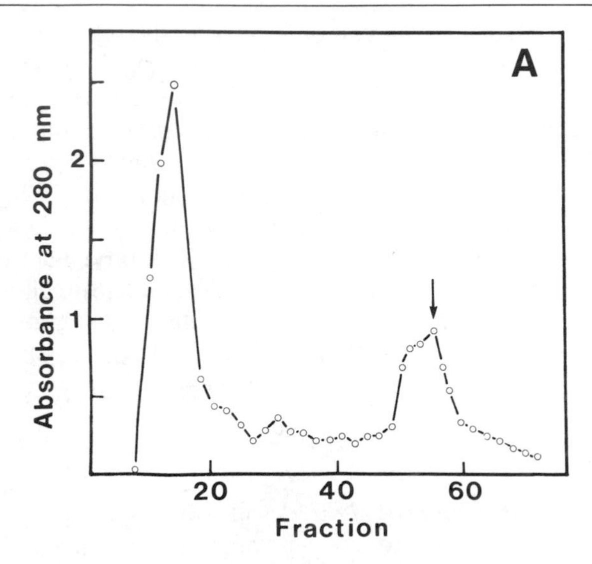

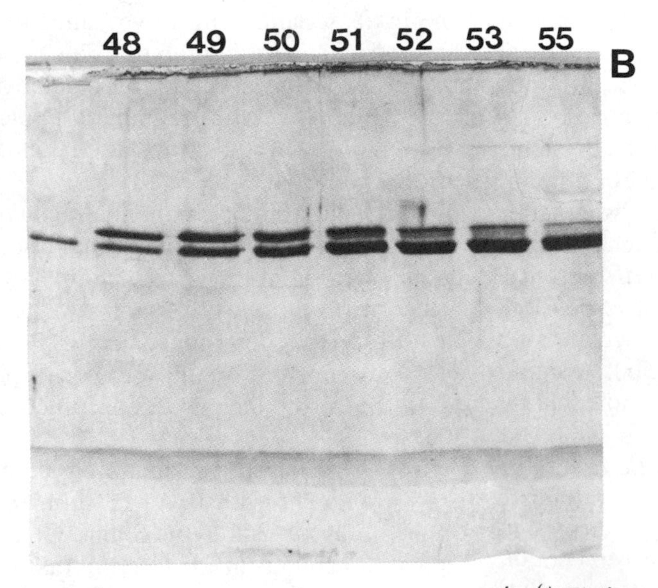

FIG. 1. Hydroxyapatite chromatography of toxin and protoxin. (A) Elution profile from a hydroxyapatite column (1.6 by 15 cm). Desalted ammonium sulfate precipitate (20 ml) was applied and eluted as described in the text. Only half of the profile is shown; the other half contained no material with an absorbance at 280 nm. The arrow marks the fraction with the highest hemolytic titer. (B) SDS–PAGE of column fractions. The labels refer to fraction numbers. The unlabeled lane contains purified mature aerolysin. Reproduced from Howard and Buckley.[3]

TABLE II
AMINO ACID COMPOSITION OF MATURE AEROLYSIN[a]

Amino acid	Number of residues[b]	Amino acid	Number of residues[b]
Lysine	22	Alanine	28
Histidine	7	Cysteine	ND[c]
Arginine	16	Valine	37
Aspartic acid	64	Methionine	3
Threonine	31	Isoleucine	17
Serine	38	Leucine	25
Glutamic acid	37	Tyrosine	24
Proline	24	Phenylalanine	14
Glycine	49	Tryptophan	ND

[a] Determined as described in J. T. Buckley, L. N. Halasa, and S. MacIntyre, *J. Biol. Chem.* **257,** 3320 (1982).
[b] Moles of amino acid per mole of enzyme.
[c] ND, not determined.

similar amounts of thermolysin and chymotrypsin as well as with an *Aeromonas* protease, but not by staphylococcal V8 protease, which destroys both forms of the toxin.

Stability. Like many other hole-forming toxins, aerolysin aggregates spontaneously in concentrated solutions. The process is greatly accelerated at temperatures above 0°. Aggregation also occurs during freezing and thawing. The process is essentially irreversible (aggregates are not dissociated by boiling in SDS–PAGE sample buffer, although they are disrupted by 6 *M* urea) and results in loss of activity. Because aggregated protein may interfere in some experiments, samples of the mature protein are routinely centrifuged (20,000 *g*, 30 min) immediately before use.

In contrast to aerolysin, proaerolysin, once purified and freed of protease, is stable at temperatures above 0° and to repeated freezing and thawing. Preliminary experiments indicate that the protoxin is incapable of forming aggregates and that this may be the reason for its very low hole-forming activity.

Amino Acid Composition and Sequence of the Amino Terminus. The amino acid composition of aerolysin is presented in Table II. It is very similar to the amino acid composition predicted by the nucleotide sequence of the gene (unpublished observations). The amino-terminal amino acid is alanine.[3] The amino acid sequence at the amino terminus, translated from the DNA sequence, confirms the published amino acid sequence at this end of aerolysin, except at position 19, where serine was reported and the nucleotide sequence is translated as cysteine.

[28] Preparation of Hemolysin from *Aeromonas*

By ÅSA LJUNGH and TORKEL WADSTRÖM

Aeromonas hydrophila, a gram-negative bacterium of the family Vibrionaceae is ubiquitously found in nature and commonly isolated from cold- and warm-blooded animals, various waters and sewage, and may act as a primary human pathogen.[1] It causes furunculosis and septicemia in fish. In contrast to most gram-negative organisms *A. hydrophila* produces several extracellular enzymes and toxins during growth, namely hemolysins, enterotoxin, proteases, phospholipases, lipase, elastase, phosphatase, aminopeptidases, and ribonuclease.[2]

One hemolysin, called aerolysin by Bernheimer and Avigad[3] and Buckley *et al.*,[4] has been shown to be a potent membrane-damaging toxin with a broad cell spectrum,[5] and has been the object of study of several investigators.[2-4,6]

Assay Method

Test solutions were diluted 2-fold in sodium phosphate-buffered saline, pH 7.2 (PBS), in microtiter trays (Linbro Sc. Comp., Hamden, CT). To each well holding 100 μl of test solution was added an equal volume of washed rabbit erythrocytes in PBS (1%). One hemolytic unit (HU) was defined as the inverted value of the dilution of toxin which lysed 50% of the erythrocytes after incubation at 37° for 1 hr.[2]

An alternative method is as follows: Test solutions were diluted in 0.145 *M* NaCl–0.01 *M* tris(hydroxymethyl)aminomethane (Tris; pH 7.2)–0.2% gelatin. Four-fold dilutions of toxin were dispersed in tubes (1 ml). To each tube was added 1 ml of washed rabbit erythrocytes in 0.145 *M* NaCl–0.01 *M* Tris, pH 7.2.[3] The density of the erythrocyte suspension was adjusted to give absorbance $A^{1.0}_{545\text{ nm}} = 0.8$ after complete hemolysis.

[1] Å. Ljungh and T. Wadström, *J. Toxicol. Toxin Rev.* **1,** 257 (1983).

[2] Å. Ljungh, B. Wretlind, and R. Möllby, *Acta Pathol. Microbiol. Scand., Sect. B* **89,** 387 (1981).

[3] A. W. Bernheimer and L. S. Avigad, *Infect. Immun.* **9,** 1016 (1974).

[4] J. T. Buckley, L. N. Halasa, K. D. Lund, and S. MacIntyre, *Can. J. Biochem.* **59,** 430 (1981).

[5] M. Thelestam and Å Ljungh, *Infect. Immun.* **34,** 949 (1981).

[6] L. M. Riddle, T. E. Graham, and R. L. Amborski, *Infect. Immun.* **33,** 729 (1981).

The tubes were incubated at 37° for 30 min and then briefly centrifuged. One HU is defined as the amount of toxin needed to give 50% hemolysis under these conditions. This method has also been used with human erythrocytes.[4]

Purification of Hemolysin

Production of Hemolysin. Hemolysin is produced during growth in various media, and released into the culture medium during the logarithmic growth phase. Bernheimer and Avigad[3] used a basal salt medium consisting of yeast extract diffusate (220 ml), 20 g casamino acids, 33 μg of thiamin, 1.2 mg of nicotinic acid, and water up to 1 liter. The pH was adjusted to pH 7.2 before sterilization. Maximal hemolysin titer was obtained after 20 hr of growth at 37°.

This medium was slightly modified by Buckley *et al.* by adding M9 salts.[4] Harvesting was made after incubation at 35° for 8 hr followed by 16 hr at 25°. Riddle *et al.* described a defined medium for the accumulation of hemolysin and protease.[6] Basal salt medium was supplemented with glycerol (0.42%) and L-glutamic acid (0.1%). Addition of zinc (1.8 μg/ml) stimulated the extracellular accumulation of hemolysin whereas supplementation of iron (0.5 μg/ml) exerted an inhibitory effect.

Purification. Unless otherway stated all operations were carried out at 4°.

Step 1: Solid ammonium sulfate was added to culture supernatant to 60% saturation at 0° and allowed to stand for 4 hr. After centrifugation, the precipitate was extracted twice with 0.03 *M* sodium borate, pH 8.2 (2.5 ml/400 ml of original supernatant).

Step 2: The supernatant from the second wash was chromatographed on a Sephadex G-25 column (Pharmacia) equilibrated with 0.03 *M* sodium borate, pH 8.2.

Step 3: The void volume was applied to DEAE-Sepharose CL-6B equilibrated with the same sodium borate buffer. The column was eluted with a linear gradient from 0 to 0.3 *M* NaCl in the borate buffer.

Step 4: Fractions with hemolytic activity were pooled and precipitated with ammonium sulfate (60% at 0°). The precipitate was dissolved in a small volume of borate buffer, pH 8.2, and chromatographed on a Sephadex G-150 column.

Step 5: Fractions with hemolytic activity were pooled and adsorbed onto a hydroxyapatite column equilibrated in 0.01 *M* phosphate buffer, pH 7. The column was eluted with a linear gradient of 0.01–0.25 *M* phosphate, pH 7.

According to this purification scheme the end product is hemolysin (aerolysin) with a specific activity of 2.68×10^4 HU/mg protein. The yield was 2.2%. Proteolytic breakdown during the early steps and losses during extraction steps accounted for the low yield.[4] The initial steps were essentially as described by Bernheimer and Avigad.[3]

We used the following scheme[2]:

Step 1: Crude culture fluid (brain heart infusion broth, Difco, Detroit, MI; 100 ml broth in 1-liter Erlenmeyer flasks with indentations was shaken at 140 rps for 18 hr at 37°) was concentrated about 20 times by ultrafiltration in a hollow fiber cartridge (Amicon, Lexington, MA) used in an Amicon CH3 concentrator. The concentrate was treated with DEAE-Sephadex A-25 in a batch procedure (5 g dry weight/liter original culture fluid, pH 8.5).

Step 2: After dialysis against 0.02 *M* glycine–NaOH, pH 7.0, isoelectric focusing was performed in a density gradient of glycerol and a pH gradient of ampholine 3-10 (1%, w/v, LKB products, Bromma, Sweden). Fractions with hemolytic activity, pH 4–6, were refocused in a shallow pH gradient (pH 4–6).

Step 3: Active fractions (p*I* 5.2–5.5) were concentrated against polyethylene glycol 20,000, dialyzed against 0.02 *M* Tris–HCl, pH 7.2, and chromatographed on BioGel P-60 (column 2.5×80; Bio-Rad, Richmond, CA) in 0.25 *M* NaCl with 0.05 *M* Tris–HCl, pH 7.0.

The specific activity was 2.13×10^4 and the yield was ~1%. The low yield was attributed to proteolytic degradation, and to losses during isoelectric focusing. Alternative methods involving fast-protein liquid chromatography are under investigation.

Properties of Hemolysin

The purification schemes above both give preparations of hemolysin free from detectable protease, phospholipase and other enzymes. The M_r is 51,500 ± 700, and isoelectric points 5.39 ± 0.17 and 5.56 ± 0.17.[4] The purified hemolysin is stable to freezing and thawing but sensitive to heat (50°, 10 min). It is inactivated by dithiothreitol, cysteine, and ferrous ions, indicating a site sensitive to reduction. The hemolysin is further inactivated by urea, EDTA, magnesium, papain, and *Streptomyces griseus* protease. It is, however, stable in the presence of trypsin and pronase (1 mg/ml).[2–4]

Precursor of Hemolysin

A precursor of hemolysin has been detected in culture fluid during the logarithmic phase of growth.[7] It is converted to mature toxin by proteolytic enzymes, hence very little precursor is present in the stationary phase. It is about 250 times less hemolytic than hemolysin and possibly nonhemolytic. Work by Howard and Buckley on this precursor–hemolysin interrelationship showed that hemolysin is released as a protoxin of somewhat higher molecular weight than hemolysin (54,000). It requires trypsin digestion or degradation by a heat-stable protease from *A. hydrophila* in the culture fluid to achieve hemolytic activity. Studies with cyanogen bromide fragments of protoxin and hemolysin indicate that the processing of protoxin occurs at the C-terminal end.

Interaction of Hemolysin with Cell Membranes

Hemolysin has been shown to bind to a specific glycoprotein receptor which is present on eukaryotic cells.[8] It may also adhere via hydrophobic interaction (personal communication). The hemolysin inserts into the lipid bilayer and forms holes which are permissive for molecules of less than M_r 100,[5] or holes of 3 nm diameter.[8] It induces typical cytopathogenic effects in cultured cells like human embryonic fibroblasts.[5] Both binding to cell membranes and subsequent cytopathogenic effects are irreversible.[5,8] Brief exposure of human embryonic lung fibroblasts to 1 HU/ml of hemolysin resulted in inhibition of DNA synthesis and cell death.[5] These data indicate that there may exist an intracellular target for hemolysin.

The hemolysin has been cloned in *Escherichia coli* K12.[9] The hemolysin gene product was detected as a 54,000-Da protein in *E. coli* maxicells harboring the *aerA* plasmid. Availability of this clone will simplify purification of the hemolysin since less protease is present in *E. coli* K12 culture supernatant, and overall fewer extracellular proteins are produced. Purified hemolysin may serve as a tool for further studies on membrane interactions and protein transport in gram-negative bacteria as recently described by Howard and Buckley.[10]

[7] S. P. Howard and J. T. Buckley, *J. Bacteriol.* **163,** 336 (1985).
[8] S. P. Howard and J. T. Buckley, *Biochemistry* **21,** 1662 (1982).
[9] T. Chakraborty, B. Huhle, H. Bergbauer, and W. Goebel, *J. Bacteriol.* **167,** 368 (1986).
[10] S. P. Howard and J. T. Buckley, *J. Bacteriol.* **161,** 1118 (1985).

[29] Preparation of Diphtheria Toxin Fragment A Coupled to Hormone

By THOMAS N. OELTMANN and RONALD G. WILEY

There have been several recent reviews describing the synthesis and selective cytotoxicity of plant and bacterial toxins covalently coupled to specific binding proteins such as lectins, antibodies, and hormones toward a variety of normal and neoplastic cells.[1,2] The diphtheria toxin molecule (62,000 Da) is well suited for the construction of such toxic hybrid molecules. Its 21,167-Da fragment A is an enzyme[3,4] which inhibits protein synthesis by the NAD^+-dependent ADP ribosylation of elongation factor 2 present in the cytoplasma of eukaryotic cells as given by the equation:

$$NAD^+ + \underset{\text{(active)}}{\text{EF-2}} \rightarrow \underset{\text{(inactive)}}{\text{ADP-ribosyl-EF-2}} + \text{nicotinamide} + H^+$$

Only a few molecules of fragment A need reach the cytoplasm in order to stop protein synthesis and kill the cell.[5] The B fragment (37,195 Da) is responsible for the interaction of the toxin with specific receptors on the cell surface of toxin-sensitive cells.[6] By replacing a B fragment with a binding protein for different receptors, such as a hormone, one can create a hybrid toxin with a new, or altered, specificity. This section describes the covalent conjugation of diphtheria fragment A (DTA) to human chorionic gonadotropin (hCG), a hormone which specifically binds to, and stimulates, Leydig cells.

Preparation of Human Chorionic Gonadotropin–Fragment A Hybrid (hCG-DTA)

The general method of coupling diphtheria toxin fragment A to new binding proteins is outlined in Fig. 1 and is accomplished by a disulfide exchange[7] reaction after modification with the thiol-containing, heterobi-

[1] S. Olsnes and A. Pihl, *Pharmacol. Ther.* **15,** 355 (1982).
[2] E. S. Vitetta and J. W. Urh, *Annu. Rev. Immunol.* **3,** 197 (1985).
[3] R. J. Collier, *Bacteriol. Rev.* **39,** 54 (1975).
[4] A. M. Pappenheimer, Jr., *Annu. Rev. Biochem.* **46,** 69 (1977).
[5] M. Yamaizumi, E. Mekada, T. Uchida, and Y. Okada, *Cell* **15,** 245 (1978).
[6] J. L. Middlebrook, R. B. Dorland, and S. H. Leppla, *J. Biol. Chem.* **253,** 7325 (1978).
[7] J. Carlsson, H. Drevin, and R. Axen, *Biochem. J.* **173,** 732 (1978).

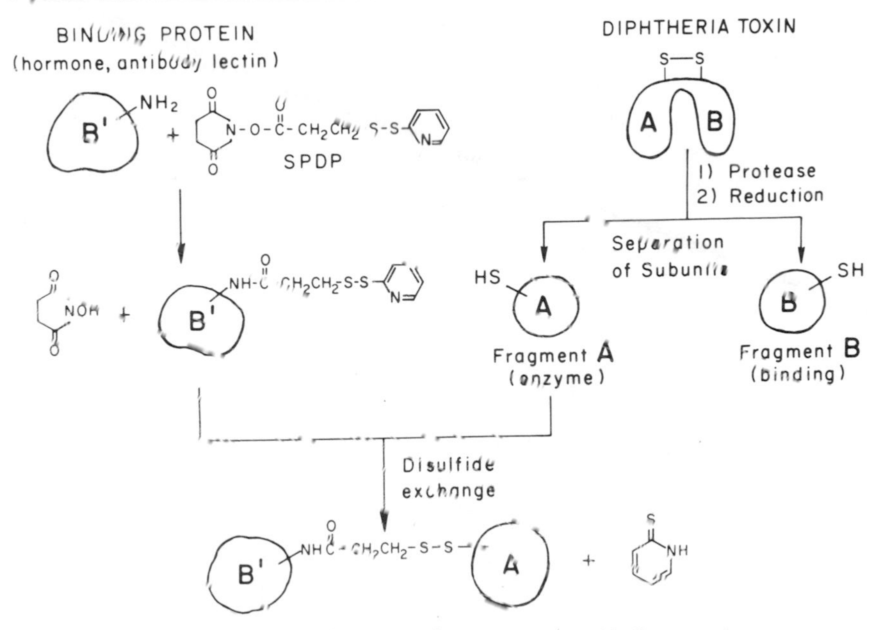

FIG. 1. Coupling of diphtheria toxin fragment A to binding proteins.

functional, cross-linking reagent *N*-succinimidyl-3-(2-pyridyldithiol) propionate (SPDP). The coupling procedure requires several steps and these are outlined below.

A. Isolation of Fragment A from Native Diphtheria Toxin

Diphtheria toxin and its constituent fragments A and B are prepared by the methods described by Chung and Collier.[8]

1. Diphtheria toxin (partially purified, Connaught Laboratories, Toronto, Canada) is purified first by chromatography on DEAE-cellulose. The toxin (20 mg/ml) is dialyzed against 5 m*M* sodium phosphate, pH 6.8, for 48 hr at 4° and then applied directly to a DEAE-cellulose column (2 × 30 cm) equilibrated in the same buffer. The column is washed with 200 ml of 5 m*M* sodium phosphate, pH 6.8, and then the toxin is eluted with a linear gradient of phosphate buffer from 5 to 20 m*M* at pH 6.8. The fractions containing diphtheria toxin are pooled and concentrated by pressure dialysis in an Amicon chamber using a PM10 membrane.

The partially purified toxin is then applied to a Sephadex G-100 column (2 × 110 cm) equilibrated with 50 m*M* Tris–HCl, 1 m*M* EDTA,

[8] D. W. Chung and R. J. Collier, *Biochim. Biophys. Acta* **483,** 248 (1977).

pH 8.2. The major peak eluting, corresponding to the pure toxin (62,000 Da), is pooled and concentrated by pressure dialysis.

2. Fragment A is prepared by limited proteolysis of diphtheria toxin by trypsin. Purified toxin (10 mg/ml in 50 m*M* Tris–HCl) is incubated with trypsin (1 μg/ml final concentration) at 25° for 45 min, after which proteolysis is stopped by the addition of soybean trypsin inhibitor (1.5 μg/ml final concentration). The nicked toxin is concentrated to 40 mg/ml by pressure dialysis in an Amicon concentrator using a PM10 membrane. The nicked toxin is reduced with 0.1 *M* dithiothreitol in the presence of 8 *M* urea. Fragment A (21,167) is isolated by gel filtration on a Sephadex G-100 column (5 × 90 cm) equilibrated with 50 m*M* Tris–HCl, pH 8.2, 1 m*M* EDTA, 2 *M* urea, and 10 m*M* dithiothreitol. Fractions containing pure fragment A are pooled and dialyzed against 0.1 *M* 2,2′-dihydroxyethyl disulfide to reversibly block the free sulfhydryl group.

B. Modification of Human Chorionic Gonadotropin with SPDP

Derivatization of proteins with SPDP involves modification of amino groups (Fig. 1) and one or more of these amino groups may be necessary for binding to specific receptors. Thus, it is important to first titrate the hormone with the cross-linking reagent to determine what ratio of SPDP to hormone can be used to give maximum derivatization of hormone with minimal loss of specific binding activity.

Human chorionic gonadotropin (National Pituitary Agency, Baltimore, MD) is labeled with ^{125}I by the chloramine-T method as modified by Bellisario and Bahl[9] for retention of biological activity. Equal aliquots of the ^{125}I-labeled hCG are added to eight culture tubes containing 0.01 μmol of unlabeled hCG (225 μl of a 1 mg/ml solution) in 50 m*M* sodium phosphate, pH 7.0, containing 150 m*M* sodium chloride.

A 10 m*M* stock solution of SPDP (Sigma) is prepared in dimethylformamide or absolute ethanol. Aliquots of the stock solution corresponding to the following molar ratios to hCG are added at 0°: 0 μl (control), 0.5 μl (0.5 : 1), 1 μl (1 : 1), 2 μl (2 : 1), 5 μl (5 : 1), 10 μl (10 : 1), 20 μl (20 : 1), and 50 μl (50 : 1). The volumes are adjusted with appropriate amounts of dimethylformamide or ethanol and the reaction mixtures are incubated for 1 hr at room temperature. Each sample is then reduced with 10 m*M* dithiothreitol (to remove the reactive dithiolpyridyl group) and dialyzed against 50 m*M* sodium phosphate, 150 m*M* sodium chloride, pH 7.0, to remove unreacted SPDP, 2-thiopyridine, and *N*-hydroxysuccinimide.

Following the dialysis, each sample is assayed for retention of binding activity by an appropriate assay. For hCG it has been determined that a

[9] R. Bellisario and O. P. Bahl, *J. Biol. Chem.* **250**, 3837 (1975).

molar ratio of SPDP to hCG of 1 : 1 gives maximum modification with minimal loss of binding activity.[10] However, this may vary for different binding proteins such as antibodies and hormones and thus each binding protein should be titrated initially.

C. *Diphtheria Toxin Fragment A–hCG Coupling*

Once a useful ratio of SPDP to hCG has been established, coupling of hCG to diphtheria toxin fragment A is carried out as follows:

1. hCG to be coupled is prepared as a 10 mg/ml (0.2 m*M*) solution in 50 m*M* sodium phosphate, pH 7.0, containing 150 m*M* NaCl.
2. A 20 m*M* stock solution of SPDP is prepared in ethanol.
3. To 1 ml of the hCG solution (0.2 μmol) is added to 10 μl (0.2 μmol) of the stock SPDP solution. This reaction mixture is incubated for 1 hr at room temperature and dialyzed overnight against 50 m*M* sodium phosphate, 150 m*M* NaCl, pH 7.0.
4. A 2-fold molar excess of diphtheria toxin fragment A is reduced with 0.1 *M* dithiothreitol at room temperature for 1 hr and then desalted on a Sephadex G-25 column (1 × 50 cm) equilibrated in 50 m*M* sodium phosphate, 150 m*M* NaCl, pH 7.0.
5. The reduced, desalted, fragment A is immediately mixed with the dialyzed SPDP-derivatized hCG and allowed to stand at room temperature for 1 hr.

D. *Purification of the Fragment A–hCG Hybrid (hCG–DTA)*

The disulfide conjugate prepared above is easily purified from unreacted starting material, first by affinity chromatography on NAD^+-Sepharose, followed by lectin chromatography on concanavalin A (Con A)-Sepharose.

1. Preparation of NAD^+-Hexyl Sepharose 4B. Sepharose 4B (100 g packed, wet) is washed with 1 liter of distilled water, resuspended in 100 ml of water, and the pH adjusted to 11. The suspension is cooled to 20° and 30 g of fresh CNBr is added. The reaction mixture is maintained at pH 11 with the addition of NaOH as needed. The reaction will subside in 15–20 min whereupon a large volume of ice is added to halt the reaction completely. The activated Sepharose is immediately washed on a glass funnel with 1 liter of ice-cold water followed by 500 ml of ice-cold 0.1 *M* $NaHCO_3$. The washed Sepharose is then suspended in 100 ml of 0.1 *M* $NaHCO_3$ containing 10 g of 6-aminohexanoic acid, the pH adjusted to 8.5, and the suspension stirred at room temperature for 24 hr. The Sepharose is collected on a glass filter and washed in succession with 1 liter of 0.1

[10] T. N. Oeltmann, *Biochem. Biophys. Res. Commun.* **133**, 430 (1985).

$NaHCO_3$, 1 liter of 0.01 *M* HCl, 1 liter of 0.5 *M* NaCl, 1 liter of water, and finally 500 ml of 80% pyridine. The resin is transferred to a beaker and 1 g of NAD^+ (Sigma grade III) in 24 ml of water is added followed by a solution of 40 g dicyclohexylcarbodiimide in 100 ml pyridine. The mixture is slowly stirred for a week at room temperature, then collected on a glass funnel and washed in succession with 1 liter distilled water, 1 liter ethanol, 1 liter butanol at 40°, 1 liter ethanol, and finally 1 liter water. The resin is then suspended in 500 ml 1 m*M* HCl for 30 min, collected on a glass filter, and washed, within 5 min, with 500 ml of ice-cold 0.1 *M* $NaHCO_3$ followed in succession with 1 liter 0.5 *M* NaCl, 1 liter water. The resin is then resuspended in water and stored at 4°.

2. *Affinity Chromatography of the Diphtheria Toxin Fragment A-hCG Conjugate on NAD^+-Sepharose*. The reaction mixture obtained in Section C above is dialyzed against 20 m*M* *N*-tris(hydroxymethyl)methyl-2-aminoethanesulfonic acid (TES, Sigma) containing 1 m*M* $CaCl_2$ and 1 m*M* $MnCl_2$ (TES-CM buffer). Following dialysis, the mixture was applied to the NAD^+-Sepharose column (1 × 10 cm) equilibrated with TES-CM buffer. Unreacted hCG is not bound to the column while the hCG–fragment A conjugate, as well as unreacted fragment A, is retained by virtue of the interaction of the Sepharose-bound NAD^+ and the NAD^+-binding site of fragment A. The conjugate as well as free fragment A are subsequently eluted with TES-CM buffer containing 10 m*M* adenine and 0.5 *M* NaCl. The eluted fractions are pooled and dialyzed against PBS containing 5 m*M* $CaCl_2$ and 5 m*M* $MgCl_2$, pH 7.5 (PBS-CM).

3. *Lectin Chromatography on Con A-Sepharose*. The partially purified hybrid obtained from the NAD^+-Sepharose column is applied to a Con A–Sepharose (Sigma) column (1 × 10 cm) equilibrated with PBS-CM. Unreacted fragment A is not bound to the lectin column whereas the hybrid was retained by virtue of the interaction between the mannose residues in the oligosaccharide moiety of hCG and the Sepharose-bound Con A. The hybrid, now free of unreacted hCG and fragment A, is subsequently eluted with PBS-CM buffer containing 0.25 *M* α-methyl mannoside. The fractions containing hybrid are pooled, dialyzed against PBS, and concentrated by pressure dialysis to 2 mg/ml and stored at −20°.

Characterization of the Diphtheria Toxin Fragment A–hCG Hybrid (hCG–DTA)

Characterization of the conjugate is by structural and functional tests which demonstrate the covalent linkage of the fragment A to the hormone, retention of the binding ability of the hybrid to Leydig cells, and the selective killing of cells expressing the hCG receptor (Leydig cells).

A. SDS–Polyacrylamide Gel Electrophoresis

The hCG–DTA conjugates can be determined by the presence of higher molecular weight products upon 10% polyacrylamide electrophoresis in the presence of 0.1% SDS.[11,12] These higher molecular weight bands appear at 21,000 higher than uncoupled hCG (45,000). Upon reduction, lower molecular weight bands corresponding to the β chain of hCG (30,000), the α chain of hCG (15,000), and fragment A (21,000) are resolved. Only conjugates that retain the binding activity of the uncoupled hCG will be useful as specific cytotoxic reagents, thus the binding specificity of the conjugate should also be examined.

B. Binding of hCG–DTA to Leydig Cells

In order to demonstrate the retention of specific binding of hCG–DTA as well as the presence of fragment A covalently bound to hCG, one can carry out a simple radioimmunoassay using anti-diphtheria toxin (Connaught Laboratories, Swiftwater, PA) or anti-hCG (Miles Laboratories, Elkhart, IN). In addition, anti-fragment A can be obtained from rabbits or goats immunized with highly purified fragment A. The hCG–DTA is allowed to bind to target cells, as well as control cells which lack the hCG receptor. The presence of covalently bound fragment A can then be demonstrated by the binding of ^{125}I-labeled anti-diphtheria toxin or ^{125}I-anti-fragment A. Briefly, 1×10^6 cells in 1 ml PBS containing 1% bovine serum albumin at pH 7.0 is incubated with hCG–DTA for 1 hr at 4° followed by five washings with PBS. The cells are then resuspended in medium containing ^{125}I-labeled anti-fragment A or ^{125}I-labeled anti-diphtheria toxin ($\sim 10^6$ cpm) and again incubated for 1 hr at 4°. The cells are then harvested, washed extensively, and the radioactivity bound to target cells and control cells are determined.

C. In Vitro Cytotoxicity of hCG–DTA

The inactivation of EF-2 in cells by diphtheria toxin requires the release of free fragment A into the cytosol. Thus hCG–DTA cytotoxicity can be determined by the inhibition of the incorporation of radiolabeled amino acids into cellular protein as a function of hCG–DTA concentration, time, temperature, and specific inhibitors (i.e., hCG, anti-hCG, anti-DTA). In a typical dose–response experiment 10^4 Leydig cells are freshly plated 24 hr prior to use. Cells are incubated at 37° with various concen-

[11] U. K. Laemmli, *Nature (London)* **227,** 680 (1970).
[12] P. O'Farrell, H. Goodman, and P. O'Farrell, *Cell* **12,** 1133 (1977).

trations of hCG–DTA (10^{-6} to 10^{-11} *M*) for 22 hr. Following this incubation, radiolabeled amino acids are added and the incubation continued for two additional hours. The cells are then harvested and the radioactive, trichloroacetic acid-insoluble content determined.[13]

N. Oeltmann and E. C. Heath, *J. Biol. Chem.* **254,** 1028 (1979).

Section II

Assay of Toxins

[30] Assay of Hemolytic Toxins

By ALAN W. BERNHEIMER

Cytolytic toxins are produced by a variety of living entities. Prominent among these are bacteria, stinging insects, poisonous reptiles, stinging marine invertebrates such as cnidarians (coelenterates), and basidiomycetes.[1] Erythrocytes are the most convenient cells for assaying cytolytic toxins because they are readily available, because they contain a built-in marker in the form of hemoglobin, and because agents lytic for other types of mammalian cells are usually lytic for erythrocytes. Moreover, it seems reasonable to assume that knowledge of the primary or basic reaction leading to lysis of erythrocytes is applicable to other types of eukaryotic cells.

Degree of hemolysis or "percentage hemolysis" can be estimated (1) by measuring hemoglobin released in a fixed period of time after mixing toxin and erythrocytes, (2) by measuring changes in the optical properties of the whole reaction mixture as a function of time either by following light absorbance or light scattering or both, or (3) by counting in a hemocytometer the number of erythrocytes remaining after a period of time following addition of lytic agent. The choice of method is determined primarily by the precision one wishes to attain, by the amount of time at one's disposal, and by the available equipment. Because of their convenience only methods under category (1) will be described.

Ponder[2] wrote: ". . . it has been shown to everybody's satisfaction that once the red cell begins to lose its pigment, the loss proceeds rapidly and can be considered as being 'all-or-none.' " The kinetics of hemolysis have been described in detail by the same author albeit in a somewhat unconventional manner based on time–dilution curves, meaning curves resulting from plotting time required for 100% or some other degree of hemolysis as a function of dilution of lytic agent. Other methods of describing the kinetics of hemolysis are based on rate of lysis as a function of concentration of one of the reactants as illustrated in Fig. 1. The sigmoid curves that are commonly obtained when percentage hemolysis is plotted against time are interpreted as reflecting the distribution in variation of sensitivity to lysis of individual cells in the population of erythrocytes under study.

[1] A. W. Bernheimer and B. Rudy, *Biochim. Biophys. Acta* **864,** 123 (1986).

[2] E. Ponder, "Hemolysis and Related Phenomena." Grune & Stratton, New York, 1948.

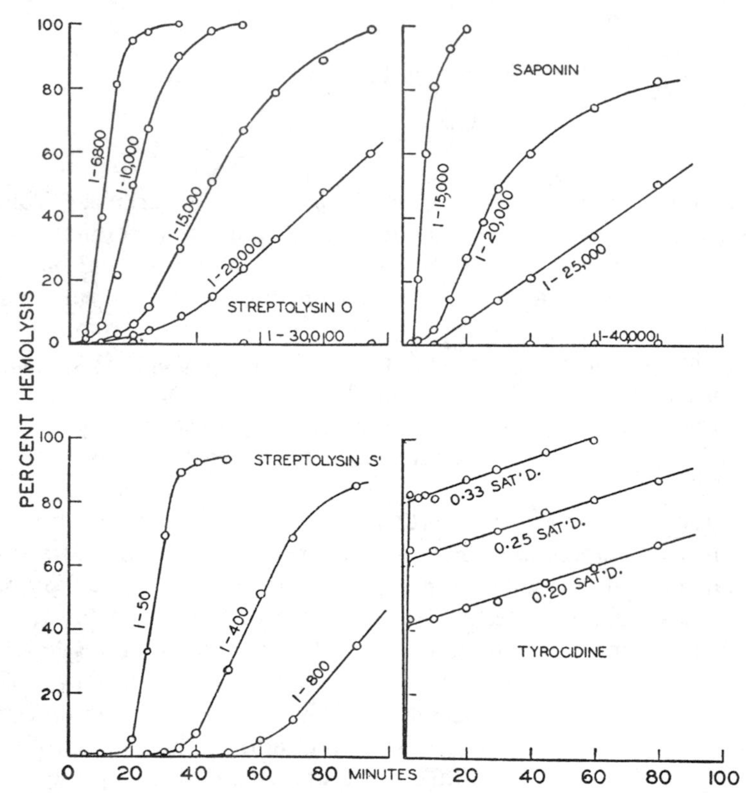

FIG. 1. Course of hemolysis induced by various concentrations of streptolysin O, streptolysin S′, saponin, and tyrocidine, all at 20°. From A. W. Bernheimer [*J. Gen. Physiol.* **30,** 337 (1947)].

Relatively sophisticated methods for following the course of hemolysis can be used, for example, that of Keilin and Hartree,[3,4] which depends on the conversion of liberated oxyhemoglobin to acid methemoglobin by potassium ferricyanide. The course of hemolysis is followed continuously by comparing the spectral absorbance of oxyhemoglobin and methemoglobin in various proportions. The writer prefers a method permitting assay of as many as six or eight samples simultaneously. A simple method applicable with slight modification to a wide variety of cytolytic toxins is described.

[3] D. Keilin and E. F. Hartree, *Nature* (*London*) **148,** 75 (1941).
[4] D. Keilin and E. F. Hartree, *Nature* (*London*) **157,** 210 (1946).

Principle

Decreasing volumes of test solution are added to a constant volume of erythrocyte suspension. After a period of incubation at constant temperature the mixtures are briefly centrifuged and the quantity of hemoglobin in the supernatants is estimated. One unit of hemolytic activity is defined as that amount of test solution liberating half of the total hemoglobin in the erythrocytes.

Reagents

Toxin diluent: 0.15 *M* NaCl in 10 m*M* Tris (pH 7.2) containing 10 m*M* $MgCl_2$, 10 m*M* $CaCl_2$, and 0.2% gelatin (see Note 1 below)

Erythrocyte Suspension. A convenient volume, as 1 to 10 ml of citrated, oxalated, or heparinized blood (see Note 2 below), is diluted in 20 vol of 10 m*M* Tris- or 10 m*M* phosphate-buffered (pH 7.2) 0.85% NaCl solution and centrifuged at 3000 *g* for 5 min. The supernatant is discarded, and the sedimented cells are washed in an additional 20 vol of buffered NaCl solution. The sedimented cells are suspended in 6 vol of buffered NaCl solution. One-half milliliter of water is added to a 0.5-ml sample of erythrocyte suspension, a small amount of saponin powder is added to cause complete lysis, and absorbance is read at 545 nm. The volume of the cell suspension is adjusted to give a 545 nm absorbance of 0.80, corresponding to 100% lysis in the test system.

Assay

(See Note 3 below.) Decreasing volumes of a suitable dilution of test toxin are added to eight 12 × 75 mm tubes, as 1.0, 0.67, 0.5, 0.4, 0.3, 0.2, 0.15, and 0.10 ml (see Note 4 below). The volumes of all tubes are brought to 1 ml with toxin diluent. To each tube is added 1 ml erythrocyte suspension, the contents of the tubes are mixed, and the tubes are placed in a 37° water bath. After 30 min (See Note 5 below) they are centrifuged at 120 *g* for 7 min.

The concentration of the hemoglobin in the supernatants is estimated either (1) by visual comparison to standards representing 10, 20, 30%, etc., hemolysis, or (2) by reading $A_{545\ nm}$ in a spectrophotometer (see Note 6 below). The tube in which 50% of the total hemoglobin has been released is determined directly or by interpolation. The dilution in that tube, prior to addition of erythrocyte suspension, is the number of hemolytic units per milliliter of test solution.

Note 1. In the assay of many cytolytic toxins it is essential to preve[illegible] denaturation by including in the system gelatin, bovine serum albumin other stabilizing agent. Magnesium ions or calcium ions are required for maximal activity of some cytolytic toxins, namely phospholipases. For this reason a Tris- rather than phosphate-buffered NaCl solution is preferable in some instances. However, divalent cations are inhibitory to at least one cytolytic toxin, namely staphylococcal alpha toxin,[5] where phosphate or other buffer is preferable. An additional advantage of a phosphate- over Tris-buffered NaCl solution is that erythrocytes stored cold in the former maintain their integrity longer than in the latter.

Note 2. The species of erythrocytes is often of great importance. For example, rabbit erythrocytes are highly sensitive to staphylococcal alpha toxin whereas sheep erythrocytes are about 4% and human erythrocytes 1% as sensitive as rabbit cells.[6] With respect to *Vibrio damsela* cytolysin, mouse erythrocytes are more than 10,000 times as sensitive as sheep or human erythrocytes.[7] With few exceptions the basis for these dramatic differences is unknown.

Erythrocytes from different individuals of the same species may or may not vary in sensitivity to a particular toxin. In the former instance there are three ways to circumvent the difficulty: (1) Always use cells from the same individual (rabbit, human, or other) on the assumption that sensitivity does not fluctuate with aging of the individual; (2) use a pool of erythrocytes from a number of individuals on the assumption that sensitivity will even out statistically; (3) assay a toxin preparation and designate that toxin preparation as a standard. Assay the standard every time a new preparation is tested, and correct the latter for fluctuation in values of the former on the assumption that the standard is stable under the conditions of storage.

The erythrocyte suspension prepared as described will contain leukocytes and platelets. So far as is known the presence of these entities can be ignored.

Note 3. For maximal activity, some cytolysins need to be activated by a thiol. This requirement can be met by incubating for 10 min at 20° equal volumes of test solution and 0.1% cysteine, dithiothreitol, or other SH compounds at neutral pH. This is done immediately before assay.[8]

Note 4. The recommended volumes are selected for convenience in pipetting. Smaller or larger decrements can be used if desired.

Note 5. In order for lysis to occur, some cytolysins, most notably phospholipases C, require a 5- to 10-min chilling of the reaction mixture in an ice bath following the 37° incubation.

[5] S. Harshman and N. Sugg, *Infect. Immun.* **47,** 37 (1985).
[6] A. W. Bernheimer, *Ann. N.Y. Acad. Sci.* **128,** 112 (1965).
[7] H. Kothary and A. S. Kreger, *Infect. Immun.* **49,** 25 (1985).
[8] A. W. Bernheimer and L. S. Avigad, *Infect. Immun.* **1,** 509 (1970).

TABLE I
ASSAY OF HEMOLYTIC ACTIVITY OF HELIANTHIN

Tube number	Volume of 1 : 15,000 dilution (ml)	Volume of toxin diluent (ml)	Hemolysis estimated visually (%)	A_{545}	Hemolysis calculated from A_{545} (%)
1	1.0	0	95	0.757	93
2	0.67	0.33	90	0.690	86
3	0.50	0.50	85	0.628	79
4	0.40	0.60	80	0.590	74
5	0.30	0.70	65	0.513	64
6	0.20	0.80	50	0.397	50
7	0.15	0.85	43	0.328	41
8	0.10	0.90	30	0.212	27

Note 6. In either case it is desirable to prepare a "standard" representing 50% hemolysis, using saponin and incubating the mixture at 37° for 30 min. This compensates for a small decrease in absorbance at 545 nm which occurs because of partial conversion of hemoglobin to methemoglobin during the 30-min incubation. If desired, this source of error can be eliminated by first converting the hemoglobin of the test erythrocyte suspension to carboxyhemoglobin or to cyanmethemoglobin by treatment with CO or HCN, respectively.

Example

A 1 mg/ml solution of helianthin, a cytolysin obtained from the sea anemone, *Stoichactis helianthis*,[9] was assayed as described after diluting 15,000 times in toxin diluent in order to bring it into range. Percentage hemolysis was estimated both visually and by use of a Zeiss PMQ spectrophotometer. The results are shown in Table I. In both methods, 50% hemolysis occurred in tube No. 6, which contained 0.20 ml of the 1 : 15,000 dilution of the test solution. The activity of the latter is therefore (1.0/0.2)(15,000), or 75,000 hemolytic units per milliliter. If no tube had shown 50% hemolysis, then an interpolated value would have been obtained by using the nearest values above and below 50%. Results are usually reproducible to within 20%.

There has recently been described a semiautomated method of microassay utilizing microtest plates and permitting the simultaneous estimation of hemolytic activity of a large number of samples.[10]

[9] A. W. Bernheimer and L. S. Avigad, *Proc. Natl. Acad. Sci. U.S.A.* **73,** 467 (1976).

[10] J. D. Young, I. G. Leong, M. A. DiNome, and Z. A. Cohn, *Anal. Biochem.* **154,** 649 (1986).

[31] Diphtheria Toxin: Quantification and Assay

By STEPHEN F. CARROLL and R. JOHN COLLIER

Diphtheria toxin (DT) inhibits protein synthesis in susceptible cells by catalytically transferring the ADP-ribosyl moiety of NAD into covalent linkage with elongation factor 2 (EF-2), inactivating the factor and causing cell death.[1-3]

$$\text{NAD} + \text{EF-2} \rightleftharpoons \text{ADP-ribosyl-EF-2} + \text{Nic} + \text{H}^+ \quad (1)$$

Under physiological conditions, the equilibrium position of reaction (1) lies far to the right, and is essentially irreversible. In the absence of EF-2 (and at a much lower rate), DT also catalyzes the hydrolysis of NAD, with water acting as the ADP-ribosyl acceptor.

$$\text{NAD} + \text{HOH} \rightleftharpoons \text{ADP-ribose} + \text{Nic} + \text{H}^+ \quad (2)$$

Both the ADP-ribosyltransferase [Eq. (1)] and NADase (NAD^+ glycohydrolase) [Eq. (2)] activities of DT are catalyzed by the amino-terminal A fragment (DTA), which is released from the intact toxin by mild proteolysis and reduction. The carboxy-terminal B fragment (DTB) functions in receptor recognition and membrane translocation.

Various factors influence the enzymatic and cytotoxic activities of DT, such as the degree to which the toxin has been nicked,[4] the presence or absence of endogenous dinucleotides,[5] or the state of aggregation.[6] The nicked, nucleotide-free monomer is believed to be the biologically active molecule. Table I summarizes the properties of the different toxin forms.

Described here are methods for quantifying the biological activities of DT, including lethality and dermonecrosis in animals, cytotoxicity in cell culture, and *in vitro* enzymatic reactions. Methods for purification and fractionation of the toxin and its fragments are described in the companion Chapter [11] in this volume. The samples of DT discussed here [and the $E^{1\%}_{1\,\text{cm}}$ (280) used for their optical quantification] include the nucleotide-

[1] R. J. Collier, *in* "ADP-Ribosylation Reactions" (O. Hayaishi and K. Ueda, eds.), p. 575. Academic Press, New York, 1982.

[2] A. M. Pappenheimer, Jr., *Harvey Lect.* **76,** 45 (1982).

[3] T. Uchida, *in* "Molecular Action of Toxins and Viruses" (P. Cohen and S. van Heyningen, eds.), p. 1. Elsevier, New York, 1982.

[4] K. Sandvig and S. Olsnes, *J. Biol. Chem.* **256,** 9068 (1981).

[5] S. Lory, S. F. Carroll, P. D. Bernard, and R. J. Collier, *J. Biol. Chem.* **255,** 12011 (1980).

[6] S. F. Carroll, J. T. Barbieri, and R. J. Collier, *Biochemistry* **25,** 2425 (1986).

TABLE I
PROPERTIES OF DIPHTHERIA TOXIN AND ITS FRAGMENTS

Sample	M_r	Enzymatic activity: ADPr-transferase		Enzymatic activity: NADase		Receptor binding	Cyto-toxicity
		Intact	Nicked	Intact	Nicked		
T monomer							
Nucleotide bound	58,342	–	+	–	+	+	+
Nucleotide free	58,342	–	+	+	+	+	+
T Dimer							
Nucleotide bound	116,684	–	+	–	+	–	–
Nucleotide free	116,684	–	+	+	+	–	–
TA	21,164	+		+		–	–
TB	37,198	–		–		(+)	–

bound form (13.4), the nucleotide-free form (13.0), and DTA (15.0). Concentrations of dimeric DT are given in monomer equivalents.

Toxicity Assays

Lethality in Guinea Pigs

DT is toxic for a wide range of animals, including man, monkeys, rabbits, guinea pigs, and various fowl.[7] That amount of toxin required to kill a 250-g guinea pig at 120 hr (ca. 22 ng) is defined as the minimum lethal dose, or MLD. On a weight basis, the other animals listed above appear equally susceptible. Mice and rats are much more resistant to the action of DT.

Materials

Guinea pigs: 200–350 g
PBS–BSA: Phosphate-buffered saline containing 200 μg/ml bovine serum albumin
DT: 3 to 400 n*M* in PBS–BSA
1-cc tuberculin syringes with 25-gauge × 5/8 inch needles
Diethyl ether

Procedure. Guinea pigs are injected subcutaneously in duplicate with 100 μl of DT (20–2000 ng). The animals may be anesthetized by brief

[7] R. J. Collier, *Bacteriol. Rev.* **39**, 54 (1975).

exposure to diethyl ether prior to injection. Survival is then monitored for up to 5 days, and the time of death in hours postinjection is determined. Once normalized to 250-g animals, plots of survival time vs log (log MLD) empirically give a linear relationship,[6] allowing determinations of lethality for survival times between 13 and 120 hr.

Dermonecrotic Lesions

Toxin-induced tissue necrosis is readily visualized by intradermal injection of DT into the shaved backs of rabbits or guinea pigs. Sensitivity of detection is roughly 0.1 pg, or five orders of magnitude lower than the guinea pig MLD.

Materials

Rabbits: The backs of New Zealand White rabbits are carefully shaved and ruled into 2.5-cm squares with a permanent ink marking pen

DT: 10^{-9} to 10^{-5} M in PBS–BSA

1-cm^3 tuberculin syringes with 25-gauge × 5/8 inch needles

Procedure. Duplicate 50-μl aliquots of DT (0.05 to 500 pg) are injected intradermally into the shaved backs of rabbits by pinching the skin between thumb and forefinger and angling the needle into the surface layer. The contents of the syringe are ejected, creating a bleb on the skin surface. The needle is then removed slowly to avoid loss of sample. After 48 hr, the diameters of the duplicate lesions are measured and averaged. A linear relationship is obtained between lesion diameter and the log of the quantity injected.[6]

Cytotoxic Activity

Although numerous tissue culture cell lines are susceptible to the action of DT (see Table II), cells derived from African Green monkey kidneys (Vero, BS-C-1, CV-1) appear the most sensitive.[8] The assay described measures the ability of DT to inhibit the incorporation of radiolabeled amino acids into acid-insoluble protein in BS-C-1 cells. That concentration of toxin which inhibits protein synthesis by 50%, relative to untreated controls, is termed the ID_{50}. BS-C-1 cells were chosen because of their sensitivity, stability, and consistency in routine assays. They contain ca. 32,000 DT receptors per cell, with a K_d of 0.7 nM.[9] An alternative method based on changes in the pH of culture medium as a function of cell death has also been described.[10]

[8] J. L. Middlebrook and R. B. Dorland, *Can. J. Microbiol.* **23,** 183 (1976).

[9] J. T. Barbieri and R. J. Collier, unpublished observations.

[10] J. R. Murphy, P. Bacha, and M. Teng, *J. Clin. Microbiol.* **7,** 91 (1978).

TABLE II
INHIBITION OF PROTEIN SYNTHESIS IN VARIOUS CELL LINES BY DT MONOMER AND DIMER[a]

Cell line	ID_{50}	
	Monomer	Dimer
Vero	3.6×10^{-14}	9.1×10^{-13}
BS-C-1	3.2×10^{-14}	4.4×10^{-13}
CV-1	3.8×10^{-14}	6.3×10^{-13}
WM-9	9.3×10^{-13}	2.9×10^{-11}
SW-620	1.9×10^{-12}	1.7×10^{-11}
HeLa	3.3×10^{-12}	8.1×10^{-11}
CHO	1.8×10^{-11}	6.3×10^{-10}

[a] Data from Ref. 6.

Reagents

Growth medium: Modified Eagle's medium (MEM, Irvine Scientific) containing 10% fetal calf serum

BS C-1 cells; ATCC CCL 26

Pulsing medium: MEM containing 1/20 the normal amount of L-leucine and 2 μCi/ml L-[^{3}H]leucine (TRK.510, Amersham, Arlington Heights, IL; or equivalent)

TCA–leucine: 10% trichloroacetic acid containing 0.1 mg/ml L-leucine

0.1 *N* NaOH

DT: 10^{-7} to 10^{-14} *M*, serial 10-fold dilutions in PBS–BSA

24-Well tissue culture plates (Falcon 3047)

Procedure. BS-C-1 cells in growth medium are seeded into 24-well tissue culture plates at a density of 5×10^4 cells/well and incubated at 37°. Twenty-four to 48 hr later the medium is removed by aspiration, and the cell lawns are rinsed once with PBS (2 ml) and covered with fresh growth medium (1 ml) containing 10^{-9} to 10^{-16} *M* DT (10 μl of the appropriate dilution/well). Each sample is tested in duplicate, and control wells receive 10 μl of PBS–BSA. The trays are incubated at 37°. After 23 hr, the toxin-containing medium is removed and replaced with pulsing medium (0.5 ml), and incubation is continued for 1 hr. The pulsing medium is then removed, and protein is precipitated by washing the cell lawns twice with ice-cold TCA–leucine (1.5 ml each wash). The precipitated cell material is washed once with PBS (1.5 ml) and dissolved in 0.5 ml of 0.1 *N* NaOH. Four-hundred microliters is removed and counted by liquid scintillation. The percentage of protein synthesized in toxin-treated cells is determined by comparison with the no-toxin controls.

Comments. Control cells incorporate 10,000–20,000 cpm into TCA-precipitable material; duplicate wells generally agree within 10%. Shorter periods of incubation require greater concentrations of DT to inhibit protein synthesis by 50%.

In Vitro Enzymatic Reactions

ADP-Ribosyltransferase Assay

DT catalytically transfers the ADP-ribosyl moiety to a modified histidine residue (diphthamide) on EF-2.[11] This residue, highly conserved in all eukaryotic cells, is also found in the Archaebacteria.[12] Other bacterial elongation factors are resistant to the action of DT. The nucleotide specificity of this reaction is also quite high.[13] Of the analogs tested, only thionicotinamide adenine dinucleotide (and to some extent, acetylpyridine adenine dinucleotide) can substitute for NAD; NADH, NADP, NADPH, and NMN are essentially inactive.

The intact toxin is unable to catalyze the ADP-ribosyltransferase reaction *in vitro,* and must be nicked prior to its introduction into the assay where the concentration of reducing agent (DTT) is sufficiently high (40 m*M*) to release the enzymatically active A fragment.

Maximal activity is observed at pH 8.2–8.5, and is markedly influenced by ionic strength; doubling the ionic strength of the reaction mixture decreases the reaction rate by ca. 50%. Inasmuch as NAD binding to DTA is relatively insensitive to salt concentrations up to 0.2 *M*,[13] high ionic strengths may interfere with the binding of EF-2 to DTA–NAD complexes. No requirements for specific anions or cations have been detected.

The Michaelis constants for the *in vitro* reaction catalyzed by DTA are 1.4 μM for NAD and 0.15 μM for EF-2. Under standard conditions, the turnover number is ca. 200 mol ADPr-EF-2 formed/mol enzyme/min. Since EF-2 does not bind DTA to any significant extent alone or in the presence of toxin ligands,[14] the reaction apparently follows an ordered sequential mechanism; the binding of NAD precedes that of EF-2. The reaction mechanism does not apparently proceed via a covalent ADP-ribosyl-DT intermediate.[14]

[11] B. G. Van Ness, J. B. Howard, and J. W. Bodley, *J. Biol. Chem.* **255,** 10710 (1980).

[12] A. M. Pappenheimer, Jr., P. C. Dunlop, K. W. Adolph, and J. W. Bodley, *J. Bacteriol.* **153,** 1342 (1983).

[13] J. Kandel, R. J. Collier, and D. W. Chung, *J. Biol. Chem.* **249,** 2088 (1974).

[14] D. W. Chung and R. J. Collier, *Biochim. Biophys. Acta* **483,** 248 (1977).

The following assay measures the incorporation of radiolabel from [*adenine*-^{14}C]NAD into acid-insoluble EF-2. Linear rates of incorporation are obtained up to 1.0 pmol of DTA or nicked DT per assay (10 n*M*).

Reagents

Assay buffer R: 0.5 *M* Tris–HCl, 10 m*M* EDTA, 0.4 *M* dithiothreitol (DTT), pH 8.2

[*adenine*-^{14}C]NAD: 4 μM in water, >400 mCi/mmol, frozen in small aliquots at −70° (NEC-743 diluted ca. 1 : 5 with water, New England Nuclear, Boston, MA, or equivalent)

EF-2: 2–3 μM, extracted from wheat germ and stored in small aliquots at −70° (see below)

TCA paper: Whatman 3MM filter paper is ruled into 1-in. squares, numbered in pencil, and impregnated with trichloroacetic acid (TCA) by immersion in diethyl ether containing 10% (w/v) TCA.[15] The TCA-impregnated paper is then air dried in a fume hood

EF-2

Numerous procedures are available for the preparation of crude EF-2 from erythrocytes,[16] yeast,[17] or wheat germ[18] which yield material suitable for use in the ADP-ribosyltransferase assay. The method outlined below is simple and reliable.[19]

Thirty grams of raw, uncooked wheat germ is suspended in 240 ml of cold buffer containing 50 m*M* Tris–HCl (pH 8.0), 5 m*M* magnesium acetate, 50 m*M* KCl, 4 m*M* $CaCl_2$, and 5 m*M* 2-mercaptoethanol, and allowed to soak for 2 min on ice. The suspension is then homogenized at top speed in a Waring blender for 50 sec (five blendings, each of 10-sec duration). Ice is added to maintain the temperature at 4°. The homogenate is first filtered through gauze, and cellular debris is then removed by two successive centrifugations at 21,000 *g* for 15 min. After the first centrifugation, the pH of the supernatant fluid is adjusted to 7.6 with acetic acid. After the second spin, solid KCl is added to a final concentration of 0.1 *M*, and ribosomes are removed by centrifugation at 250,000 *g* for 1 hr. The postmicrosomal supernatant is fractionated with ammonium sulfate; material precipitating between 30 and 50% saturation at 4° contains roughly 70% of the total EF-2 present in the supernatant. The crude EF-2 is

[15] F. J. Bollum, this series, Vol. 12, p. 169.

[16] B. Hardestry and W. McKeehan, this series, Vol. 20, p. 330.

[17] J. W. Bodley, P. C. Dunlop, and B. G. VanNess, this series, Vol. 106, p. 378.

[18] A. B. Legocki, this series, Vol. 60, p. 703.

[19] D. W. Chung and R. J. Collier, *Infect. Immun.* **16,** 832 (1977).

dissolved in approximately 30 ml of buffer containing 50 m*M* Tris–HCl (pH 7.5), 1 m*M* EDTA, 1 m*M* DTT, 5% glycerol, and dialyzed twice (1 liter each time) against the same buffer at 4°. Any precipitate that forms during dialysis is removed by centrifugation, and the EF-2 is aliquoted (ca. 250 μl) into small tubes and stored frozen at −70°. Ten microliters of this solution usually contains between 15 and 30 pmol of acceptor protein. EF-2 prepared in this manner is stable for several years in the freezer, at least several hours on ice, and is relatively unaffected by repeated freeze-thawing.

Procedure. Reaction mixtures contain the following volumes (and final concentrations): 10 μl assay buffer R (50 m*M* Tris–HCl, 40 m*M* DTT, 1 m*M* EDTA, pH 8.2), 10 μl EF-2 (0.3 μ*M*), 10 μl [*adenine*-^{14}C]NAD (0.4 μ*M*), and a combination of sample (0.1 to 2.0 pmol; 1.0 to 20 n*M*) and water in a final volume of 100 μl. Routinely, a mixture of assay buffer R (10 μl), EF-2 (10 μl), and water (30 μl) sufficient for all samples is prepared, and 50 μl is distributed into 0.5-ml microfuge tubes. Samples and water are then added to a final volume of 90 μl, and the reaction is initiated by the addition of NAD (10 μl). Incubation is at 25° for 15 min, and the reaction is terminated by applying duplicate 40-μl aliquots to TCA paper. The filter paper is washed twice with 5% TCA (ca. 10 ml/square, 10 min each wash), rinsed, and then washed once with methanol (5 min), and dried under a stream of hot air. Individual squares are excised and counted by liquid scintillation.

Comments. For each assay, both negative (no toxin) and positive (0.5 pmol DTA) controls are included for determination of background and standard levels of incorporation, respectively. Background counts are subtracted from all assays, and the amount of toxin present in each sample is then derived by comparison with the standard. For kinetic studies, the concentrations of NAD and EF-2 are increased 2-fold, and the reaction time is shortened to 5 min. If necessary, sensitivity of the assay can be increased several orders of magnitude by employing [^{32}P]NAD and highly purified EF-2.[20]

NAD⁺ Glycohydrolase Assay

Several methods are available for analyzing the rate of nicotinamide release from NAD catalyzed by DTA and nucleotide-free DT. In the method described below, hydrolysis of NAD is terminated by covalent modification of ribosyl hydroxyls with borate, creating highly charged NAD and ADP-ribosyl derivatives. The liberated nicotinamide, now un-

[20] K. Reich and R. J. Collier, unpublished observations.

charged, is then extracted into ethyl acetate in which it is freely soluble. Alternatively, the charged reaction by-products can be removed by adsorption to Dowex AG1-X2 ion-exchange resin, either by batch[21] or column[22] techniques.

Reagents

Assay buffer G: 0.1 *M* Tris–HCl, pH 7.8, 0.2 mg/ml BSA [carbonyl-^{14}C]NAD: 45 μM in water, ca. 50 mCi/mmol, stored in small aliquots at −70° (CFA.372, Amersham, Arlington Heights, IL, or equivalent)
Termination buffer: 1 *M* boric acid titrated to pH 9.0 with 10 *N* NaOH
Water-saturated ethyl acetate

Procedure. Reaction mixtures (100 μl) contain 50 m*M* Tris–HCl, pH 7.8, 100 μg/ml BSA, 4.5 μM [carbonyl-^{14}C]NAD, and toxin samples up to 10 μM. Fifty microliters of assay buffer G is aliquoted into 1.5-ml microfuge tubes on ice; samples (up to 1 nmol) and water are then added to a final volume of 90 μl. [carbonyl-^{14}C]NAD (10 μl) is added to initiate the reaction, and the tubes are incubated at 37° for 1.5 hr. Hydrolysis is terminated by adding termination buffer (25 μl) with vortexing. One milliliter of water-saturated ethyl acetate is added, the samples are vortexed for 10 sec, and then centrifuged for 10 sec in an Eppendorf microfuge to enhance phase separation. Duplicate 0.4-ml aliquots are removed and counted by liquid scintillation.

Comments. As in the ADP-ribosyltransferase assay, negative (no toxin) and positive (0.25 nmol DTA) controls are included for determination of background and reference levels of incorporation, respectively. For greater sensitivity, incubation times can be extended to ca. 24 hr and compared to appropriate controls. Smaller reaction volumes[23] can also be used to minimize sample consumption.

[21] S. F. Carroll, S. Lory, and R. J. Collier, *J. Biol. Chem.* **255,** 12016 (1980).
[22] J. Moss, V. C. Manganiello, and M. Vaughan, *Proc. Natl. Acad. Sci. U.S.A.* **73,** 4424 (1976).
[23] C. M. Collins and R. J. Collier, *J. Biol. Chem.* **259,** 15159 (1984).

[32] Use of Exotoxin A to Inhibit Protein Synthesis

By CATHARINE B. SAELINGER

The importance of *Pseudomonas* exotoxin A (PE) as one of the prime virulence factors of *Pseudomonas aeruginosa* has been well established. The toxin has been characterized on both the structural and biochemical level. *Pseudomonas* exotoxin A has a molecular weight of 66,583. It is composed of three domains, one having ADP-ribosyltransferase activity, and the other two involved in binding toxin to sensitive cells and translocation of the toxin across a cellular membrane to reach the cytoplasm.[1,2] PE exerts its toxic effect by specifically stopping protein synthesis both *in vitro* and *in vivo*.[3,4] At the biochemical level it acts in a manner identical to fragment A of diphtheria toxin, i.e., by the ADP ribosylation of cytoplasmic elongation factor 2 (EF-2). The resultant ADPribose–EF-2 complex is nonfunctional in protein synthesis, EF-2 is depleted, and cell death ensues.[3] The enzymatic reaction is

$$\text{NAD} + \text{Elongation factor 2} \rightleftarrows \text{ADP-ribose}\sim\text{elongation factor 2} + \text{nicotinamide}$$

The pH optimum for the reaction is 8.0. The ADP-ribosylation reaction can be reversed if the ribosylated EF-2 is incubated with an excess of either PE or fragment A of diphtheria toxin and excess nicotinamide at low pH (5.2). NAD^+ is produced in this reaction.

PE is produced by *P. aeruginosa* as a proenzyme and must be activated to express enzyme activity.[5,6] While enzyme active peptides of PE have been described,[7] active toxin is usually formed by conformational changes in the intact native toxin molecule.[5] This can be achieved by incubating PE with reducing agent in the presence of urea.

ADP Ribosylation of EF-2

ADP ribosyl transferase activity of activated toxin is easily assayed by measuring the incorporation of radioactivity from [*adenine*-^{14}C]NAD into

[1] J. Hwang, D. J. Fitzgerald, S. Adhya, and I. Pastan, *Cell* **48,** 129 (1987).
[2] C. Guidi-Rontani and R. J. Collier, *Molec. Microbiol.* **1,** 67 (1987).
[3] B. H. Iglewski and D. Kabat, *Proc. Natl. Acad. Sci.* **72,** 2284 (1975).
[4] K. Snell, I. A. Holder, S. H. Leppla, and C. B. Saelinger, *Infect. Immun.* **19,** 839 (1978).
[5] S. H. Leppla, O. C. Martin, and L. A. Muehl, *Biochem. Biophys. Res. Commun.* **81,** 532 (1978).
[6] M. L. Vasil, D. Kabat, and B. H. Iglewski, *Infect. Immun.* **16,** 353 (1977).
[7] D. W. Chung and R. J. Collier, *Infect. Immun.* **16,** 832 (1977).

acid-insoluble material in the presence of crude EF-2, prepared from extracts of wheat germ as described by Chung and Collier.[7] Both native and activated toxin are assayed. PE is activated by incubating PE with an equal volume of 8.0 *M* urea containing 2% DTT for 20 min at 25°. As a control native PE is incubated with an equal volume of 0.1 *M* Tris–HCl, pH 8.1. The final reaction mixture contains 50 m*M* Tris–HCl, pH 8.1, 0.1–2.0 μg PE, 0.0125 μCi [^{14}C]NAD, and 500 μg wheat germ extract. Activated or native PE (5.0–10.0 μl) is added to 50 μl wheat germ extract (diluted in 0.1 *M* Tris, pH 8.1) at 4°. Nicotinamide [U-^{14}C]adenine dinucleotide (ammonium salt, 100–250 mCi/mmol; Amersham, Arlington Heights, IL) is diluted to 2.5 μCi/ml in cold 0.1 *M* Tris, pH 8.1. To initiate the reaction, 5.0 μl of [^{14}C]NAD is added, the mixture is vortexed, and incubated 10 min at 25°. The reaction is stopped by the addition of 0.2 ml 12% (w/v) cold trichloroacetic acid (TCA). After 15 min on ice, the precipitate is collected on Metricel membrane filters (0.8 μm, Gelman Sciences, Inc., Ann Arbor, MI) by vacuum filtration, the precipitates are washed three times with 5% cold TCA (3.0 ml/wash), and the filters are subjected to liquid scintillation counting without solubilization. Toxin activity is determined by subtracting counts obtained with unactivated toxin from those obtained with activated PE.

Increasing the length of incubation with [^{14}C]NAD will increase the sensitivity of the assay; we do not recommend incubating longer than 60 min, as we have seen loss of incorporation after that time. The assay can detect as little as 100 ng activated PE. The assay can be used to quantitate toxin in body fluids (e.g., serum), in Percoll gradient fractions, in culture supernatants and during toxin purification. It should be noted that the overall protein concentration in the reaction mixture should be kept low (less than 5 mg/ml) to obtain maximal enzyme activity.

Level of Functional EF-2 in Tissues

A modification of the ADP-ribosylation assay measures the level of active, functional EF-2 in tissues[4] or in mammalian cells in culture,[8] and thus provides an assay of the cytotoxic activity of PE *in vivo* and *in vitro*.

Tissues are pooled, weighed, minced, and homogenized in 4 vol of 0.25 *M* sucrose at 4°. The homogenates (3 ml) are treated with 0.5 ml of 4.0 *M* NH_4Cl to release EF-2 from ribosomes and with 0.8 g of a pellet of washed, activated charcoal (Norit neutral) to remove excess endogenous NAD. Homogenates are centrifuged at 100,000 *g* for 60 min, and the supernatant fluid is collected and assayed. Aliquots of superna-

[8] M. Michael and C. B. Saelinger, *Curr. Microbiol.* **2**, 103 (1979).

tant fluid are incubated in a reaction mixture containing 50 mM Tris–HCl (pH 8.2), 0.07 M dithiothreitol (DTT), 1 mg of bovine serum albumin/ml, and 0.1 mg of diphtheria toxin/ml (List Biological Laboratories, Inc., Campbell, CA). The mixture is adjusted to pH 8.0. The control tube contains the reaction mixture without toxin. Tissue extract is added to each tube; the mixtures are equilibrated at 37°, and 0.12 μCi of [^{14}C]NAD is added. After incubation for 15 min, the reaction is stopped by addition of an equal volume of cold 12% trichloroacetic acid. Precipitates are collected as described above and washed by filtration, and radioactivity is measured by liquid scintillation counting. Protein content of tissue homogenates is determined by the method of Lowry *et al.*[9] Routinely, all assays are run in triplicate. The difference between the cpm with and without toxin represents the ADPribosyl-EF-2 that is formed. Data are expressed as cpm/mg of tissue protein.

Inhibition of Protein Synthesis *in Vitro*

A wide range of cells are susceptible to the cytotoxic action of PE, with mouse fibroblasts (L929, LM, 3T3) being the most sensitive cell lines described to date.[10] The mouse fibroblast LM is our standard test cell line. One simple assay to measure PE activity is to assess inhibition of protein synthesis.

1. Mouse LM fibroblasts (ATCC-CCC 1.2, LM, a derivative of L929) are maintained in stock culture in 75-cm^2 flasks in McCoy's 5A (Gibco Laboratories, Grand Island, NY) medium containing 5% heat-inactivated fetal calf serum (Gibco Laboratories), penicillin (200 U/ml), and streptomycin (0.2 mg/ml). For experimental use, cells are seeded in 24-well culture dishes at a concentration of 5×10^5 cells/ml, and incubated 18 hr at 37°. This results in an approximately 80% monolayer.
2. Prior to use, monolayers are washed one time with complete medium, and cooled to 4° for 15 min.
3. PE is added to wells in 1 ml of medium, and cells are incubated at 4° for 60 min to allow binding of toxin. In a typical experiment PE over a concentration range of 10^{-8} to 10^{-10} M is used. All dilutions are assayed in triplicate.
4. Cell monolayers are washed three times with medium and cells are reincubated in medium for 5 hr at 37° to allow expression of toxin activity.
5. To measure protein synthesis, cells are incubated for 60 min in

[9] O. H. Lowry, N. J. Rosebrough, A. L. Farr, and R. J. Randall, *J. Biol. Chem.* **193,** 265 (1951).

[10] J. L. Middlebrook and R. B. Dorland, *Can. J. Microbiol.* **23,** 183 (1977).

medium containing 2 μCi/ml [^{3}H]leucine (L-[4,5-^{3}H]leucine; activity 45–70 Ci/mmol, Amersham; Arlington Heights, IL).

6. Monolayers are washed three times in phosphate-buffered saline, pH 7.2, and dissolved in 1.0 ml of 0.1 *N* NaOH. After 5 min at 37° the NaOH is collected in centrifuge tubes, precipitated with 1.5 ml of cold 12% TCA, and allowed to sit on ice for at least 60 min.

7. *Processing:* The acid-insoluble material is washed two times in 6% TCA (3 min, 1500 *g*) and digested in 1.5 ml 0.1 *N* NaOH for 30 min at 56°. Aliquots are assayed for protein by the method of Lowry *et al.*[9] For radioactivity, aliquots are put in Ready-solv HP/b (Beckman Instruments, Inc., Irvine, CA) and counted in a scintillation counter.

8. Data are expressed as incorporation of [^{3}H]leucine in counts per minute per microgram of protein, or as percentage of inhibition of protein synthesis in toxin-treated cells as compared to control cells. From these data the concentration of toxin required to inhibit protein synthesis by 50% (TCD_{50}) can be calculated. A representative experiment is given in Table I.

9. If the investigator wishes to base data on counts per minute of [^{3}H]leucine incorporated per well, steps 6 and 7 may be modified. Cells are washed three times with phosphate-buffered saline, and 1 ml 6% TCA is added to each well. After 10 min at room temperature the TCA is removed, and one additional milliliter of TCA is added per well. After 5 min, this is removed and 0.2 ml NaOH (0.1 *N*) is added to each well to dissolve the cells. The samples are transferred to counting vials, Ready-solv HP/b is added, and the samples are counted in a scintillation counter.

TABLE I
INHIBITION OF PROTEIN SYNTHESIS BY *Pseudomonas* EXOTOXIN A

Toxin	LM[a]		Henle[b]	
(ng/ml)	(cpm/μg)	(% inhibition)	(cpm/μg)	(% inhibition)
0	163.2 ± 3.7	0	74.0 ± 3.7	0
10	125.6 ± 4.2	23	nd[c]	
50	68.3 ± 2.6	58	65.6 ± 0.4	11
100	27.1 ± 3.8	83	55.9 ± 4.3	24
500	3.5 ± 0.6	98	22.9 ± 2.7	69
1000	nd[c]		13.9 ± 0.9	81

[a] PE prebound for 60 min at 4°, cells washed and incubated at 37° for 5 hr prior to measurement of protein synthesis. TCD_{50} 30 ng/ml.

[b] Cells incubated for 5 hr at 37° in the presence of PE prior to measurement of protein synthesis. TCD_{50} 250 ng/ml.

[c] nd, Not done.

Results are expressed as percentage of the incorporation in control samples.

10. *Modifications:* When less sensitive cell lines are used, it is necessary to incubate cells continuously in the presence of PE and/or to extend the length of incubation at 37° (see Table I).

11. *Comments:* Specific antibody to PE has been detected in calf, horse, and human serum but not in fetal calf serum[11]; therefore it is recommended that studies with this toxin be performed with fetal calf serum. Binding of PE can be carried out in Hanks' balanced salt solution (Gibco Laboratories) or in phosphate-buffered saline. Calcium, however, is required for internalization of PE.[12] The protocol can be manipulated by addition of agents or shifts in temperature at different time points[13–15]; this allows the investigator to monitor intracellular events.

Inhibition of Protein Synthesis *in Vivo*

Inhibition of protein synthesis also can be measured *in vivo* in the intact animal. This can be approached in two ways: directly, by monitoring the decrease in incorporation of radiolabeled amino acids into tissue, or indirectly, by measuring the level of functional EF-2 in tissues (see above).

In Vivo Protein Synthesis

Incorporation of [^{3}H]leucine (L-[4,5-^{3}H]leucine; 30 to 50 Ci/mmol) during an *in vivo* exchange period is measured.[4,16] Animals are injected intraperitoneally 90 min to 2 hr before sacrifice (e.g., 150 μCi/mouse or 500 μCi/guinea pig). After sacrifice animals are perfused with Ringer's solution and 3% sucrose, the tissues excised, quick frozen on dry ice, and stored at −70° until processed. At a convenient time, the tissues are homogenized in saline, the protein precipitated with 12% TCA, washed twice in 6% TCA, and the precipitates extracted by the following protocol: 70% ethanol, 95% ethanol, absolute ethanol, ethanol and ether (1 : 1) at 55°, and ether (twice). The precipitates are dried overnight, dissolved in 1.0 *N* NaOH; aliquots are taken for protein determination and for radioactivity determination in a scintillation counter.

[11] J. L. Middlebrook and R. B. Dorland, *Can. J. Microbiol.* **23,** 175 (1977).

[12] D. FitzGerald, R. E. Morris, and C. B. Saelinger, *Infect. Immun.* **35,** 715 (1982).

[13] C. B. Saelinger, R. E. Morris, and G. Foertsch, *Eur. J. Clin. Microbiol.* **4,** 170 (1985).

[14] R. E. Morris and C. B. Saelinger, *Infect. Immun.* **52,** 445 (1986).

[15] K. Sandvig, A. Sundan, and S. Olsnes, *J. Cell. Physiol.* **124,** 54 (1985).

[16] P. F. Bonventre and J. G. Imhoff, *J. Exp. Med.* **124,** 1107 (1966).

Data can be expressed as percentage of leucine incorporation in tissues of experimental animals as compared with uninfected control animals.

This methodology can be used to measure the effect of purified toxin (injected im, ip, or iv) or of toxin produced *in vivo* following challenge with viable organisms.

[33] Shiga Toxin as Inhibitor of Protein Synthesis

By ARTHUR DONOHUE-ROLFE, MARY JACEWICZ, and GERALD T. KEUSCH

Shiga toxin, a protein produced by all strains of *Shigella dysenteriae*, is a potent inhibitor of eukaryotic protein synthesis.[1-5] The molecule is composed of two distinct polypeptide subunits, both of which play an essential role in mediating inhibition of cellular protein synthesis caused by the addition of toxin to intact mammalian cells. The B subunit (M_r 7691) of the toxin is responsible for the binding of holotoxin to the eukaryotic cell surface.[6] Subsequently, the A subunit itself (M_r 32,225) or the smaller A_1 portion (M_r 28,000) must reach the cell cytoplasm where it inhibits protein synthesis, by an enzymatic effect.[1] In cell-free *in vitro* systems, however, the A subunit or the activated A_1 fragment alone is sufficient to inhibit protein synthesis.

Two bacterial toxins, pseudomonas exotoxin A[7] and diphtheria toxin,[8] are known to inhibit protein synthesis by catalyzing the transfer of the adenosine diphosphate moiety from NAD to the protein synthesis elongation factor, EF-2. Unlike these two toxins, Shiga toxin acts directly on the ribosome and functionally inactivates the 60S subunit.[1] Shiga toxin inhibits eukaryotic protein synthesis by cleaving the *N*-glycosidic bond of

[1] R. Reisbig, S. Olsnes, and K. Eiklid, *J. Biol. Chem.* **256,** 8739 (1981).

[2] J. E. Brown, M. A. Ussery, S. H. Leppla, and S. W. Rothman, *FEBS Lett.* **117,** 84 (1980).

[3] J. E. Brown, S. W. Rothman, and B. P. Doctor, *Infect. Immun.* **29,** 98 (1980).

[4] M. R. Thompson, M. S. Steinberg, P. Gemski, S. B. Formal, and B. P. Doctor, *Biochem. Biophys. Res. Commun.* **71,** 783 (1976).

[5] K. Eiklid and S. Olsnes, *J. Receptor Res.* **1,** 199 (1980).

[6] A. Donohue-Rolfe, G. T. Keusch, C. Edson, D. Thorley-Lawson, and M. Jacewicz, *J. Exp. Med.* **160,** 1767 (1984).

[7] T. Honjo, Y. Nishizuka, O. Hayaishi, and I. Kato, *J. Biol. Chem.* **243,** 3553 (1968).

[8] B. H. Iglewski and D. Kabat, *Proc. Natl. Acad. Sci. U.S.A.* **72,** 2284 (1975).

the adenine at nucleotide position 4324 in the 28S rRNA.[9] The end result is the inhibition of nascent polypeptide chain elongation.[1,2,4] This biochemical mode of action is identical to that of the plant toxin, ricin, and in fact significant homology has been observed in the amino sequences of the ricin A chain and the A chain of Shiga-like toxin which is a molecule nearly identical to Shiga toxin.[10,11,12]

Toxicity to Intact Cells

Shiga toxin is a lethal cytotoxin to a limited number of tissue culture cell lines.[5,13,14] Susceptible cells are primate epithelial cells, including the human cervical carcinoma-derived HeLa line, and the African Green monkey kidney Vero cell line. Many epithelial and nonepithelial cell lines are resistant to Shiga toxin.[5,13,16] In addition, not all strains of a cell line are necessarily equally sensitive. For example, if HeLa S3 or the American Type Culture Collection CCL-2 HeLa cells are cloned by limiting dilution methods, lines can be isolated differing in sensitivity to toxin from exquisitely sensitive to highly resistant, with over a million-fold difference in the LD_{50} dose.[5,16]

Holotoxin is required to cause lethal cytotoxicity in intact cells; isolated A or B subunits are without effect. As studied in HeLa cells, toxin binding to cell surface toxin receptors via the B subunit is a necessary prerequisite for toxin to be translocated across the cell membrane to reach its ribosomal locus of action. Evidence has been presented that this translocation proceeds by the mechanism of receptor-mediated endocytosis in HeLa cells.[17,18] As a consequence, there is a delay of 30 min to several hours after toxin is inoculated until inhibition of protein synthesis is detectable.[3,5,19] Methods for the purification of Shiga toxin are discussed in Chapter [22] in this volume.

[9] Y. Endo, K. Tsurugi, T. Yutsudo, Y. Takeda, T. Ogasawara, and K. Igarashi, *Eur. J. Biochem.* **171,** 45 (1988).

[10] Y. Endo, K. Mitsui, M. Motizuki, and K. Tsurugi, *J. Biol. Chem.* **262,** 5908 (1987).

[11] S. B. Calderwood, F. Auclair, A. Donohue-Rolfe, G. T. Keusch, and J. J. Mekalanos, *Proc. Natl. Acad. Sci. USA* **84,** 4364 (1987).

[12] N. A. Strockbine, M. P. Jackson, L. M. Sung, R. K. Holmes, and A. D. O'Brien, *J. Bact.* **170,** 1116 (1988).

[13] G. T. Keusch, *Trans. N.Y. Acad. Sci.* **35,** 51 (1973).

[14] G. Vicari, A. L. Olitzki, and Z. Olitzki, *Br. J. Exp. Pathol.* **41,** 179 (1960).

[15] G. T. Keusch and S. T. Donta, *J. Infect. Dis.* **131,** 58 (1975).

[16] G. T. Keusch, A. Donohue-Rolfe, and M. Jacewicz, *in* "Microbial Toxins and Diarrhoeal Disease" (D. Evered and J. Whelan, eds.), p. 193. Pittman, London, 1985.

[17] M. Jacewicz and G. T. Keusch, *J. Infect. Dis.* **148,** 844 (1983).

[18] G. T. Keusch and M. Jacewicz, *Biochem. Biophys. Res. Commun.* **121,** 69 (1984).

[19] G. T. Keusch, *in* "Receptor Mediated Binding and Internalization of Toxins and Hormones" (J. Middlebrook and L. Kohn, eds.), p. 95. Academic Press, New York, 1981.

Inhibition of *in Vitro* Protein Synthesis

Activation of Shiga Toxin. The toxin A subunit is responsible for the inhibition of protein synthesis caused by the intact toxin.[1] There is a trypsin-sensitive region on the A subunit which, upon cleavage and reduction with dithiothreitol (DTT) or 2-mercaptoethanol, yields two polypeptide fragments, A_1 (M_r 28,000) and A_2 (M_r 4000).[20] Reduction is necessary to separate the two A subunit peptides, indicating that the toxin A_1 and A_2 fragments are held together by a disulfide bridge. The A_1 fragment by itself is capable of inhibiting protein synthesis in a cell-free system, and it is at least 6-fold more active than the unnicked intact A subunit.[1]

Activation of holotoxin for use in a cell-free system can be accomplished with trypsin, urea, and DTT treatment as described by Reisbig *et al.*[1] Purified Shiga toxin (1–2 μg) is incubated at 37° for 60 min in 10 μl PBS buffer (140 m*M* NaCl, 10 m*M* sodium phosphate, pH 7.4) containing 5 μg/ml trypsin. Phenylmethylsulfonyl fluoride is added to a final concentration of 20 ng/ml and then the toxin is treated with 8 *M* urea and 12 m*M* DTT for 60 min at room temperature. Finally, the treated toxin mixture is diluted 25-fold in PBS/0.1% Triton X-100 and 50 μg/ml rabbit hemoglobin.

The A subunit can be separated from the B subunit and isolated in biologically active form.[6] To separate A and B subunits, 100 μg of purified toxin is dissolved in 200 μl of 5% formic acid and applied to a column (0.9 × 18 cm) containing BioGel P-60 (Bio-Rad Laboratories, Richmond, CA) at a column flow rate of 2 ml/hr. Fractions of 0.4 ml are eluted in 5% formic acid and absorbance at 280 nm is determined. Two peaks should be obtained, the first of which is the A subunit, as demonstrated by SDS–polyacrylamide gel electrophoresis (Fig. 1). Fractions containing the A subunit are pooled and lyophilized. Isolated A subunit (50 μg) is resuspended in 1 ml of 10 m*M* Tris–HCl, pH 7.4, containing 8 *M* urea. The solution is then transferred to dialysis tubing and dialyzed initially against the same buffer. The concentration of urea is gradually diluted to 0.8 *M* over a 6-hr period by the addition of 10 m*M* Tris–HCl, pH 7.4. For long-term storage of the A subunit, we dialyze the final solution against 20 m*M* ammonium bicarbonate buffer, lyophilize, and store the lyophilized A subunit at −70°. For activation of the purified A subunit, the procedure outlined above for the holotoxin is suitable.

Inhibition of Protein Synthesis in a Reticulocyte Lysate System. Shiga toxin effects on protein synthesis can be conveniently assayed in a cell-free rabbit reticulocyte system.[1] Reticulocyte lysate can be prepared from blood obtained from rabbits rendered anemic by acetylphenylhydrazine injections to increase the reticulocyte count by the procedure of

[20] S. Olsnes, R. Reisbig, and K. Eiklid, *J. Biol. Chem.* **256,** 8732 (1981).

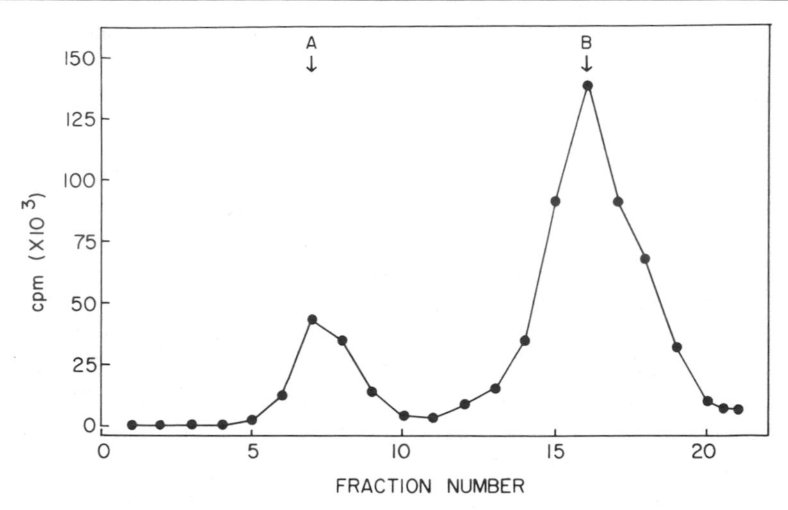

FIG. 1. BioGel P-60 chromatography of formic acid-dissociated[125] I-labeled toxin. The identity of peaks A and B was determined by SDS–polyacrylamide gel electrophoresis. From Donohue-Rolfe *et al.*[6]

Lingrel[21] or it can be purchased commercially (Bethesda Research Laboratories, Gaithersburg, MD). The erythrocyte lysate is supplemented according to Pelham and Jackson.[22] Using stock solutions of 5 mg/ml creatine kinase in 50% aqueous glycerol and 1 m*M* hemin in 90% ethylene glycol, 20 m*M* Tris–HCl (pH 8.2), 50 m*M* KCl, the lysate is made 50 μg/ml in creatine kinase and 25 μM with respect to hemin. To 0.8 ml of the supplemented reticulocyte lysate, 50 μl creatine phosphate (0.2 *M*) and 50 μl KCl/$MgCl_2$ (2 *M* and 10 m*M*, respectively) are added. To this is added 50 μl of a mixture of leucine-free unlabeled amino acids at the following concentrations (m*M*): Ala, 3.0; Arg, 0.5; Asn, 0.5; Asp, 2.0; Cys, 0.5; Gln, 0.5; Glu, 2.0; Gly, 2.0; His, 2.0; Ile, 0.5; Lys, 2.0; Met, 0.5; Phe, 1.5; Pro, 1.0; Ser, 2.0; Thr, 1.5; Trp, 0.5; Tyr, 0.5; Val, 3.0.

The supplemented lysate is stored in small aliquots in liquid nitrogen. Protein synthesis with and without toxin is measured in 50 μl of the supplemented lysate to which 0.5 μCi of [^{14}C]leucine (30 mCi/mmol) is added. The reaction mix is incubated at 30° for 10 min when a 10-μl aliquot is withdrawn from each sample and transferred to tubes containing 1 ml of 0.1 *M* KOH/0.5 *M* H_2O_2 to stop the reaction and to decolorize the sample. After incubation at room temperature for 30 min, 0.2 ml of 50% (w/v) trichloroacetic acid (TCA) is added to each tube to precipitate protein. The precipitates are collected by filtration on glass fiber filters

[21] J. B. Lingrel, *in* "Methods in Protein Biosynthesis" (A. E. Laskin and J. A. Last, eds.), Vol. 2, p. 231. Dekker, New York, 1972.

[22] H. R. B. Pelham and R. J. Jackson, *Eur. J. Biochem.* **67,** 247 (1976).

(Whatman GF/C), washed with 5% TCA, and dried at 100° for 1 hr. Radioactivity is measured by scintillation counting.

In Vitro Inactivation of Ribosomes. Quantitative measurement of Shiga toxin-mediated ribosome inactivation is performed as described by Reisbig *et al.*[1] Rabbit reticulocyte ribosomes (300 n*M*) are incubated with either activated holotoxin (3 n*M*) or A subunit (500 p*M*) in 60 μl of 50 m*M* Tris–HCl (pH 7.4), 25 m*M* KCl, 2 m*M* $MgCl_2$, and 6 m*M* 2-mercaptoethanol. The samples are incubated at 37° and at varying time intervals up to 30 min, 10-μl samples are withdrawn and transferred to tubes containing 3 μl of rabbit polyclonal antitoxin antiserum at 0°.[6] Ribosomal function is then assayed by determining polyuridylic acid [poly(U)]-directed polymerization of phenylalanine. Each sample is therefore brought to 100 μl in buffer (50 m*M* Tris–HCl, pH 7.4; 60 m*M* KCl; 4 m*M* $MgCl_2$; 1.2 m*M* spermidine; 9 m*M* 2-mercaptoethanol) containing 10 μg of poly(U), 500 μ*M* GTP, 10 μl of pH 5 supernatant prepared by the method of Felicetti and Lipmann,[23] and 5 nCi of [^{14}C]- or 92 nCi of [^{3}H]phenylalanyl-tRNA. Incubation proceeds for 15 min at 37°, and the tubes are placed in ice. First, 100 μl of 1% bovine serum albumin is added, followed by 2 ml of 5% TCA. The samples are then incubated at 90° for 10 min, and the heat-stable acid-precipitable material is collected by filtration on glass fiber filters. The filters are washed with 5% TCA, dried, and radioactivity is measured by scintillation counting.

Acknowledgments

Our work has been supported by Grants AI-20325 and AI-16242 from the National Institute of Allergy and Infectious Diseases, NIH, Bethesda, Maryland, Grant 82008 from the Programme for Control of Diarrhoeal Diseases, World Health Organization, Geneva, and a Grant in Geographic Medicine from the Rockefeller Foundation.

[23] L. Felicetti and F. Lipmann, *Arch. Biochem. Biophys.* **125**, 548 (1968).

[34] [^{32}P]ADP-Ribosylation of Proteins Catalyzed by Cholera Toxin and Related Heat-Labile Enterotoxins

By D. MICHAEL GILL and MARILYN WOOLKALIS

Cholera toxin and the heat-labile enterotoxins of some enteric bacteria are related multimeric proteins that catalyze identical ADP-ribosylations of the form[1,2]

$$NAD^+ + \text{protein} \rightarrow \text{ADP-ribose–protein} + \text{nicotinamide} + H^+$$

[1] D. M. Gill and R. Meren, *Proc. Natl. Acad. Sci. U.S.A.* **75**, 3050 (1978).
[2] D. M. Gill and S. Richardson, *J. Infect. Dis.* **141**, 64 (1980).

The physiologically relevant, and most readily modified, substrates are various forms of the alpha subunits of Gs (= N_s), the membrane-bound positive regulator of adenylate cyclase. However, the toxins also ADP-ribosylate a variety of particulate and soluble proteins more slowly. The ADP-ribose residues are probably always attached to arginine residues.

For its maximal activity, cholera toxin needs to interact with an active S site[3] also known as ARF.[4] S/ARF is a membrane protein that binds guanyl nucleotides but is distinct from G_s and G_i. In many cases, the predominant form of S/ARF is a soluble protein known as cytosolic factor[5] (CF) described in chapter [35] in this volume. Thus the ADP-ribosylation of G_s in membranes requires either the simultaneous presence of CF and a guanyl nucleoside triphosphate (GTP itself suffices unless the tissue sample has a very active GTPase activity) or prior exposure of membranes to CF and a nonhydrolyzable analog of GTP such as GTPγS, Gpp(NH)p, or Gpp(CH_2)p.

Conditions for ADP-Ribosylation *in Vitro*

Generation of Fragment A_1 of Cholera Toxin

The entire toxin multimer (A5B) allows the ADP-ribosylation to take place inside whole cells but for work with cell fractions its fragment A_1 is required and this must be separated from the rest of the molecule. A_1 is separated from fragment A_2 by reduction of the linking disulfide. Its activity is increased further by the presence of SDS which separates A_1 from noncovalent association with other components of the toxin and probably also prevents aggregation.

Prepare a solution of 2 mg/ml toxin in 50% glycerol and store at $-20°$. The ADP-ribosyltransferase activity is stable. On the day of use, mix 10 μl of this stock solution with 90 μl of 0.5% SDS, 5 m*M* dithiothreitol, 130 m*M* NaCl, 10 m*M* HEPES, pH 7.3. Stopper and incubate at 37° for 10 min. Do not cool on ice or the SDS will precipitate. Use the activated toxin at a ratio of 1 to 20 so that the membranes are exposed to about 0.025% SDS, which is tolerated well. If a diluent is necessary, use the same activating buffer plus 0.1% ovalbumin. Discard any excess toxin because it loses potency by oxidation.

Alternatively, prepare carboxymethyl toxin which is stably activated. Dissolve 1 mg of toxin in 4 ml of 100 m*M* Tris–HCl, pH 8.8, 5 m*M* dithio-

[3] D. M. Gill and R. Meren, *J. Biol. Chem.* **258,** 11908 (1983).
[4] R. Kahn and A. G. Gilman, *J. Biol. Chem.* **261,** 7906 (1986).
[5] D. M. Gill and J. Coburn, *Biochemistry* **26,** 634 (1987).

threitol, 0.5% SDS, 1 m*M* EDTA. Incubate 10 min at 37°. Add 7.4 mg iodoacetamide (giving 10 m*M*) and incubate for 30 min more. Dialyze against two changes of 0.1% SDS, 130 m*M* NaCl, 10 m*M* HEPES, pH 7.3. Bring up to 5 ml and store at −20° in 0.5 ml portions.

One-Step Method for ADP-Ribosylation

Wash membranes by diluting them in 10 to 20 vol of buffered saline (130 m*M* NaCl, 10 m*M* HEPES, pH 7.3, 0.01% NaN_3, 0.01 trypsin inhibitor units (TIU)/ml aprotinin) and centrifuging. Avoid greater dilutions. Additional protease inhibitors can be added if appropriate. Resuspend the pellet in 2 vol of NAD-free cell cytosol, or a more purified source of CF such as the DEAE-Sephadex-purified material described in this volume [35]. If no CF is available, compensate by using more guanyl nucleotide, toxin, and NAD, e.g., 3× more of each.

For one gel slot, incubate 15 μl of the membrane suspension for 30 min at 25 to 30° (avoid 37°) with 5 μl of the following mixture: one part of 200 μg/ml activated toxin, two parts of 50 μM [^{32}P]NAD (e.g., 10,000–50,000 cpm/pmol), one part of guanyl nucleotide, and one part of 200 m*M* thymidine (warmed to 37° to dissolve). The final concentrations are about 10 to 15 mg/ml of membrane protein, 10 μg/ml toxin, 5 μM NAD, up to 1 m*M* GTP, GTPγS, or Gpp(NH)p, 10 m*M* thymidine, and 100–200 m*M* NaCl.

Two-Step Method for ADP-Ribosylation

It is often advantageous to activate the S protein before adding the toxin and NAD. One reason is that activation of S is promoted at 37° but the toxin reaction is impaired at 37°.[3] In this case incubate 15 μl of the membrane-CF suspension with up to 1 m*M* Gpp(NH)p for 10 min at 37°. Then add 4 μl of the other ingredients (toxin, NAD, and thymidine) and continue for 30 min at 25 to 30°.

Gel Analysis and Interpretation

End the incorporation by adding 1 ml of ice-cold buffered saline, vortex, and recover the membranes by centrifugation. Before discarding the washings, 1 μl portions can be spotted onto TLC plates for analysis of the soluble products (see below for details).

If it is necessary to digest DNA before loading the gel, suspend the pellet in one drop of buffered saline containing 2 m*M* $CaCl_2$ and 3 U/ml of micrococcal nuclease. Incubate for 15 min at 37° and wash again.

Make the membrane pellet into a slurry by vortexing vigorously before adding SDS gel sample buffer. Fractionate on a polyacrylamide slab gel containing SDS (7.5–15%). Start at 100 V: the voltage can be increased when the dye front is about 1 cm past the stack. Stain with Coomassie blue. Remove unincorporated [^{32}P]NAD and other nucleotides by including in the destain bath a nylon net bag containing about 5 g of a mixed bed resin such as Dowex MR3. Dry the gel and prepare an autoradiogram.

The migration rates of ADP-ribosyl G_s vary with the gel system. On a 7–15% gradient gel the fastest band occupies a position equivalent to 42,000 M_r. In erythrocytes this is the only band (Fig. 1a) but many tissues have an additional band at 46,000–48,000 M_r. This is often resolved into a

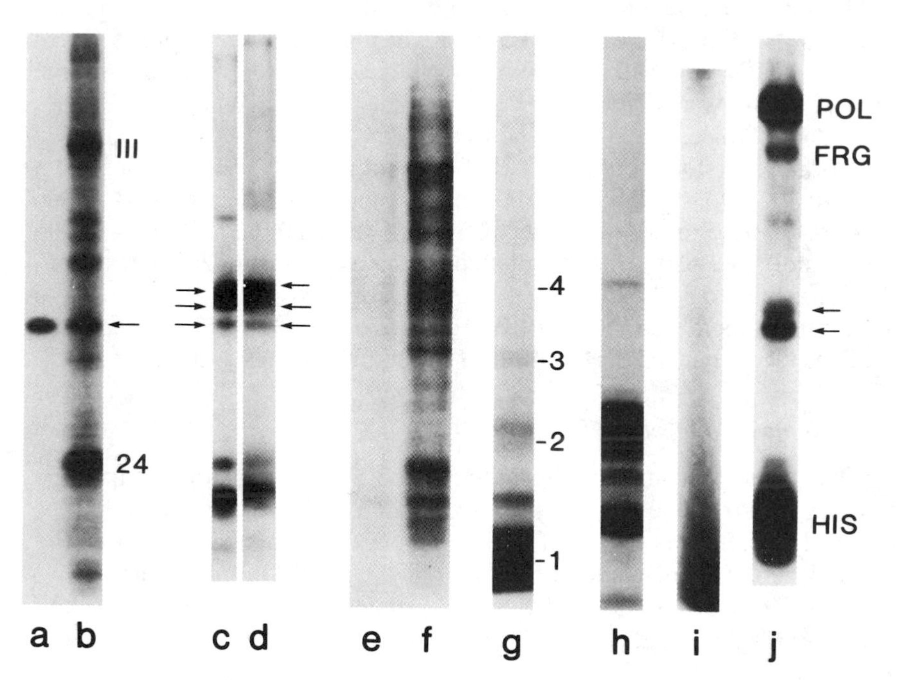

FIG. 1. Various [^{32}P]ADP-ribosylation patterns. The forms of G_s are indicated by arrows. (a) Pigeon erythrocyte membranes with only G_s (42,000 M_r) labeled; (b) the same, with secondary target proteins labeled as well (III, band III; 24, ca. 24,000 M_r); (c) and (d): calf brain membranes in which the 46,000/48,000 M_r form of G_s is labeled as well as the 42,000 M_r form. The two lanes show different arrangements of the larger bands; (e) no GTP, and (f) with 5 μM GTP, represent cyc$^-$ S49 lymphoma cell membranes. They lack G_s but the secondary target proteins are labeled when GTP is available. (g) The nonenzymatic labeling of globin monomer, dimer, trimer and tetramer in erythrocyte cytosol; (h) the labeling of tRNA molecules and half-molecules in bovine testis cytosol. GTP was supplied. (i) The radioactive smear obtained by running [^{32}P]ADP-ribose through an acrylamide gel; (j) the particulate fraction of wild-type S49 lymphoma cells which shows, in addition to G_s, some labeling of poly(ADP-ribose) polymerase (POL), a fragment of the polymerase (FRG), and core histones (HIS) despite the presence of thymidine.

doublet and its migration is sometimes anomalous (Fig. 1c and d). The 42K and 46/48K bands straddle the prominent stained band of actin (M_r 45,000) which is invariably present. On nongradient gels the G_s bands migrate relatively more slowly and may have apparent sizes of 44,000–45,000 and 49,000–52,000, respectively.

For quantitation, cut out gel bands after autoradiography and count them in Omnifluor. Depending on the situation, estimate the background from an adjacent area of the same gel track or the equivalent area of a companion track generated without toxin. When it is present at all, the 46K/48K species is usually more abundant than the 42K species but it is ADP-ribosylated an order of magnitude more slowly. The ratio of the radioactivities in the two bands therefore indicates the overall efficiency of the ADP-ribosylation.

ADP-Ribosylation of Secondary Substrates

The many other target proteins are ADP-ribosylated considerably more slowly than G_s but are often much more abundant and can dominate the labeling pattern (Fig. 1b). Although they are unrelated to G_s, GTP is required for their ADP-ribosylation (Fig. 1e and f). A constant finding is an ADP-ribosylated protein of 22,000–25,000 M_r which is at least 10 times more abundant than G_s. In erythrocytes this appears to be labeled in the soluble phase and then sticks to the membranes. Most other minor bands, particularly those that are smaller than 40,000 M_r, are less predictable and vary in labeling intensity according to conditions. A high-mobility group nuclear protein, M_r ca. 28,000, is enhanced in its labeling by high salt or SDS. In erythrocyte membranes band III is generally ADP-ribosylated to a small extent (Fig. 1b).

Tests for ADP-Ribosyl G_s

Criteria by which the various forms of [^{32}P]ADP-ribosyl-G_s can be recognized against a background of other toxin-dependent [^{32}P]ADP-ribosylated proteins include the following: (1) molecular size; (2) G_s is the most readily labeled and so is the first protein saturated as the toxin or NAD concentration is increased; (3) its labeling is blocked more readily by pretreatment of the membranes with the protein-modifying reagents diethyl pyrocarbonate[5] or MPU (3-chloromercuri-2-methoxypropylurea),[5,6] (4) in certain cases its labeling is reduced by GTPγS and a hormone, as described below.

[6] D. M. Gill and M. Woolkalis, *Ciba Found. Symp.* **112,** 57 (1985).

GTP Analog

G_s is a poorcr substrate for cholera toxin when it is fully activated to support adenylate cyclase activity. This occurs on binding guanyl nucleotides and involves a conformational rearrangement of the subunits which, at least in detergent solution, can be reflected in an actual physical dissociation of the α subunit from the β and γ subunits. The dissociation occurs more readily with GTPγS than with Gpp(NH)p and most readily of all with GTPγS and a suitable hormone. In the case of pigeon erythrocyte membranes, GTPγS alone somewhat reduces the number of G_s molecules that can be ADP-ribosylated [i.e., the extent of labeling with Gpp(NH)p plus GTPγS may be lower than with Gpp(NH)p alone] and the application of GTPγS with the beta adrenergic agonist isoproterenol can reduce the labeling of G_s by 90%. Thus, although GTPγS is fundamentally the most potent agonist available for activating S, under certain circumstances it will support less ADP-ribosylation of G_s than does Gpp(NH)p. Also, GTPγS is slowly hydrolyzed to GDP but Gpp(NH)p is not.

Ethylenediaminetetraacetic Acid

The binding of guanine nucleotides to G_s, which can be detrimental to the labeling, has a much more stringent magnesium requirement than does their binding to S.[3] Probably for this reason, labeling is sometimes increased by EDTA. Even very high concentrations of EDTA (e.g., 300 m*M*) can be tolerated. See below for its effect on poly(ADP-ribose) formation.

Salt. The toxin reaction requires some salt. The optimal amount of NaCl or KCl is approximately isotonic but higher concentrations of phosphate or sulfate ions stimulate the reaction further. They affect the toxin itself, not the substrate.[7,8] Avoid ammonium ions and Tris.

NADase

It is commonly difficult to ADP-ribosylate G_s in a membrane preparation that exhibits a high endogenous NADase (NAD^+ glycohydrolase) activity. NADase (NAD $\rightarrow$ nicotinamide + ADPribose + H^+) is an ectoenzyme of the plasma membrane, particularly of macrophages,[9] and is therefore very active in membranes made from lymphoid tissues, brain, and gut. In certain species (e.g., the rabbit) it is also very active on

[7] J. Moss and M. Vaughan, *J. Biol. Chem.* **252,** 2455 (1977).
[8] G. Soman, K. B. Tomer, and D. J. Graves, *Anal. Biochem.* **134,** 101 (1983).
[9] M. Artman and R. J. Seeley, *Arch. Biochem. Biophys.* **195,** 121 (1979).

erythrocytes.[10] There is not usually a significant NAD^+ pyrophosphatase activity (NAD → NMN + AMP) which can in any case be controlled by EDTA.

Detection of NAD Degradation. To assess the degradation of NAD during the toxin incubation, analyze 1-μl portions of the final washings by thin-layer chromatography. A suitable system is cellulose TLC plates (Eastman) developed with 60% ethanol : 0.3 *M* ammonium acetate buffer, pH 5.0. Typical R_f values are ATP, 0.08; NAD, 0.15; ADPR, 0.40; AMP, 0.49; orthophosphate, 0.60; nicotinamide, 0.94. Avoid the use of the polyethyleneimine cellulose (PEI) plates because traces of [^{32}P]ADP-ribose react with the PEI itself to generate a radioactive smear that cannot be quantitated and is easily overlooked. Expose X-ray film overnight with an intensifying screen.

ADP-Ribosylation in the Face of High NADase Activity

A combination of adjustments may be necessary.

1. Reduce the membrane concentration.
2. Increase the NAD concentration.
3. Increase the toxin concentration.
4. Supply 20 m*M* isonicotinic acid hydrazide (INH, isoniazid) and 1 m*M* 3-acetylpyridine adenine dinucleotide (3-APAD) to inhibit the NADase. This combination is most successful against the NADases of birds and ruminants, which are particularly sensitive to INH and in which 3-APAD acts synergistically with INH to preserve NAD.[11] The combination is less effective against the INH-insensitive NADases, for which higher concentrations of the inhibitors may be tested. 3-APAD is a partial substrate for cholera toxin (ca. 1.5% as efficient as NAD) so for quantitative work it is necessary to allow for the accumulation of nonradioactive ADP-ribosyl G_s from the 3-APAD.[11] The 3-APAD should be added at the same time as the NAD.

INH-adenine dinucleotide (INHAD), which is a side product of vertebrate NADases but is not a good substrate for the toxins, runs with NAD on TLC unless benzaldehyde is added to form the hydrazone. Use 50% ethanol, 0.3 *M* ammonium acetate buffer, pH 5.0, 10% benzaldehyde. Typical R_f values on cellulose plates are NAD, 0.41; ADPR, 0.49; orthophosphate, 0.63; INHAD, 0.72.

5. Supply 10 m*M* dithiothreitol as a further inhibitor of INH-sensitive and INH-insensitive NADases.[11]

[10] S. G. A. Alivasatos and O. F. Denstedt, *Science* **114,** 281 (1951).
[11] D. M. Gill and J. Coburn, *Biochim. Biophys. Acta* **954,** 65–72 (1988).

6. ADP-ribose, 20 mM, is occasionally beneficial. It inhibits NADase but is also an inhibitor of the toxin reaction.

7. Use a two-step incubation so that S is already activated by the time that the NAD is supplied.

Example: ADP-Ribosylation of Calf Brain Membranes

The following allows complete ADP-ribosylation with calf brain membrane preparations which have an active INH-sensitive NADase. As with all brain membrane preparations, these contain many closed vesicles and a nonionic detergent must be supplied to allow access. Since the detergent tends to inactivate CF, the latter is generally omitted.

Wash membranes and resuspend them in buffered saline. Preincubate at 25° (sic) for 10 min with 0.2% Triton X-100 and 1 mM Gpp(NH)p. To 10 μl of membrane suspension (100 μg protein), immediately add 10 μl of a second mixture and incubate for another 60 min at 25°. The final concentrations are 20 μg/ml activated toxin, 20 μM [^{32}P]NAD (5000–15,000 cpm/pmol), 20 mM INH, 1 mM 3-APAD, 10 mM thymidine, 10 mM dithiothreitol, 0.1% Triton X-100, 0.5 mM Gpp(NH)p (Fig. 1c and d).

Comparison with the ADP-Ribosylation of Brain G_i by Pertussis Toxin

This is much simpler than the ADP-ribosylation of G_s. S is not involved so only one step has to be considered. ATP or an ATP analog is required instead of GTP.[12] Guanyl nucleotides are best omitted for they inactivate the substrate.[13] Conversely EDTA is beneficial, possibly because it maintains G_i in the toxin-reactive, but functionally inactive, configuration. The devices for limiting NADase action are the same as those developed for cholera toxin. The following concentrations seem to allow the total ADP-ribosylation of G_i in calf brain: washed membranes, 5 mg/ml of protein, 10 μg/ml of activated pertussis toxin (preincubate at 37° for 30 min with 100 mM dithiothreitol; 0.1% SDS may also help), additional dithiothreitol for a final concentration of 10 mM, 100 μM [^{32}P]NAD (1000–5000 c/m/pmol), 50 mM EDTA, 20 mM INH, 1 mM 3-APAD, 1 mM ATPγS, 0.1% Triton X-100. Incubate at 37° for 30 min, then process as for G_s. Correct the observed [^{32}P]ADP-ribose incorporation for additional nonradioactive ADP-ribosylation from the 3-APAD, which is about 3% as efficient a substrate as is NAD.[11]

[12] L.-K. Lim, R. D. Sekura, and H. R. Kaslow, *J. Biol. Chem.* **260,** 2585 (1985).

[13] S. K. F. Wong, B. R. Martin, and A. M. Tolkovsky, *Biochem. J.* **232,** 191 (1985).

Competing Reactions and Artifactual Results

Nonenzymatic Addition of Free ADP-Ribose to Proteins and Other Macromolecules.[14] [^{32}P]ADP-ribose may be present in the starting [^{32}P]NAD sample (check by TLC) and more may be generated from the NAD by tissue NADase (plasma membrane) or by the combined action of poly(ADP-ribose) polymerase (nuclear) and poly(ADP-ribose) glycohydrolase (soluble). There is some specificity but in principle ADP-ribose can form an adduct with any protein and the reaction is usually noticed only for abundant proteins. In several tissues the main target is glyceraldehyde-3-phosphate dehydrogenase (subunit M_r 36,000). The labeling of globin and its multimers is shown in Fig. 1g. The nonenzymatic reaction is largely reversible and in our hands comes to equilibrium after about 24 hr at 37° with 1–2% of the ADP-ribose attached to protein. It cannot be blocked by adding an excess of unlabeled ADP-ribose. It is recognizable by these criteria: (1) it continues for many hours, (2) it increases with increasing temperature whereas the reaction of cholera toxin is faster at 25° than at 37°, (3) it is not prevented by digesting the [^{32}P]NAD to [^{32}P]ADP-ribose.

To prepare [^{32}P]ADP-ribose, weigh 2 mg of pig brain acetone powder (sold as NADase) into an Eppendorf tube. Soak it in water and wash several times by centrifugation. Discard any granules that float. Add 100 μl of 50 μM [^{32}P]NAD in 2 mM HEPES buffer, pH 7.3. Incubate at 37° with constant agitation for 10 min. Boil 2 min. Centrifuge. A portion should be analyzed by TLC (cellulose, not PEI) to ensure that no NAD remains and that the breakdown to [^{32}P]AMP and ortho[^{32}P]phosphate is modest. Other available mammalian NADases such as the one from bovine spleen may also be used but the material sold as NADase from *Neurospora crassa* is very crude and should be avoided.

Generation of [^{32}P]tRNA from [^{32}P]ADP-Ribose

Most tissues contain ADP-ribose phosphodiesterase, which rapidly converts [^{32}P]ADP-ribose to [^{32}P]AMP when magnesium is present.[15] The AMP may be dephosphorylated to yield ortho[^{32}P]phosphate or converted to [α-^{32}P]ATP at the expense of another nucleoside triphosphate. In the latter case, the radioactivity is not conjugated to proteins but is readily incorporated into the 3′-CCA terminal positions of tRNA molecules that have lost their final A. During gel electrophoresis, the mixture of end-labeled [^{32}P]tRNA species gives a cluster of radioactive bands in the

[14] E. Kun, A. C. Y. Chang, M. L. Sharma, A. M. Ferro, and D. Nitecki, *Proc. Natl. Acad. Sci. U.S.A.* **73,** 3131 (1976).

[15] J. M. Wu, M. B. Lennon, and R. J. Suhadolnik, *Biochim. Biophys. Acta* **520,** 588 (1978).

21,000 M_r region and another cluster in the 12,000-M_r region. The latter represents half-molecules of tRNA split in their anticodon loops (Fig. 1h).

Such incorporation has the following properties: (1) It depends on a nucleoside triphosphate such as GTP or UTP, or best of all CTP. ATP is less effective since it reduces the specific activity of the product. Nonhydrolyzable triphosphates do not work. (2) It is blocked by EDTA, which inhibits ADP-ribose phosphodiesterase. (3) Depending on the mechanism of producing the [^{32}P]ADP-ribose, the incorporation is blocked by inhibitors either of poly(ADP-ribose) synthesis or of NADases. (4) The product is alkali labile.

ADP-Ribosylation of the Acrylamide Gel

If the incubated membranes are now washed well, any residual [^{32}P]ADP-ribose reacts with the polyacrylamide itself during gel electrophoresis and cannot be removed. Over 0.1% of the input counts may attach to the gel. This effect generates a radioactive smear which is most intense toward the bottom of the gel where the migration of [^{32}P]ADP-ribose is slowest (Fig. 1i). The smear is absent when the tissue samples contain ADP-ribose phosphodiesterase and magnesium ions.

ADP-Ribosylation Catalyzed by Endogenous Transferases

Moss and colleagues have described several tissue enzymes that catalyze nonspecific ADP-ribosylations of proteins in broken cells.[16] The activity of these enzymes is particularly high in frog and turkey erythrocyte lysates where they can mask toxin-dependent labeling. While it is evident that these enzymes must not be as active in the intact cell as they are *in vitro,* neither the natural inhibitory mechanism nor a sufficiently selective artificial inhibitor has been identified. The best cure is to further purify the target membranes.

Poly(ADP)-Ribosylation

Membrane preparations usually contain nuclear fragments that contain both poly(ADP-ribose) polymerase (NAD^+ ADP-ribosyltransferase) and the broken DNA that this enzyme requires. It is advisable to add thymidine to inhibit this polymerase. Even so, there remains some ADP-ribose incorporation at 115,000 M_r (this is the polymerase itself) and often also in the region below 20,000 apparent M_r, which represents core histones ADP-ribosylated as a result of poly(ADP-ribose) formation. If further inhibition is needed, supply 10 mM 3-aminobenzamide and 10 mM EDTA or 100 mM phosphate. EDTA both inhibits polymer formation and

[16] D. A. Yost and J. Moss, *J. Biol. Chem.* **258,** 4926 (1983).

stimulates the activity of poly(ADP-ribose) glycohydrolase. In a two-step incubation, include EDTA only in the second step.

Alternatively, extract the ADP-ribosylated products with a detergent that dissolves G_s but not the products of poly(ADP-ribose) polymerase. For example, use 0.5–0.7% Lubrol, 15 min at room temperature[17] or 60 min on ice. For more complete extraction also add 0.15 *M* NaCl and 0.15% cetyltrimethylammonium bromide.

Toxin Activity Assays

The most sensitive assay involves measuring the rise in adenylate cyclase activity as a consequence of ADP-ribosylation. Freshly lysed pigeon erythrocytes are very sensitive and have a low basal cyclase activity. Wash the erythrocytes to remove serum and leukocytes, then suspend packed cells in 1 vol of HEPES-buffered saline using a glass tube. Immerse the tube in ethanol–dry ice, then thaw in room temperature water. Supplement the lysate with 5 m*M* NAD and 5 m*M* GTP. Incubate 40-μl portions with activated toxin for 30 min at 25°. Wash the membranes in 1 ml of saline and measure their adenylate cyclase activity, with GTP, by any standard method (this series, Vol. 38). The cyclase activity is approximately proportional to the logarithm of the toxin concentration between 1 and 100 ng/ml.

A less sensitive but more rapid method involves TCA precipitation of all the ADP-ribosylated proteins that are formed in brain particles. We use the total microsomal fraction from bovine cortex. Under the conditions described, G_s is greatly outnumbered by the secondary products. First pellet out calf brain particles sufficient for 200 μg/assay tube. Just before the assay, add the following ingredients to the pellet and then distribute 20-μl aliquots. Final concentrations are 100 μ*M* [^{32}P]NAD (1000 cpm/pmol), 1 m*M* 3-APAD, 20 m*M* INH, 10 m*M* thymidine, 0.1% Triton X-100, 1 m*M* Gpp(NH)p, 10 m*M* dithiothreitol, 10 m*M* EDTA, 10 m*M* magnesium diacetate, 1 *M* potassium phosphate, pH 7.6 (added last). The final pH is about 7.1. Add 1 μl of activated toxin or 1 μl of the toxin activation buffer for a control, and immediately mix. Incubate at 25° for 60 min, dilute with about 200 μl of ice-cold saline, then precipitate with ice-cold 10% TCA, sit 10 min on ice, and collect the protein by filtration on glass fiber, washing with 2.5% TCA. At 10 μg/ml, cholera toxin gives an incorporation of 5–15 pmol ADP-ribose which is about 10 times the background level.

Acknowledgment

This work was supported by NIH Grant AI 16928.

[17] G. L. Johnson, H. R. Kaslow, and H. R. Bourne, *J. Biol. Chem.* **253,** 7120 (1978).

[35] Assay and Purification of Cytosolic Factor Required for Cholera Toxin Activity

By MARILYN WOOLKALIS, D. MICHAEL GILL, and JENIFER COBURN

A cytosolic protein widespread in vertebrates is required in addition to a guanyl nucleotide triphosphate for the formation of a membrane complex that, in its turn, raises the ADP-ribosyltransferase activity of cholera toxin. The two steps can be separated in time. The cytosolic factor has been called CF[1] and its membrane-bound form S[2,3,4] or ARF.[5]

Assaying the Cytosolic Factor

Pigeon Erythrocyte Membranes for CF Assay

Pigeon erythrocyte membranes are particularly dependent on CF. They also have negligible NADase activity. Membranes can be prepared in bulk and stored until needed.

Draw blood from the wing veins of pigeons into syringes containing heparin. Wash the cells several times with buffered saline [130 m*M* NaCl, 10 m*M* HEPES, pH 7.3, 0.01% NaN_3, 0.01 trypsin inhibitor units (TIU)/ml aprotinin]. Remove surface white cells by aspiration after each centrifugation and by passing the suspension once through a column of washed lambswool. Suspend the erythrocytes to a hematocrit of 50% and lyse them by immersing the tube in ethanol–dry ice. Thaw, dilute with 20 vol of buffered saline, and centrifuge. To digest the DNA in the crude ghost pellet, add an equal volume of 4 m*M* $CaCl_2$, 6 U/ml micrococcal nuclease, and incubate at 37° for 15 min or until a small sample remains fluid when made 1% with SDS. Without washing, dispense the red slurry into 0.3-ml aliquots and store at −70°.

Standard CF Assay

On the day of use, thaw a tube of nuclease-treated erythrocyte ghosts, dilute with 1.5 ml of saline, vortex well, and recover the membranes by centrifugation (12,000 *g*, 5 min). Wash again. Resuspend the pink membrane pellet in one pellet volume of buffered saline. Include 0.1% ovalbu-

[1] K. Enomoto and D. M. Gill, *J. Biol. Chem.* **255,** 1252 (1980).
[2] D. M. Gill and R. Meren, *J. Biol. Chem.* **258,** 11908 (1983).
[3] D. M. Gill and M. Woolkalis, this volume [34].
[4] D. M. Gill and J. Coburn, *Biochemistry* **26,** 6364 (1987).
[5] R. Kahn and A. G. Gilman, *J. Biol. Chem.* **261,** 7906 (1986).

min in order to swamp differences in protein concentration between various samples of CF. On ice, add Gpp(NH)p or GTPγS to 200 μM and quickly dispense 10-μl portions to cold tubes that already contain 10 μl, including the test CF and appropriate buffer to standardize the salt concentration and volume. Incubate 15 min at 37° and return to ice. To each tube add 10 μl of a mixture of 20 μg/ml activated cholera toxin,[3] 20 mM thymidine, 15 μM [^{32}P]NAD, and 15% (w/v) polyethylene glycol 8000. Incubate at 25° for 20 min, then dilute with 1 ml of buffered saline. Recover the membranes by centrifuging at 15,000 g for 10 min. Discard the supernatant, make the pellet into a slurry by vigorous vortexing, and then dissolve it in 40 μl of gel sample buffer. For maximum sensitivity fractionate the entire pellet solution on an SDS–polyacrylamide gel (7.5–15% acrylamide) and quantify the [^{32}P]ADP-ribosyl G_s.[3] For quantitation, compare the test CF with a standard preparation over a range of concentrations. Since crude CF often demonstrates inhibition as well as activation, it is important to use several concentrations of the unknown and to compare rising phases of the response curves. As a negative control use a CF solution that has been heated to 80° for 15 min.

Rapid Minigel Assay

For rapid assay of column fractions, use 10 μl of each fraction per assay but fractionate only 10 μl of the sample buffer solution of membranes. Use minigels (10 × 8 cm, 12.5% acrylamide; IDEA Scientific, Corvallis, OR). Include positive and negative controls. Set the supply to 100 V until the sample clears the stacking gel, then 250 V for about 1 hr. Stain for 5 min, destain for 1 to 2 hr, dry the gel (30 min), and prepare an autoradiograph overnight.

More Rapid TCA Assay

An even faster alternative is to count the trichloroacetic acid-insoluble material. At the end of the 25° incubation, wash the membranes in buffered saline, resuspend in 100 μl of 1% serum albumin, precipitate with 10% TCA, filter on Whatman GFA glass fiber circles, and count. Beware, however, of using the TCA assay for crude CF which contains poly(ADP-ribose) polymerase.

Partial Purification of CF from Bovine Testis

Supernatants

Testis is a good source. Cool the testes of freshly slaughtered bulls on ice and conduct all subsequent steps of the purification at 0–4°. Discard the tunica albuginea. Mince testicular tissue (500 g) and homogenize in aliquots in a cooled Waring blender in 1 liter of buffer A (see below for recipe). Centrifuge at 30,000 *g* for 15 min. Aspirate the floating lipid. Centrifuge the combined supernatants at 100,000 *g* for 2 hr. Again aspirate any residual lipid off the surface. Collect the clarified supernatants, avoiding the cloudy layer above the pellet, and store at −70°.

Column Fractionation: General

Use polypropylene tubes to collect the fractions from all of the columns. The purification can be improved by analyzing fractions by SDS–polyacrylamide minigels (15% acrylamide, Coomassie blue staining) and selecting and pooling those fractions that minimize protein complexity. CF has a molecular size of 20,000–21,000 M_r. In the later stages when the protein concentration becomes very low, CF may be stabilized by mixing in additional protein, such as insulin, that can be readily resolved from CF. The recoveries from step to step are difficult to quantitate because CF of different purities gives response curves of different shapes and different plateaus.

DEAE-Sephadex

Apply 200 ml of the 100,000 *g* supernatant to a DEAE-Sephadex column (2.6 × 40 cm) equilibrated in buffer A. This column removes material that interferes with the resolution of later columns; batch procedures have not proved effective and DEAE-Sephacel cannot be substituted. Develop the column by pumping buffer A at 15 ml/hr, collecting 7-ml fractions. Assay portions of each sample for CF activity, using a mini-gel assay. The TCA assay should not be used at this stage because the earliest peak of incorporation represents fractions rich in poly(ADP-ribose) polymerase. This peak should be discarded. Pool the fractions with CF activity (200–300 ml) that represent the majority of the UV-absorbing material. CF purified to this stage is adequate for routine work.

QAE-Sephadex

For further purification, concentrate the DEAE-Sephadex pool to 25 ml using an Amicon TCF10 unit with a YM10 filter, dialyze extensively against buffer B, and centrifuge at 30,000 *g*. Apply to a QAE-Sephadex column (2.6 × 40 cm) equilibrated in buffer B. Pump at 30 ml/hr with buffer B, collecting 7-ml fractions. CF elutes at the trailing shoulder of the second major protein peak. Concentrate the active fractions to 5–10 ml as above.

Gel Filtration

Apply the concentrate to a BioGel P-30 column (2.5 × 85 cm) equilibrated in buffer A. Elute at a constant 40 cm pressure head. Collect 3-ml fractions. Precalibrate the column with standard proteins: CF elutes between soybean trypsin inhibitor (20,000 M_r) and cytochrome *c* (13,000 M_r). The TCA assay may be used: if there are two peaks the first peak represents poly(ADP-ribose) polymerase.

Hydroxylapatite

Concentration is not usually necessary. Dialyze the P30 pool extensively against buffer C and apply to a BioGel HTP column (0.9 × 11 cm) equilibrated in the same buffer. Pump at 40 ml/hr and collect 3-ml fractions. Develop the column with 60 ml of buffer C, 60 ml of buffer D, and then a linear gradient formed from 95 ml each of buffers D and E. CF represents 10 to 50% of the protein eluted between 120 and 165 m*M* phosphate. Unfortunately, however, some of the main contaminants may have molecular sizes close to that of CF. Before assaying the fractions, neutralize and supplement with 100 m*M* NaCl. Pool the active fractions, concentrate 10-fold and change the medium to buffer A. Store aliquots at −70°.

Buffers Employed in the CF Purification

Buffer A: 50 m*M* HEPES, 10 m*M* sodium phosphate, 100 m*M* NaCl, 2 m*M* magnesium diacetate, 1 m*M* EGTA, 1 m*M* dithiothreitol, 0.01 TIU/ml aprotinin, 5 μg/ml leupeptin, pH 7.6
Buffer B: 50 m*M* AMPSO, 10 m*M* sodium phosphate, 50 m*M* NaCl, 2 m*M* magnesium diacetate, 1 m*M* EGTA, 1 m*M* dithiothreitol, 0.01 TIU/ml aprotinin, 5 μg/ml leupeptin, pH 9.0
Buffer C: 20 m*M* sodium phosphate, 1 m*M* dithiothreitol, pH 5.8
Buffer D: 80 m*M* sodium phosphate, 1 m*M* dithiothreitol, pH 5.8
Buffer E: 200 m*M* sodium phosphate, 1 m*M* dithiothreitol, pH 5.8

Use degassed distilled water for preparing buffers C, D, and E.

Acknowledgment

This work was supported by NIH Grant AI 16928.

[36] Selective Destruction of Vagal Sensory Neurons Using *Shigella* Cytotoxin

By Ronald G. Wiley, Arthur Donohue-Rolfe, Gerald T. Keusch, and Thomas N. Oeltmann

Selective destruction of neurons based on retrograde axonal transport of cytotoxins is an elegant experimental technique and a possible mechanism of neurological disease. As an experimental tool, this "suicide transport" strategy has been successful in producing anatomically restricted destruction of neurons sending axons through a peripheral nerve injected with cytotoxin. The toxic lectins, ricin, abrin, and modeccin[1–4] have been used to selectively destroy *all* sensory and motor neurons projecting axons through a particular nerve by pressure microinjecting the toxin subepineurally into the appropriate peripheral nerve. These toxins are taken up by axons within the injected nerve and retrogradely transported to the parent cell bodies where they act to kill just those neurons. Such lesions can be valuable in neurochemical,[5] neuroanatomical,[6] and neurophysiological[7] studies. Recently, we have used the cytotoxin from *Shigella dysenteriae* (SdT) in similar fashion to destroy sensory neurons while sparing motor neurons sending axons through the injected nerve.[8] Animals prepared in this manner are uniquely valuable in experiments requiring selective sensory dennervation.

Materials and Methods

Toxin Production and Purification

Shigella dysenteriae 1 strain 60R[9] (Chapter 22 in this volume) is grown in modified syncase medium containing 1% casamino acids, 0.004% tryp-

[1] R. G. Wiley, W. W. Blessing, and D. J. Reis, *Science* **216,** 889 (1982).
[2] T. Yamamoto, Y. Iwasaki, and H. Konno, *J. Neurosurg* **60,** 108 (1984).
[3] C. J. Helke, C. G. Charlton, and R. G. Wiley, *Brain Res.* **328,** 190 (1985).
[4] R. G. Wiley and T. N. Oeltmann, *J. Neurosci. Meth.,* **17,** 43 (1986).
[5] C. J. Helke, C. G. Charlton, and R. G. Wiley, *Neuroscience,* **19,** 523 (1986).
[6] L. E. Westrum, M. A. Henry, L. R. Johnson, and R. C. Canfield, *Abstr. Soc. Neurosci.* **15,** 559 (1985).
[7] J. T. Wall, R. G. Wiley, and C. G. Cusick, *Abstr. Soc. Neurosci.* **15,** 904 (1985).
[8] R. G. Wiley, A. Donohue-Rolfe, and G. T. Keusch, *J. Neuropathol. Exp. Neurol.* **44,** 496 (1985).
[9] R. J. Dubos and J. W. Geiger, *J. Exp. Med.,* **84,** 143 (1946).

tophan, 0.2% glucose, and no added iron.[8] Cultures are grown aerobically with shaking at 300 rpm at 37° and harvested in early stationary growth (A_{600} = 3–3.5). Unless otherwise stated, all purification steps are at 4°. Cells are pelleted by centrifugation at 10,000 $g \times 10$ min and washed twice by resuspension in 10 m*M* Tris–HCl, pH 7.4. The final pellet is resuspended in the same buffer at 2% of the original volume and sonicated to achieve >95% cell lysis as indicated by A_{600}. Unbroken cells are spun out at 5000 $g \times 20$ min and the crude supernatant is applied to a column (2.5 × 50 cm) of Cibacron Blue F3G-A coupled to Sepharose CL-6B (Pharmacia) equilibrated with 10 m*M* Tris–HCl, pH 7.4. After 12 hr of continuous recycling, the column is washed with 10 column volumes of the same buffer followed by elution of bound material with 0.5 *M* NaCl in the same buffer. The protein peaks are pooled and dialyzed against 25 m*M* Tris–acetate, pH 8.3 before application to a column (0.9 × 20 cm) of Polybuffer exchanger 94 (Pharmacia) equilibrated with the dialysis buffer. Bound SdT is eluted with a degassed solution of Polybuffer 96 diluted 1 : 13 in H_2O and adjusted to pH 6.0 with acetic acid. Fractions (1.5 ml) are collected and assayed for toxicity on HeLa cells.[10] Toxin-containing fractions are then pooled, transferred to a dialysis bag, and concentrated to 2 ml final volume by incubation at room temperature with dry polyethylene glycol. Concentrated SdT is next applied to a column (1.5 × 100 cm) of BioGel P-60 equilibrated with 20 m*M* ammonium bicarbonate. The toxin-containing fractions are then pooled and lyophilized. The overall yield of toxin is 47%, the increase in specific activity is 1360-fold, and the final product gives two bands (32000 and 65000 M_r) on PAGE under reducing conditions.[11]

Production of Antiserum

SdT is converted to toxoid by a 3-day incubation at 37° in 0.1 *M* sodium phosphate, pH 8, containing 1% formalin. After dialysis of toxoid against PBS, 0.5 ml PBS containing 100 μg toxoid is mixed with 0.5 ml of Freund's complete adjuvant and injected subcutaneously at four sites into an adult New Zealand White rabbit. At 5-week intervals, 50 μg toxoid booster injections in Freund's incomplete adjuvant are given. Animals are bled for antiserum 8–10 days after each booster. After centrifugation to remove erythrocytes, aliquots of antiserum are stored at −70°. The antiserum was high titer; 10 μl of a 1 : 40,000 dilution immunoprecipitated 1 ng of SdT.

[10] G. T. Keusch, M. Jacewicz, S. Z. Hirschmann, *J. Infect. Dis.* **125,** 539 (1972).

[11] A. Donohue-Rolfe, G. T. Keusch, C. Edson, D. Thorley-Lawson, and M. Jacewicz, *J. Exp. Med.* **160,** 1767 (1985).

Peripheral Nerve Injection

Rats or mice are anesthetized and the desired nerve is surgically exposed, taking care not to damage the nerve. A dissecting or operating microscope is essential. After placing a 1-cm-wide strip of thin plastic film under the nerve the tip of a glass micropipet is inserted into the nerve trunk and threaded coaxially along the nerve for 1–3 mm. The micropipet tip diameter should be 25–50% of the nerve trunk diameter. SdT dissolved in PBS (9 g/liter NaCl in 5 m*M* sodium phosphate, pH 7.6) containing 0.1% Fast Green dye is loaded into the pipet prior to inserting it into the nerve. We prefer drawing the micropipet from 5-μl capillary pipets that have 1-μl graduations, which permit accurate measurement of injection volumes as small as 50 nl. Ejection pressure is applied to the micropipet using a water-filled 10-μl Hamilton syringe coupled via water-filled polyethylene tubing. A small bubble within the micropipet separates water from toxin solution. To obtain 90% destruction of vagal sensory neurons and little or no damage to motor neurons, we recommend injecting 0.1–0.2 μg of SdT in a 0.1-μl vol into the cervical vagus nerve of mice. For rats, doses of 0.5–1.0 μg in 0.3 μl is suggested. With these doses of SdT, most animals will die within a few days of systemic toxicity unless protected by subcutaneous injection of 0.5–1.0 ml of rabbit anti-SdT antiserum 30–60 min after injecting SdT into the nerve.[6] Delayed injection of antiserum in this fashion has no apparent effect on the ability of SdT to destroy vagal sensory neurons but it does uniformly protect against lethal systemic neurotoxicity (ascending paralysis) in mice and enterotoxicity in rats. Satisfactory vagal sensory neuron lesions with sparing of vagal motor neurons can often be produced without antiserum protection using 0.05–0.1 μg SdT in mice and rats but some animals may die after several days. Nerves larger than the vagus may require correspondingly larger doses of SdT to obtain satisfactory sensory neuron lesions and it is our impression that vagal neurons with small unmyelinated axons are more readily destroyed by subepineurally injected SdT. Although not specifically tested to date, injection of large nerves such as the sciatic would likely require larger doses of SdT with antiserum protection and would be best suited to experimental designs requiring destruction of the small sensory fibers.

Assessment of Lesion

The action of SdT on neurons is readily demonstrated within 24 hr using routine Nissl stains (see Fig. 1A). To ascertain the extent and anatomic distribution of the sensory neuron lesion produced by subepineural SdT injection, the animal is reanesthetized and perfused transcar-

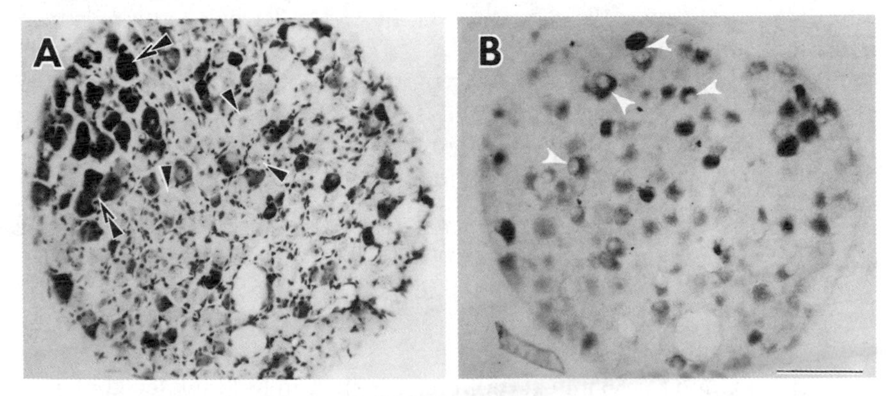

FIG. 1. Mouse nodose ganglion neurons 54 hr after injection of the ipsilateral cervical vagus nerve with 0.4 μg of SdT. (A) Cresyl violet (Nissl) stain demonstrating numerous severely chromatolytic neurons poisoned by SdT (single arrowheads) along with some normal-appearing neurons (double arrowheads); (B) indirect immunoperoxidase staining of a section from same ganglion as in (A). Many neurons contain numerous dark granules but nuclei are unstained (arrowheads). Primary rabbit anti-SdT antiserum was used at 1 : 20,000 dilution. Sections processed as described in text. Magnification bar is 100 μm. In the medulla of this animal, vagal motor neurons were well preserved.

dially with 4% formaldehyde (freshly prepared from paraformaldehyde) in 100 m*M* sodium phosphate buffer, ph 7.6. After 5–10 min of fixative, the perfusion is changed to 30% sucrose in 10 m*M* sodium phosphate, pH 7.6, for another 10 min. The appropriate tissues (ganglia and brainstem or spinal cord) are then removed by dissection and sectioned in a cryostat. Ganglia can be sectioned at 16 μm and brainstem or spinal cord at 32 μm. Sections are mounted on gelatin-coated slides and dried. Ganglia are best picked up directly off the microtome knife by contact with a room temperature slide, but brainstem or spinal cord sections can be lifted off with a fine brush and collected in PBS for subsequent mounting on gel-coated slides.

Dried slides are soaked briefly in water and stained for 5 min in cresyl violet (0.125% in H_2O with one drop glacial acetic acid/100 ml). Differentiation is begun by a brief water dip followed by 3 min each in 70 and 95% ethanol. Careful immersion in 95% ethanol containing two drops of glacial acetic acid/100 ml completes the differentiation process. Dehydration is accomplished by a brief rinse in 95% ETOH followed by 3 min each in two baths of 100% ethanol then clearing in xylene for 5–10 min. Slides are mounted and coverslipped using any of the standard media (Histoclad, Permount, etc.). Neurons poisoned by SdT will appear devoid of rough endoplasmic reticulum (Nissl substance) within 24–36 hr and will gradually disintegrate over the next several days. By 2 weeks, the poisoned neurons will be gone.

To demonstrate that SdT is transported to the appropriate neurons, cryostat sections dried in a vacuum desicator at room temperature can be processed for indirect peroxidase immunohistochemistry using the rabbit antiserum (see Fig. 1B). The steps include the following:

1. Presoak for 30 min in 1% normal goat serum in PBS containing 1% Triton X-100
2. Overnight incubation in humidified Petri dishes at 4° with rabbit anti-SdT antiserum at 1 : 5000–1 : 20,000 dilution in PBS
3. Two 5-min washes in PBS
4. Incubation for 60 min at room temperature in biotinylated goat anti-rabbit IgG (Vector Labs, Burlingame, CA)
5. Wash twice in PBS, 5 min each
6. Incubate for 60 min in Avidin-DH (Vector) in PBS
7. Wash twice in PBS, 5 min each
8. Incubate in peroxidase reaction mixture containing 0.5 mg/ml 3,3′-diaminobenzidine and 0.01% H_2O_2 in 0.05 *M* Tris–HCl, pH 7.2, for 10 min
9. Wash in distilled water for 2–3 min, then dehydrate through graded alcohols; clear, mount, and coverslip

In sections processed using the above protocol, SdT-poisoned neurons will contain numerous brown cytoplasmic granules.

[37] Measurement of Neurotoxic Actions on Mammalian Nerve Impulse Conduction

By JONATHAN J. LIPMAN

Neurotoxins act to impair the proper ionic and electrical activity of, and communication between, the excitable tissues of nerve and muscle. The sites at which neurotoxins act differ for each toxin and most toxins act at more than one site. Certain toxins act on neuronal membranes or those of their support cells, while others act on axonal translocation processes. Still others act on the nerve terminal processes responsible for neurotransmitter release. Measurement of the conduction properties of nerve does not supply information on synaptic or postsynaptic toxicity. On the other hand, measurement of the postsynaptic response properties of muscle following nerve stimulation identifies toxicity at the nerve and the neuromuscular junction in addition to identifying myotoxic effects. Neuromuscular junction preparations may thus be used to screen for neurotoxic effect of microbial and other neurotoxins. Data obtained from such preparations require further examination in other assay systems to identify the specific locus of toxicity. These other assay systems include intracellular recording, which is used to identify toxicity at the nerve axon membrane, and extracellular recording, which provides information on changes in conduction velocity in isolated nerves and in nerve bundles.

For the purpose of this chapter, the use of intracellular recording techniques is first described to illustrate the information which can be gleaned regarding the actions of toxins on nerve membrane processes. Extracellular recording techniques are then described with regard to the measurement of toxic effects on conduction velocity in isolated nerves and nerve bundles. Finally, neuromuscular junction preparations are described, to show how these identify toxic actions at perisynaptic motoneuron and effector tissue sites. Since a working understanding of the physiological processes deranged by toxin action is required, cursory descriptions of key aspects of these processes are given.

Toxicity at Axon Membrane Sites of Action Potential Propagation

The essential properties of the nerve cell membrane are common in principle and mechanism to all excitable membranes, including those of muscle. The intracellular microelectrode technique is used to assess toxicity at this level of organization. This uses a glass microelectrode with which the isolated nerve is impaled.[1] These microelectrodes are 50 to 0.5 μm at the tip and are filled with conducting solution, usually 3 M KCl. The electrical potential difference (PD) across the membrane can then be measured between the intracellular electrode and the bathing solution. In the resting state, this PD across the membrane is due to selective segregation of ionic charges largely due to Na^+, K^+, Cl^-, and large organic anions. This segregation, maintained against the ionic concentration gradients by active processes, is illustrated in general terms in Fig. 1. The electrostatic PD across the membrane (E_m) is approximately predicted from the Nernst equation, stated as:

$$E_m = \frac{RT}{Fn} \ln \frac{[\text{external concentration of permeating ion}]}{[\text{internal concentration of permeating ion}]}$$

where R is the thermodynamic gas constant (8.3 $JK^{-1} \cdot mol^{-1}$); T, the absolute temperature; F, the Faraday constant (96,500 C/mol); and n, the

[1] J. Del Castillo and B. Katz, *J. Physiol.* (*London*) **124,** 560 (1954).

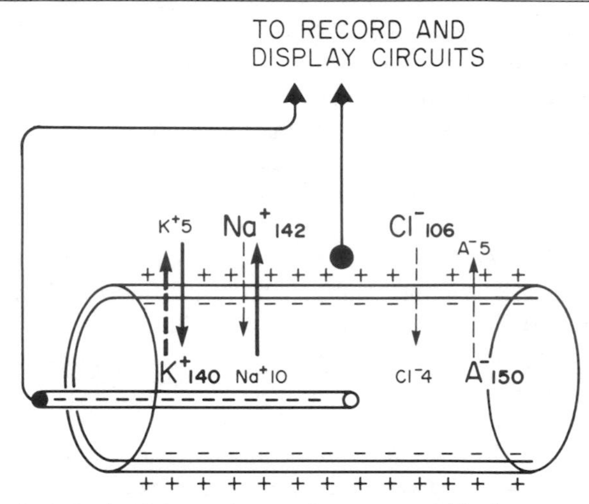

FIG. 1. The ionic basis of the resting membrane potential ionic concentrations is given (mmol/liter). Broken lines indicate the direction of ionic concentration gradients. Solid lines indicate directions of active transport. The membrane is impermeable to large intracellular organic anions (A^-) which contribute to the potential difference. Intracellular recording (as shown) would measure a membrane potential of about −85 mV (inside negative with respect to outside). Modified from W. C. Bowman and M. J. Rand, "Textbook of Pharmacology," 2nd Ed. Blackwell, London, 1980.

valency of the permeant ion. A sophistication of the intracellular recording technique, the voltage clamp method developed by Hodgkin and Huxley,[2] allows the toxicologist to change or control the membrane potential and to prevent the all-or-nothing action potential from occurring. By using this method to separately examine sodium and potassium currents, the actions of toxins affecting one or the other may be studied. The voltage clamp technique necessitates the introduction of two electrodes into the axon, one of which monitors the membrane potential in the usual way; the other being connected to the output of a feedback amplifier producing just sufficient current to hold the potential at the desired value.

The sites of toxin action on nerve membrane excitability and function are explicable with reference to the ionic basis of nerve membrane electrical events. At rest, active ionic charge separation gives the nerve membrane a resting potential of about 40 to 60 mV measured between the intracellular electrode and the external bathing fluid. The inside is negative with reference to the outside (Fig. 2). The nerve axon may be stimulated experimentally by producing a localized reduction in membrane PD of about 15 mV smaller than the resting potential (termed the *depolariza-*

[2] A. L. Hodgkin and A. F. Huxley, *J. Physiol. (London)* **117,** 500 (1952).

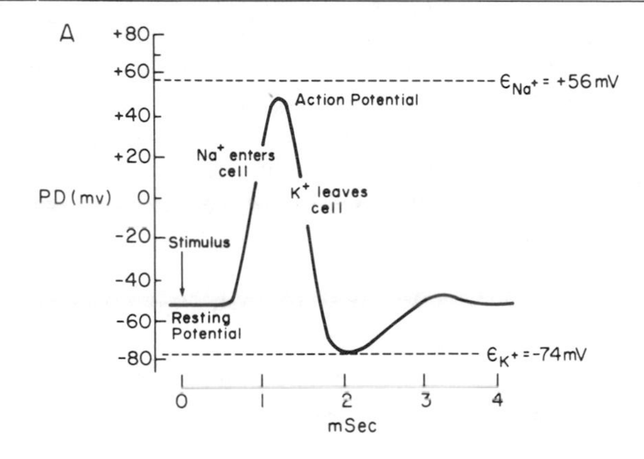

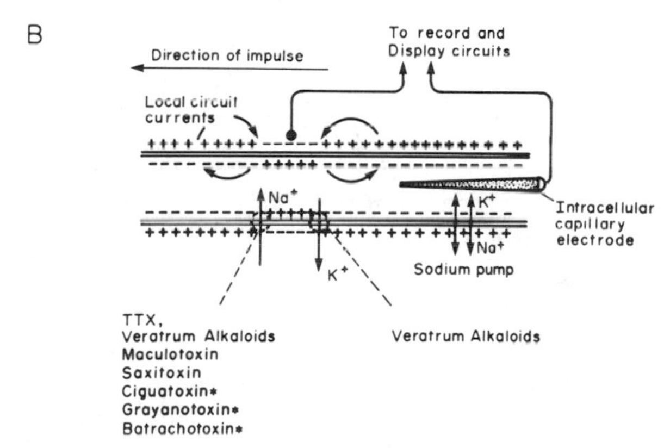

FIG. 2. Intracellular recording. (A) Illustrating the changes in membrane potential (PD, mV) measured between an intracellular capillary electrode and an electrode outside the membrane, as would be observed during the generation and propagation of an action potential illustrated in (B). The equilibrium potentials for sodium (E_{Na^+}) and potassium (E_{K^+}), as calculated from the Nernst equation, are shown. Data are idealized findings from the squid giant axon. (B) Showing the ionic and electrical events occurring during the generation of an action potential. Dotted lines indicate sites of neurotoxic action of some toxins. TTX, Tetrodotoxin. Those toxins shown block sodium channel activation, except where marked with asterisks (*), which depolarize by impairing inactivation.

tion threshold) which triggers an action potential (AP). Although a sequence of subthreshold depolarizations may summate to trigger an action potential, these must occur within a brief interval, known as the period of latent addition. Experimentally, the depolarization is conveniently initiated by a suprathreshold pulse of current passed between closely adjacent electrodes on the outside of the nerve. The voltage drop across the membrane resistance under the cathode locally depolarizes the membrane to initiate subsequent events. A large and sudden increase in the sodium permeability (termed *activation*) follows (see Fig. 2), and the membrane potential momentarily becomes reversed as sodium flows down its concentration gradient. This depolarization overshoots and reverses in polarity, approaching but not reaching the equilibrium potential for Na^+ before the Na^+ channels close (*inactivation*). Tetrodotoxin (TTX) obtained from the ovaries of the Japanese puffer fish (suborder Tetradontiae) has a cationic guanidium ion at physiological pH. This substitutes for Na^+, blocks the increase in Na^+ conductance (blocks activation), and arrests AP recruitment at this stage. Local anesthetic agents likewise prevent Na^+ activation by a membrane-stabilizing action. TTX also blocks conduction in skeletal muscle cells where the inward depolarizing current is carried by sodium ions, but not in smooth muscle where this is carried by other ions. A related toxin, saxitoxin, is produced by unicellular dinoflagellates (*Gonyaulax catanella*) which are ingested by a variety of shellfish. It was first isolated from the Alaskan butterclam (*Saxidomus giganteus*). The saxitoxin molecule possesses two guanidium ions and blocks neuromuscular transmission (see below) as well as nerve conduction. It is clinically responsible for "paralytic shellfish poisoning," a distinct clinical entity whose symptoms resemble tetrodotoxin poisoning. The veratrum alkaloids (obtained from *Veratrum alba*) exert a part of their toxic effect by delaying activation. Batrachotoxin (from the skin of the frog *Phallobates aurotaenia*), conversely, produces irreversible depolarization by increasing the resting Na^+ permeability and opening Na^+ conduction gates, blocking inactivation. Ciguatoxin, from ciguateric moray eels and grayanotoxin, from rhododendron leaves, exert a similar neurotoxic action to batrachotoxin. Their effect is antagonized, therefore, by tetrodotoxin. As the positive overshoot peaks and sodium channels inactivate, the restoration of the resting membrane potential is facilitated by a secondary increase in potassium permeability along its concentration gradient. A component of the neurotoxic actions of the veratrum alkaloids is exerted by inhibition of these potassium channels. Following these ionic exchanges which constitute the action potential, the resting state is recovered by an increase in the activity of the Na^+/K^+ pump, during which time the nerve is refractory to further impulse transmission. A band of refractoriness therefore allows the AP as it propagates along the nerve, the local circuit current generated by the AP triggering further depolarization of the resting region ahead in the direction of propagation.

Toxic Effects on Conduction

The significance of the local circuit current in initiating the AP is critical as we consider the properties of myelinated neurons. The velocity

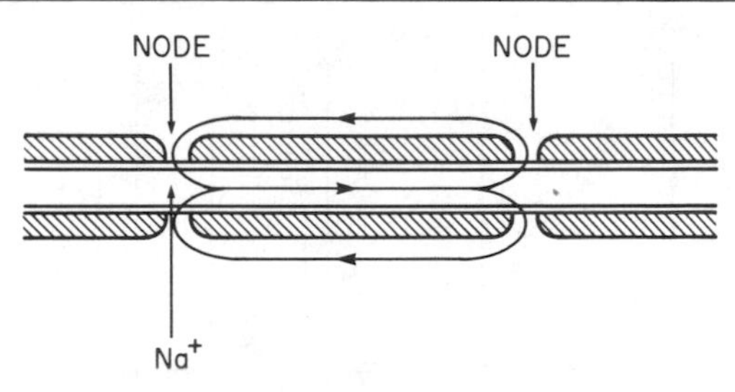

FIG. 3. Local circuit currents in myelinated nerve emanating from a node of Ranvier at which an action potential occurs travel to the next node to initiate an impulse there. Conduction along the nerve as a whole is termed saltatory since it jumps from one node to another. The myelin sheath is shown shaded.

of an AP's travel along a nerve is invariate for a particular fiber; it is determined by fiber size (which also determines the excitability or threshold for depolarization) and the presence or absence of myelination (see Table I). Myelination is a biological strategy for increasing the conduction velocity (CV) of a nerve. Nerve fibers are functionally classified on the basis of CV and fiber size. Class A fibers are the fastest conducting of the myelinated fibers. B fibers are also myelinated but are of smaller diameter than A, and hence slower in conduction. C fibers are nonmyelinated and hence are the slowest. The myelin sheath of a myelinated nerve develops from the surface membranes of the Schwann cells which envelope the developing nerve. The sheath is electrically of high resistance, and is discontinuous along the length of the nerve, gaps interrupting successive Schwann cell-derived lengths of myelin. Those gaps, termed nodes of Ranvier, expose the underlying axon (Fig. 3). Nerve conduction velocity in myelinated fibers is so much faster than in nonmyelinated axons because APs occur only at the nodes. Local circuit currents generated by the AP pass at the speed of electrical conduction through the axoplasm and extracellular fluid to the next available node, where depolarization is triggered. The greater the diameter of the axon the lower is the axoplasmic resistance to local circuit current and the faster is the electrotonic spread of current from one node to the next. This type of conduction, where the impulse jumps from one node to the next, is termed saltatory conduction.

Neurotoxic demyelination such as that produced by diphtheria toxin (from *Corynebacterium diphtheriae*) can thus be detected in myelinated nerve as a reduction in CV. This has been observed in a variety of preparations.[3-5] In isolated nerves, Rasminsky and Sears[6] have demonstrated

[3] B. H. Waksman, R. D. Adams, and H. C. Mansmann, *J. Exp. Med.* **105,** 591 (1957).

[4] H. E. Kaeser and E. H. Lambert, *Electroencephalogr. Clin. Neurophysiol. Suppl.* **22,** 29 (1962).

TABLE I
PROPERTIES OF PERIPHERAL MAMMALIAN NERVE

Property	A: Fiber group α	A: Fiber group β	A: Fiber group γ	B	C: Fiber group, Sympathetic	C: Fiber group, Dorsal root
	Motor to striated muscle and sensory afferents	Sensory afferents	Muscle spindle efferents and sensory afferents	Preganglionic autonomic efferents	Postganglionic sympathetic	Sensory afferents
Fiber diameter (μm)		← 1–25 →		3	← 1–1.3 →	
Conduction velocity (m/sec)	100–200	60	40	3–14	← 1–2.01 →	
Spike (AP) duration (m/sec)		← 0.4–0.5 →		1.2	← 2.0 →	
Absolute refractory period duration (m/sec)		← 0.4–0.5 →		1.2	← 2.0 →	

increased internodal conduction times of 500 μsec in ventral nerve root fibers demyelinated by diphtheria toxin compared with 19.7 μsec in control fibers.

Conduction velocity is most conveniently measured by extracellular recording techniques. Nerves occur *in vivo* in mixed bundles containing both myelinated and nonmyelinated axon embedded in a ground substance, the endoneurium. The bundles themselves are organized into nerve trunks bounded by perineural connective tissue. When extracellular recording electrodes are placed on the surface of a nerve trunk at some distance from the stimulating electrodes, a monophasic record of the compound action potential (CAP) of the whole trunk is obtained.[7] Extracellular recordings of this type differ in meaning and appearance from the intracellular recordings referred to previously. The monophasic record indicates the passage of AP depolarization under the recording electrode relative either to a widely separated or indifferent electrode. The CAP is the summated current due to individual APs of conducting nerves within the nerve trunk. Since most nerve trunks contain fibers of several different types and each type conducts at a different velocity (Table I), the impulses arrive temporally separated to some measurable degree at the recording electrode, and the shape and form of the CAP reflects these individual contributions. Each potential wave in the CAP reflects the population response of a single fiber type. The AP travelling along the fast A_α fibers are detected first, followed by A_β, A_γ, B, and finally C fiber potentials. The stimulation thresholds for the thicker fibers are lower than for the finer ones, so that it is possible to stimulate the fast conducting fibers preferentially, although the converse is not the case; any stimulus adequate to trigger slow fibers is necessarily suprathreshold for faster fibers. The process of selective stimulation of the fibers in a nerve bundle is illustrated in Fig. 4. In practice, the CVs of a number of nerves of a single fiber type vary with a distribution that is typical for the bundle. Several factors conspire to complicate the interpretation of CAPs of a nerve bundle. These factors include a rather complex recovery cycle attending the passage of each AP as it is generated during the sequence of repetitive impulse trains.[8] The CAP of nerve bundles may nevertheless be analyzed in terms of the distribution of its component conduction velocities (DCV). A formal analysis of the factors responsible for and the analy-

[5] J. A. Morgan-Hughes, *J. Neurol. Sci.* **7,** 157 (1968).

[6] M. Rasminsky and T. A. Sears, *J. Physiol.* (*London*) **227,** 323 (1972).

[7] H. S. Gasser and J. Erlanger, *Am. J. Physiol.* **62,** 496 (1922).

[8] S. G. Waxman, *in* "Conduction Velocity Distributions" (L. J. Dorfman, K. L. Cummins, and L. J. Leifer, eds.). Liss, New York, 1981.

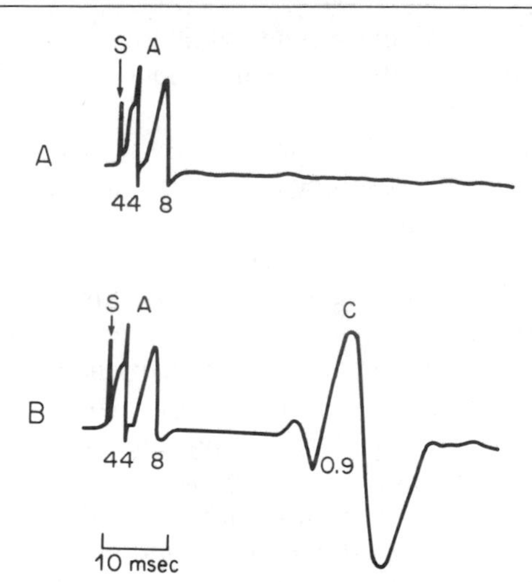

FIG. 4. The compound action potential (CAP) measured from rabbit saphenous nerve, antidromically stimulated 4 cm from the measuring electrodes. S, Stimulus artifact. (A) Following rectangular pulse stimulation of 100-μsec duration, two groups of fast and slow A fiber potentials conducting at 44 and 8 m/sec. (B) When the stimulus amplitude was doubled using 500-μsec duration, the higher threshold "C" fibers are also excited, conducting at 0.9 m/sec. The rabbit saphenous nerve does not contain B fibers, classified on the basis of threshold. Adapted from P. J. Watson, *Eur. J. Pharmacol.* **1,** 407 (1967).

sis of these distributions is reviewed by Dorfman *et al.*[9] Using the DCV technique, Hernández-Cruz and Munoz-Martinez[10] find that Tullidora toxin (from buckthorn, *Karwinskia humboldtiana*) preferentially affects larger rather than smaller diameter fibers. These authors provide a plausible explanation, therefore, for the ability of the toxin to produce clinical quadraplegia with intact sensorium via preferential motoneuron blockade.

Conduction velocity studies thus provide information on a nerve's ability to generate and propagate an AP. Factors affecting generation, such as sodium conduction blockade, and propagation, such as demyelination, interfere with the DCV measured from single nerves or bundles. Clinically, these principles form the basis of the practice of electromyography.[11] CV studies do not directly provide information on toxic effects mediated peri- or postsynaptically, however, for which nerve–muscle preparations must be used.

[9] L. J. Dorfman, K. L. Cummins, and L. J. Leiffer (eds.), "Conduction Velocity Distributions: A Population Approach to Electrophysiology of Nerve." Liss, New York, 1981.

[10] A. Hernández-Cruz and E. J. Munoz-Martinez, *Neuropathol. Appl. Neurobiol.* **10,** 11 (1984).

[11] M. Cherington and R. D. Snyder, *N. Engl. J. Med.* **278,** 95 (1968).

Presynaptic Toxic Sites

The intracellular microtubular system of the neuron is in active use and turnover, translocating cytoplasmic components to and from nerve cell body to presynaptic terminal. Toxins that inhibit the metaphase of mitotic division also disrupt this microtubular integrity and block nerve function. Examples of such toxins include colchicine (from the autumn crocus), the vinca alkaloids (vinblastine and vincristine from *Vinca rosea,* the periwinkle plant), podophyllotoxin (from *Podophyllum* resin), and griseofulvin (from the molds, *Penicillium griseofulvum* Dierckx and *P. janczewskii*). The neurotoxic effect of these toxins is to disrupt microtubular translocation processes and hence to block neurotransmitter delivery and availability at presynaptic nerve endings, both central[12,13] and peripheral.[14]

The AP on arrival at the nerve terminus evokes the release of neurotransmitter substances which transit the cleft between presynaptic membrane and effector cell. Acetylcholine (ACh) is the neurotransmitter at the neuromuscular junction striated muscle, often used to assay for these neurotoxic effects. A large number of neuromuscular junction preparations are available to the toxicologist, prominent among these being the rat (or mouse) phrenic nerve diaphragm which has been used, for instance, to assay β-bungarotoxin[15,16] and the rat hemidiaphragm preparation. This has been used for assaying the effects of lophotoxin.[17] The most widely used system is probably the frog nerve–sartorius muscle preparation described below.

At the resting neuromuscular junction prior to the arrival of an AP, a steady, nonstimulated, low-frequency release of quanta of ACh molecules occurs. These quanta, representing the entire contents of single vesicles, transit the synaptic cleft and provoke miniature end-plate potentials (MEPPs) in the muscle. Individually these are inadequate to induce a muscle membrane depolarization. On receipt of an AP at the nerve terminus, concerted exocytotic release of many ACh quanta occurs. Unlike the AP itself, this quantal release is not blocked by tetrodotoxin, indicating that sodium conductance is not critical for this presynaptic process. Present evidence suggests that vesicle migration and ACh release require Ca^{2+} conductance changes and, indeed, Ca^{2+}-binding agents (chelating

[12] P. R. Howe, P. F. Rogers, and W. W. Blessing, *Neurosci. Lett.* **52,** 287 (1984).

[13] R. E. Davis, B. E. Schlumpf, and P. D. Klinger, *Toxicol. Appl. Pharmacol.* **80,** 308 (1985).

[14] E. M. Volkov and V. N. Frosin, *Neirofiziologia* **16,** 231 (1984).

[15] R. B. Kelly and F. R. Brown, *J. Neurobiol.* **5,** 135 (1974).

[16] C. C. Chang and M. C. Huang, *Navnyn-Schmiedeberg's Arch. Pharmacol.* **282,** 129 (1974).

[17] P. Culver and R. S. Jacobs, *Toxicon* **19,** 825 (1981).

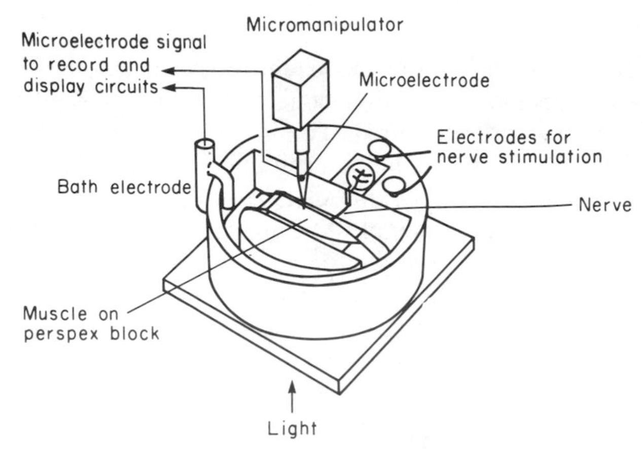

FIG. 5. A nerve–muscle chamber suitable for toxicity studies on the frog nerve–sartorius muscle preparation, after Fatt and Katz.[18] Muscle cells in the end-plate region are impaled by the intracellular microelectrode for examination of end-plate potentials arising from nerve stimulation. In use, the bath is fluid filled to about 3 mm above the muscle. See text for details.

agents) exert neurotoxic action at this locus. Postsynaptic receptor activation, in response to an excitatory neurotransmitter, simultaneously increases the membrane permeability to all small ions, including Na^+, K^+, and Ca^{2+}, causing localized depolarization of an extent and degree proportional to the rate of neurotransmitter–receptor interaction. The isolated frog nerve–sartorius muscle preparation of Fatt and Katz[18] has been used to examine these miniature end-plate potentials and excitatory postsynaptic potentials (EPSPs). In this preparation (Fig. 5), an intracellular recording electrode impales the muscle cells of the end-plate region. A buffer containing an elevated Ca^{2+} concentration (3.6 m*M* instead of 1.8 m*M*) is used to raise the threshold of the muscle fiber and to raise the end-plate potential by about 25%. Measurement of the end-plate potential in this preparation is facilitated by treating the muscle with tubocurarine (explained below).[19,20] The effect of this treatment is to reduce the amplitude of the EPP below the threshold of the muscle fiber so that no muscle impulse arises and a local subthreshold potential change remains. Kao and Nishiyama[21] have used this preparation to explore saxitoxin action and Abe *et al.*[22] have used it to assay β-bungarotoxin.

[18] P. Fatt and B. Katz, *J. Physiol.* (*London*) **115,** 320 (1951).

[19] J. C. Eccles, B. Katz, and S. W. Kuffler, *J. Neurophysiol.* **4,** 362 (1941).

[20] S. W. Kuffler, *J. Neurophysiol.* **5,** 18 (1942).

[21] C. Y. Kao and A. Nishiyama, *J. Physiol.* (*London*) **180,** 50 (1965).

[22] T. Abe, A. R. Limbrick, and R. Miledi, *Proc. R. Soc. London, Ser. B* **194,** 545 (1976).

The excitatory postsynaptic potential (EPSP) differs from the AP in four ways. First, it is a simple depolarization rather than a reversal in polarity; second, it is a graded response depending on the amount of neurotransmitter released; third, it has no refractory period such that successive responses may summate; and fourth, it is nonpropagated and attenuates rapidly with distance from the receptor. Summation of EPSPs to a critical value triggers the postsynaptic event. α-Latrotoxin (from black widow spider venom) is believed to inactivate the presynaptic membrane by forming ion channels in the membrane which permit the passage of Na^+, K^+, Ca^{2+}, and Mg^{2+}.[23-25] At the neuromuscular junction, this provokes spontaneous release of ACh quanta, thereby increasing the frequency of postsynaptic MEPPS to 500–1000 times normal. This is followed by a gradual decline over 30–60 min to zero and to a condition of presynaptic neurotransmitter depletion and inactivation. Observation of this sequence of events has been made for the concentrated venom itself.[26] In contrast to α-latrotoxin, botulinum toxin initially produces a flaccid paralysis when administered to the whole animal. This appears to be due to the inhibition of evoked release of ACh by combination with membrane release sites. Postsynaptically reduced MEPP frequency is observed, eventually leading to complete inhibition.[27,28] β-Bungarotoxin from the Taiwan banded krait, and Australian red–black spider venom are believed to disrupt synaptic vesicles and deplete the nerve endings of acetylcholine. The effect of tetanus toxin, in contrast to that of botulinus toxin, is to produce a spastic paralysis by preventing the release of ACh and other neurotransmitters selectively in spinal inhibitory processes of the central nervous system (CNS).[29] However, at the neuromuscular junction, tetanus toxin inhibits ACh release in a botulinus toxin-like manner.[30,31] Tetanus toxin is not delivered to the CNS via the circulation but by internalization within and retrograde transport by the peripheral nervous system in sensory, motor, and autonomic axons.[32,33]

[23] W. P. Hurlbut and B. Ceccarelli, *Adv. Cytopharmacol.* **3,** 87 (1979).

[24] H. E. Longenecker, Jr., W. P. Hurlbut, A. Mauro, and A. W. Clark, *Nature* (*London*) **225,** 701 (1970).

[25] A. Finkelstein, L. L. Rubin, and M. C. Tzeng, *Science* **193,** 1009 (1976).

[26] J. E. Smith, A. W. Clark, and T. A. Kuster, *J. Neurocytol.* **6,** 519 (1977).

[27] E. J. Schantz and H. Sugiyama, *Essays Toxicol.* **5,** 99 (1974).

[28] A. S. V. Bergen, F. Dickens, and L. J. Zatman, *J. Physiol.* (*London*) **109,** 10 (1949).

[29] V. B. Brooks, D. R. Curtis, and J. C. Eccles, *J. Physiol.* (*London*) **135,** 655 (1957).

[30] G. N. Kryzhanovsky, *Navnyn-Schmiedeberg's Arch. Pharmacol.* **276,** 247 (1973).

[31] N. Ambache, R. S. Morgan, and G. P. Wright, *J. Physiol.* (*London*) **107,** 45 (1948).

[32] K. Stöckel, M. Schwab, and H. Thoenen, *Brain Res.* **99,** 1 (1975).

[33] W. Dimpfel and E. Habermann, *Navnyn-Schmiedeberg's Arch. Pharmacol.* **280,** 177 (1973).

A number of polypeptide neurotoxins derived from snake venoms and snails act at the neuromuscular junction. So-called alpha-type venom toxin, e.g., α-bungarotoxin and α-conotoxin,[34] act postsynaptically to bind the ACh receptor (see below); others act presynaptically to prevent ACh release. Both effects are measurable in isolated nerve–muscle preparations. Examples of these presynaptic toxins include the aforementioned β-bungarotoxin (from the banded krait, *Bungarus multicinctus*), crotoxin (from the South American rattlesnake, *Crotalus terrificus*), notexin (from the Australian tiger snake, *Notechis scutatus scutatus*), and taipoxin (from *Oxyuranus scutellatus scutellatus,* the Australian taipan).[35] The mechanisms of neurotoxic action have been fairly well explored, particularly for β-bungarotoxin, which disrupts presynaptic neurotransmitter vesicles. The neuromuscular blockade produced by this agent occurs in three stages.[36,37] Initially, a slight reduction in the stimulated EPP is observed, followed by an increase which is attributed to a toxic influx of Ca^{2+} lasting 30–60 min. During the third stage, MEPP frequency diminishes to complete neuromuscular blockade. Atraxotoxin from the Australian funnel-web spider (*Atrax robustus*) similarly evokes an initial increase of ACh release followed by depletion of ACh from somatic motor nerve endings. Unlike black widow spider venom, atratoxin does not affect adrenergic neurons, however. A different toxic effect at this site, induced by preferentially inactivating Ca^{2+} entry into the nerve terminal, is responsible for the toxicity of the polypeptide ω-conotoxins from fish-hunting cone snails of the genus *Conus*.[34]

Intrasynaptic Sites of Toxicity

Between the presynaptic membrane and the receptors of the effector cells lies the synaptic cleft, which neurotransmitters must cross to evoke receptor and postreceptor responses. Neurotransmitter synthesis, storage and inactivation by either reuptake, metabolism, or both, each provide vulnerable sites for neurotoxic action. A large variety of neurotransmitter and neuromodulator substances are used by the nervous system subserving various functions in different types of synapses. Although a review of such substances and their susceptibility to toxic derangement lies outside the scope of the present chapter, the cholinergic synapse at the striated neuromuscular junction may usefully be considered. Acetylcholinester-

[34] B. M. Olivera, W. R. Gray, R. Zeikus, J. M. McIntosh, J. Varga, J. Rivier, V. DeSantos, and L. J. Cruz, *Science* **230,** 1338 (1985).

[35] B. D. Howard and C. B. Gundersen, Jr., *Annu. Rev. Pharmacol. Toxicol.* **20,** 307 (1980).

[36] T. Abe, S. Alema, and R. Miledi, *Eur. J. Biochem.* **80,** 1 (1977).

[37] R. B. Kelly, R. J. von Wedel, and P. N. Strong, *Adv. Cytopharmacol.* **3,** 77 (1979).

ase (AChE) is the principal enzyme responsible for terminating the action of acetylcholine (ACh) by hydrolysis at this site. AChE is located both pre- and postjunctionally. Prejunctionally, it is formed in the cholinergic cell body and transported by axonal translocation processes to the presynaptic nerve terminus. Postjunctionally, AChE is formed in muscle but its formation arises as a consequence of presynaptic ACh release and effect. Neurotoxic inhibitors of AChE include physostigmine from calabar beans, the seeds of the vine *Physostigma venenosum,* and galantamine, an alkaloid extracted from the bulbs of the caucasian snowdrop *Galanthus woronowii.* The effects of intrasynaptic toxin actions are best revealed by neuromuscular junction assay systems, as described above. AChE inhibition at the neuromuscular junction greatly enhances postsynaptic cholinergic receptor stimulation, evoking spontaneous muscle fasciculation and enhanced and prolonged muscle contraction. Certain central nervous system (medullary) synapses subserving respiratory movements are cholinergic, and central AChE toxicity is usually manifest, therefore, as respiratory paralysis.

Postsynaptic Sites of Toxicity

Neurotoxins which act at the postsynaptic ACh receptor at the neuromuscular junction block this receptor by either a depolarizing or nondepolarizing action. They therefore have a muscle relaxing effect. The "α-type" toxins exert such postsynaptic ACh receptor blockade either directly at the ACh-binding site or at functionally linked membrane structures. Included among these are polypeptide constituents of the venoms of certain snakes in the family Elapidae (cobras, kraits, corals, mambas, tiger snakes, death adders, black snakes, and taipans) and the sea snakes, Hydrophiidae. These polypeptide α-toxins bind strongly, specifically, and irreversibly to the ACh receptor. Curare, a generic term for various alkaloids extracted from plants of the species *Strichnos* and *Chondrodendon,* is a nondepolarizing blocker of the receptor. Presently, the most important of these alkaloids is (+)-tubocurarine isolated from *Chondrodendon tomentosum* by King in 1935.[38] It is so named from the practice of preparing this drug as a blowgun arrow poison by packing it into tubes of bamboo by certain Amazonian aboriginals from the rain forests of Peru's eastern border. Another curare-type neurotoxin is C-toxiferine, isolated from the bark of *Strychnos toxifera.* The prefix C indicates the practice of the aboriginals, who isolated this, of packing the prepared toxin in calabashes (or gourds). C-Toxiferine is about 25 times more potent than tubocurarine and has a longer duration of action. Nondepolarizing blockade is

[38] H. King, *Nature (London)* **135**, 469 (1935).

TABLE II
PRIMARY SITES OF ACTION OF, AND DETECTION METHODS FOR, SOME NEUROTOXINS ON MAMMALIAN PERIPHERAL NERVOUS SYSTEMS

		Detection method		
Toxin	Primary site of action	Intracellular recording from nerve	Extracellular recording from nerve	Neuromuscular junction
Tetrodotoxin	Nerve membrane	⋆	⋆	⋆
Batrachotoxin	Nerve membrane	⋆	⋆	⋆
Ciquatoxin	Nerve membrane	⋆	⋆	⋆
Grayanotoxin	Nerve membrane	⋆	⋆	⋆
Veratrum alkaloids	Nerve membrane	⋆	⋆	⋆
Saxitoxin	Nerve membrane	⋆	⋆	⋆
Vinca alkaloids	Intraneuronal microtubles		⋆	⋆
Colchicine	Intraneuronal microtubles		⋆	⋆
Tullidora toxin	Nerve fiber		⋆	⋆
Diphtheria toxin	Myelin sheath		⋆	⋆
β-Bungarotoxin	Presynaptic terminal			⋆
α-Latrotoxin	Presynaptic terminal			⋆
Black widow spider venom	Presynaptic terminal			⋆
Botulinum toxin	Presynaptic terminal			⋆
Tetanus toxin	Presynaptic terminal			⋆
Crotoxin	Presynaptic terminal			⋆
Notexin	Presynaptic terminal			⋆
Atratoxin	Presynaptic terminal			⋆
ω-Conotoxin	Presynaptic terminal			⋆
Physostigmine	Intrasynaptic			⋆
Galantamine	Intrasynaptic			⋆
(+)-Tubocurarine	Postsynaptic membrane			⋆
Venom α-toxins	Postsynaptic membrane			⋆
C-toxiferine	Postsynaptic membrane			⋆
Nicotine	Postsynaptic membrane			⋆
Lobeline	Postsynaptic membrane			⋆
Coniine	Postsynaptic membrane			⋆
α-Conotoxin	Postsynaptic membrane			⋆

competitive, that is reversible, and prevents the response of skeletal muscle to ACh or other nicotinic agonists. The classical studies of Bernard[39] showed that when curare was applied to the neuromuscular junction of the frog, stimulation applied to the nerve failed to evoke a muscular

[39] C. Bernard, *C. R. Seances Soc. Biol. Ses Fil.* **2,** 195 (1851).

response, whereas stimulation applied directly to the muscle evoked muscle twitches.

Recently, lophotoxin, isolated from several species of sea whips (genus *Lophogorgia*) has been shown to exert a curare-like block at the ACh receptors of the rat hemidiaphragm and frog rectus abdominus muscle preparations.[17,40]

Depolarizing blocking agents compete with ACh for the ACh receptor and, since they exert ACh agonistic activity, cause an initial depolarization before blocking further stimulation by endogenous ACh. Depolarization block is thus preceded by muscle fasciculation and augmentation of maximum evoked muscle twitch amplitude. Naturally occurring depolarizing blockers include nicotine (from *Nicotiana*, tobacco), coniine from hemlock (*Conium maculatum*), lobeline (from *Lobelia inflata*), murexine (also called urocanylcholine) from certain mollusks and leptodactyline from the skin of a lizard. Depolarizing agents may be distinguished from other types of blocking drugs by their effects on multiply innervated muscle of frogs and fowl. Thus, the rectus abdominus muscle of the frog and the gastrocnemius and neck muscles of the fowl respond to depolarizing agents with a sustained contracture, whereas other agents produce a flaccid paralysis. Tubocurarine, of course, antagonizes their action.

In summary (see Table II), techniques for the detection of neurotoxicity on isolated nerve and nerve–muscle junction have shown that neurotoxins acting on central and peripheral nervous tissue act at the nerve membrane to block ion conduction and propagation, act axonally to interfere with axonal translocation processes, act presynaptically to derange presynaptic transmitter storage and release, and act intra- and postsynaptically to derange synaptic-effector tissue responsiveness.

Acknowledgment

I wish to acknowledge, with gratitude, the influence of Professor P. S. J. Spencer (UWIST, Wales) and Professor W. C. Bowman (Strathclyde) upon my thoughts and interests. Editorial assistance in the production of this manuscript was given to me by numerous colleagues and I especially thank Drs. P. E. Teschan and Andrew Fingret for their comments.

[40] W. Fenical, R. K. Okuda, M. M. Bandurraga, P. Culver, and R. S. Jacobs, *Science* **212,** 1512 (1981).

[38] Electroencephalogram as a Tool for Assaying Neurotoxicity

By JONATHAN J. LIPMAN

Toxins are substances that occur in nature, biosynthesized or elaborated by animals, plants, or microbes, which induce deleterious effects when administered to an individual to which they are exogenous. Functionally, toxic agents may be classified in a variety of ways. Neurotoxins are describable as agents which exert their toxic effect on nervous tissue. Neurotoxins may be further described in terms of the site of their induced biological lesion within intoxicated nervous tissue. Thus, tetrodotoxin acts on the nerve fiber; β-bungarotoxin, black widow spider venom, and scorpion toxin may be described as having presynaptic action; and α-bungarotoxin, cobratoxin, and histrionicotoxin may be described as having postsynaptic action. It is also generally useful to be able to describe the central or peripheral nervous system predilection of the neurotoxin, and on this basis, for instance, apamin and physalemin can be described as excitatory in their effects on the central nervous system (CNS).[1]

Subcellularly, neurotoxic effects are mediated by a plethora of mechanisms in nervous tissue, including interference with receptor function, with neurotransmitter release or inactivation, and with activation and inactivation of ion channels. Textrodotoxin and saxitoxin, for example, block sodium channels.

Regardless, however, of the site or mechanism of the neurotoxic effect, neurotoxins by their very definition derange the function of the central nervous system, and the output of this system, behavior, is changed by their action.

It has been estimated that a major portion, perhaps 60%, of the energy metabolism of the brain as a whole is devoted to electrical events. These include the maintenance of ion fluxes of sodium and potassium gradients required for the function of the organ.[2] At a molecular level, the electrical potentials measured by the electroencephalogram (EEG) derive from the summative activity of these graded potentials' ion fluxes in nerve cell bodies and dendrites, largely the axon hillock potentials rather than axonal action potentials.

[1] H. P. Von Han and C. G. Honegger, *Experientia* **30,** 207 (1974).

[2] R. Hawkins, *in* "Cerebral Energy Metabolism and Metabolic Encephalopathy" (D. W. McCandless, ed.), Chap. 1. Plenum, New York, 1985.

Neurotoxicity mediated at any functional level therefore perturbs the EEG either focally or generally, and the purpose of this chapter is to describe how this may be measured in laboratory animals and to describe how recent advances have rendered this technique routinely useful.

The cortical electroencephalogram, as this is represented at the scalp, is used for human clinical studies since it is convenient and noninvasive. Animal studies do not suffer this limitation, however, and electrodes may be emplaced at specific intracerebral loci with stereotaxic ease. Such focal depth electrodes monitor regional electrical activity in brain areas of interest to the investigator desiring to examine the electrical function of specific brain loci believed to be involved in the mediation of neurotoxic effects. Due, however, to the organizational, anatomical, and functional arrangement of the brain, the cortical EEG is a useful screening tool for use in laboratory animals. The electrocorticogram (ECoG), as this is called, is easily measurable by means of simple amplification circuits or commercially available equipment using electrodes readily constructed in the laboratory.

The paper polygraph trace generated by the EEG or ECoG provides the experienced eye with considerable information regarding metabolic and autoregulatory activity as this is represented in the background rhythm of the ECoG. This rhythm is depressed when metabolic substrates are limiting, as occurs in hypoglycemia and hypoxia,[3] and when regional cerebral blood flow (rCBf) is reduced by hypercapnia,[4] or metabolic rate is limited by hypothermia.[5] Normally, the background rhythm of the EEG and the cerebral metabolic rate of oxygen ($CMRO_2$) are related and result in an EEG–$CMRO_2$ coupling that is reflected in an EEG–rCBf coupling. These couplings are illustrated by the changes which occur in the EEG, $CMRO_2$, or rCBf during mental activity, during seizure states, or during pharmacologic manipulation of the CNS.[4-7]

Historically, visual estimation of the intensity of electrical features and their frequences seen in the polygraph paper trace has been the sole method of clinical EEG analysis for both focal and background rhythm analysis. To an extent, this is still the clinical analytical technique of choice for focal and episodic feature description. The advent of computer methods of analysis has recently placed reliable techniques for measure-

[3] E. Gellhorn and M. Kessler, *Am. J. Physiol.* **136,** 1 (1942).

[4] C. J. Wehr, J. J. Lipman, J. W. Hammon, Jr., H. W. Bender, and W. H. Merrill, *Surg. Forum* **35,** 252 (1984).

[5] C. J. Wehr, W. H. Merrill, J. J. Lipman, J. W. Hammon, Jr., and H. W. Bender, *Surg. Forum* **34,** 240 (1983).

[6] W. J. Ray and H. W. Cole, *Science* **228,** 750 (1985).

[7] O. B. Paulson and F. Sharborough, *Acta Neurol. Scand.* **50,** 194 (1974).

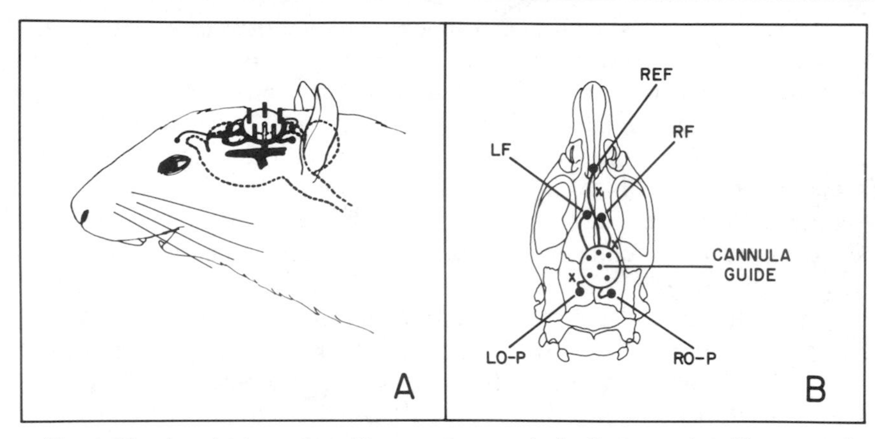

FIG. 1. The chronic electrode and intracerebroventricular (icv) cannula guide preparation for the rat. (A) The relationship between the implant, the brain ventricular system, and the head of the rat is shown. (B) The position of the implant in relation to skull landmarks is shown in dorsal plan view. Stainless steel extradural anchoring screws are emplaced at "×." The acrylic bone cement mantle securing the implant to the anchoring screws and the skull is not shown. Electrodes are labeled: reference (REF), left frontal (LF), right frontal (RF), and left and right occipitoparietal (LO-P and RO-P). When not in use the cannula guide carries an occluding sylette (not shown).

ment of background frequency of the EEG (or ECoG) in the hands of any investigator having access to a microcomputer. This development in technology revolutionizes the field of EEG measurements in general, and the assay of neurotoxicity in particular, and renders the quantitative electroencephalogram both achievable and useful to the research laboratory.

In this laboratory, we have developed an automated technique for intermittent or continuous ECoG monitoring of the conscious, ambulatory freely responding rat which is particularly useful for neurotoxicity studies on the central nervous system of this animal. Neurotoxins or agents under investigation for neurotoxic effect are administered peripherally or centrally, the latter by means of a chronically indwelling intracerebroventricular (icv) cannula guide which is incorporated into the design of the cortical electrode array permanently mounted to the animal's head. This electrode and cannula guide arrangement is illustrated in Fig. 1. The method for construction and implantation of this device is described elsewhere.[8] By means of this apparatus, suitably connected via a strong, light, and flexible tether carrying the electrode wires to an EEG machine (we use a clinical eight-channel model, Grass model 89C), the ECoG paper trace can be inspected for electroencephalographic conse-

[8] J. J. Lipman and S. Harshman, *Toxicon* **23,** 325 (1985).

quences of neurotoxin administration. By selecting a monopolar rather than a bipolar configuration of electrode connection to the amplifier circuits of the EEG machine, each electrode trace represents the electrical activity of the cortex directly beneath each of the electrodes and is therefore regional (see Fig. 2). This is important when assessing the spread of electrocortical derangements such as seizures since it is usually desirable to identify both the originating foci and also the order and progression of spread or generalization, when this should occur.

We have identified both focal and general abnormalities in the ECoG measured from rats injected intracerebroventricularly with various doses of α-staphylotoxin. Focal, isolated features of the ECoG are illustrated in Fig. 2 arising in single cortical areas or occurring episodically in a general fashion over the whole brain. The tracings of Fig. 2 clearly possess differ-

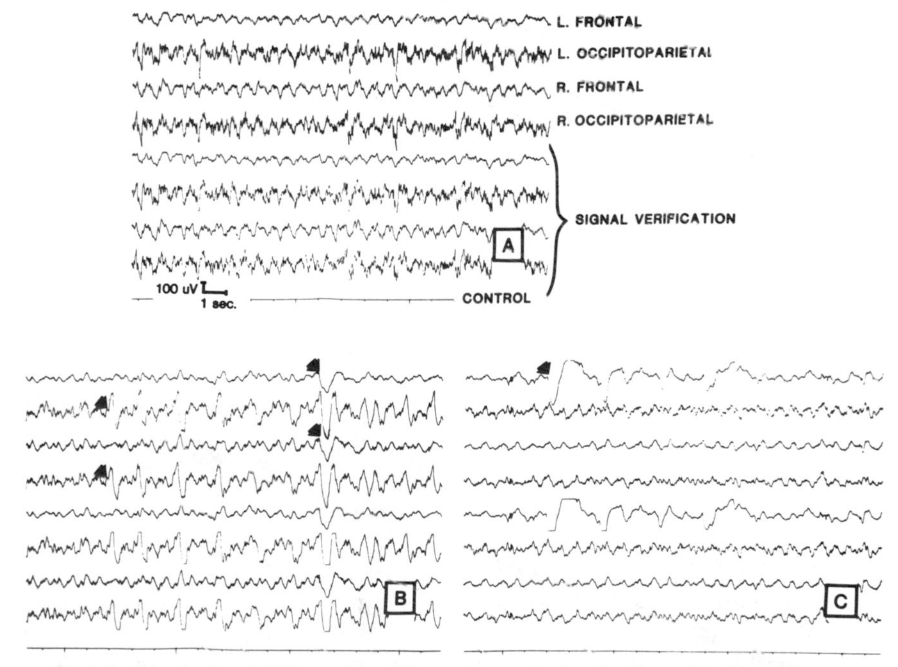

FIG. 2. Monopolar EEG tracings illustrating typical episodic and regionally isolated features arising in the ECoG of the rat injected icv with α-staphylotoxin. (A) Control EEG illustrating regional (monopolar) recording. The lower four traces labeled "signal verification" are echoes of the upper four traces after FM signal processing for computer analysis. These echo traces are used for quality control purpose to ensure veracity of the analysis of the recorded signal. (B) Isolated delta range wave forms restricted to bilateral O-P areas are shown at left arrows. Right arrows show occasional generalization to frontal electrodes. (C) Isolated left frontal derangement of the ECoG trace resulting from focal high amplitude slow wave activity.

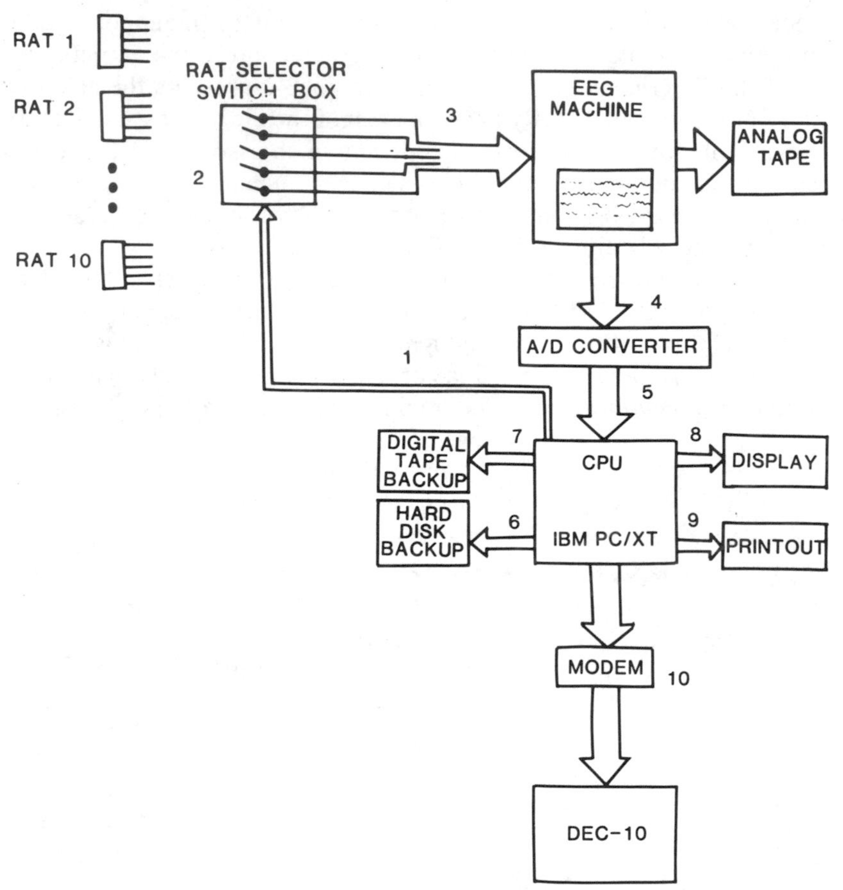

FIG. 3. The automated ECoG acquisition system for sequential measurement of ECoG signals from up to 10 animals. See text for explanation.

ent frequencies in the background rhythm of the ECoG, yet these would be very difficult if not impossible to analyze manually with any accuracy. The task of manually analyzing such records from the large numbers of animals required by behavioral studies would be overwhelming and impractical. We have therefore developed a system by which 10 electrode-implanted animals may be simultaneously connected to a single EEG machine and microcomputer for analysis of the recorded signal. The acquisition of multiple ECoG samples from several animals may thus proceed automatically. The automated acquisition system is illustrated in Fig. 3. A digital switching box is employed to which all animals' electrode tethers are connected. The switchbox (2, in Fig. 3) is controlled by a software-generated control sent from the computer's parallel port (1, in Fig. 3). The selected rat's ECoG signal passes from the switchbox (2) to the EEG machine (3), producing a paper record for inspection and analysis. The amplified signal is additionally sent to an analog-to-digital converter (4) where the digitized signal is read into a standard IBM PC/XT minicomputer (5) and stored on the hard disk of the system (6). We have equipped the system with a digital tape backup unit for permanently archiving data (7). Following an ECoG recording session the data are displayed on a monochrome monitor (8) and, where required, a hard copy is produced on the attached dot-matrix printer (9).

The software that we have developed[9] permits the automatic selection of ECoG signals, one at a time. The ECoG of each rat is recorded for a selectable time at a sample rate of 64 Hz (software selectable). Each successive 100-sec data sample is separated into 25 $\times$ 4-sec epochs and each epoch is transformed into the frequency domain using a 256-point fast Fourier transform (FFT) algorithm.[10] The power in the frequency bands (delta, δ, 1–3 Hz; theta, θ, 3–7 Hz; alpha, α, 7–13 Hz; and beta, β, 13–20 Hz) are calculated. The process is illustrated in Fig. 4.

Using this technique we have subjected the ECoG of rats treated with icv α-staphylotoxin to analysis of the background rhythm, data that were not readily obtainable from the polygraph trace of Fig. 2.

Typical data illustrating the dose-dependent and time-dependent depression of $\theta + \alpha$ (3–13 Hz) power in the ECoG frequency spectrum are shown in Fig. 5. Notice that the power in these frequency bands is greater in occipitoparietal than frontal regions of the brain. It is also clear from the illustration that 10 μg of α-staphylotoxin produces a rapid decline in ECoG ($\theta + \alpha$) power within 100 sec of injection, which further declines until death. The effect of a 1-μg dose is clearly less intense and of slower progression, nevertheless continuing to decline inexorably to near isoelectric levels before death ensues, apparently from a lethal, centrally mediated mechanism. We consistently find in these and other studies that the $\theta + \alpha$ power integrals are a *nonspecific* indicator of cerebral depression generally indicative of impaired states of arousal. Under resting conditions, we have found that the diurnal rhythm of $\theta + \alpha$ power varies in a characteristic pattern which differs marginally for each rat and provides a cerebral monitor of arousal in daily behavioral activity cycles. The depression illustrated in Fig. 5 is several orders of magnitude greater than the normal minima encompassed by the bounds of the diurnal rhythm.

[9] D. DeBoer, J. J. Lipman, P. Lawrence, and P. E. Teschan, *Proc. Am. Assoc. Med. Instrum.*, 1985.

[10] L. R. Rabiner and B. Gold, "Theory and Application of Digital Signal Processing," p. 58. Prentice-Hall, Englewood Cliffs, New Jersey, 1975.

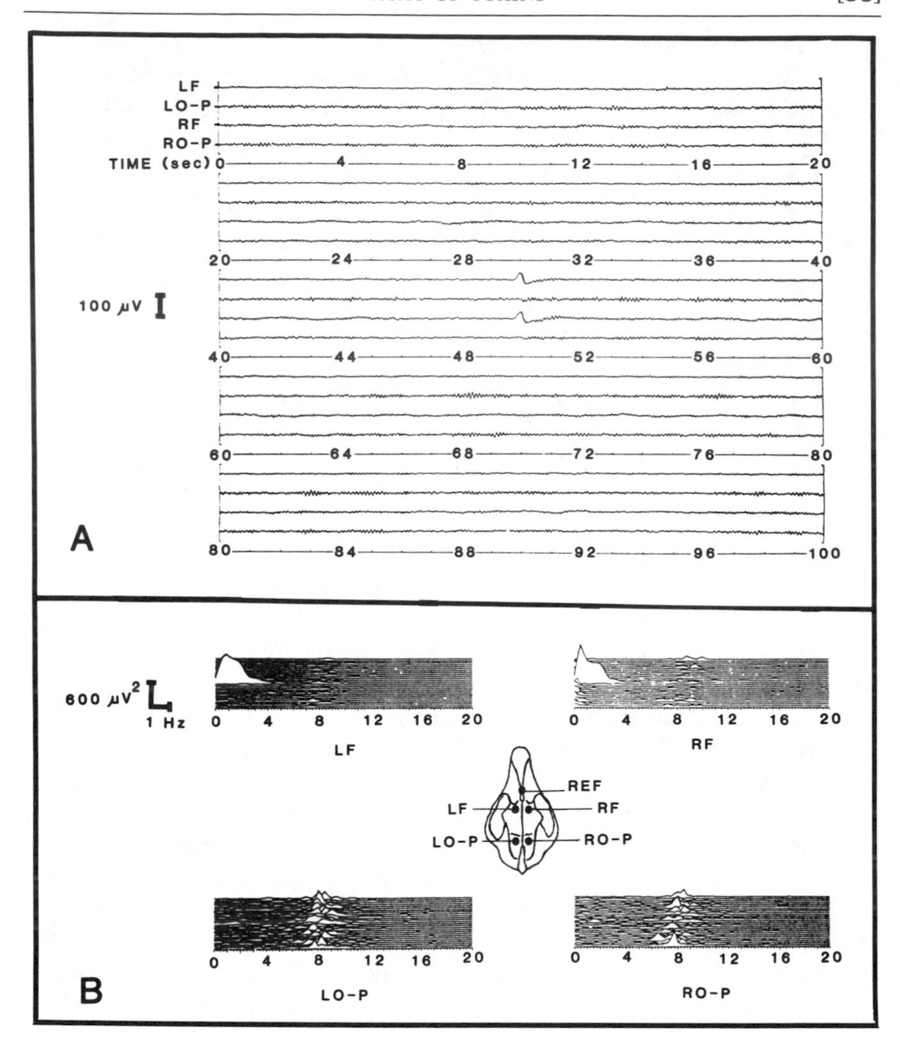

FIG. 4. Fast Fourier transformation (FFT) of the digitized ECoG signal for regional frequency analysis. Upper frame (A) illustrates the process by which 100 sec of the signal is divided into 4-sec epochs. An artifact has been deliberately inserted into the LF and RF signals at the fiftieth second. The FFT yields four compressed spectral arrays or analyses (CSA), one for each electrode (lower frame, B). Each line of the CSA represent the power in μV^2 at each frequency measured during each 4-sec epoch. Note the irregularity seen between 0 and 4 Hz of the CSAs derived from the LF and RF electrodes arising from the artifact. This appears in the line of the CSA originating from the 48- to 52-sec epoch. Modified from J. R. Bourne, B. Hamel, D. Giese, G. M. Woyce, P. L. Lawrence, J. Ward, and P. E. Teschan, *IEEE Trans. Biomed. Eng.* **27,** 656 (1980) with permission.

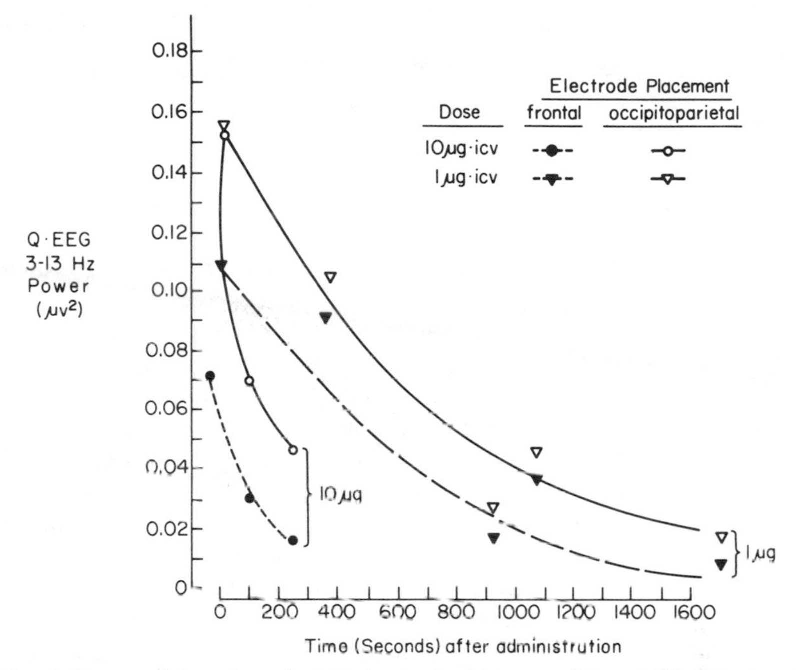

FIG. 5. Dose- and time-dependent depression in the power of the rat ECoG background rhythm measured in $\theta + \alpha$ bands (3–13 Hz) following icv injection of α-staphylotoxin. Frontal electrode data are shown by the broken lines, occipitoparietal data by the solid lines. Note that this depression develops generally throughout the cortex.

The electroencephalogram (electrocorticogram in our hands since our electrodes are cortically placed) provides a wealth of information on the viability of the brain under treatment by putative neuroactive agents. Since EEG activity is physiologically coupled to rCBf and $CMRO_2$, this renders its use as an index subject to false positive errors where vasoactivity is a prominent feature of the agent under test. Such properties require identification in other test systems. As an adjunctive test in a behavioral screening process, however, the advent of quantitative methods of analysis commend the electroencephalogram as a valuable assay of neurotoxicity. As a technique for identifying intracerebral loci of behavioral neurotoxicity, the combined use of stereotaxic techniques renders the EEG exquisitely useful in predicting and assaying neurotoxic actions.

Acknowledgments

This work was supported in part by a Vanderbilt University Biomedical Sciences Research Grant RR-5424. The expert assistance of Patricia Lawrence and the collaboration of Paul E. Teschan, M.D., is gratefully acknowledged.

[39] Assay of Pore-Forming Toxins in Cultured Cells Using Radioisotopes

By MONICA THELESTAM

Introduction

Pore-forming toxins belong to a group of protein toxins that associate with cell membranes, unspecifically or to specific receptors, and form transmembrane pores of discrete sizes.[1] The pores are thought to consist of oligomeric aggregates of the toxins. The resulting membrane permeabilization leads to a disturbed electrolyte balance, and to cellular leakage of various substances of larger molecular size. The final cellular effects depend on the diameter of the toxin-induced pores and the type of target cell. All membrane-damaging toxins (hemolysins, cytolysins) are potential pore formers but so far pore formation has actually been established only in a few cases (Table I). Other membrane-damaging toxins may disorganize the plasma membrane more unspecifically, leading to the same final cellular consequences as with pore formers.

The assays described below are based on the concepts that (1) toxin-induced pore formation can be detected as a leakage of cytoplasmic constituents from toxin-treated cells, and (2) the molecular size of the leaked material is indicative of the size of the functional pores induced in the membranes. These leakage assays utilize cells cultivated *in vitro* to confluent monolayers, and prelabeled with radioisotopes to obtain cytoplasmic markers of defined molecular sizes. Labeled cells are exposed to toxin and the leaked radioactivity scored and expressed as percentage of the total radioactivity present in the cytoplasm at the beginning of the experiment.

Cultivation of Cells

Procedure for Cell Cultivation

1. Prepare a uniform suspension of freshly trypsinized cells, containing 3×10^4 cells/ml of the desired cell culture medium.
2. Add exactly 1 ml of this suspension to each well in 24-well polystyrene culture plates.
3. Incubate the plates at 37° in a humid atmosphere containing 5% CO_2 until labeling with radioactive substance (see below).

[1] S. Bhakdi and J. Tranum-Jensen, *Trends Biochem. Sci.* **8**, 134 (1983).

TABLE I
PORE-FORMING BACTERIAL TOXINS

Toxin	Producing organism	Ref.
Alpha toxin	*Staphylococcus aureus*	*a*
Delta toxin	*Staphylococcus aureus*	*b*
Streptolysin O	*Streptococcus pyogenes*	*c*
Aerolysin	*Aeromonas hydrophila*	*d, e*
Enterotoxin	*Clostridium perfringens*	*f*

[a] R. Füssle, S. Bhakdi, A. Sziegoleit, J. Tranum-Jensen, T. Kranz, and H. J. Wellensiek, *J. Cell Biol.* **91,** 83 (1981).
[b] J. H. Freer, T. H. Birkbeck, and M. Bhakoo, *in* "Bacterial Protein Toxins" (J. E. Alouf, F. J. Fehrenbach, J. H. Freer, and J. Jeljaszewicz, eds.), p. 181. Academic Press, Orlando, Florida, 1984.
[c] S. Bhakdi, J. Tranum-Jensen, and A. Sziegoleit, *Infect. Immun.* **47,** 52 (1985).
[d] M. Thelestam and Å. Ljungh, *Infect. Immun.* **34,** 949 (1981).
[e] S. P. Howard and J. T. Buckley, *Biochemistry* **21,** 1662 (1982).
[f] B. McClane, *Biochim. Biophys. Acta* **777,** 99 (1984).

Comments

Human diploid embryonic lung fibroblasts (line WI-38 and MRC-5) were used in the leakage assays developed originally in this laboratory in 1975.[2] More recently also other cell lines have been successfully used (see final section).

One basic prerequisite for reliable results is that the number of cells per well is constant in each experiment. If this requirement is fulfilled, quantitative results are obtained without the need for counting the cells or estimating some other parameter of cell growth (e.g., the amount of cellular protein). The absolute amount of radioactivity incorporated (and released) per culture varies with the cell type and may vary between experiments with the same cell type. However, the values for maximal and spontaneous release of markers (see below) will serve as internal standards, enabling a comparison between different experiments.

As a routine we use double or triplicate samples of cultures for toxin treatment, triplicates of the spontaneous release, and six parallel cultures for the maximal release in each experiment to obtain reliable values.

[2] M. Thelestam, Doctoral dissertation. Karolinska Institute, Stockholm, Sweden, 1975.

TABLE II
CYTOPLASMIC MARKERS[a]

Substance used for labeling cells	Designation of cytoplasmic marker	Approximate M_r	Mean spontaneous release[b] (% of maximal release)
α-Amino[1-^{14}C]isobutyric acid	AIB label	103	22.5 (15–30)
[5-^{3}H]Uridine	Nucleotide label	<1000	5 (3–7)
[5-^{3}H]Uridine	RNA label	>200 000	5 (3–7)

[a] From Thelestam.[2]
[b] In TBS after incubation of the labeled cells for 30 min at 37°.

Cytoplasmic Markers

Assay systems with three different cytoplasmic markers have been developed in this laboratory, to be used singly or in combination (Table II).

AIB Label

α-Aminoisobutyric acid (AIB) enters cells by means of sodium-coupled active transport.[3] This amino acid is nonmetabolizable and therefore remains in the cytosol as a low-molecular-weight marker of defined size (M_r 103).[4]

Labeling Procedure

1. Prepare a solution of 1 μCi/ml α-amino[^{14}C]isobutyric acid in Hanks' balanced salt solution (HBSS) containing 10 m*M* TES and buffered to pH 7.4 with 1 *M* NaOH. Warm this AIB solution to 37° before addition to cells.
2. Rinse confluent monolayers of cells three times with HBSS at 22°.
3. Incubate the cultures for 60 min at 37° with 0.3 ml of the AIB solution.
4. Replace the AIB solution by 1 ml of Eagle's minimal essential medium (MEM), pH 7.4, and incubate the cultures for 30 min at 37° in 5% CO_2 atmosphere.
5. Rinse the labeled cultures rapidly three times with 22° HBSS.
6. The cells are ready for immediate use.

[3] N. N. Christensen and M. Liang, *Biochim. Biophys. Acta* **112,** 524 (1966).
[4] M. Thelestam and R. Möllby, *Infect. Immun.* **12,** 225 (1975).

Nucleotide Label

Uridine added to cell cultures in stationary growth phase is taken up by facilitated transport, rapidly phosphorylated and incorporated into di- and trinucleotides, but not into larger molecules under the standardized conditions described below. The molecular weight of the ensuing marker, referred to as nucleotide label, will be <1000.[2,5] More than 90% of the total cell-associated radioactivity will be found in the cytoplasmic fraction of human lung fibroblasts among other cell types (Thelestam, unpublished).

Labeling Procedure

1. Rinse confluent monolayers of cells three times with HBSS at 22°.
2. Incubate the cultures for 60 min at 37° with 0.3 ml of 1 μCi/ml [5-^{3}H]uridine in Eagle's MEM with 10% calf serum, pH 7.6–7.8.
3. Rinse the labeled cultures three times with HBSS at the desired temperature.
4. The cells are ready for immediate use.

This labeling method is simplified from the originally described more time-consuming procedure.[5] The molecular size of the generated cytoplasmic marker has been confirmed to agree exactly with that of the originally reported nucleotide label.[6]

RNA Label

Uridine added to cell cultures in logarithmic growth phase will be incorporated into various RNA species. When pulse-labeled cells are grown out to confluency in medium free of radioactive uridine the label will gradually accumulate in high-molecular-weight rRNA (M_r >200,000) because it is more slowly turned over than the other RNA species.[5]

Labeling Procedure

1. Incubate cell cultures in logarithmic growth phase (24 hr after inoculation) for 24 hr at 37° with 1 ml of 1 μCi/ml [5-^{3}H]uridine in Eagle's MEM with 10% calf serum, pH 7.6–7.8.
2. Rinse the cells once with sterile HBSS and continue the incubation for 3–5 days at 37° in medium free of radioactive uridine.
3. When the cultures are confluent rinse three times with HBSS at the desired temperature.
4. The cells are ready for immediate use.

[5] M. Thelestam and R. Möllby, *Infect. Immun.* **11,** 640 (1975).

[6] M. Thelestam, *Biochim. Biophys. Acta* **762,** 481 (1983).

Assay

1. Add toxin diluted in medium or appropriate buffer to the labeled cell cultures. The volume of toxin solution can be varied between 0.3 and 1 ml/well.
2. Incubate at 37° for the desired time period.
3. Transfer the toxin solutions to tubes and centrifuge 1000 *g* for 5 min to remove cell debris.
4. Add 0.1 ml of the supernatant to 10 ml of scintillation liquid and measure the radioactivity.

Maximal, Spontaneous, and Specific Release of Markers

The *maximal release* (*MR*) is determined after labeling the cells. The cell membranes are ruptured completely by a procedure leaving the nuclei intact. Thus the *MR* is equal to the total leakable radioactivity present exclusively in the cytoplasmic fraction at the beginning of the experiment.[5]

Procedure

1. Incubate labeled and rinsed cells for 60 min at 37° in 0.3–1 ml of 0.06 *M* sodium borate buffer, pH 7.8. This treatment causes the cells to swell but not disrupt.
2. Disrupt the cell membranes by gently scraping the bottom of the culture well with a rubber policeman.
3. Centrifuge the cell lysate 1000 *g* for 5 min to remove the nuclei.
4. Measure the radioactivity in 0.1 ml of the supernatant by liquid scintillation as above.

The *spontaneous release* (*SR*) is the radioactivity released into toxin-free buffer or medium. The *SR* varies with the marker (Table II). Using the AIB label, 45 min can be considered as a maximum assay time since about 80% of the AIB label is spontaneously released within 2 hr in TBS or Eagle's MEM.[4] In contrast both the nucleotide and the RNA label are efficiently retained by cells as long as the plasma membranes are intact. Thus both markers can be used in assays extended for up to 24 hr. A maximum of 20% (nucleotide) and 12% (RNA) of these labels will spontaneously be released by human lung fibroblasts in Eagle's MEM with 10% calf serum in 24 hr.

The *specific toxin-induced release* of cytoplasmic markers (*Q*) is calculated according to the following expression:

$$Q = 100[(ER - SR)/(MR - SR)] \quad (1)$$

where *ER* is the experimental value for released radioactivity. The *SR* should be as low as possible since the *Q* value increases with increasing *SR*. Based on various theoretical considerations[2] we have set the limit for the highest tolerable *SR* as 30% of the *MR*.

Relative Sensitivities of the Assays with Different Markers

Toxin-induced leakage of the cytoplasmic markers described above has been compared to leakage of ^{51}Cr and related to uptake of trypan blue and to morphological changes of cells.[7] The following order of sensitivity for detection of small transmembrane pores was established: leakage of AIB label > leakage of nucleotide label > leakage of ^{51}Cr label = uptake of trypan blue = morphological changes > leakage of RNA label. When large membrane lesions were induced, e.g., with detergents, we obtained coinciding curves for release of the four markers upon incubation for 30 min with increasing detergent concentrations. When the detergent-induced release was measured as a function of time, however, the markers were released in the order indicated above.[7] When more subtle membrane lesions were induced, e.g., with pore-forming toxins, the higher molecular weight markers were not released at all.[8]

Thus the three markers could be used in combination to obtain so-called leakage patterns after exposure to toxins at standard conditions.[9] The appearance of these patterns give a notion about the relative sizes of pores induced by different toxins. The leakage patterns can be transformed to numerical expressions, so-called ED_{50} ratios. Based on ED_{50} ratios 38 membrane-damaging toxins were classified into 5 groups, established with the aid of reference substances with known mechanisms of membrane-damaging action.[9]

Utility of the Different Markers

The AIB label, although giving the most sensitive detection of pore formation, is of a more limited use than the other two markers. Its high spontaneous release in short time makes it unsuitable for long-term studies of membrane damage. Moreover the labeling of cells with AIB may be tricky since its uptake is extremely sensitive to changes of environmental conditions, e.g., temperature and pH.[4]

The pulse-labeling procedure with [^{3}H]uridine to obtain the RNA label involves an increased risk of contamination of the cells due to labeling and

[7] M. Thelestam and R. Möllby, *Med. Biol.* **54,** 39 (1976).
[8] M. Thelestam and R. Möllby, *Toxicon* **21,** 805 (1983).
[9] M. Thelestam and R. Möllby, *Biochim. Biophys. Acta* **557,** 156 (1979).

TABLE III
USE OF LEAKAGE ASSAYS WITH VARIOUS CELL TYPES

Cells	Markers	Ref.
Human U-amnion	Nucleotide label	*a*
Green monkey kidney (Vero)	AIB label	*b*
	Nucleotide label	
	RNA label	
Human intestinal epithelial cells (Henle 407)	Nucleotide label	*c*
Human skin fibroblasts	Nucleotide label	*d*
	RNA label	

[a] M. Degré and T. Hovig, *Acta Pathol. Microbiol. Scand. Sect. B* **84,** 437 (1976).

[b] B. A. McClane and J. L. McDonel, *Biochim. Biophys. Acta* **600,** 974 (1980).

[c] A. Buxser and P. F. Bonventre, *Infect. Immun.* **31,** 929 (1981).

[d] R. Linder and A. W. Bernheimer, *Toxicon* **22,** 641 (1984).

TABLE IV
CELL LINES IN WHICH NUCLEOTIDE LEAKAGE HAS BEEN APPLIED[a]

Cell type/origin	Designation of cell line[b]
Mouse adrenocortical tumor cells	Y 1 (two sublines)
Mouse embryo fibroblasts	3T3
Transformed mouse embryo fibroblasts	3T3-SV40
Chinese hamster ovary cells	CHO-Kl
Chinese hamster ovary mutant	RPE-44[c]
Baby hamster kidney fibroblasts	BHK-21
Rabbit kidney cells	LLC-RK_1
Human cervical carcinoma cells	HeLa
Mouse myeloma cells	SP2/O
Human intestinal fibroblasts	Flow 11 OOO
Human glia cells[d]	Two lines of different origin
Human glioma cells[d]	U-251-MG and TP-483-MG
Human meningioma cells[d]	Five lines from different tumors

[a] In the author's laboratory.

[b] Obtained from Flow Laboratories (Irvine, Scotland) unless otherwise stated.

[c] Kindly supplied by J. M. Moehring, University of Vermont, Burlington, Vermont.

[d] Established at the Department of Tumour Pathology, Karolinska Hospital, S-104 01 Stockholm, Sweden.

rinsing several days before the assay is performed. However, both the AIB and the RNA label are useful for the establishment of leakage patterns.

For routine assay of pore-forming toxins the nucleotide label is excellent. The labeling procedure is simple and rapid and can be performed under nonsterile conditions. The *SR* is low over extended time periods although the marker is small enough to afford a rather sensitive test.

All three markers have been applied in other laboratories using other cell lines than human fibroblasts (Table III). During the past 15 years we have successfully applied the nucleotide leakage assay to a number of still other cell types (Table IV). In fact this assay has worked well with every type of cell tested hitherto. Despite the varying origin of the cells (Table IV), and their different degrees of malignancy, the molecular weight of the nucleotide label appears consistently to be <1000. This has been verified by gel filtration on Sephadex G-25 of cell lysates from nucleotide-labeled Vero, CHO, Y1, 3T3, 3T3-SV40, LLC-RK_1, glia, and glioma cells[6] (unpublished observations).

Acknowledgments

I thank Inger Florin for helpful suggestions. The investigations carried out in the author's laboratory were supported by the Swedish Medical Research Council (Grant No. 16X-2562).

[40] Incorporation of Toxin Pores into Liposomes

By ROSWITHA FÜSSLE, JØRGEN TRANUM-JENSEN, ANDREAS SZIEGOLEIT, and SUCHARIT BHAKDI

Introduction

Many cytolytic toxins primarily attack the lipid bilayer of target cells, whereby membrane damage can be effected through two basically distinct mechanisms. The first is a direct enzymatic degradation of lipids by phospholipases.[1] Alternatively, membrane perturbation can occur through a nonenzymatic, physical derangement of the lipid bilayer, effected either through a detergent-like action of the toxin as described for mellitin,[2] or through the insertion of the toxin molecules into the membrane with

[1] R. Möllby, *in* "Bacterial Toxins and Cell Membranes" (J. Jeljaszewicz and T. Wadström, eds.), p. 367. Academic Press, New York, 1978.

[2] E. Habermann and H. Kowallek, *Hoppe-Seyler's Z. Physiol. Chem.* **351,** 884 (1970).

formation of transmembrane pores. The latter mechanism was first described in context with the immune cytolysis of cells by complement.[3,4] Subsequently, it was found that cytolysis by staphylococcal α-toxin and by streptolysin O follow a basically similar pattern of events.[5,6] Binding to membranes is accompanied by a hydrophilic–amphiphilic transition which occurs during oligomerization of native protomers into macromolecular pore structures. Penetration of these oligomers into the apolar membrane domain is due to the exposure of hydrophobic regions that are primarily hidden within the native molecules.[7] Transmembrane pore formation results in free diffusion of ions and small molecules across the bilayer. If unhalted, this process ultimately leads to cellular swelling and rupture of the membrane.

Staphylococcal α-toxin is a single-chain cytolytic exotoxin of M_r 34,000 that is produced by most pathogenic staphylococci.[8–11] Early studies indicated that the primary target of toxin attack is the lipid bilayer.[12–14] Thus, the toxin binds to and damages artificial liposomes, causing release of low-molecular-weight markers[13,15,16]; it induces an increase in electrical conductance of lipid monolayers[17]; and it lyses nucleated cells as well as erythrocytes, exhibiting a broad reactivity toward cells from different species.[18,19] Specific "receptors" are probably not basically required for toxin binding,[8,13,20,21] but it is possible that certain cells possess high-affinity binder molecules that render them particularly sensitive to toxin

[3] M. Mayer, *Proc. Natl. Acad. Sci. U.S.A.* **69**, 2954 (1972).

[4] S. Bhakdi and J. Tranum-Jensen, *Proc. Natl. Acad. Sci. U.S.A.* **75**, 5655 (1978).

[5] R. Füssle, S. Bhakdi, A. Sziegoleit, J. Tranum-Jensen, T. Kranz, and H. J. Wellensiek, *J. Cell Biol.* **91**, 83 (1981).

[6] S. Bhakdi, J. Tranum-Jensen, and A. Sziegoleit, *Infect. Immun.* **47**, 52 (1985).

[7] S. Bhakdi and J. Tranum-Jensen, *Rev. Physiol. Biochem. Pharmacol.* **107**, 147 (1987).

[8] C. McCartney and J. P. Arbuthnott, *in* "Bacterial Toxins and Cell Membranes" (J. Jeljaszewicz and T. Wadström, eds.), p. 89. Academic Press, New York, 1978.

[9] J. P. Arbuthnott, *in* "Microbial Toxins" (T. C. Montie, S. Kadis, and S. A. Ajl, eds.), Vol. 3, p. 189. Academic Press, New York, 1970.

[10] A. W. Bernheimer, *Biochim. Biophys. Acta* **344**, 27 (1974).

[11] M. Rogolsky, *Microbiol. Rev.* **43**, 320 (1979).

[12] J. P. Arbuthnott, J. H. Freer, and B. Billcliffe, *J. Gen. Microbiol.* **75**, 309 (1973).

[13] J. H. Freer, J. P. Arbuthnott, and A. W. Bernheimer, *J. Bacteriol.* **95**, 1153 (1968).

[14] J. H. Freer, J. P. Arbuthnott, and B. Billcliffe, *J. Gen. Microbiol.* **75**, 321 (1973).

[15] P. Cassidy, H. R. Six, and S. Harshman, *Biochim. Biophys. Acta* **332**, 413 (1974).

[16] G. Weissman, G. Sessa, and A. W. Bernheimer, *Science* **154**, 772 (1966).

[17] A. R. Buckelew and G. Colaccico, *Biochim. Biophys. Acta* **233**, 7 (1971).

[18] A. W. Bernheimer, *Science* **159**, 847 (1968).

[19] A. W. Bernheimer, K. S. Kim, C. C. Remsen, J. Antanavage, and S. W. Watson, *Infect. Immun.* **6**, 636 (1972).

[20] S. Bhakdi, M. Muhly, and R. Füssle, *Infect. Immun.* **46**, 318 (1984).

[21] G. M. Phimister and J. H. Freer, *J. Med. Microbiol.* **18**, 197 (1984).

action.[21–24] In the electron microscope, ring-shaped structures have been observed on toxin-treated membranes[13,14,19] and the original proposal that these represented membrane-bound toxin hexamers has recently been verified.[5,25,26] The functional size of the lesions has been approximated by marker release experiments using resealed erythrocyte ghosts and nucleated cells[5,27] as well as by osmotic protection experiments[20] to be in the order of 2–3 nm.

Streptolysin O (SLO) is an oxygen-labile, thiol-activated bacterial exotoxin produced by Lancefield group A β-hemolytic streptococci.[28,29] It consists of a single polypeptide chain of M_r 69,000.[30] All thiol-activated toxins initially bind to cholesterol molecules in the target membranes and exert cytolytic effects on a broad spectrum of mammalian cells.[28,29] In the electron microscope, cytolysis is often accompanied by the appearance of circular and semicircular structures on the membranes that can be visualized by electron microscopy.[10,31,32] Until recently, it was thought that membrane damage ensues as the result of cholesterol disarrangement and destabilization of the lipid bilayer, and that the structures viewed in the electron microscope represented complexes of the toxin with tightly bound cholesterol.[10,28,32] However, extensively delipidated SLO recovered after sucrose density gradient centrifugation of membrane solubilizates in the presence of detergent was reincorporated into liposomes of pure phosphatidylcholine and shown to retain the pore structures.[6] Thus cholesterol, although playing a key role in the initial binding of SLO to membranes, does not directly participate in the formation of the membrane-perturbing toxin channels. SLO-damaged membranes have been shown to exhibit large functional holes that allow for the passage of very large molecules with diameters exceeding 15 nm.[33,34]

Following the isolation of the complement pore and its reconstitution

[22] P. Cassidy and S. Harshman, *Biochemistry* **15,** 2348 (1976).

[23] I. Kato and M. Naiki, *Infect. Immun.* **13,** 289 (1976).

[24] S. J. Anderson, S. R. Carlsen, and S. Harshman, *Toxin* **3** (*Suppl.*), 13 (1983).

[25] S. Bhakdi, R. Füssle, and J. Tranum-Jensen, *Proc. Natl. Acad. Sci. U.S.A.* **78,** 5475 (1981).

[26] N. Tobkes, B. A. Wallace, and H. Bagley, *Biochemistry* **24,** 1915 (1985).

[27] M. Thelestam and R. Möllby, *Infect. Immun.* **12,** 225 (1975).

[28] J. E. Alouf, *Pharmacol. Ther.* **11,** 661 (1980).

[29] C. J. Smyth and J. C. Duncan, *in* "Bacterial Toxins and Cell Membranes" (J. Jeljaszewicz and T. Wadström, eds.), p. 130. Academic Press, New York, 1978.

[30] S. Bhakdi, M. Roth, A. Sziegoleit, and J. Tranum-Jensen, *Infect. Immun.* **46,** 394 (1984).

[31] R. R. Dourmashkin and W. F. Rosse, *Am. J. Med.* **41,** 699 (1966).

[32] J. L. Duncan and R. Schlegel, *J. Cell Biol.* **67,** 160 (1975).

[33] L. Buckingham and J. L. Duncan, *Biochim. Biophys. Acta* **729,** 115 (1983).

[34] J. L. Duncan, *Infect. Immun.* **9,** 1022 (1974).

into liposomes, we have adapted analogous approaches in the study of toxin pores. Reincorporation of isolated protein complexes into artificial lipid bilayers has generated valuable structural information on the pores and the simple procedures used for these purposes will be detailed in this chapter.

Methodology

Production of Toxin-Lysed Erythrocyte Membranes

Rabbit or human blood is drawn into a solution of 200 m*M* EDTA (pH 7.4). Erythrocytes are sedimented by centrifugation and washed three times in phosphate-buffered saline (PBS), pH 7.2, to remove the bulk of white cells. The washed erythrocytes are suspended to 10^9 cells/ml in PBS.

The erythrocyte suspensions are incubated either with purified bacterial toxins or, if these are not available, with partially purified proteins. Heavily loaded membranes are obtained, for example, by incubating 10^9 cells with 100 μg purified α-toxin for 15 min, 37°, or with 20–30 μg SLO for 1–2 min, 37°. The latter incubation should be performed in the presence of 5 m*M* dithiothreitol. At the given toxin concentrations, hemolysis occurs within seconds. Binding of α-toxin approaches 40% at high toxin concentrations, whereas SLO binds almost quantitatively. The hemolysed cell membranes are then pelleted by centrifugation (16,000 rpm for 15 min, in a Sorvall SS 34 rotor, 4°) and washed three times with ice-cold 5 m*M* phosphate buffer, pH 8.0.

A sodium dodecyl sulfate-polyacrylamide gel electrophoresis (SDS–PAGE) analysis of α-toxin-treated and nontreated control membranes is shown in Fig. 1. Typically, pore formers do not induce any noticeable degradation of membrane constituents; hence, the toxin bands simply superimpose themselves upon the original membrane protein patterns.

Isolation of Membrane-Bound Toxin Oligomers

Detergent Solubilization. The washed, packed membranes are solubilized by addition of a mild, nondenaturing detergent. We currently employ sodium deoxycholate (DOC, Merck AG, Darmstdt, FRG) at a final concentration of 125–250 m*M*. A 250 m*M* stock solution of the detergent may be prepared (dissolved in water) and 1 vol added to 1 vol of packed, washed erythrocyte ghosts. Solubilization is quantitative and occurs within seconds at room temperature. The DOC-solubilization procedure has replaced the method utilizing 2% Triton X-100 that was initially employed in the α-toxin studies.[5]

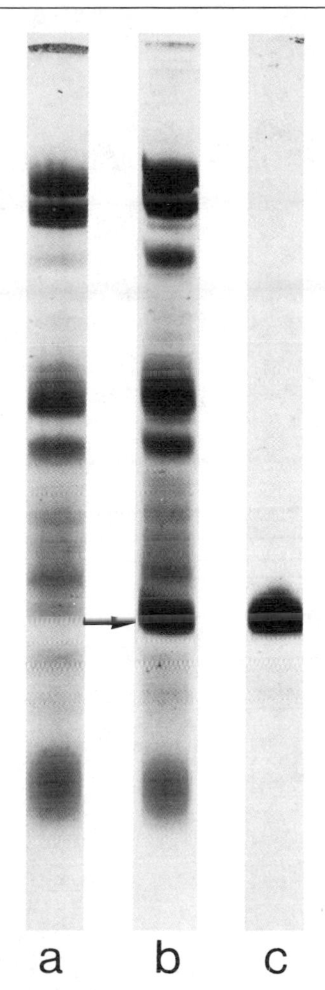

FIG. 1. SDS gel electrophoresis of (a) hypotonically lysed rabbit erythrocytes (control), (b) α-toxin-treated rabbit erythrocyte membranes, (c) native α-toxin. Samples applied to the gels were boiled in 2% SDS before electrophoresis. Toxin-lysed erythrocyte membranes showed the appearance of a 34-kDa protein band that migrated identically as native α-toxin (arrow). Direction of electrophoresis: top to bottom. Reproduced from Ref. 5 with permission.

Protein Isolation. Pores formed by large molecular aggregates such as found with SLO may be purified by a single centrifugation of membrane detergent extracts in a sucrose density gradient. Smaller oligomers such as formed by α-toxin cannot be purified in this manner because of cosedimentation of several membrane proteins. For isolation of these complexes, gel chromatography, e.g., over a Sepharose column is recom-

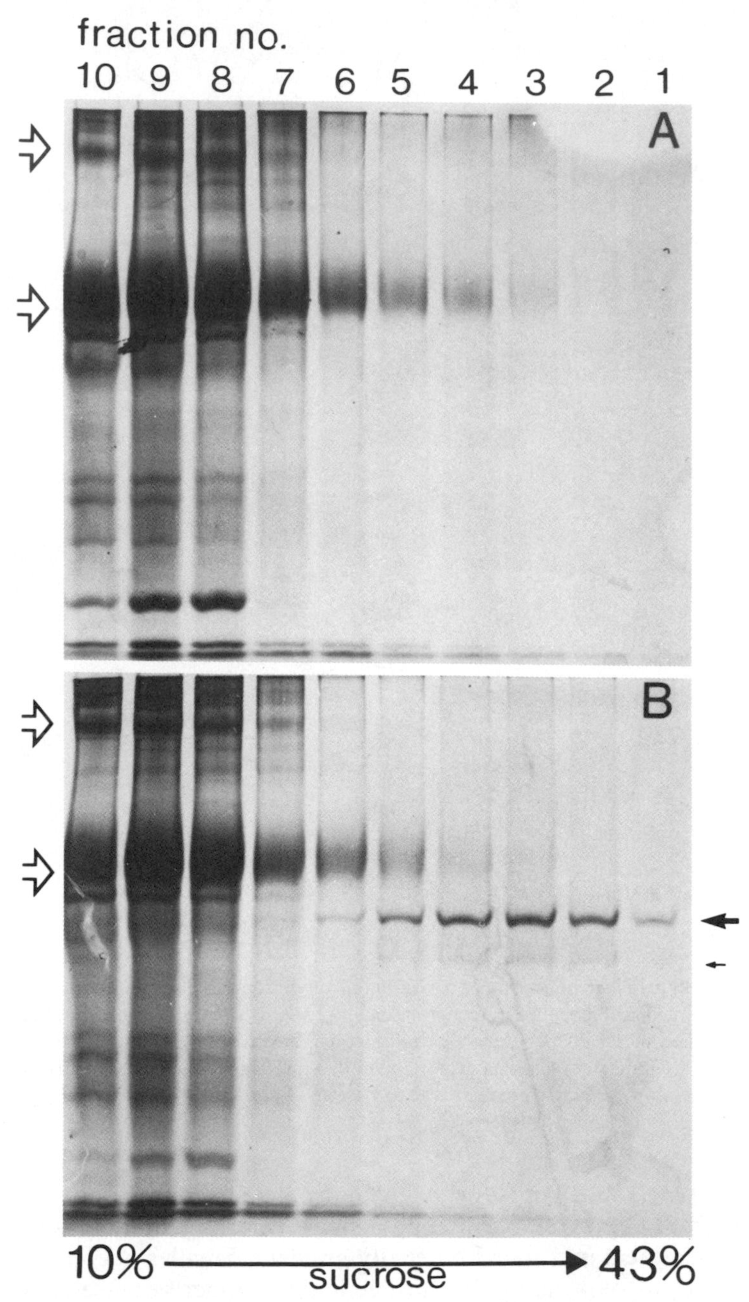

FIG. 2. Isolation of SLO oligomers from rabbit erythrocyte membranes. Control (A) and SLO-treated (B) membranes were solubilized in 250 mM DOC and centrifuged through

mended as a first step. The toxin-containing fractions may then be subjected to sucrose density gradient centrifugation, which we generally perform in linear 10–43% (w/v) gradients in 10 m*M* Tris, 50 m*M* NaCl, 7.5 m*M* NaN_3, pH 8.2, containing a low concentration of detergent (e.g., 6.25 m*M* DOC). For preparative isolations, a Beckman SW 41 Ti rotor is loaded with six gradients, each of 12-ml total volume. Centrifugation is performed at 150,000 *g* for 16 hr, 4°. Thereafter, the bottom of each tube is punctured and the gradients dropped out in 10 or 20 equal fractions. The individual fractions may be analyzed by SDS–PAGE, immunoelectrophoresis, and by electron microscopy. Figure 2 depicts the results of SDS–PAGE of fractions obtained after centrifugation of SLO-treated and control membranes. It is apparent that the bulk of membrane proteins is recovered in the upper half of the gradient. SLO protein bands are seen to have sedimented faster than other membrane proteins and are recovered in high-molecular-weight fractions covering a broad range corresponding to 20–40S. Fractions containing the toxin should be pooled and can be used for immunization and for membrane reconstitution experiments.

Membrane Reconstitution

Phosphatidylcholine, 100 mg/ml (type V-E, Sigma Chemicals Co., Munich, FRG) in chloroform : methanol (9 : 1) is evaporated to form a thin film on a conical evaporation flask. Subsequently the lipid film is dissolved under vigorous shaking into a solution of 10 ml 20 m*M* Tris, 50 m*M* NaCl, 15 m*M* NaN_3, pH 8.0, containing 50 m*M* DOC. Fractions containing SLO or α-toxin from sucrose density gradients are admixed with this solution of lecithin at approximate protein : lipid ratios (w/w) of 1 : 20 to 1 : 30. The lecithin–toxin mixture is dialyzed for 48 hr at 4° against 400 vol of 20 m*M* Tris, 100 m*M* NaCl, 7.5 m*M* NaN_3, pH 8.0, with one buffer change after 24 hr. Removal of detergent during this period of dialysis leads to formation of lipid vesicles (liposomes). During the dialysis a variable fraction of the toxin becomes lipid associated. The liposomes are concentrated over Amicon PM10 membranes and supplemented with sucrose to a concentration of 50% (w/v). Aliquots of 1.5 ml are overlayered with 3.5 ml of sucrose, 30% (w/v), in 20 m*M* Tris, 100 m*M* NaCl, 7 m*M* NaN_3, pH 8.0, in Beckman SW 50.1 ultracentrifugation tubes (total vol-

---

linear, DOC-containing sucrose density gradients (direction of sedimentation: left to right). Ten fractions were collected and anlayzed by SDS–PAGE. The SLO bands (solid arrows) are seen in the high-molecular-weight fractions 1–6, separated from the bulk of the contaminating erythrocyte membrane proteins (open arrows point to the spectrins and band 3 protein). Reproduced from Ref. 6 with permission.

ume 5 ml). Centrifugation at 35,000 rpm for 4 hr at 15° leads to flotation of liposomes and their separation from nonincorporated protein.

Figure 3 depicts a fused rocket immunoelectrophoresis in an experiment performed with staphyloccal α-toxin. Fifty microliters of fractions 1–7 was filled in each well and electrophoresed for 16 hr at 1.5 V/cm against specific antiserum to α-toxin (2 μl/cm^2). Liposome-bound toxin was found at the top of the sucrose gradient (B). In the control, where α-toxin was dialyzed in the absence of lecithin, all toxin remained at the bottom of the tube during centrifugation (A).

Figure 3C shows an SDS-polyacrylamide gel electrophoresis of the top fraction from Fig. 3B, containing the α-toxin-bearing liposomes. As described for membrane-derived staphylococcal α-toxin, the 34-kDa band

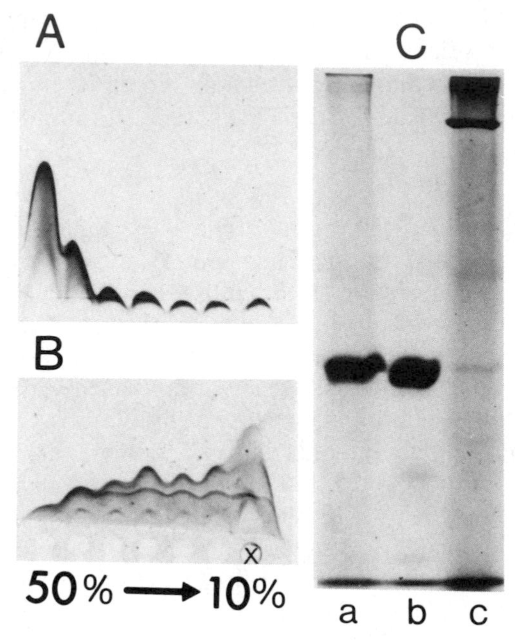

FIG. 3. Preparations of membrane-derived α-toxin were dialyzed in the absence (A) or presence (B) of phosphatidylcholine/DOC. Removal of the detergent during dialysis led to the formation of liposomes associated with the toxin, as demonstrated by subsequent centrifugation in sucrose: most of the toxin floated with lipid to the top of the sucrose gradient (B). In the control experiment without lipid, the protein remained at the bottom of the gradient (A). (C) SDS-polyacrylamide gel electrophoresis of the top fractions recovered from experiment (B). When toxin-carrying liposomes were boiled in the presence of SDS before electrophoresis (gel a), the relative mobility of the protein was identical to that of unheated native α-toxin (gel b), while electrophoresis of unheated, toxin-carrying liposomes (gel c) led to the appearance of a high-molecular-weight band (200,000) representing the toxin hexamer. Reproduced from Ref. 5 with permission.

corresponding to the native toxin is observed only when the liposomes are boiled in SDS (2%, v/v) before electrophoresis. If samples are not boiled in SDS, the 200-kDa band corresponding to the hexameric membrane-bound α-toxin is observed.

Conclusion

The described procedures require no special expertise or equipment, and should be of use for the isolation and characterization of other membrane-damaging proteins in the future. Oligomerizing pore formers that attack erythrocytes should be particularly easy to study, even when the native proteins are not available in purified form. Methods for ultrastructural analyses of biological target membranes and reconstituted liposomes are detailed in chapters [50] and [51] of this volume, and electron microscopy should provide useful complementary information on the orientation of the protein pores in the lipid bilayer.

Acknowledgments

These studies were supported by the Deutsche Forschungsgemeinschaft (Bh 2/1-5 and 2/2) and the Verband der Deutschen Industrie.

[41] Assay Methods for Alpha Toxin from *Clostridium perfringens:* Phospholipase C

By COLETTE JOLIVET-REYNAUD, HERVÉ MOREAU, and JOSEPH E. ALOUF

Phosphatidylcholine → phosphorylcholine + 1,2-diacylglycerol
Sphingomyelin → phosphorylcholine + ceramide

Numerous assay methods have been developed to determine the enzymatic activity of phospholipase C (phosphatidylcholine cholinephosphohydrolase, EC 3.1.4.3): physical methods (turbidimetric assays[1–4]), acidimetric methods (titrimetric assays[5–7]), chemical methods (acid-soluble

[1] W. E. van Heyningen, *Biochem. J.* **35,** 1246 (1941).
[2] A. C. Ottolenghi, *Anal. Biochem.* **5,** 37 (1963).
[3] R. Murata, A. Yamamoto, S. Soda, and A. Ito, *Jpn. Med. Sci. Biol.* **18,** 189 (1965).
[4] R. Möllby, C. E. Nord, and T. Wadström, *Toxicon* **11,** 139 (1973).
[5] M. Rodbell, *J. Biol. Chem.* **241,** 130 (1966).

phosphorus assays[8–10] with alternate methods including solvent extraction assay[11] or phospholipase–phosphatase coupled assays[12,13]), a spectrophotometric method (*p*-nitrophenylphosphorylcholine assay[14]), radiochemical methods (release of radioactive phosphorylcholine assay[15,16]), a bioluminescence method,[17] and a fluorescence-quenching method.[18]

Furthermore, in contrast to most bacterial phospholipases C, *Clostridium perfringens* phospholipase C has toxic and membrane-damaging activities: lethal, necrotizing, and hemolytic effects have been used for assays of the toxin activity.[4,7,19–22]

Most of these methods have been reported and discussed in earlier volumes of this series by Ottolenghi,[23] Takahashi *et al.*,[24] and Krug and Kent.[25] However, none of them is strictly specific for *Clostridium perfringens* phospholipase C and must be used with caution with impure preparations of enzyme.

In this chapter we describe three assay methods which are modified procedures of the general methods listed above. These assays have been selected with regard for their simplicity and sensitivity.

Hemolytic Assay

Principle

The lytic activity of *Clostridium perfringens* phospholipase C is determined by spectrophotometric measurement of hemoglobin release from rabbit erythrocytes.

---

[6] R. F. A. Zwaal, B. Roelofsen, P. Comfurius, and L. L. M. van Deenen, *Biochim. Biophys. Acta* **233,** 474 (1971).
[7] R. Möllby and T. Wadström, *Biochim. Biophys. Acta* **321,** 569 (1973).
[8] M. G. MacFarlane and B. C. J. Knight, *Biochem. J.* **35,** 884 (1941).
[9] A. C. Ottolenghi, *Biochim. Biophys. Acta* **106,** 510 (1965).
[10] Y. Yamakawa and A. Ohsaka, *J. Biochem. (Tokyo)* **81,** 115 (1977).
[11] J. H. Kleiman and W. E. M. Lands, *Biochim. Biophys. Acta* **187,** 477 (1969).
[12] A. Ohsaka and T. Sugahara, *J. Biochem. (Tokyo)* **64,** 335 (1968).
[13] E. L. Krug, N. J. Truesdale, and C. Kent, *Anal. Biochem.* **97,** 43 (1979).
[14] S. Kurioka and M. Matsuda, *Anal. Biochem.* **75,** 281 (1976).
[15] B. A. Diner, *Biochim. Biophys. Acta* **198,** 514 (1970).
[16] R. D. Mavis, R. M. Bell, and P. R. Vagelos, *J. Biol. Chem.* **247,** 2835 (1972).
[17] S. Ulitzur and M. Heller, *Anal. Biochem.* **91,** 421 (1978).
[18] R. F. Chen, *Anal. Lett.* **10,** 787 (1975).
[19] K. Mitsui, N. Mitsui, and J. Hase, *Jpn. J. Exp. Med.* **43,** 65 (1973).
[20] C. J. Smyth and J. P. Arbuthnott, *J. Med. Microbiol.* **7,** 41 (1974).
[21] M. G. MacFarlane, *Biochem. J.* **47,** 270 (1950).
[22] T. Takahashi, T. Sugahara, and A. Osaka, *Biochim. Biophys. Acta* **351,** 155 (1974).
[23] A. C. Ottolenghi, this series, Vol. 14, p. 188.
[24] T. Takahashi, T. Sugahara, and A. Ohsaka, this series, Vol. 71, p. 710.
[25] E. L. Krug and C. Kent, this series, Vol. 72, p. 347.

Reagents

Buffer A: Tris–HCl, 20 mM, pH 7.2; NaCl, 0.158 M
Buffer B: Tris–HCl, 20 mM, pH 7.2; NaCl, 0.158 M; $Ca(CH_3COO)_2$, 10 mM; $ZnCl_2$, 0.1 mM
Cholesterol: 20 mg/ml in absolute ethanol

Standardized Rabbit Erythrocyte Suspension (RRBC). Rabbit blood is drawn from the ear vein and collected in Alsever's solution. After centrifugation at 1500 g for 10 min, the plasma and buffy layer are removed by aspiration. Erythrocytes are washed three times with Tris–HCl-buffered saline. The packed erythrocytes are standardized to 2.5% (v/v) in buffer A as follows: 0.5 ml of the suspension lysed with 14.5 ml of 0.1% Na_2CO_3 gives an absorbance of 0.2 at 541 nm. The standard RRBC suspension (~3×10^8 cells/ml) is kept at 4° and used within 4 days.

Procedure

Serial dilutions of phospholipase C in 1 ml of buffer B containing 0.1% BSA are incubated at 37° for 10 min before adding 0.5 ml of standardized RRBC. Test tubes are incubated at 37° for 45 min and then centrifuged at 1500 g for 5 min. The percentage of hemolysis is estimated by the optical absorbance of the hemoglobin released in the supernatant at 541 nm compared with the absorbance of standards.

In routine assays of enzyme activity during purification or with impure toxin preparations, the hemolytic activity of theta toxin could interfere in results. In this case, test samples are incubated previously for 10 min, at 37°, with 10 μl of an ethanolic solution of cholesterol which inhibits theta toxin activity. RRBC suspension is then added as described above.

Unit. Values are estimated graphically by plotting, on a log probit graph, the percentage of lysis versus the volume of diluted toxins used in the assay. One hemolytic unit (HU) is defined as the amount of toxin which causes hemolysis of 50% of the cells.

Comments. This assay is rapid, reproducible, and very sensitive. In these conditions, 1 HU corresponds to about 15 ng of enzyme (3.10^{-4} nmol).

Most erythrocytes except those of horse and goat are sensitive to *Clostridium perfringens* phospholipase C and in many laboratories, sheep erythrocytes are used for the hemolytic assay of this toxin. However, rabbit erythrocytes are 5-fold more sensitive to the toxin than sheep erythrocytes and the hot–cold phenomenon is not observed (sheep erythrocytes do not lyse when exposed to alpha toxin at 37°, lysis occurs only after cooling to 4°).

The difference in sensitivity of these two erythrocyte species to alpha toxin characterizes the specific hemolytic effect of this toxin. In contrast, theta toxin lyses erythrocytes of sheep and rabbit to the same extent whereas rabbit erythrocytes are insensitive to delta toxin.

Turbidimetric Method

Principle

This method, based on that described by van Heyningen,[1] measures the increase in turbidity of egg yolk suspension after incubation with phospholipase C. The turbidimetric method, with modifications, has been used in the assay of phospholipase C from different sources. For the assay of *Clostridium perfringens* phospholipase C, Möllby *et al.*[4] replaced egg yolk substrate by L-α-lecithin from soybean emulsified in 0.1 *M* Tris–HCl buffer, pH 7.2, and 10 m*M* $CaCl_2$, by ultrasonic treatment. In this chapter we develop a modified method with two alternate procedures by using egg lecithin dispersion in sodium cholate.

Substrate

Solution A (stock substrate): 0.9 g of egg lecithin (Merck) is dispersed in 25 ml of buffer containing 20 m*M* Tris–HCl, pH 7.2, 0.158 *M* NaCl, and 2.4% sodium cholate

This mixture is equilibrated at room temperature for 2 hr under stirring and is stable for 4 days at room temperature.

Solution B: 16 ml of solution A are diluted to 100 ml with buffer containing 20 m*M* Tris–HCl, pH 7.2, 0.158 *M* NaCl, 10 m*M* $Ca(CH_3COO)_2$, and 0.2 m*M* $ZnCl_2$ in final concentration

This solution is prepared daily and equilibrated at room temperature under agitation.

Procedure 1

One milliliter of substrate suspension (solution B) is added into each of a set of six cuvettes in a thermostatted spectrophotometer (Beckman DU7) and allowed to equilibrate at 37° for 10 min. Different toxin dilutions in a volume of 0.1 ml are then added into each cuvette except in one taken as a control where 0.1 ml of 20 m*M* Tris–HCl buffer, pH 7.2, is added. Increase of turbidity in each cuvette is measured at 510 nm and recorded continuously.

After subtracting the increase of turbidity in the control, the enzymatic activity is determined in the linear portion of the curve.

Procedure 2

The reaction mixture described in procedure 1 is carried out in test tubes. After 15 min of incubation at 37°, the reaction is stopped by adding

10 μl of 0.2 *M* EDTA. The increase of turbidity in each tube is measured at 510 nm.

Unit. One unit of activity is defined arbitrarily as the amount of enzyme which causes an increase of turbidity corresponding to an optical absorbance of 0.2/min.

Comments. Egg yolk preparations with uniform quality are difficult to obtain. This modified method with egg lecithin suspension in the presence of a detergent is simple to perform and gives reproducible results.

Within limits of enzyme concentration, a good proportionality is obtained between the increase in turbidity of the reaction mixture and the enzyme quantity.

Procedure 2, which is simple to perform, is used for routine assay of enzyme activity in column chromatography fractions.

Titrimetric Method

Principle

By the action of phospholipase C, the phosphate bond ester of phospholipids is hydrolyzed with the liberation of H^+ ions. The measurement of the H^+ liberated is the basis of this assay, which has been described by Rodbell,[5] Zwaal *et al.*,[6] and Möllby and Wadström.[7] Some modifications in the substrate preparation have been incorporated.

Substrate. Egg lecithin (Merck), 6.5 g, is dispersed in 100 ml 0.2 *M* NaCl, 5 m*M* $Ca(CH_3COO)_2$, 0.1 m*M* $ZnCl_2$ containing 3% of sodium cholate. This dispersion is allowed to equilibrate under stirring for 2 hr at room temperature before using.

Procedure

The reaction is performed at 37° by adding enzyme sample to 1.5 ml of substrate. The pH is maintained at 6.7. Titratable H^+ is followed in a titration unit in connection with a titrigraph TT11 (Radiometer, Copenhagen, Denmark) by addition of 0.1 *N* NaOH.

Unit. One unit (U) liberates 1 μmol of H^+ ions per minute. The specific enzymatic activity is expressed as units per milligram of protein.

Comment. Detergent is of primary importance in the level of enzyme activity. Dispersion of egg lecithin with Triton X-100 instead of sodium cholate increases the enzymatic activity of phospholipase C. However, results are less reproducible.[26]

[26] H. Moreau, C. Jolivet-Reynaud, and J. E. Alouf, manuscript in preparation (1988).

[42] Phospholipase Activity of Bacterial Toxins

By LARRY W. DANIEL, LYNN KING, and MARTHA KENNEDY

Bacterial toxins may possess phospholipase activity that contributes to their membrane-damaging properties. These bacterial phospholipases have a variety of substrate specificities and include phospholipases A, C, and D.[1] The products of these reactions are summarized below:

$$\text{Phospholipid} \xrightarrow{\text{phospholipase A}} \text{lysophospholipid} + \text{fatty acid}$$
$$\text{Phospholipid} \xrightarrow{\text{phospholipase C}} \text{diacylglycerol} + \text{phospho base group}$$
$$\text{Phospholipid} \xrightarrow{\text{phospholipase D}} \text{phosphatidic acid} + \text{base group}$$

Since the biological activities of various, previously described bacterial toxins are a function of their phospholipase activity it is of interest to test newly isolated toxins for phospholipase activity. In this chapter we describe methods for the preparation of radiolabeled phospholipid substrates and methods for determining phospholipase activity and specificity. The assay of a recently recognized phospholipase A_2 (PLA_2) produced by *Vibrio vulnificus* and the phospholipase D (PLD) activity of a recently isolated toxin produced by *Vibrio damsela* will be used as representative examples of the phospholipase determinations.

Preparation of Radiolabeled Phospholipid Substrates from Cell Culture

Substrates are prepared which contain predominately [1-^{14}C]palmitic acid (^{14}C 16 : 0) in the *sn*-1 position and [5,6,8,9,11,12,14,15-^{3}H]arachidonic acid (^{3}H 20 : 4) in the *sn*-2 positions of the phospholipids by incubating cultured cells with the radiolabeled fatty acids in medium containing 10% fetal bovine serum. We use Madin Darby canine kidney cells since their uptake and incorporation of [^{3}H]20 : 4 and [^{14}C]16 : 0 has been previously characterized[2]; however, other cultured cell types are suitable. Alternatively, ^{32}P-labeled phospholipids are obtained by incubating the cells with [^{32}P]P_i; 10 μCi/ml culture medium. In addition, phospholipids doubly labeled with [^{3}H]20 : 4 and ^{32}P may be prepared by these procedures.

[1] H. van den Bosch, *in* "Phospholipids" (J. N. Hawthorne and G. B. Ansell, eds.), p. 313. Elsevier, Amsterdam, 1982.

[2] L. W. Daniel, L. King, and M. Waite, *J. Biol. Chem.* **256,** 12830 (1981).

Extraction and Purification of Individual Phospholipids

After labeling cells for 24 hr, the medium is discarded and the cells are scraped from the culture dishes into 2 ml methanol. The lipids are then extracted as described by Bligh and Dyer.[3] The solvent is evaporated under a stream of N_2; the phospholipid classes are separated by thin-layer chromatography (TLC) on silica gel 60 plates (E. Merck, Darmstadt, West Germany) developed in a solvent system consisting of chloroform–methanol–acetic acid–water (75 : 48 : 12.5 : 4, v/v/v/v). The phospholipid classes are identified by comparing their migration to known phospholipid standards (Serdary Research Laboratories, London, Ontario, Canada). The standards, but not radiolabeled lipids, were visualized by exposure to iodine vapors. The appropriate phospholipid-containing areas were then scraped from the plates and reextracted as described above. Alternatively, high-performance liquid chromatography (HPLC) can be used for the separation of phospholipid classes. We isolated substrates by a modification of the HPLC method of Geurts Van Kessel[4] using a Waters μPorasil column (10 μm, 4 mm × 30 cm). The phospholipids were applied to the column in $CHCl_3$ and eluted with a solvent gradient of 1.5 to 9% water in a mixture of 2-propanol and hexane (8 : 6, v/v) at 1 ml/min. The change in gradient concentration was accomplished in 18 min and was followed by an additional 15 min at 9% water. The elution of phospholipids was monitored by absorbance of the sample at 208 nm. It should be noted that the absorbance is not quantitative since the phospholipid species vary in unsaturation. Phospholipid substrates may be quantitated by the method of Rouser *et al.*[5] and the specific activity of the substrates can be determined. It should be noted that some ^3H- or ^{14}C-labeled phospholipids are commercially available.

Phospholipase Assays

Phospholipase A_2

Phospholipases A_2 specifically hydrolyze the *sn*-2-acyl moiety of phospholipids and therefore preferentially release ^3H-labeled fatty acids from the $^3H/^{32}P$-labeled phospholipids prepared as described above. We have used the following procedure to assay the phospholipase A_2 of *Vibrio vulnificus:* Reaction mixtures containing samples (0.1 ml) of the phospho-

[3] E. G. Bligh and W. J. Dyer, *Can. J. Biochem. Physiol.* **37,** 911 (1959).

[4] W. S. M. Guerts Van Kessel, W. M. A. Hax, R. A. Demel, and J. DeGier, *Biochim. Biophys. Acta* **486,** 524 (1977).

[5] G. Rouser, A. N. Siakotos, and S. Fleischer, *Lipids* **1,** 85 (1966).

TABLE I
HYDROLYSIS OF [^{3}H,^{32}P]PHOSPHATIDYLCHOLINE[a]

	Percentage of radiolabel recovered as[b]					
	PC		LPC		FFA	
Enzyme source	^{3}H	^{32}P	^{3}H	^{32}P	^{3}H	^{32}P
Vibrio vulnificus	50.2	78.0	4.4	21.1	44.5	.3
Crotalus adamanteus phospholipase A_2	5.0	18.2	3.7	72.2	87.7	1.5
No enzyme	91.0	97.0	1.4	2.5	4.2	0.1

[a] Modified from Ref. 6.
[b] PC, Phosphatidylcholine, LPC; lysophosphatidylcholine; FFA, free fatty acid.

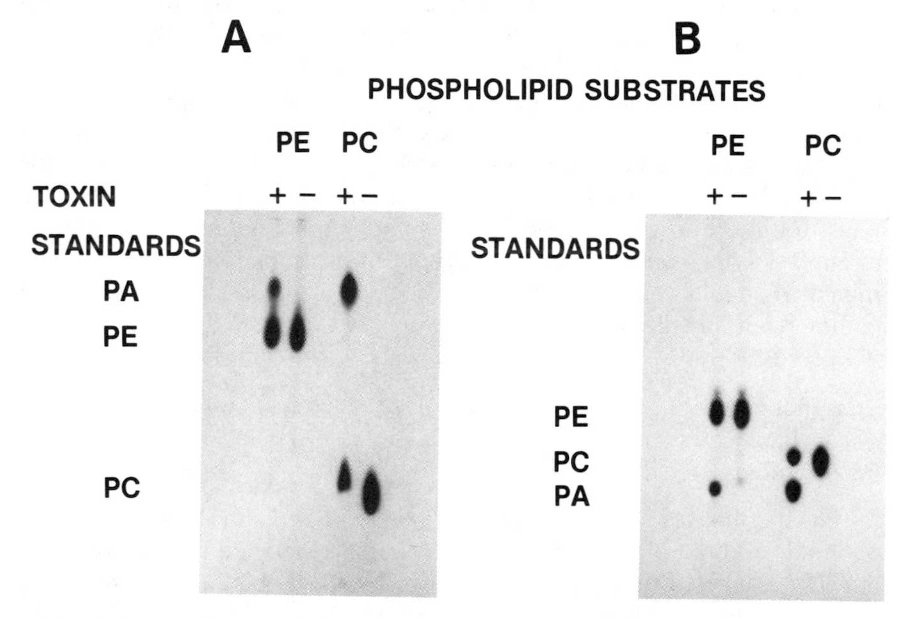

FIG. 1. Phospholipase activity of *Vibrio damsela* toxin. ^{32}P-Labeled phospholipid substrates were prepared and incubated with *Vibrio damsela* toxin as described. The products of the reactions were separated by thin-layer chromatography in an acid solvent system consisting of chloroform–methanol–acetic acid–water (75/48/12.5/4.5, v/v) (A) or a basic solvent system consisting of chloroform–methanol–ammonium hydroxide (65/35/8, v/v) (B). The products were visualized by autoradiography and identified by comigration with known standards. Two solvent systems were used to confirm the identification.

lipase peak fraction obtained by isoelectric focusing as described by Testa *et al.*[6] and 0.9 ml of a suspension of labeled phosphatidylcholine substrate in PBS (0.067 *M* Na_2HPO_4–HCl, 0.077 *M* NaCl, pH 7) were incubated at 37° for 2 hr. The lipids were then extracted and separated by TLC as described above. The labeled lipids were located by fluorography using ENHANCE spray (New England Nuclear, Boston, MA) and Kodak SB-5 film. The areas containing radiolabeled products were then scraped from the plates and quantitated by scintillation counting. When [^{3}H]arachidonic acid and ^{32}P-labeled phosphatidylcholine (PC) were used as substrate, we found that [^{3}H]arachidonic acid was released as free fatty acid and there was an increase in ^{32}P-labeled lysophosphatidylcholine (Table I). Enzymatic activity is calculated from the percentage hydrolysis and the known specific activities of the substrates.

Phospholipase D

We have also assayed the recently purified toxin from *Vibrio damsela*[7,8] for phospholipase activity and found no release of fatty acids from [^{3}H]20 : 4- and [^{14}C]16 : 0-labeled substrates (data not shown). However, when the toxin was assayed using ^{32}P-labeled phospholipids as substrates, we observed a conversion of PC and (PE) to phosphatidic acid (Fig. 1). Therefore, the toxin of *Vibrio damsela* is a phospholipase D with a rather unique specificity, since most bacterial phospholipases do not degrade intact phosphatidylcholine or phosphatidylethanolamine.[1]

Conclusions

The methods described herein provide a simple and sensitive method for determining the phospholipase activity and phospholipid class specificity of bacterial phospholipases. These procedures are easily modified to determine the ion requirements and pH optima of the enzymes. However, these procedures are not recommended for kinetic analysis of the enzymes due to problems of substrate insolubility in aqueous solutions.[9]

[6] J. Testa, L. W. Daniel, and A. S. Kreger, *Infect. Immun.* **45,** 458 (1984).
[7] M. H. Kothary and A. S. Kreger, *Infect. Immun.* **49,** 25 (1985).
[8] A. S. Kreger, M. H. Kothary, and L. D. Gray, this volume [25].
[9] L. W. Daniel, *in* "Biochemistry of Arachidonic Acid Metabolism" (W. E. M. Lands, ed.), p. 175. Nijhoff, Boston, 1985.

[43] Detection of Endotoxin by Enhancement with Toxic Shock Syndrome Toxin-1 (TSST-1)

By G. A. BOHACH and P. M. SCHLIEVERT

Introduction

Toxic shock syndrome (TSS) is an illness characterized by an array of clinical symptoms including fever, scarlatiniform rash, desquamation, multisystem involvement, and hypotension or shock. The pathogenesis of TSS is attributed to an exotoxin,[1] TSST-1, produced by *Staphylococcus aureus*. The mechanism by which TSST-1 causes hypotension and shock is incompletely understood. However, it is known that TSST-1, like other pyrogenic toxins, enhances host susceptibility to endotoxin shock.[2]

There are at least three mechanisms by which TSST-1 can enhance the biological effects of endotoxin. First, TSST-1 is immunosuppressive,[3] thereby allowing increased growth of gram-negative opportunistic organisms. This results in greater than normal amounts of endotoxin entering the systemic circulation. Second, the interaction of TSST-1 with the reticuloendothelial system hinders the clearance of endotoxin from the circulation.[3] Third, TSST-1 can directly enhance the cytotoxicity of endotoxin.[4] The combination of these effects demonstrated *in vivo* and *in vitro* results in lethal endotoxin shock in experimental animals.

This chapter is devoted to describing two systems that exploit the enhancing effect of TSST-1 for detecting small amounts of endotoxin in a sample. One technique, an *in vivo* assay, is based upon the ability of sublethal doses of TSST-1 to cause up to (and possibly greater than) a 50,000-fold increase in susceptibility of rabbits to lethal endotoxin shock. We also describe a sensitive *in vitro* cytotoxicity assay for detection of less than 0.1 ng of endotoxin per milliliter of sample.

Toxin Preparation

Endotoxin-free TSST-1 is prepared from cultures of toxigenic strains of *Staphylococcus aureus* grown in dialyzable beef heart medium. Usu-

[1] Y. B. Kim and D. W. Watson, *J. Exp. Med.* **131,** 611 (1970).
[2] P. M. Schlievert, *Infect. Immun.* **36,** 123 (1982).
[3] P. M. Schlievert, *J. Infect. Dis.* **147,** 391 (1983).
[4] W. F. Keane, G. Gekker, P. M. Schlievert, and P. K. Peters *Am. J. Pathol.* **122,** 169 (1986).

ally strain MN8,[5] a high TSST-1 producer, is used. Purification of the toxin by isoelectric focusing after ethanol precipitation is described elsewhere in this volume.[5]

Endotoxin from a variety of gram-negative organisms can be used as standards. We use either commercially available endotoxin (Sigma Chemical Co., St. Louis, MO, or List Biological Laboratory, Inc., Campbell, CA) from *Salmonella typhimurium, Salmonella typhi,* or *Salmonella minnesota* Re595 or prepare it by the method of Westphal *et al.*[6]

In Vivo Assay

This assay, described by Schlievert,[2] is performed in rabbits. Most of the symptoms typical of human TSS, excluding rash and desquamation, can be produced in rabbits by injection of TSST-1 and endotoxin. The animals are first given a sublethal dose of TSST-1 and rested. A second dose of a sample containing endotoxin is administered 4 hr later, and the animals are observed for mortality. A critical feature of the assay involves preconditioning of the animals since TSST-1 is highly lethal to unconditioned rabbits. This is likely due to high levels of endogenous endotoxin originating from opportunistic *Pasteurella multocida* infections in nonconditioned animals.[2]

Animal Preparation

American Dutch-belted rabbits [1.0–1.5 kg of body weight (kg BW)] are conditioned before the assay. The animals are restrained for 3 hr on a pyrogen test rack and then allowed to adjust to the laboratory environment for at least 1 week. If necessary, pulmonary infections are treated with tetracycline. Routine screening for bacteremia is performed by plating 0.1 ml of peripheral blood on agar plates. Only healthy rabbits should be used.

Endotoxin Screening

The rabbits (including controls) are restrained in pyrogen test racks with rectal thermometers inserted for monitoring TSST-1 induced fever. The rabbits are allowed to rest 1 hr on the test rack prior to receiving injections to allow their temperatures to return to normal. Animals whose temperatures remain above 39° are not used. At least three rabbits per group are then given a pretreatment intravenous (iv) injection of TSST-1

[5] D. Blomster-Hautamaa and P. M. Schlievert, this volume [6].

[6] O. Westphal, O. Lüderitz, and F. Bister, *Z. Naturforsch., B* **7,** 148 (1952).

(0.1–10 μg/kg BW) in a standard volume of 1 ml/kg BW. The toxin diluent is pyrogen-free phosphate-buffered saline (PBS, 0.05 *M* sodium phosphate, pH 7.0, 0.15 *M* NaCl). The temperature of the animals typically rises linearly in response to TSST-1 and peaks at 4 hr. At this time, an iv injection of sample containing endotoxin should be administered. Control rabbits receiving only TSST-1 should be included in each assay to ensure that conditioning of the animals was successful. The animals are returned to their cages and observed for lethal shock over a 48-hr period.

After receiving TSST-1 and endotoxin, the rabbits show a fever response, become hypothermic with evidence of vascular collapse, and develop diarrhea and labored breathing. High doses of endotoxin (>1.0 μg/kg BW) cause death within 3 hr whereas animals receiving lower doses (0.01–1.0 μg/kg BW) may survive for up to 24 hr. The symptoms observed are not typical of endotoxin shock, but rather are greatly accelerated. It is important to note that the LD_{50} of endotoxin alone in these animals is between 500 and 1000 μg/kg BW.

Endotoxin Quantitation

If desired, the assay described above may be expanded to quantitate endotoxin in a sample. The sensitivity of rabbits to lethal shock after injection of known amounts of TSST-1 and endotoxin standards must be determined as shown in Table I. Appropriately diluted unknown samples can then be tested for lethality in rabbits pretreated with TSST-1 and their effects compared to those of the endotoxin standards.

In Vitro Assay

Endotoxin causes necrosis of rat renal tubular epithelial cells (RTC) *in vitro*. Keane *et al.*[4] showed that the cytotoxic effect is augmented at least 1000-fold by pretreatment of the cells with TSST-1. Thus, TSST-1-treated RTC are sensitive to endotoxin concentrations of less than 0.1 ng/ml.

Preparation of RTC

The procedure used is a modification[4] of that described by Cade-Treyer and Schigeru.[7]

Adult male Sprague-Dawley rats are anesthetized with Nembutal (50 mg/kg BW). The kidneys are exposed and perfused *in situ* with 30 ml of PBS containing 0.04% EDTA and then with 30 ml of PBS. A final perfusion with 0.05% collagenase and 0.125% trypsin in PBS (12 ml) is per-

[7] D. Cade-Treyer and T. Schigeru, *Cell Tissue Res.* **163,** 15 (1975).

TABLE I
LETHALITY OF ENDOTOXIN IN RABBITS PRETREATED WITH TSST-1

Pretreatment with TSST-1 (μg/kg BW)	Results (dead/total) of different endotoxin doses (μg/kg BW) given 4 hr after TSST-1						
	0	0.001	0.01	0.1	1.0	10.0	100.0
0				0/3	0/3	0/3	0/3
0.1				0/3	1/3	2/3	2/3
1.0	0/3	0/3	0/3	1/3	2/3	3/3	
5.0	0/3	0/3	1/3	1/3	2/3	2/3	
10.0	0/3	0/3	1/3	3/3	3/3	3/3	
100.0	0/3	0/3	2/3	2/3	3/3		

formed. After removal of the capsule and fat tissue, the outer cortex is cut into 5-mm fragments. The tissue is placed in a PBS–trypsin–collagenase solution and incubated at 37° with shaking. After 20 min of digestion, the tissue is filtered through gauze and the filtrate is discarded. The tissue is recovered and subjected to two additional digestions (20 min each). After each of the last two digestions, the tissue is filtered through gauze and the filtrates are collected and pooled in a 50-ml conical centrifuge tube. RTC are pelleted by centrifugation for 5 min (180 *g*) and resuspended in Hanks' balanced salts solution (Grand Island Biological Company, Grand Island, NY) with 0.1% gelatin (GHBSS). The cells are filtered through nylon mesh (40 μm), pelleted by centrifugation (180 *g*), and resuspended in GHBSS to a concentration of 1×10^6 cells/ml. The cell suspension is stored at 4° and used within 3 hr of preparation.

Assay Procedure

RTC are first incubated with nontoxic levels of TSST-1 for 20 min, washed, and exposed to endotoxin. Cytotoxicity can be assessed directly by the inability of damaged cells to exclude trypan blue. An alternative method employs chemiluminescence (CL) to detect reactive oxygen species generated in cells irreversibly injured by endotoxin.[4]

Trypan Blue Method. TSST-1 is first tested at various concentrations (0.1–10.0 ng/ml) to determine its maximum nontoxic level. RTC (1×10^5) in 0.1 ml of GHBSS are mixed with 1 ml of the endotoxin. After incubating for 30 min (37°), ice-cold GHBSS is added. The cells are pelleted by centrifugation at 200 *g* (5 min) and resuspended in PBS containing 0.1% trypan blue. Uptake of the dye by RTC is evaluated by microscopic examination and compared to control cell suspensions that were not exposed to toxin.

For use in assays to detect endotoxin in a specimen, indicator RTC are pretreated with the highest concentration of TSST-1 that causes baseline cytotoxicity (usually 1.0 ng/ml). The cells are preincubated with TSST-1 for 20 min (37°) and washed with GHBSS by centrifugation at 200 *g*. The RTC pellet is resuspended in a sample containing endotoxin and incubated for an additional 30 min at 37°. The cells are pelleted by centrifugation in cold GHBSS, resuspended in 0.1% trypan blue, and observed by microscopic examination for assessment of cytotoxicity.

Chemiluminescence (CL) Method. A preliminary experiment is performed to determine the baseline CL response of RTC pretreated with nontoxic concentrations of TSST-1. Suspensions of RTC (0.1 ml) containing 1×10^5 cells are mixed with 3.6 μl of a stock luminol solution (0.275 m*M* 5-amino-2,3-dihydro-1,4-phthalazinedione) and dark adapted (10 min) in a luminometer. TSST-1 is added at various concentrations (0.1–10 ng/ml) and the CL response is monitored for 60 sec.

A similar technique is used to detect the presence of endotoxin in a sample. RTC are pretreated with the highest concentration of TSST-1 that stimulates only baseline CL. After 20 min the cells are washed with GHBSS, mixed with luminol, and dark adapted for 10 min in a luminometer. The cells are mixed with the endotoxin sample and the CL response is monitored for 60 sec. Endotoxin-induced cell damage is indicated by CL levels higher than controls treated with GHBSS.

[44] Ileal Loop Fluid Accumulation Test for Diarrheal Toxins

By Merlin S. Bergdoll

I. Introduction

Many different organisms have been identified with the production of diarrhea in humans with new ones continually being associated with this illness. The major test that has been and is being used in the study of diarrhea-associated organisms is the ileal loop fluid accumulation test. De and Chatterje[1] used the ileal loop test in rabbits in their search for a test method for studying the biological action of *Vibrio cholerae,* one of the earliest diarrheal diseases associated with a specific organism. Initially it was used to study the organism but later was used as an aid in the identification and purification of the toxin responsible for the diarrhea. Since its

[1] S. N. De and D. N. Chatterje, *J. Pathol. Bacteriol.* **66,** 559 (1953).

initial development it has been used in the study of a number of diarrheal agents: *Escherichia coli, Clostridium perfringens, Bacillus cereus, Shigella dysenteriae, Salmonella, Vibrio parahaemolyticus, Aeromonas hydrophila, Yersinia enterocolitica,* and *Campylobacter.*

This chapter is written primarily to give specific information about the ileal loop test as it has been and is being used with rabbits. Other animals have been used, such as lambs[2-4] and mice[5] for *C. perfringens,* mice[6] and pigs[7] for *E. coli,* mice for *B. cereus,*[8] and rats for *C. jejuni.*[9,10] It is not possible to review here all of the research that has been done in which the ileal loop has been used; however, brief mention will be made of some of the diarrheal diseases in which the test has been used to study the disease and to aid in the elucidation of the causative factor.

The specific agents responsible for the symptoms of the diseases that are produced by the organisms are proteins to which antibodies can be prepared in most cases. Once specific antibodies are available, the need for the ileal loop test is decreased, specifically for analytical purposes, for example, testing specific microorganisms for the production of the toxic agent. The name given to the diarrheal toxins is enterotoxin, preceded by the name of the organism, for example, cholera enterotoxin, *E. coli* enterotoxin, etc. This created some confusion because the term enterotoxin had been applied to the staphylococcal toxin responsible for staphylococcal food poisoning by Dack *et al.*[10a] in 1930 and was the only toxin so called for many years. The staphylococcal enterotoxins do not produce ileal loops even though they elicit diarrhea (sometimes profuse) when ingested. Because they do not produce ileal loops some investigators do not consider them to be true enterotoxins even though they do produce the intragastric symptoms of vomiting and diarrhea. The staphylococcal enterotoxins are performed in the food before ingestion and produce a reaction within a few hours (1–6) while the so-called diarrheal enterotoxins normally are produced in the intestinal tract by the organisms growing there.

[2] A. H. W. Hauschild, L. Niilo, and W. J. Dorward, *Appl. Microbiol.* **16,** 1235 (1968).

[3] A. H. W. Hauschild, L. Niilo, and W. J. Dorward, *Can. J. Microbiol.* **16,** 339 (1970).

[4] L. Niilo, *Infect. Immun.* **3,** 100 (1971).

[5] K. Yamamoto, I. Ohishi, and G. Sakaguchi, *Appl. Environ. Microbiol.* **37,** 181 (1979).

[6] K. Punyashthiti and R. A. Finkelstein, *Infect. Immun.* **4,** 473 (1971).

[7] M. M. R. ElDeib, C. R. Dove, C. D. Parker, T. L. Veum, G. M. Zinn, and A. A. White, *Infect. Immun.* **51,** 24 (1986).

[8] K. Shinagawa, N. Matsusaka, H. Konuma, and H. Kurata, *Jpn. J. Vet. Sci.* **47,** 557 (1985).

[9] F. A. Klipstein and R. F. Engert, *Infect. Immun.* **45,** 314 (1984).

[10] F. A. Klipstein, R. F. Engert, H. Short, and E. A. Schenk, *Infect. Immun.* **50,** 43 (1985).

[10a] G. M. Dack, W. E. Woolpert, and H. Wiggers, *J. Prevent. Med.* **4,** 167 (1930).

II. Ileal Loop Test Method

The ileal loop test method as described by De and Chatterje[1] provides the basis for all subsequent work done with this method; it is described in detail below as they presented it. Certain improvements and/or modifications that other investigators have applied to it will be described.

A. *Original Ileal Loop Procedure*

Rabbits weighing 1200–1500 g were denied all food and water for 24 hr prior to use. With aseptic precautions and under local procaine anaesthesia a midline incision about 2 in. long was made just below the middle of the abdomen. A segment of the small intestine about midway between the upper and lower ends was isolated with two silk ligatures, carefully avoiding any blood vessels. One milliliter of Dunham's peptone–water medium inoculated with one loopful of a 24-hr culture of *V. cholerae* was injected slowly into the lumen of the isolated loop; previous experience showed that 1 ml was the most suitable dose. The abdomen was closed in two layers with thread. Animals were not allowed food or water and were killed after 24 hr by rapidly injecting intravenously 5 cm^3 of air. The loop and the sections above and below were carefully examined. The fluid was aspirated with a sterile syringe and measured. The *V. cholerae* were determined by plating on MacKonkey plates and in Dunham's medium. Experiments were conducted in 10 rabbits and 15 rats with 6 additional rabbits and 5 additional rats as controls in which uninoculated sterile medium was used. To determine leakage three test rabbits and two controls were injected in the ear with 5 ml of 2% Evan's blue in normal saline 4 hr after the start of the experiment. Evidence of leakage was looked for the following day. The part of the intestine proximal to the test loop was distended with fluid, which was tested for contamination. The part distal to the test loop collapsed as well as the loops in control rabbits. The test loops were the size of the thumb and contained 14–20 ml of fluid which was positive for *V. cholerae*. The rats were not usable because only about 0.4 ml of fluid had accumulated in their loops.

De *et al.*[11] modified the original procedures by tying off two loops 12 in. apart with the second loop acting as a control. They were able to produce fluid accumulation of 20 ml of *Bacterium coli* (*E. coli*). They were unable to produce fluid accumulation in the large intestine. It was essential that the 1 ml of inoculum be injected slowly (45–60 sec) because a more rapid injection into the controls may result in false-positive reactions.[12] All control loops (4 in.) in the 14 animals used by De *et al.*[13] were

[11] S. N. De, K. Bhattacharya, and J. K. Sarkar, *J. Pathol. Bacteriol.* **71,** 201 (1956).
[12] S. N. De, *Nature (London)* **183,** 1533 (1959).

collapsed and all of those with the sterile *V. cholerae* filtrates contained 10 ml of fluid. This was the first effort to show that the fluid accumulation was due to a substance produced by the organisms.

B. Modifications of the Ileal Loop Method by Other Investigators

McNaught and Roberts[14] used 2 to 3-kg rabbits and allowed them to have water but not food for 18 hr before surgery. They used loops 4–5 in. long and 9 in. apart tied off with one silk ligature. They injected 2 ml of inoculum of *E. coli* in both loops with the intermediate area as a control. The rabbits were sacrificed after 24 hr and examined for fluid accumulation. They observed a good correlation between fluid accumulation in the rabbits and the diarrheal strains they tested.

Jenkin and Rowley[15] used six different breeds of rabbits 12–24 weeks of age; no differences in results were noted with the different breeds. They tied off three or four loops 6 in. apart with two ligatures 3 in. apart with colored cotton thread so the different loops could be easily identified. After injecting the *V. cholerae* inoculum the muscle layers were sewn up with nylon thread, using a running stitch. The skin was sutured with three or four Michel clips and the abdomen gently manipulated to redistribute the small intestine. The entire operation required 15–20 min. The rabbits were standing within 2 hr. They obtained variable results with the duodenum and concluded that the ileum was best.

Taylor and Wilkins[16] used Copenhagen or New Zealand albino rabbits that weighed 1.25 to 1.75 kg; they stated that this size must be used for best results. They used two test loops with a blank loop in between. They sacrificed the rabbits 24 hr after the injection of *S. typhimurium* and *Shigella sonnei* and *S. flexneri* cultures. They obtained fluid accumulation with all of these organisms, the first demonstration that these organisms were reactive in the ileal loop test.

Burrows and Musteikis[17] purchased rabbits 8 weeks old and challenged them at 9–11 weeks of age at a weight of 2 kg. They reported that there was no evidence that the age of the rabbit was a significant factor within relatively wide limits in the experimental infection (*V. cholerae*) but they concluded from their data that rabbits as old as 16 weeks were appreciably more resistant. It was essential that rabbits be enteritis free. Later they found that rabbits of 9 weeks of age at a weight of 1.8 kg were best. They starved their rabbits for 48 hr with water given *ad libitum*.

[13] S. N. De, M. L. Ghose, and A. Sen, *J. Pathol. Bacteriol.* **79,** 373 (1960).
[14] B. S. McNaught and B. S. Roberts, *J. Pathol. Bacteriol.* **76,** 155 (1958).
[15] C. R. Jenkin and D. Rowley, *Br. J. Exp. Pathol.* **40,** 474 (1959).
[16] J. Taylor and M. P. Wilkins, *Indian J. Med. Res.* **49,** 544 (1961).
[17] W. Burrows and G. M. Musteikis, *J. Infect. Dis.* **116,** 183 (1966).

Results were irregular if the animals were starved for only 24 hr. They tied off the small bowel 60 cm above the appendix and washed it out with sterile physiological saline (10 ml) with injection just below the tie. A second tie was made to isolate the point of injection. Bowel contents were washed down with minimum manipulation and the bowel tied off above the appendix. Washing did not appear to affect the results significantly. Washing was resorted to only when the bowel was not emptied by starving. Four loops, 12–14 cm, were tied off and 2 ml of inoculum was injected adjacent to the tie nearest the appendix. A second tie was made to isolate the site of injection, thus separating the 12- to 14-cm loops by the double ties. They found the operating technique to be critical; it was necessary to carry it out with facility or inconsistent results would be obtained. Their tests showed that the position of the loop in which the sample was placed was not important; this was shown by what they called shams (2 ml of saline). Water was available to the animals but food was withheld. Allowing the experiment to extend beyond 30–36 hr was doubtful although the results with shams were negative with healthy animals. The small segments between the double tie often showed increase in fluid and hemorrhagic contents due to local trauma. There was no evidence of leakage between loops, determined by the injection of Pontamine sky blue 6BX injected in the second and fifth loops (0.1 ml of 2%). The animals were sacrificed at 18 hr to get full blown and intermediate reactions with inocula of 10^1 to 10^4. Later, six loops were used with the first and last injected with saline.[18]

Smith and Halls[19] reported that the first and last 18 in. (45 cm) of the small intestine was unusable; they used the in-between section (40–50 in.) to make 3- to 4-in. loops. They used the H1 strain of *E. coli* in the first and last loop for testing other *E. coli* strains.

Coleman *et al.*[20] reported the purification of the cholera toxin and used the rabbit ileal loop method for quantitating the toxin (ml/cm basis). They used four dilutions of the toxin in at least four rabbits for each dilution. They determined the unit of activity to be 10 μg although they did get positive reactions with less toxin. They found the antibody they prepared against their toxin to neutralize the action of the toxin in the ileal loop test.

Duncan *et al.*[21] used New Zealand White rabbits of both sexes, acclimated to the laboratory at least 1 week; they were 7–10 weeks old and

[18] G. N. Kasai and W. Burrows, *J. Infect. Dis.* **116,** 606 (1966).
[19] H. W. Smith and S. Halls, *J. Pathol. Bacteriol.* **93,** 499 (1967).
[20] W. H. Coleman, J. Kaur, M. E. Iwert, G. J. Kasai, and W. Burrows, *J. Bacteriol.* **96,** 1137 (1968).
[21] C. L. Duncan, H. Sugiyama, and D. H. Strong, *J. Bacteriol.* **95,** 1560 (1968).

weighed 1.4–2.2 kg at the time of testing. Water was supplied *ad libitum* at all times but food was withheld 24 and usually 48 hr before surgery. The operative procedures were those used by Jenkin and Rowley.[15] The ileum was ligated approximately 90 cm anterior to the mesoappendix. Intestinal contents were washed down with 10 ml of physiological saline injected just below the tie, and the site of injection was isolated by a second ligature 5 cm below the first tie. Segments of 10 cm were tied off caudally from this ligature. Saline packs were used to keep the exposed intestine moist during the operative procedure. Intraluminal injections of 2 ml of *C. perfringens* inoculum were made into alternate loops so that a blank loop separated the test segments. A control loop injected with sterile culture medium was included in each rabbit. The animals were sacrificed after 20–24 hr (only water available) by intravenous injection of Diabutal. The length and fluid volume of the loops were measured to calculate the loop volume–length ratio (ml/cm), which was used to compensate for the unavoidable variability in the length of the loops.[17] The results from any rabbits in which either the control or uninjected loops were filled with fluid were considered invalid. The *C. perfringens* was enumerated to determine the relationship between the fluid accumulation and the growth of the organisms in the loops.

Moon *et al.*[22] used 9-week-old rabbits weighing 1.8 kg for testing diarrheal *E. coli* strains from swine. They used six experimental loops 10 cm in length for each rabbit. They injected 2 ml of inoculum per loop and used one loop for each strain tested in at least six rabbits.

Pierce and Wallace[23] used 3.5- to 4.0-cm loops in 2-kg rabbits with a single tie with surgical silk between adjacent segments to test *V. cholerae* strains. One milliliter of isotonic solution was used in two loops as a negative control and 1 ml of isotonic crude cholera toxin was injected in two loops as a positive control. The rabbits were sacrificed after 6 hr and the volume–length ratios determined. Results were excluded if either the controls were positive or the cholera toxin was negative. Fourteen segments in 7 rabbits were used. Six hours of development was sufficient for *E. coli* enterotoxin but 18 hr gave the best results for the cholera enterotoxin.

Spira and Goepfert[24] used the same procedures employed by Duncan *et al.*[21] for testing strains of *B. cereus* for activity in rabbit ileal loops. They used 10-cm loops separated by 5-cm blank loops. The volume–length ratios were calculated and plate counts were made to determine the

[22] H. W. Moon, S. C. Whipp, G. W. Engstrom, and A. L. Baetz, *J. Infect. Dis.* **121,** 182 (1970).

[23] N. F. Pierce and C. K. Wallace, *Gastroenterology* **63,** 439 (1972).

[24] W. M. Spira and J. M. Goepfert, *Appl. Microbiol.* **24,** 341 (1972).

growth–fluid accumulation ratio. They found that although there were variations from test loop to test loop, the average of three test loops proved to be a reasonably accurate indication of the fluid-inducing ability of a particular strain. The age of the rabbits proved to be important because rabbits weighing less than 1200 g gave consistently positive responses while the responses in older rabbits were sporadic or negative. Subsequent work was done with rabbits 8 weeks or younger and weighing under 1 kg. They found that the position of the loop made no difference in the results obtained. Initially they held the animals 24 hr before sacrificing but most of the rabbits died within 10 hr. None died in the first 6 hr, hence, this was selected as the standard time for holding the rabbits. They were able to show that *B. cereus* strains (19 of 22) as well as *C. thurengiensis* strains (4 of 6) were positive in the ileal loop test; they also showed that culture filtrates were positive in the loops.

Thompson *et al.*[25] employed the rabbit ileal loop method in the purification of the *B. cereus* enterotoxin. They used only female New Zealand White rabbits weighing 700 to 1000 g. Ligations were made at least 6 cm from the appendix and spaced to yield up to seven test loops (averaging 6 cm in length) per animal; the test loops were separated by interloops of approximately 4 cm in length. Each set of samples was tested in at least seven animals, and the position of each sample was randomized for each animal. Animals were held for 6 hr postsurgery before they were sacrificed. A positive response was defined as a volume/length ratio greater than or equal to 0.50 in at least 50% of the test animals.

C. *Summary of Ileal Loop Test Method Modifications (Table I)*

The basic principles of the ileal loop method as De and Chatterje[1] reported them have been retained for the most part. Although the breed of the rabbits did not appear to make a difference in the results,[15] essentially all investigators use albino rabbits, usually New Zealand White.The size of the rabbits used varied but it would appear that the smaller ones, less than 2 kg, are most suitable with even smaller sizes needed for *B. cereus* investigations.[24] The time of starving the rabbits was increased from 24 to 48 hr, which appears to give the more consistent results.[17] It is generally agreed that water should be provided at all times even though De and Chatterje[1] did not provide it. The small intestine proved to be the only usable area as it was shown that the large intestine[11] and the duodenum[15] did not respond. The major change was in the increase in the number of loops per rabbit used, from the original 1 to 6–20. The position of the

[25] N. E. Thompson, M. J. Ketterhagen, M. S. Bergdoll, and E. J. Schantz, *Infect. Immun.* **43,** 887 (1984).

TABLE I
CURRENT RABBIT ILEAL LOOP TEST METHOD PROCEDURES

Rabbits:	Albino (New Zealand White)
Weight:	1.2 to 18 kg[a]
Starving:	48 hr with only water
Ileal loops:	Six to 20 with at least 1 negative control and 1 positive control; blank controls optional
Sample replicates:	At least four to six rabbits
Holding:	18 to 24 hr[a]
Results:	Usable if blanks and negative controls are negative and if positive control is positive
Positive:	Volume (ml)/length (cm) ratio of 0.5 or greater

[a] Less than 1 kg and holding period of less than 10 hr for *B. cereus*.

loops in the intestine did not appear to affect the results[17] although there was a difference in the distance from either end of the intestine avoided by the different investigators. Most investigators used 2 ml of inoculum although De and Chatterje[1] used only 1 ml. The replicates for each sample varied with the investigator but four to seven appeared to be a desirable number. The type of controls and the number also varied, some using a positive as well as a negative control.[23] All were agreed that a positive reaction in the negative control rendered the results from that animal useless. The time of holding the animals before sacrificing varied from 5 to 18–24 hr, with 18–24 hr being most suitable; however, for *E. coli* heat-stable enterotoxin, a period of 6 hr was necessary as the reaction decreased with longer holding periods.[26] Also, a shorter holding period was necessary for *B. cereus* because the strains were lethal at 10 hr.[24] Some variability in the procedures used will be necessary because of the particular organism the investigator is working with.

III. Diarrheal Disease Organisms

A. *Vibrio cholerae*

Cholera was known for many years before it received much attention, primarily because it appeared to be endemic with Pakistan and India. It was not until an outbreak in Egypt in 1947 and another one in Thailand in 1958–1959[27] that very much attention was given to it. It is probably the most serious diarrheal disease with many deaths resulting from the ill-

[26] D. G. Evans, D. J. Evans, Jr., and N. F. Pierce, *Infect. Immun.* **7,** 873 (1973).
[27] R. A. Finkelstein, *Crit. Rev. Microbiol.* **2,** 553 (1973).

ness. Deaths were greatly decreased when it was learned that proper treatment with fluids could prevent most of the deaths; apparently the organism has little effect on the intestinal tract other than to stimulate excessive diarrhea. To study the mechanism of the disease De and Chatterje[1] employed the use of the rabbit ileal loop test. De[12] was successful in showing that culture filtrates could produce fluid accumulation in rabbit ileal loops, indicating that a product of the *V. cholerae* was responsible for the diarrhea; he applied the term enterotoxic to the toxic activity[13] which was the forerunner of the name enterotoxin now applied to all of the diarrheal toxins.

Burrows and associates[17,18,20] did extensive studies on the cholera problem, also using the rabbit ileal loop as their biological test system. They reported partial purification of the cholera toxin with approximately 10 μg required to produce fluid accumulation in the rabbit ileal loop.[20] Finkelstein and LoSpalluto[28] claimed to have purified the toxin, which they labeled cholargen; less than 1 μg was required to produce fluid accumulation in the rabbit ileal loop. Later these same investigators[29] reported the crystallization of the cholera toxin; they changed the name to exo-enterotoxin. The amount required to produce fluid accumulation in the rabbit ileal loop was approximately 1 μg. Antibodies that had neutralizing capacity were prepared against the purified toxin. Cholera is still a subject of much interest as well as are other species of *Vibrio;* for example, Spira and Fedorka-Cray[30] have purified an enterotoxin produced by *V. mimicus,* which appears to be identical to the cholera enterotoxin. The rabbit ileal loop was one of the biological tests used in their studies.

B. *Escherichia coli*

Escherichia coli can be a harmless commensal; however, for many years it has been considered to be a potential pathogen for man and animals.[31] Although diarrheal outbreaks in nurseries had occurred, it was not until 1967 that the first recognized outbreak caused by *E. coli* which was related to nonchlorinated water occurred in the United States.[32] The first well-documented outbreak of foodborne illness in the United States caused by *E. coli* due to the consumption of Camembert cheese from France occurred in 1971.[33] Since that time a number of foodborne related

[28] R. A. Finkelstein and J. LoSpalluto, *J. Exp. Med.* **130,** 185 (1969).

[29] R. A. Finkelstein and J. LoSpalluto, *Science* **175,** 529 (1970).

[30] W. M. Spira and P. J. Fedorka-Cray, *Infect. Immun.* **45,** 679 (1984).

[31] J. L. Kornacki and E. H. Marth, *J. Food Protect.* **45,** 1051 (1982).

[32] S. A. Schroeder, R. J. Caldwell, T. M. Vernon, P. C. White, S. I. Granger, and J. V. Bennett, *Lancet* **1,** 737 (1968).

[33] R. Marrier, J. G. Wells, R. C. Swanson, W. Callahan, and I. J. Mehlman, *Lancet* **2,** 1376 (1973).

TABLE II
FLUID ACCUMULATION IN RABBIT ILEAL LOOPS IN RESPONSE TO ENTEROTOXINS[a]

Time after injection (hr)	*E. coli* strain	Heat-labile enterotoxin	Heat-stable enterotoxin	Crude cholera enterotoxin
2	0.29[b]	0.10	0.19	0.01
4	0.65	0.15	0.48	
6	1.06	0.42	0.61	0.38
10	1.43	1.00	0.42	0.68
18	1.90	1.58	0.30	1.30

[a] Data taken from Fig. 1 in Evans *et al.*[26]
[b] Fluid volume (ml)/loop length (cm) ratio (average of 12 to 25 loops).

outbreaks have occurred around the world. The symptoms generally are mild to severe diarrhea with profound dehydration and shock and fever.

Bacterium coli strains isolated from acute and chronic enteritis were reported by De *et al.* in 1956[11] to produce fluid accumulation in rabbit ileal loops and by others later. The ileal loop reaction was interpreted to be due to an enterotoxin produced by the organisms. It was reported in 1971[34] that *E. coli* can produce both heat-stable and heat-labile enterotoxins. Both types were able to produce fluid accumulation in rabbit ileal loops[26] (Table II). A number of investigators have attempted to purify the heat-labile enterotoxin with some success; however, the procedures used by Clements and Finkelstein[35] appeared to give the best results. Their purified preparation did produce fluid accumulation in rabbit ileal loops. Several different forms of the heat-stable enterotoxin have been reported; the preparations tested in the rabbit ileal loop did stimulate fluid accumulation, however, the infant mouse test is normally used for testing the heat-stable enterotoxins. The heat-labile enterotoxin is antigenic and antibodies to it have been prepared[35] while the heat-stable enterotoxins are small-molecular-weight peptides that are not antigenic.[36] Many papers have been published over the past few years on the involvement of *E. coli* in human disease as it is an on-going problem.

C. *Clostridium perfringens*

The possibility that *C. perfringens* was involved in food poisoning was noted by Knox and Macdonald in 1943[37]; the predominant symptoms were abdominal cramps, nausea, and diarrhea developing 10–12 hr after

[34] C. L. Gyles, *Ann. N.Y. Acad. Sci.* **176,** 314 (1971).
[35] J. D. Clements and R. A. Finkelstein, *Infect. Immun.* **24,** 760 (1979).
[36] M. R. Thompson and R. A. Giannella, *Infect. Immun.* **47,** 843 (1985).
[37] R. Knox and E. K. Macdonald, *Med. Off.* **69,** 21 (1943).

TABLE III
RABBIT ILEAL LOOP RESPONSE TO THE INJECTION OF *Bacillus cereus*[a]

Injection[b]	Response[c]
0	Negative
2.2×10^2	Negative
1.6×10^4	Negative
1.6×10^6	0.14
1.6×10^8	0.30
1.6×10^{10}	0.81
Culture filtrate	0.83

[a] Data from Spira and Goepfert.[24]
[b] Two milliliters of washed cells.
[c] Average of two tests in two different rabbits.

ingestion of food containing large numbers of *C. perfringens* organisms. Initially, human volunteers were used to show that this organism was involved in food poisoning with the probability that the production of a toxin in the intestinal tract was the causative agent. Duncan *et al.*[21] and Strong *et al.*[38] reported on the use of the ileal loop test in rabbits and Hauschild *et al.*[2] reported on using it in lambs for the testing of strains of *C. perfringens* for their involvement in food poisoning. Duncan and Strong[39] were the first to show that cell extracts from *C. perfringens* could produce fluid accumulation in the rabbit ileal loop; activity was observed as soon as 3 hr. Heating the extract to 60° for 10 min destroyed the ileal loop activity. The ileal loop reaction was an important adjunct in the purification of the *C. perfringens* enterotoxin.[40,41] Antibodies specific for the enterotoxin were prepared, thus essentially eliminating the need for the ileal loop test.

D. *Bacillus cereus*

Bacillus cereus was shown to be responsible for food poisoning in the early 1950s when Hauge[42,43] reported illnesses in individuals 10–12 hr after they consumed foods that contained large numbers of these organisms. The typical symptoms were abdominal pain, profuse watery diar-

[38] D. H. Strong, C. L. Duncan, and G. Perna, *Infect. Immun.* **3,** 171 (1971).
[39] D. L. Duncan and D. H. Strong, *J. Bacteriol.* **100,** 86 (1969).
[40] R. L. Stark and C. L. Duncan, *Infect. Immun.* **6,** 662 (1972).
[41] A. H. W. Hauschild and R. Hilsheimer, *Can. J. Microbiol.* **17,** 1425 (1971).
[42] S. Hauge, *Nord. Hyg. Tidskr.* **31,** 189 (1950).
[43] S. Hauge, *J. Appl. Bacteriol.* **18,** 591 (1955).

TABLE IV
ILEAL LOOP RESPONSE OF FRACTIONS FROM PURIFICATION OF *B. cereus*[a]

Fraction	Major antigens	Minor antigens	Ileal loop response
1	580		0/5[b]
2	580	577	0/5[b]
3	575, 577	580	7/12
4	575	577	0/5
2 + 4	575, 580	577	9/12

[a] Data from Thompson *et al.*[25]
[b] Number positive (ml/cm = ≥0.5) versus number of rabbits used.

rhea, rectal tenesmus, and moderate nausea. Essentially all of the outbreaks reported occurred in Europe and it was not until 1969 that the first well-documented outbreak occurred in the United States.[44] Spira and Goepfert[24] reported the use of the rabbit ileal loop test for the examination of *B. cereus* organisms (Table III). They were able to identify strains that produced the ileal loop activity and associate them with specific food-poisoning outbreaks. This was the only test that was specific for testing the food-poisoning organisms and confirming their involvement in food poisoning.

An all-out effort was carried on in the Food Research Institute to purify the enterotoxin responsible for this type of food poisoning. Spira and Goepfert[45] did report some of the properties of the enterotoxin but it was not until 1984 that the factor responsible for the diarrheal action was identified by Thompson *et al.*[25] (Table IV). It was necessary to combine three proteins to accomplish the ileal loop activity as singly or a combination of any two did not produce fluid accumulation. Antibodies to two of the proteins were prepared and used in identifying strains that were involved in outbreaks. Shinagawa *et al.*[8] reported the purification of the *B. cereus* enterotoxin using both the rabbit and mouse ileal loop test to follow the purification. They claimed to have a single protein entity but this has not been confirmed. Activity in this field is at a minimum because few cases of this type of food poisoning occur in the United States. The more common symptom observed is the emetic reaction which occurs

[44] T. Midura, M. Gerber, R. Wood, and A. R. Leonard, *Public Health Rep. Wash.* **85,** 45 (1970).
[45] W. M. Spira and J. M. Goepfert, *Can. J. Microbiol.* **21,** 1236 (1975).

within 2–4 hr after ingestion of food containing the toxin; in this case the ileal loop test is not applicable.[46]

E. Salmonella

Salmonellosis is one of the leading foodborne diseases in the United States and many countries of the world. The major symptom of this disease is diarrhea with abdominal cramps along with nausea and vomiting during the first day. Mild fever (38 to 39°) sometimes follows and headache, chills, and prostration may occur. The incubation period in most cases is between 20 and 48 hr with the symptoms lasting from 2 to 55 days. Originally, salmonellosis was related to the invasiveness of the organisms rather than being due to a toxin produced by them. Taylor and Wilkins[16] were able to stimulate fluid accumulation in rabbit ileal loops by *S. typhimurium,* indicating toxin involvement. Later, Giannella *et al.*[47] also were able to evoke fluid accumulation in the ligated rabbit ileum by the live organisms but were unable to induce fluid response by culture filtrates. However, Koupal and Deibel[48] were able to demonstrate fluid accumulation in rabbit loops by culture supernatant fluids as well as cell extracts of *Salmonella enteritidis*. Deibel's group[49] was successful in the partial purification of an enterotoxin produced by a strain of *S. typhimurium* as determined by the rabbit ileal loop test. Apparently the major enterotoxin produced by the *Salmonella* is heat labile, however, Jiwa and Mansson[50] reported that some of the strains they worked with produced both a heat-labile and a heat-stable enterotoxin; both were not produced by all strains. They found that the heat-labile enterotoxin activity in the rabbit ileal loop was neutralized by cholera antitoxin. Finkelstein *et al.*[51] reported the purification and characterization of the heat-labile enterotoxin which was isolated from *S. typhimurium*. They found that the toxin was closely related to the heat-labile *E. coli* enterotoxin and to cholera enterotoxin. Singh *et al.*[52] reported the purification of a heat-labile enterotoxin from a strain of *S. saint-paul* isolated from a calf in India. Their toxin produced fluid in the rabbit ileal loop; however, the antibody to their enterotoxin did not cross-react with cholera toxin. The ileal loop test has

[46] J. Melling, B. J. Capel, P. C. B. Turnbull, and R. J. Gilbert, *J. Clin. Pathol.* **29,** 938 (1976).
[47] R. A. Giannella, S. B. Formal, G. J. Dammin, and H. Collins, *J. Clin. Infect.* **52,** 441 (1973).
[48] L. R. Koupal and R. H. Deibel, *Infect. Immun.* **11,** 14 (1975).
[49] D. M. Sedlock, L. R. Koupal, and R. H. Deibel, *Infect. Immun.* **20,** 375 (1978).
[50] S. F. H. Jiwa and I. Mansson, *Vet. Microbiol.* **8,** 443 (1983).
[51] R. A. Finkelstein, B. A. Marchlewica, R. J. McDonald, and M. Boseman-Finkelstein, *FEMS Microbiol. Lett.* **17,** 239 (1983).
[52] S. P. Singh, V. D. Sharma, and I. P. Singh, *FEMS Microbiol. Lett.* **26,** 301 (1985).

been used by a number of investigators to study the mode of action of *Salmonella* in salmonellosis. Duebbert and Peterson[53] found that prostaglandins synthesized by the epithelial cells were indicated to be involved in the pathogenesis of both experimental cholera and salmonellosis. Their results were consistent with an enterotoxin-mediated mechanism for both diarrheal diseases.

F. *Shigella dysenteriae*

The Centers for Disease Control started keeping records on the food-related outbreaks of shigellosis in 1965[54,55]; however, the number of outbreaks is only a small percentage of the total number of foodborne outbreaks recorded each year. Only 11 outbreaks were reported in 1980, which was 5% of the total number of outbreaks recorded. Shigellosis is characterized by diarrhea, abdominal pain, fever, and often vomiting and tenesmus developing 7 to 36 hr after the ingestion of food containing the *Shigella* organisms. Very little has been reported on this type of food poisoning but studies with the rabbit ileal loop test by Taylor and Wilkins[16] showed that *S. sonnei* and *S. flexneri* could produce fluid accumulation in the ileal loop. Later, Asnani and Kaur[56] and Ramachandran *et al.*[57] were able to show fluid accumulation in the rabbit ileal loop with *S. dysenteriae,* which indicated that an enterotoxin may be responsible for the disease. Ramachandran *et al.*[57] used this test to distinguish between two types of toxin produced by *Shigella,* an enterotoxin and a cytotoxin; the latter did not cause fluid accumulation in the rabbit ileal loop. No attempts on the purification of the enterotoxin have been reported.

G. *Vibrio parahaemolyticus*

The involvement of *V. parahaemolyticus* in foodborne disease was discovered in 1950 when 272 individuals experienced acute gastroenteritis after consuming half-dried sardines in Osaka, Japan.[58] Fujino and colleagues[58,59] reported the main symptoms to be abdominal pain, vomiting, and diarrhea developing between 2 and 6 hr after the ingestion of the

[53] I. E. Duebbert and J. W. Peterson, *Toxicon* **23,** 157 (1985).
[54] Centers for Disease Control, *Morbid. Mortal. Weekly Rep.* **27,** 207 (1978).
[55] Centers for Disease Control, *Morbid. Mortal. Weekly Rep.* **27,** 231 (1978).
[56] P. J. Asnani and S. Kaur, *Acta Microbiol. Hung.* **30,** 154 (1983).
[57] V. G. Ramachandran, A. Varghese, and D. S. Agarwal, *Indian J. Med. Res.* **78,** 161 (1983).
[58] T. Fujino, *Saishin Igaku* **6,** 263 (1951).
[59] T. Fujino, Y. Okuno, D. Nakasa, A. Aoyama, K. Fukai, T. Mukai, and T. Ueho, *Med. J. Osaka Univ.* **4,** 299 (1953).

sardines. These investigators identified the organism as *Pasteurella parahaemolytica* which was later changed to *Vibrio parahaemolyticus*.[60] This disease has become the leading foodborne disease in Japan and has been identified as the cause of illness in other countries, including the United States.[61,62] This organism is of marine origin and is associated with salt water fishes. Several Japanese workers showed that the *V. parahaemolyticus* strains that were positive in humans were also positive in the rabbit ileal loop test. Initially a thermostable direct hemolysin that was purified from *V. parahaemolyticus* was thought to be the most important enteropathogenic factor when it was found by Zen-Yogi *et al.*[63] to cause fluid accumulation in the rabbit ileal loop. However, Sakazaki *et al.*[64] reported that strains that did not produce the hemolysin also produced fluid accumulation in rabbit ileal loops. Honda *et al.*[65] showed that antiserum against the hemolysin did not prevent fluid accumulation in the rabbit ileal loop by strains that were loop positive, indicating that this was not the toxin responsible for the disease. As of 1986 no further work has been reported on the identification of the enterotoxigenic factor.

H. Aeromonas hydrophila

Aeromonas hydrophila has attracted attention because of its involvement in diarrheal disease. Von Gravenitz and Mensch[66] reported 30 cases of this type of illness in 1968 and since that time a number of papers have been published on the subject. Wadström *et al.*[67] used the rabbit ileal loop test to indicate that this organism did produce an enterotoxin. Attempts have been made to identify the enterotoxigenic factor but it is uncertain whether the hemolysin purified by Asao *et al.*[68] is involved in the illnesses even though it did produce fluid accumulation in the rabbit ileal loop.

[60] R. Sakazaki, S. Iwanami, and H. Fukumi, *Jpn. J. Med. Sci. Biol.* **16,** 161 (1963).

[61] J. R. Molenda, W. G. Johnson, M. Fishbein, B. Wentz, I. J. Mehlman, and T. A. Dadisman, *Appl. Microbiol.* **24,** 444 (1972).

[62] W. H. Barker, Jr., *in* "International Symposium of *Vibrio parahaemolyticus*" (T. Fujino, G. Sakaguchi, R. Sakazaki, and Y. Takesa, ed.), p. 47. Saikon, Tokyo, 1974.

[63] H. Zen-Yogi, H. Hitokoto, S. Morozumi, and R. A. LeClair, *J. Infect. Dis.* **127,** 237 (1971).

[64] R. Sakazaki, K. Tamura, A. Nakamura, T. Kurata, A. Ghoda, and Y. Kazuno, *Jpn. J. Med. Sci. Biol.* **27,** 35 (1974).

[65] T. Honda, S. Taga, T. Takeda, M. A. Hasibuan, Y. Takeda, and T. Miwatani, *Infect. Immun.* **13,** 133 (1976).

[66] A. Von Gravenitz and A. H. Mensch, *N. Engl. J. Med.* **278,** 245 (1968).

[67] T. Wadström, Å Ljungh, and B. Wretland, *Acta Pathol. Microbiol. Scand., Sect. B* **84,** 112 (1976).

[68] T. Asao, Y. Kinoshita, S. Kozaki, T. Yemura, and G. Sakaguchi, *Infect. Immun.* **46,** 122 (1984).

Chakraborty *et al.*[69] concluded from cloning experiments that *A. hydrophila* produced a cytotonic enterotoxin that was distinct from cytotoxin and hemolysin. The rabbit ileal loop test was one of the biological tests used in their experiments. The so-called hemolysin enterotoxin apparently is not the cause of human diarrhea as Morgan *et al.*[70] were able to produce diarrhea in only 2 of 57 human volunteers with a strain of *A. hydrophila* that produced cytotoxin and hemolysin enterotoxin even though the strain produced fluid accumulation in rabbit ileal loops. It appears from these studies that accumulation of fluid in the rabbit ileal loop does not necessarily indicate that the substance responsible will cause diarrhea in humans.

I. Yersinia enterocolitica

Yersinia as the possible cause of foodborne disease did not attract attention until the early 1970s when diarrheal illnesses appeared to be associated with the presence of *Y. enterocolitica* in the foods consumed. The most frequent symptoms reported are diarrhea, fever, vomiting, abdominal pain, nausea, and headaches developing 24 to 37 hr after the ingestion of food containing relatively large numbers of *Yersinia*. Recovery is generally complete within 1 to 2 days. Sakazaki *et al.*[71] demonstrated as early as 1974 that cell-free filtrates of cultures of *Y. enterocolitica* could produce fluid accumulation in rabbit ileal loops. Although others had demonstrated the production of a heat-stable enterotoxin by *Yersinia,* Pai and Mors[72] showed that this toxin was active in the rabbit ileal loop test. Robbins-Browne *et al.*,[73] with the aid of the rabbit ileal loop test, showed that the heat-stable enterotoxin may be similar to the action of the *E. coli* heat-stable enterotoxin. The heat-stable enterotoxin was essentially purified by Okamoto *et al.*[74] with determination of its activity by the suckling mouse assay, the method that is currently being used for this toxin. There is doubt that this toxin is the causative factor in yersin-

[69] T. Chakraborty, M. A. Montenegro, S. C. Sanyal, R. Helmuth, E. Bulling, and K. N. Timmis, *Infect. Immun.* **46,** 435 (1984).

[70] D. R. Morgan, P. C. Johnson, H. L. DuPont, T. K. Satterwhite, and L. V. Wood, *Infect. Immun.* **50,** 62 (1985).

[71] R. Sakazaki, K. Tamura, A. Nakamura, T. Kurata, A. Ghoda, and Y. Kazuno, *Jpn. J. Med. Sci. Biol.* **27,** 45 (1974).

[72] C. H. Pai and V. Mors, *Infect. Immun.* **19,** 908 (1978).

[73] R. M. Robbins-Browne, C. S. Still, M. D. Miliotis, and H. J. Koornhof, *Infect. Immun.* **25,** 680 (1979).

[74] K. Okamoto, T. Inoue, H. Ichikawa, Y. Kawamoto, and A. Miyama, *Infect. Immun.* **31,** 554 (1981).

iosis as Schiemann[75] has reported the isolation of a *Yersinia* strain that does not produce the heat-stable enterotoxin but is capable of producing diarrhea in mice. Tests were not done in the rabbit ileal loop.

J. Campylobacter

The first outbreaks of enteritis attributed to *Campylobacter* in the United States occurred in 1978.[54,55] The illnesses were characterized by abdominal pain or cramps, diarrhea, malaise, headache, and fever. The first outbreak was attributed to the water supply and the second one was related to the consumption of raw milk. Gastroenteritis caused by *Campylobacter* has become one of the most common causes of this disease and frequency of isolation even exceeds that of *Salmonella* in many instances.[76] Ruiz-Palacios *et al.*[77] reported that an isolate from a patient with severe diarrhea produced fluid accumulation in rat ileal loops but not in rabbit ileal loops; however, McCardell *et al.*[78] found that a 20-fold concentrate of the supernatant fluid from a human fecal isolate was able to produce fluid accumulation in rabbit ileal loops. Klipstein and Engert[79] purified a heat-labile enterotoxin that gave a reaction of identity with *E. coli* heat-labile enterotoxin. They did show that this enterotoxin could produce fluid accumulation in rat ileal loops,[9] but did not report on its action in rabbit ileal loops.

K. Staphylococcus

Koupal and Deibel[80] did report on rabbit intestinal fluid accumulation by an enterotoxigenic factor produced by a *Staphylococcus aureus* strain that produced enterotoxin A (SEA); however, they were unable to demonstrate an ileal loop reaction with purified SEA. The reaction was neutralized with the antibody specific for SEA and the partially purified factor had a migration pattern in polyacrylamide gel electrophoresis nearly identical to that of SEA. However, two strains that were nonenterotoxigenic in monkeys also gave a positive ileal loop reaction while one SEA-positive strain and several enterotoxin B-positive strains gave negative ileal loop reactions. No additional work was done on this factor but it apparently is not the same as the toxins responsible for the emetic and diarrheal

[75] D. A. Schiemann, *Infect. Immun.* **32,** 571 (1981).

[76] M. P. Doyle, *J. Food Protect.* **44,** 480 (1981).

[77] G. M. Ruiz-Palacios, J. Torres, N. I. Torres, E. Escamilla, B. R. Ruiz-Palacios, and T. Tamayuo, *Lancet* **2,** 250 (1983).

[78] B. A. McCardell, J. M. Madden, and E. C. Lee, *Lancet* **1,** 448 (1984).

[79] F. A. Klipstein and R. F. Engert, *Lancet* **1,** 1123 (1984).

[80] A. Koupal and R. H. Deibel, *Infect. Immun.* **18,** 298 (1977).

action in staphylococcal food poisoning. It is possible that this factor is closely related to the enterotoxins but would have to be produced in the intestinal tract to result in a diarrheal reaction. This is unlikely to occur as the staphylococci are poor competitors and do not grow in the intestinal tract in the presence of the natural flora.

IV. Conclusions

It is obvious from the information presented here that the rabbit ileal loop fluid accumulation method has been widely used and is still being used today in those cases where alternate procedures are not available. Also, identification of organisms with diarrheal diseases is on-going and the ileal loop method is the one of choice for initial investigation of the involvement of the organisms in the disease. Care needs to be taken, however, in examining particular products of the organisms as to their involvement in any particular disease because the production of fluid in the ileal loop does not always correlate with the production of diarrhea in human subjects.[64,65,71]

The impression one receives from the descriptions of the method is that it is a relatively simple method to use. Our experience with it in the study of the *B. cereus* enterotoxin[25] was to the contrary as the results from a number of rabbits were unusable. Anyone inexperienced with the method should be aware of this and not be overly frustrated if their initial results are less than perfect. Note should be taken that Burrows and Musteikis[17] found that the operating technique was critical. In spite of the difficulties, the ileal loop method has proved to be an invaluable tool in the investigation of the diarrheal diseases; without it, progress in resolving them would have been very difficult.

Acknowledgments

The writing of this chapter was supported by the College of Agricultural and Life Sciences of the University of Wisconsin-Madison.

[45] Monkey Feeding Test for Staphylococcal Enterotoxin

By MERLIN S. BERGDOLL

I. Introduction

The first report that staphylococci can produce a toxin that causes emesis and diarrhea was by Barber in 1914.[1] He, along with two other volunteers, became ill after drinking milk from a cow with mastitis but only after the milk had stood at 28–30° for several hours. His attempts to reproduce the illness in monkeys was unsuccessful. Dack and associates in 1930[2] confirmed that the illness experienced by Barber was indeed due to a toxin (called enterotoxin) produced by staphylococci. They isolated staphylococci from a Christmas cake that had made several individuals ill with vomiting and diarrhea. The culture supernatant fluid from growing the staphylococci in laboratory medium produced vomiting and diarrhea in human volunteers. Their attempts to reproduce the illness in monkeys were unsuccessful, even though they fed as much as 90 ml of the culture supernatant fluid.

The first indication that monkeys may be useful for testing this toxin was from the work of Jordan and McBroom.[3] They described feeding experiments with juvenile South American monkeys in which five animals developed diarrhea with one also developing vomiting. The onset of symptoms was about 2 hr after the administration of the staphylococcal filtrates by stomach tube. Later Woolpert and Dack[4] reported that 20 of 21 rhesus monkeys (*Macaca mulatta*) they tested developed the symptoms of staphylococcal food poisoning. They used improved methods for the growth of the staphylococci in the laboratory, hence, they apparently were able to produce much higher concentrations of enterotoxin than they had used in their earlier experiments. The symptoms observed were as follows: about 1.5 hr after feeding, the animals became somewhat pale with increased salivation and swallowing movements; some of the animals bent over in the corner of their cages with forearms folded across

[1] M. A. Barber, *Philipp. J. Sci., Sect. B* **9,** 515 (1914).
[2] G. M. Dack, W. E. Cary, O. Woolpert, and H. Wiggers, *J. Prevent. Med.* **4,** 167 (1930).
[3] E. O. Jordan and J. McBroom, *Proc. Soc. Exp. Biol. Med.* **29,** 161 (1931).
[4] O. C. Woolpert and G. M. Dack, *J. Infect. Dis.* **52.** 6 (1933).

their abdomen; there was a short premonitory period marked by regurgitation; at the beginning of the third hour vomiting set in which was mild in some cases but profuse and projectile in others. Paroxysms of vomiting often recurred over a period of an hour or so. During this time the animals were pallid and abject. Diarrhea often was a marked feature; in two cases in which large amounts of toxic material were fed there was profuse diarrhea without vomiting; similar reactions had been noted in humans. Recovery set in rather abruptly after several hours, and on the following day the monkeys appeared to be normal except for loss of weight. All animals that were fed recovered, although in a few instances diarrhea persisted for several days. Woolpert and Dack[4] concluded that the results were reproduced so consistently by food poisoning strains of staphylococci, and the period of incubation and symptoms were so similar to those seen in human feeding, that there was little doubt that the same toxic principle was involved in both cases.

Subsequent research on staphylococcal enterotoxin conducted by Dack and graduate students at the University of Chicago utilized rhesus monkeys as the major test animal for their studies. Davison and Dack[5] reported that 25 of 29 and later 41 of 50 monkeys[6] they used in their research reacted positively when given enterotoxin-containing samples. They concluded that the monkey was the animal of choice because tests with kittens showed them to give inaccurate results. Thus the monkey feeding test proved to be the only specific test available for testing whether any given staphylococcal strain was enterotoxigenic. Any substance produced by the staphylococci that was resistant to the peptic action of the stomach and was capable of producing an emetic reaction was considered to be an enterotoxin. The basis for this conclusion was that early experiments with both humans and monkeys showed that samples that were emetic in humans were also emetic in monkeys and vice versa. Tests in other animals such as the kitten required that the enterotoxin samples be given intravenously or intraperitoneally which necessitated special treatment to assure that the only active substance present was the enterotoxin. Thus the monkey became the accepted animal model for use in staphylococcal enterotoxin research, particularly by the University of Chicago research group. Surgalla and Hite,[7] Surgalla,[8] and Segalove[9] used the monkey feeding test in their research on the production of enterotoxin by various staphylococcal strains.

[5] E. Davison and G. M. Dack, *J. Infect. Dis.* **64,** 302 (1939).
[6] E. Davison, Ph.D. thesis. University of Chicago, Chicago, Illinois, 1940.
[7] M. J. Surgalla and K. E. Hite, *J. Infect. Dis.* **76,** 78 (1945).
[8] M. J. Surgalla, *J. Infect. Dis.* **81,** 97 (1947).
[9] M. Segalove, *J. Infect. Dis.* **81,** 228 (1947).

II. Enterotoxin Identification

A. Monkey Feeding Test

1. Administration of Enterotoxin. Attempts to detect the presence of enterotoxin in foods that had been implicated in staphylococcal food poisoning were unsuccessful[2] even though the ingestion of the same foods by humans resulted in illness. To address this problem a 5-year project was initiated in the Food Research Institute in 1948 by Surgalla and coinvestigators.[10] Funding was supplied by the food industry as it was important to them to be able to determine whether enterotoxin was present in suspect foods. The monkey feeding test was the logical method for following the production and purification of the enterotoxin, hence, a colony of 30 monkeys was set up. Later the colony was expanded to as many as 130 monkeys.

Based on Surgalla's previous experience, the following procedures were used: samples, usually 50 ml, were injected intragastrically by stomach tube to monkeys that had been starved overnight[10]; after administration of the samples the monkeys were fed and then were watched continuously for 5 hr. Although a number of symptoms may be observed, it was decided that the emetic reaction was the most reliable and only if the animal vomited was a positive reaction recorded. Normally, at least two positive reactions were required because on rare occasions an animal may vomit without provocation.

a. Hand Holding of Monkeys: Two methods of injection of the enterotoxin samples into the stomach have been employed. The one that was used for many years in the Food Research Institute was to have the animal caretaker hold the animal between his knees while holding its head with one hand and its mouth open with a mouth gag with the other hand. This allowed the stomach tube to be passed through the mouth down the esophagus and into the stomach. This required an individual with considerable strength, especially with larger animals. Some employers of the monkey feeding test injected the enterotoxin samples by passing a small tube through the nostrils into the stomach but this worked best when only small volumes (25 ml or less) were being injected.

b. Anaesthesia: Melling[11] reported in 1977 the use of anesthesia with ketamine–HCl in the administration of enterotoxin samples. He found that doses of 7.5 to 15.0 mg/kg of ketamine–HCl injected intramuscularly into the thighs of rhesus monkeys did not result in an emetic action in the animals at those doses even though ketamine is an emetic. The ketamine

[10] M. J. Surgalla, M. S. Bergdoll, and G. M. Dack, *J. Lab. Clin. Med.* **41,** 782 (1953).

[11] J. Melling, *Br. J. Exp. Pathol.* **58,** 40 (1977).

TABLE I
MONKEY FEEDING METHOD

Staphylococcal culture no.	Volume of culture supernatant fluid		
	50 ml	300 ml	1000 ml
560	5/6[a]		
569	1/6	3/6	
547	0/6	5/6	
553	0/6	1/6	6/6
548	0/6	0/6	4/6
562	0/6	0/6	0/6, 0/6

[a] Number positive/number fed.

had no effect on the results he obtained with enterotoxin B (SEB) but it did make the injection of enterotoxin samples much easier and safer. Our use of this method with cynamologous monkeys (*Macaca fascicularis*) revealed that an occasional animal vomited with the recommended dosages for rhesus monkeys. Also, Adesiyun and Tatini[12] reported that 10 of 41 cynamologous monkeys showed an emetic response to 2.5 to 20 mg/kg of ketamine injected intramuscularly within the 5-hr observation period. They determined that any animal showing an emetic response to ketamine could still be used for enterotoxin assay if an experimentally determined nonemetic dose for individual monkeys is employed for sedation. This is an important observation because at the present time only cynamologous monkeys are readily available for such studies unless a conveniently located primate center is available.

2. Dosage. To access the potency of any given sample in the Food Research Institute, 5-fold levels of sample were given to three groups of six monkeys each. In the beginning, crude culture supernatant fluids were fed at an equivalent of 40, 200, and 1000 ml; a negative reaction at the 1000 ml level was the dividing line between enterotoxin-positive and enterotoxin-negative staphylococcal strains. In later testing the amounts were changed to 50, 300, and 1000 ml (Table I). For the testing of dry materials such as purified enterotoxin, different amounts were dissolved in 50 ml of water for feeding. One of the problems encountered with the monkey feeding test was that the animals became resistant to the enterotoxin after a few feedings, hence, it was necessary to replace them. To extend their

[12] A. A. Adesiyun and S. R. Tatini, *Br. J. Exp. Pathol.* **63,** 330 (1982).

TABLE II
REPRESENTATIVE VOMITING REACTIONS OBTAINED IN UNUSED AND USED RHESUS MONKEYS[a]

Sample no.	Milliliters	Unused	Used
4	40	1/3[b]	0/3
	200	3/3	1/3
	1000	2/3	2/2
7	40	1/3	0/2
	200	2/3	2/2
	1000	2/3	0/2
9	200	1/3	0/3
	1000	3/4	0/2
	Totals	15/25 (60%)	5/19 (26%)

[a] From Surgalla *et al.*[10]
[b] Number positive/number fed.

usefulness three unused and three used animals were included in each group of six to which any given sample was fed (Table II).

3. Temporary Resistance. Initially the frequency of using any given monkey in the feeding experiments was determined by the rate of recovery from the previous feeding and at times monkeys were challenged again within 1 week. However, this was before Sugiyama *et al.*[13] discovered that monkeys develop an early resistance to rechallenge, particularly if this is done within 24 hr; the animals were resistant to 10–15 times the initial dose. The resistance gradually decreased with time so that after 1 week this early resistance had essentially disappeared. Subsequent to this finding no animals were rechallenged within 7 days and usually they were held at least 2 weeks before rechallenging.

4. Susceptibility to Second Enterotoxin. It was logical to assume that only one toxin was responsible for staphylococcal food poisoning as the symptoms observed in humans and monkeys were the same with all enterotoxin-producing strains. It was only after the purification of the first enterotoxin that it was discovered that more than one enterotoxin was produced by the staphylococci. The enterotoxins were identical in their activity but each reacted with a different antibody.[14] It was discovered that monkeys resistant to one enterotoxin were only slightly resistant to an antigenically different one,[15] thus the usefulness of the monkeys could

[13] H. Sugiyama, M. S. Bergdoll, and G. M. Dack, *J. Infect. Dis.* **111,** 233 (1962).
[14] M. S. Bergdoll, M. J. Surgalla, and G. M. Dack, *J. Immunol.* **83,** 334 (1959).
[15] M. S. Bergdoll, C. R. Borja, and R. M. Avena, *J. Bacteriol.* **90,** 1481 (1965).

TABLE III
FEEDING OF ENTEROTOXIN FROM HETEROLOGOUS TYPES TO MONKEYS RESISTANT TO A SINGLE TYPE

	Enterotoxin A		Enterotoxin B		Enterotoxin C_1	
Monkeys	MED[a]	Results[b]	MED	Results	MED	Results
Enterotoxin A resistant	10	3/6	2	2/6	Not fed	
	30	3/6	Not fed		3	5/6
Enterotoxin B resistant	12	9/16	200	4/16	Not fed	
	Not fed		100	3/12	2	4/12

[a] Minimum emetic dose
[b] Number positive/number fed.

be extended (Table III). Over the period of years five antigenically different enterotoxins were identified; however, others remain to be identified. The monkey feeding test was particularly useful in identifying a food-poisoning strain as enterotoxigenic when testing against the different specific antibodies of the identified enterotoxins produced negative results. Twenty-four staphylococcal strains that were isolated from food-poisoning outbreaks and did not produce one of the identified enterotoxins were found to be enterotoxigenic by the monkey feeding test: 12 positive at 50 ml, 7 positive at 300 ml, and 5 positive at 1000 ml of culture supernatant fluid per animal.

B. Purification of Enterotoxins

The monkey feeding test was essential in the purification of new enterotoxins as a means of identifying enterotoxin-containing fractions generated in the purification procedures. Once an enterotoxin was purified and specific antibodies were prepared to it monkey feedings were no longer necessary for that enterotoxin. This procedure was followed in the purification of all seven of the enterotoxins, SEA, SEB, the SECs, SED, and SEE. No work is being done today on the purification of additional enterotoxins but if such work is undertaken in the future, the monkey feeding test will be an essential tool; no other specific biological activity is available for use in this type of work.

1. Dosage: Intragastrically. The amount of purified enterotoxin required to produce a reaction in 50% of the monkeys (ED_{50}) was estimated to be approximately 5 to 10 μg/animal for all of the enterotoxins except SED; 10 to 20 μg/animal of SED was required in the Food Research Institute to produce emesis in sensitive animals. Schantz *et al.* calculated

TABLE IV
EFFECT OF ENTEROTOXIN A
ON RHESUS MONKEYS

Route of administration			
Intragastric		Intravenous	
μg/animal	Result[a]	μg/animal	Result[a]
5	4/10	0.017	2/6
10	16/30	0.035	2/9
20	9/12	0.070	4/9
		0.140	5/6

[a] Number positive/number fed.

the ED_{50} for both SEA[16] and SEB[17] to be approximately 1 μg/kg in 3-kg animals; 100 unused animals were employed in the testing of each enterotoxin type. These investigators used both emesis and diarrhea to determine toxic activity while we used only the emetic reaction in our testing. Our results were based on the use of 12 animals for each enterotoxin. The 1000 ng/kg dose in monkeys when compared to the estimated 2 to 3 ng/kg dose required to produce illness in humans indicates why it is almost impossible to produce an emetic reaction in monkeys from enterotoxin-containing food.

2. Dosage: Intravenously. Intravenous injections of the enterotoxin also were used in the testing of purified enterotoxins for their potency as well as the effect of the neutralizing capacity of heterologous antibodies. Enterotoxin A was the most potent by this route although all of the enterotoxins were 10 to 100 times more potent by this route than by the oral route. Enterotoxin A produced an emetic reaction in monkeys at the level of 0.02 μg/kg (Table IV). Schantz *et al.*[17] reported that the intravenous injection of SEB at the level of 0.1 μg/kg gave an ED_{50} reaction while we found that 0.5 μg/kg/animal gave a positive reaction although we did not determine the minimum dose.

3. Neutralization. The results of testing the effect of heterologous antibodies in the neutralizing of enterotoxins are given in Table V. Some cross-reactions were observed such as the neutralizing of SEE by antibodies to SEA but this was not understood until it was discovered that

[16] E. J. Schantz, W. G. Roessler, M. J. Woodburn, J. M. Lynch, H. M. Jacoby, S. J. Silverman, J. C. Gorman, and L. Spero, *Biochemistry* **11,** 360 (1972).

[17] E. J. Schantz, W. G. Roessler, J. Wagman, L. Spero, D. A. Dunnary, and M. S. Bergdoll, *Biochemistry* **4,** 1011 (1965).

TABLE V
TREATMENT OF ENTEROTOXINS WITH HETEROLOGOUS ANTISERA

Enterotoxin	Amount (μg)	Antisera	Results[a]
Enterotoxin C_1	0.5	None	4/6
	1.0	SEC_1	0/6
	0.5	SEA	3/3
Enterotoxin E	1.0	None	3/3
	2.5	SEE	0/6
	2.5	SEA	0/6
	2.5	SEC_1	2/3

[a] Number positive/number fed.

these two enterotoxins have a common minor antibody.[18,19] Also, the SEB antisera appeared to neutralize the SEC before it was discovered that some SEB antisera cross-react with SEC.

C. *Mode of Action*

1. Emetic Stimulation. Sugiyama and co-workers carried out many experiments on the mode of action of the enterotoxins in which monkeys were used as the test animals. They studied the stimulation of the emetic reaction as well as the inhibition of this reaction with various drugs. They found that Thorotrast (25% ThO_2), a nonemetic material, injected intravenously 18 hr before administration of the enterotoxin, increased the sensitivity of monkeys to the emetic action; the dosage of enterotoxin required was one-twentieth that required to produce emesis in control animals.[20] In a separate experiment, Sugiyama *et al.*[21] found the emetic drug, dihydroergotamine methane sulfonate, injected subcutaneously in subemetic doses 45–60 min before the administration of enterotoxin, doubled the number of monkeys reacting when compared to the control. Although the purpose of these experiments was to study the mode of action of the enterotoxins, these chemical substances could be used to enhance the results obtained with the monkey feeding test.

2. Emetic Inhibition. Other drugs such as perphenazine, a tranquil-

[18] M. S. Bergdoll, C. R. Borja, R. N. Robbins, and K. F. Weiss, *Infect. Immun.* **4,** 593 (1971).

[19] A. C.-M. Lee, R. N. Robbins, and M. S. Bergdoll, *Infect. Immun.* **21,** 387 (1978).

[20] H. Sugiyama, E. M. McKissic, Jr., and M. S. Bergdoll, *Proc. Soc. Exp. Biol. Med.* **113,** 468 (1963).

[21] H. Sugiyama, M. S. Bergdoll, and G. M. Dack, *Proc. Soc. Exp. Biol. Med.* **97,** 900 (1958).

lizer and antiemetic, reserpine, a hypotensive tranquillizer and sedative, and Trasylol, a saline solution of polypeptides with antiproteolytic enzyme activity were tested for their effect on the inhibition of the emetic action of enterotoxin in the monkey. Intravenous injections of reserpine 18 hr prior to enterotoxin administration reduced the emetic reaction by 68% when compared to controls.[22] Perphenazine given intravenously 45 min after the administration of the enterotoxin reduced the emetic reaction by 92%.[22] Neither of these drugs had an appreciable effect on the emetic action of $CuSO_4$, indicating that the effect of the drugs was not due to depression of the vomiting center. Trasylol given intravenously immediately before administration of enterotoxin reduced the incidence of vomiting significantly when compared to a control.[23] It had no apparent protective effect against the emetic action of $CuSO_4$ or Veriloid (veratrum alkaloid). These results indicated that the emetic site was at some other point than the vomiting center.

3. Site of Emetic Action. Sugiyama and Hayama[24] concluded that the site of emetic action in the monkey was in the abdominal viscera and that the sensory stimulus for vomiting reaches the vomiting center by way of the vagus and sympathetic nerves. This conclusion was based on experiments in which transthoracic vagotomy gave partial protection against the emetic action of enterotoxin given intragastrically[25] and coupled with abdominal sympathectomy rendered the monkey completely resistant to the emetic action of the enterotoxin.[24] The monkeys with differentiated abdominal viscera still vomited as normal animals when Veriloid was administered. Bilateral destruction of the area postrama on the floor of the fourth ventricle (chemoreceptor trigger zone) made the rhesus monkeys completely refractory to the emetic action of enterotoxin given intravenously or intragastrically.[26]

D. Immunization

Rhesus monkeys were used as a model for determining whether immunization of humans against enterotoxin was possible. Three injections of enterotoxoid B intramuscularly 5 weeks apart protected the monkeys against 200 minimal doses of SEB.[27] This immunity gradually decreased

[22] H. Sugiyama, M. S. Bergdoll, and R. G. Wilkerson, *Proc. Soc. Exp. Biol. Med.* **103,** 168 (1960).

[23] H. Sugiyama, *J. Infect. Dis.* **116,** 162 (1966).

[24] H. Sugiyama and T. Hayama, *J. Infect. Dis.* **115,** 330 (1965).

[25] H. Sugiyama and T. Hayama, *Proc. Soc. Exp. Biol. Med.* **115,** 243 (1964).

[26] H. Sugiyama, K. L. Chow, and L. R. Dragsted II, *Proc. Soc. Exp. Biol. Med.* **108,** 92 (1961).

[27] M. S. Bergdoll, *J. Infect. Dis.* **116,** 191 (1966).

over a period of 1 year to an estimated protection against 20 minimal doses. Precipitating antibodies were not present in the sera from all of the monkeys even though the animals were protected against several minimal doses of enterotoxin. Animals also became resistant to the continuous feeding of larger and larger doses of SEB until the animals were resistant to 200 minimal doses. The sera from these animals when given intravenously to unused monkeys before challenging with 10 minimal doses of SEB protected the animals against the emetic action.[27] Precipitating antibodies were not detectable in the sera from the animals that developed a resistance from the oral feeding of the enterotoxin.

III. Conclusions

At the present time, very little if any research is being done on the staphylococcal enterotoxins that require animal testing. Probably more work would be done if the animals were readily available at a reasonable cost. In addition, the requirements for maintaining a monkey colony adds to the expense. We maintained a monkey colony from 1948 until 1981 and after that did utilize the monkeys at the Primate Center at the University of Wisconsin for two additional years. At the present time our research does not require the use of monkeys because we are working only with the identified enterotoxins. The monkey feeding test is a valid one and would be useful at any time work was done on identifying additional enterotoxins or in the identification of staphylococcal strains that produce an unidentified enterotoxin.

Acknowledgments

The writing of this chapter was supported by the College of Agricultural and Life Sciences of the University of Wisconsin-Madison.

[46] Assays for Epidermolytic Toxin of *Staphylococcus aureus*

By JOYCE C. S. DE AZAVEDO and JOHN P. ARBUTHNOTT

The staphylococcal epidermolytic toxins A and B (ET_A and ET_B) are extracellular proteins which are responsible for the symptoms of the scalded skin syndrome. They induce splitting of skin at the level of the stratum granulosum in man, monkeys, mice, and golden hamsters. The bioassay is based on injection of toxin, or ET-producing organisms, subcutaneously into neonatal mice and subsequent examination for splitting of skin as detected by a positive Nikolsky reaction. Several *in vitro* assays for ET are available which are based on detection using anti-ET sera.

Mouse Assay

Newborn ShaSha, BALB/c, or Park strain mice approximately 3 days old and weighing between 1.8 and 2.2 g are used. Other strains of mice such as the Webster/Swiss,[1] CD-1,[2] TA,[3] and ICR[4] are also suitable. Four mice per sample are injected subcutaneously in the back with 50–100 μl of test sample using a tuberculin syringe and a 26-gauge needle; samples are diluted in phosphate-buffered saline (PBS, 76 m*M*, pH 7.2). If the ET-producing capacity of strains is being tested, 5×10^8 colony-forming units per mouse are injected. Strains may be grown in advance, washed three times in peptone water, stored at $-70°$, and diluted to the appropriate concentration on the day of the assay. Mice are examined after 2 hr and up to 6 hr by gently stroking or pinching the skin on the back. Loosening of the skin or formation of a wrinkle (Nikolsky sign) indicates ET activity (Fig. 1). The lowest quantity of purified ET which can be detected in the mouse assay is 0.3 to 0.5 μg. Although the sensitivity of the assay is fairly high, one disadvantage is that it does not allow differentiation between the serotypes.

Ouchterlony Immunodiffusion

Materials

Glass plates, 8.5 × 9.5 cm, or microscope slides
Gel punch, 2.5-mm diameter
Template for wells
0.1% agarose (w/v) in 150 m*M* NaCl
100 m*M* NaCl

[1] K. D. Wuepper, D. H. Baker, and R. L. Dimond, *J. Invest. Dermatol.* **67,** 526 (1976).

[2] A. D. Johnson, J. F. Metzger, and L. Spero, *Infect. Immun.* **12,** 1206 (1975).

[3] M. E. Melish, F. S. Chen, S. Sprouse, M. Stuckey, and M. S. Murata, *in* "Staphylococci and Staphylococcal Infections" (J. Jeljaszewicz, ed.), pp. 287–297. Fischer-Verlag, Stuttgart, Federal Republic of Germany, 1981.

[4] I. Kondo, S. Sakurai, and Y. Sarai, *Infect. Immun.* **10,** 851 (1974).

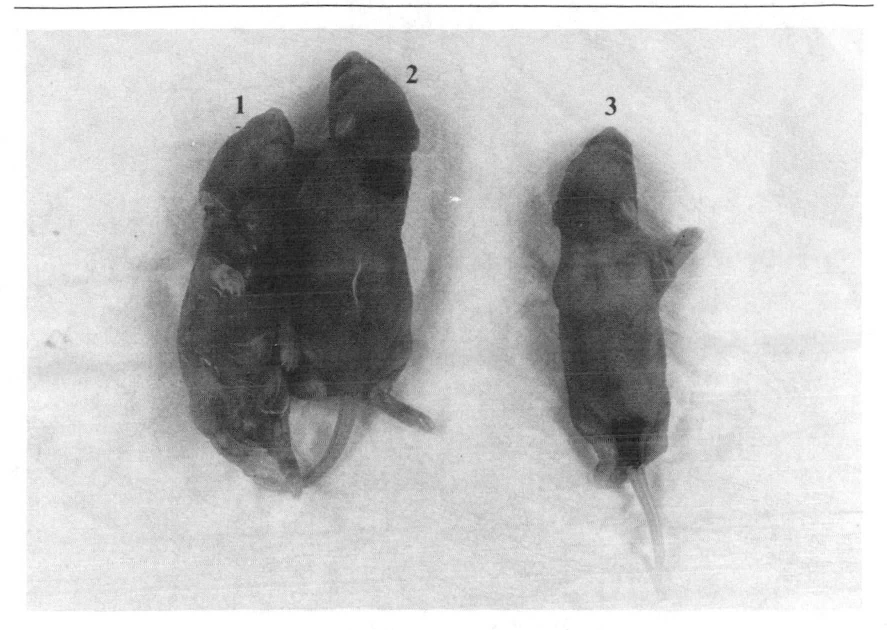

FIG. 1. Nikolsky sign in mice. (1) Frank peeling of the skin due to a large dose of ET. (2) A typical Nikolsky sign. (3) Saline-injected control.

Whatman No. 1 filter paper
0.25% (w/v) Coomassie blue R-250 in 45% (v/v) methanol, 9% (v/v) glacial acetic acid

Glass plates or microscope slides are coated with 10 or 2.5 ml agarose, respectively. Wells, 9 mm apart center to center, are punched in the agarose using the template, as shown in Fig. 2. Antiserum (5 μl) is added to the central well and doubling dilutions (5 μl) of test samples including appropriate controls are added to the surrounding wells. Plates are incubated in a moist box at room temperature overnight or at 37° for 4 hr to allow precipitin lines to develop. Plates are washed in 100 m*M* saline for 20 min, covered with a sheet of moist filter paper and several sheets (1–2 cm thick) of absorbent tissue paper, and pressed under a weight for 10 min; this process is carried out three times. After the final press, plates are rinsed with distilled water, dried in a stream of air, and stained with Coomassie Blue. The minimum quantity of ET which is detectable by this method is 60 μg/ml; the sensitivity varies depending on the potency of the antiserum. Although this method is not highly sensitive, it has an advantage in that monospecific serum is not essential if purified ET samples are used to check for lines of identity.

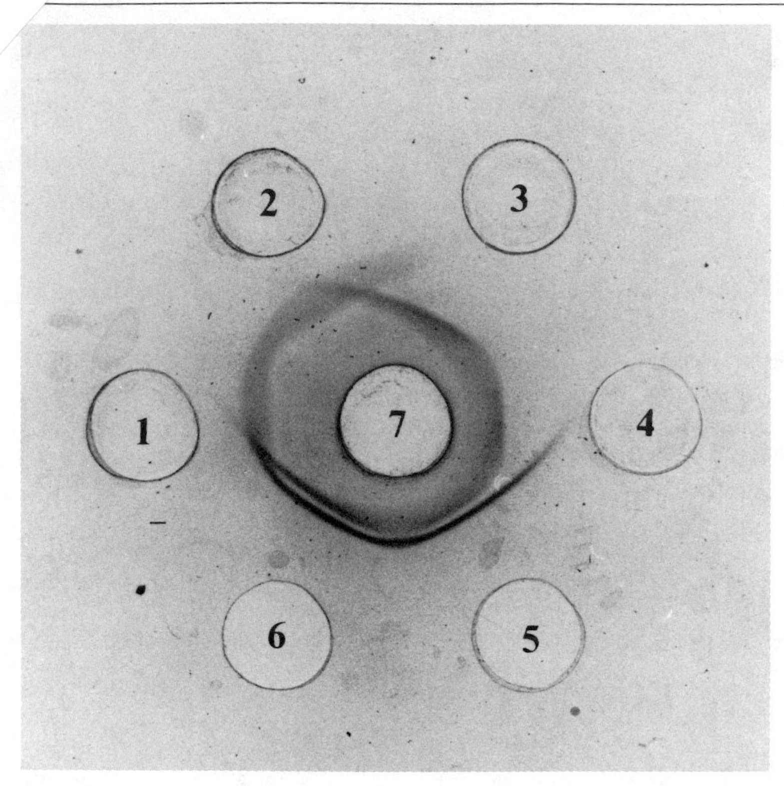

FIG. 2. Ouchterlony immunodiffusion showing a reaction of nonidentity between ET_A and ET_B. Wells 1 and 2 contain purified ET_A, wells 3 and 4 contain purified ET_B, wells 5 and 6 contain mixtures of purified ET_A and ET_B, and well 7 contains antiserum against both ET_A and ET_B.

Dot Immunoblotting

Materials

Nitrocellulose filters, 0.45 μm, 8.2-cm diameter (BA 85/22, Schleicher and Schull)
50 m*M* Tris–HCl, pH 7.4
50 m*M* Tris–HCl, pH 7.4, 150 m*M* NaCl (TBS)
TBS + 5% bovine serum albumin (BSA), fraction V
TBS + 0.5% BSA
Horseradish peroxidase-conjugated protein A
0.6% (w/v) chloronaphthol in methanol (can be stored at −20° up to 1 week)
30% H_2O_2

Test samples (5 μl) are spotted onto nitrocellulose filters and dried gently in a stream of air. Filters are incubated with 10 ml 5% BSA/TBS for

1 hr with gentle shaking followed by a 2-hr incubation with anti-ET serum (diluted 1/100 in 5% BSA/TBS). Filters are washed three times (10 min/wash) in 10 ml 0.5% BSA/TBS and incubated with conjugate (diluted 1/1000 in 0.5% BSA/TBS for 1 hr. Filters are washed three times in 0.5% BSA/TBS as before and incubated with freshly prepared substrate solution (1 ml chloronaphthol, 9 ml Tris–HCl, 5 μl H_2O_2) until the color develops. Finally, the filters are rinsed in distilled water, dried with filter paper, and stored in the dark. ET concentration is determined by end-point dilution and comparison with the end point of a known concentration of purified ET. The minimum amount of ET which is detectable by this method is 5–10 ng, i.e., 1–2 μg/ml. If this assay is used for screening of strain supernatants, it is important to ensure that the serum is monospecific since otherwise false positives are likely to arise. The use of a protein A–conjugate eliminates false positives due to protein A in culture supernatants. Ideally a monoclonal serum should be used.

Other Assays

Radial immunodiffusion,[1,5] radioimmunoassay,[1,3,6] and enzyme-linked immunosorbent assays[7] have also been used for detection of ET.

Radial immunodiffusion involves the use of agarose plates containing antiserum. Wells (2-mm diameter) are punched in the agarose containing $\frac{1}{20}$–$\frac{1}{30}$ dilution of anti-ET and filled with 3 μl test sample. Plates are incubated overnight at room temperature in a moist box and the diameters of the precipitin rings are measured. Concentration of ET is determined from a standard curve of ring diameter squared plotted against protein concentration of purified ET. Reports of sensitivity of the assay have varied from 5 μg/ml[1] to approximately 200 μg/ml.[5]

A sensitive radioimmunoassay was described by Melish *et al.*[3] using purified ET labeled with iodine-125. Test samples are incubated with 50 μl of a 10^{-5} dilution of rabbit anti-ET. Labeled ET (50 μg) is then added and the mixture incubated at 4° for 24 hr. Separation of bound and free ET is effected by addition of a second antibody (goat anti-rabbit) and centrifugation of the precipitate. Radioactivity of the precipitates is counted using a gamma counter and the ET concentration determined from a standard regression line of B/B_0 vs log concentrations of known ET (B = counts bound, B_0 = counts bound for zero dose of unlabeled ET). This assay is reported to detect ET at concentrations as low as 100 pg/ml.

Recently, Piemont *et al.*[6] described a sensitive enzyme-linked immunosorbent assay which is rapid and easy to perform. Briefly, microtiter

[5] J. P. Arbuthnott and B. Billcliffe, *J. Med. Microbiol.* **9,** 191 (1976).
[6] Y. Piemont, M. Maubensack, and H. Monteil, *J. Clin. Microbiol.* **20,** 1114 (1984).
[7] N. Harboe and A. Ingild, *Scand. J. Immunol.* **2,** 161 (1973).

wells are coated with anti-ET and incubated with varying dilutions of test sample. After washing to remove unbound antigen, bound ET is detected using anti-ET conjugated to alkaline phosphatase. The sensitivity of the method is reported to be 3 ng/ml.

Preparation of Rabbit Anti-ET Sera

Purified ET (100–200 μg) in 2 ml PBS, 76 m*M*, pH 7.2, is emulsified with 2 ml Freund's complete adjuvant. On day 0, a New Zealand White rabbit weighing 3–4 kg is injected subcutaneously on the back in multiple sites with 100 μl/site. On days 14 and 28 the procedure is repeated but the antigen is emulsified in Freund's incomplete adjuvant. The titer of the serum is checked 5 days later and if the titer is good, the animal is either bled out, or 20–30 ml drawn every fortnight. Booster doses are given when the antibody titer drops. Immunoglobulins are purified by ammonium sulfate precipitation.[7]

Comments

About 6% of coagulase-positive strains of *S. aureus* isolated from a variety of clinical sources are ET producers and of these the majority (88%) produce ET_A.[8] However, the prevalence of ET production among strains associated with exfoliative skin lesions is high (70% of strains) and although most of these strains (84%) belong to phage group II, a number belonging to other phage groups also produce ET.[9]

Production of ET *in vitro* varies considerably depending on culture conditions. The authors have found that the following protocol gives good yields: a single colony is inoculated into 30 ml yeast diffusate medium (see Chapter [5] in this volume) contained in a flanged Erlenmeyer Flask. Flasks are incubated at 37° on an orbital shaker at 150 rpm for 20 to 40 hr. Maximum yields of ET_A and ET_B are produced at 20 and 40 hr respectively. If ET is undetectable in culture supernatants, they may be concentrated up to 100-fold by precipitation with 90% saturated ammonium sulfate solution.

In screening clinical isolates for ET production, it is important to bear in mind that further serotypes of ET in addition to ET_A and ET_B may arise. Consequently, the bioassay using viable organisms should be used when *in vitro* tests for ET production are negative, particularly with strains isolated from an exfoliative skin lesion.

[8] Y. Piemont, D. Rasoamananjara, J. M. Fouace, and T. Bruce, *J. Clin. Microbiol.* **19,** 417 (1984).

[9] J. de Azavedo and J. P. Arbuthnott, *J. Med. Microbiol.* **14,** 341 (1981).

[47] Immunochemical Assays for Toxic Shock Syndrome Toxin-1

By PATRICK M. SCHLIEVERT

Toxic shock syndrome toxin-1 (TSST-1) is a low-molecular-weight single peptide toxin secreted by some strains of *Staphylococcus aureus*.[1-3] The toxin is highly associated with menstrually related TSS and is considered a likely causative toxin.[4] The toxin has a number of important biological properties including capacity to induce fever, enhancement of host susceptibility to shock by endotoxin, either of exogenous or endogenous origin, and nonspecific T-lymphocyte mitogenicity, a property which leads to suppression of immunoglobulin synthesis and enhancement of delayed hypersensitivity.[4]

Studies have been done to evaluate physical and chemical factors that affect toxin production by TSS *S. aureus*. Such studies have been undertaken to evaluate which strains make the toxin,[1,2] what conditions yield large amounts of toxin for ease of purification,[5] and why certain tampons are associated with the syndrome.[6,7] At least three different methods have been developed to achieve the above goals: polyacrylamide gel isoelectric focusing,[1] double immunodiffusion,[2,5] and most recently ELISA.[8] In this report, methods are presented which allow easy qualitative and quantitative determinations of TSST-1.

Production of TSST-1

TSST-1 can be prepared in large amounts by ethanol precipitation of culture fluids followed by electrofocusing in successive gradients of 3.5–10 and 6–8.[9,10] Staphylococcal strain MN8 (a high TSST-1 producer) is

[1] P. M. Schlievert, K. N. Shands, B. B. Dan, G. P. Schmid, and R. D. Nishimura, *J. Infect. Dis.* **143,** 509 (1981).

[2] M. S. Bergdoll, B. A. Crass, R. F. Reiser, R. N. Robbins, and J. P. Davis, *Lancet* **1,** 1017 (1981).

[3] M. S. Bergdoll and P. M. Schlievert, *Lancet* **2,** 691 (1984).

[4] P. M. Schlievert, *J. Infect. Dis.* **147,** 391 (1983).

[5] P. M. Schlievert and D. A. Blomster, *J. Infect. Dis.* **147,** 236 (1983).

[6] P. M. Schlievert, D. A. Blomster, and J. A. Kelly, *Obstet. Gynecol.* **64,** 666 (1984).

[7] P. M. Schlievert, *J. Infect. Dis.* **152,** 618 (1985).

[8] J. T. Parsonnet, J. T. Mills, Z. A. Gillis, and G. B. Pier, *J. Clin. Microbiol.* **22,** 26 (1985).

[9] D. A. Blomster-Hautamaa, B. N. Kreiswirth, R. N. Novick, and P. M. Schlievert, *Biochemistry* **25,** 54 (1986).

[10] D. A. Blomster-Hautamaa and P. M. Schlievert, this volume [6].

used as the source of toxin and is cultured in a dialyzable beef-heart medium until late stationary phase is achieved. The purified toxin may be stored lyophilized indefinitely (>2 years) or at 4° or frozen for extended periods of time. Toxin thus purified is homogenous when subjected to sodium dodecyl sulfate-polyacrylamide gel electrophoresis followed by silver staining.[9,10]

Preparation of Antisera

Hyperimmune rabbit antiserum against TSST-1 is prepared by subcutaneous injection of toxin emulsified in adjuvant (Freund's incomplete, Difco Laboratories, Detroit, MI).[11] TSST-1, 50 μg/animal/1-ml vol, is emulsified in an equal volume of adjuvant with the use of a syringe and is then injected subcutaneously in two sites (one-half volume each) in the back of the neck. Injections are given every other week for four injections and then monthly. One week after the fifth and subsequent injections, antiserum is obtained and tested for reactivity with purified TSST-1 in Ouchterlony immunodiffusion assays.

It has been our experience that approximately one-half of American Dutch-belted rabbits will make a demonstrable antibody response to TSST-1. Others have noted that New Zealand White animals will also respond. If animals to be injected have visible respiratory illness (probably due to *Pasteurella* sp.) or if the animals are older than 8 months of age at the beginning of immunization, many likely will not develop immunity.[4]

Qualitative Assay for TSST-1

Our laboratory provides a free testing service to physicians who wish to know if a potential TSS *S. aureus* is expressing TSST-1 or not. We provide this on a routine basis and have developed a simple and relatively quick assay for this purpose.

Todd–Hewitt (Difco) 1% agar medium is prepared and 20 ml/plate poured into standard Petri dishes. The test organism is then streaked onto one part of the plate and cultured at 37° overnight (culture at 37 to 40° may be used; culture at room temperature yields only low levels of TSST-1 that may not be detected). Subsequently, 2- to 4-mm-diameter wells are punched 4 mm from the growing *S. aureus* and from each other, and 20 μl hyperimmune antisera is added to one well, and 20 μl purified TSST-1 (25–50 μg/ml) is added to the other. The plate is then incubated an addi-

[11] P. M. Schlievert, K. M. Bettin, and D. W. Watson, *Infect. Immun.* **16,** 673 (1977).

tional 4–12 hr at 37° and examined for a precipitin arc which forms a line of identity with the purified TSST-1.

The results obtained (number positive) from this method agree completely with those from our quantitative test described next. We have observed that all TSS isolates tested thus far secrete enough TSST-1 to be detected by this method. As a last point, occasionally, a second precipitin arc develops which does not form a line of identity with the control toxin. This may represent a protein A–rabbit antibody reaction (it presumably occurs only when both are in high concentration since rabbit antibody does not precipitate with protein A unless it is in high concentration).

Quantitative Assays for TSST-1

An assay was developed based upon double immunodiffusion to quantify TSST-1 for the purpose of elucidating both conditions that yield high levels of toxin for ease of purification[5] and the possible mechanism of tampon involvement in TSS.[6,7]

TSST-1-producing *S. aureus* are cultured in the desired medium, usually from 1 to 100 ml. Subsequently, 4 vol 4° absolute ethanol is added (it is not necessary to remove the cells), and the cultures placed at 4° for at least 1 hr to allow the TSST-1 to precipitate. We have previously shown[5] that TSST-1 is quantitatively precipitated in the presence of 75% final concentration ethanol. After precipitation, the cultures are centrifuged (650 *g*, 10 min) and the supernatant fluid discarded. The tube or flask containing the pellet is placed on its side in a chemical hood for approximately 15 min to evaporate the remaining ethanol. The pellet is next suspended in distilled water to a final volume of 1/10 to 1/100 the original culture fluid volume. Serial 2-fold dilutions of the suspension are made in distilled water, and 20 μl of each is used in a double-immunodiffusion assay. It is not necessary to centrifuge the suspension to remove insoluble debris; we have observed that clarified suspensions and suspensions which contain insoluble material give identical results. Also, the 2-fold dilutions may be made in bacterial culture media, buffers such as phosphate or Tris, or saline.

The double-immunodiffusion (Ouchterlony) test is run as follows. Standard microscope slides are covered with 4.5 ml of 0.75% agarose (Low EEO, Sigma Chemical Company, St. Louis, MO), melted in phosphate-buffered saline (0.005 *M* sodium phosphate, 0.15 *M* NaCl, pH 7.0); the slides are not precoated with a thin layer of agarose solution. Hexagonal patterns (three/slide) of wells are punched in the solidified agarose gel. Each well is 4 mm in diameter and 4 mm from a center 4-mm-diameter well. LKB-Produkter AB (Bromma, Sweden) makes an Ouchterlony gel punch system that serves this purpose. The slide is then placed in a humidified Petri dish. Twenty microliters of dilutions of test or control purified toxin of known concentration is then added to the outer wells, and 20 μl hyperimmune antisera against TSST-1 placed in the center. The slide is then incubated 4 hr at 37° in a standard incubator or overnight at room temperature to allow development of precipitin arcs. The end point is considered the last well which gives a visible precipitin arc. The experimental test may then be compared to a similarly assayed control toxin for determination of toxin concentration. The end-point of standard preparations of TSST-1 in this assay is usually 4 μg/ml (80 ng/20 μl).

The sensitivity of the assay may be increased 2- to 4-fold by drying and staining the slides. This is accomplished by the following method. The slide is placed in approximately 500 ml of distilled water for 3 days with changes of water each day; the agarose gel is loosened from the slide after immersion in the water in order to facilitate removal of ions and soluble proteins from the gel. Subsequently, the gel is replaced on the slide and covered with wet Whatman No. 1 filter paper. The wells on the agarose should be filled with water such that air is not trapped under the filter paper since air will lead to distortion of the wells. The agarose gel is then dried at 37°, the filter paper gently peeled off, and the slide stained at least 15 min in Coomassie brilliant blue (0.25% in methanol : water : acetic acid, 5 : 5 : 1). Finally, the slide is destained in a solution of water : ethanol : acetic acid (8 : 3 : 1).

Uses of the Quantitative Assay

The double-immunodiffusion test has been used to elucidate conditions that yield good production of TSST-1.[5] These conditions include pH values of 7–8, temperatures of 37 to 40°, complex medium of animal origin, high aeration, and low glucose. Conditions which yield little, if any, TSST-1 production include pH values of 4–6.5 or 9 or greater, room temperature incubation, high glucose levels, and anaerobiasis. Toxin is made throughout cell growth, though it is maximally expressed just prior to entering stationary phase. Once made, TSST-1 is stable in culture media for days.

We have also used the assay to assess the capacity of tampons to alter TSST-1 production.[6,7]

Limitations of the Quantitative Assay

There are two limitations to the use of this assay. First, in culture media which yield high levels of TSST-1, approximately 10^7 cells/ml must be present in the original culture to detect toxin in 50× concentrated culture fluid (with strain MN8 as the test organism). The ELISA assay, recently developed,[8] is approximately 10-fold more sensitive, but is more difficult to set up. Second, the test is restricted to measuring 2-fold differences in toxin production, though we have never found the need to measure smaller differences.

Alternative Quantitative Procedure

Recently, Parsonnet *et al.*[8] have developed a competitive ELISA for measurement of TSST-1. This assay is performed according to the following procedure.

ELISA Reagents. The reagents used were described by Voller *et al.*[12] and included carbonate–bicarbonate buffer, pH 9.6, as the coating buffer. The diluent and washing buffer was phosphate-buffered saline, containing 0.05% Tween 20 and 0.1% bovine serum albumin (PBS–Tween). The substrate was *p*-nitrophenyl phosphate (No. 104, Sigma) (1 mg/ml) in 10% diethanolamine buffer, pH 9.8. IgG against TSST-1 was prepared from hyperimmune serum by precipitation three times with 35% ammonium sulfate. It was stored in PBS at −20°.

TSST-1–Enzyme Conjugation. TSST-1 was conjugated to alkaline phosphatase (type V11-S, Sigma) after modification of a previous method.[13] TSST-1 (100 μl of 10 mg/ml in PBS) was mixed with 600 μl alkaline phosphatase in 3.2 *M* ammonium sulfate. After dialyzing against PBS, 20 μl glutaraldehyde (4% solution) was added, the solution was mixed for 2 hr and then brought up to 2 ml, and finally was dialyzed against PBS. Insoluble material was removed by centrifugation.

The TSST-1–alkaline phosphatase conjugate was separated from unlinked toxin and enzyme by gel filtration (2.6 × 40 cm column, Ultragel AcA 44, LKB). Collected fractions were assayed for immune reactivity and enzyme activity.

Immunoassay. Wells of flat-bottomed microtiter plates (Immunolon-2, Dynatech Laboratories, Inc., Alexandria, VA) were coated with anti-TSST-1 antibody diluted in carbonate–bicarbonate buffer by incubating 100 μl of solution in wells for 4 hr at 37°. Plates were washed three times with PBS–Tween. Samples of culture fluids to be tested for TSST-1 or control standard toxin preparations were added to the coated plates in duplicate, followed by the addition of appropriately diluted TSST-1–alka-

[12] A. Voller, D. W. Bidwell, and A. Bartlett, *in* "Manual of Clinical Immunology" (N. R. Rose and H. Friedman, eds.), p. 506. Am. Soc. Microbiol., Washington, D.C., 1976.

[13] P. E. Kauffman, *J. Assoc. Off. Anal. Chem.* **63,** 1138 (1980).

line phosphatase conjugate. After overnight incubation at room temperature in a humidified environment, the wells were washed three times with PBS–Tween and 200 μl of prewarmed substrate solution was added; the plates were then incubated at 37°. Color development was measured at 405 nm when wells containing uninhibited TSST-1–alkaline phosphatase conjugate reached an absorbance of 1.0 (approximately 1 hr). Toxin concentrations in cultures were determined by comparison to values for standard toxin preparations.

[48] Chick Embryo Assay for Staphylococcal Toxic Shock Syndrome Toxin-1

By JOYCE C. S. DE AZAVEDO, ROGER N. LUCKEN, and JOHN P. ARBUTHNOTT

Toxic shock syndrome toxin-1 (TSST-1) is produced by strains of *Staphylococcus aureus* associated with toxic shock syndrome,[1–3] an acute, febrile disease in man leading to cardiovascular shock.[4] TSST-1 (M_r 23,000, pI 7.3)[5] is pyrogenic, causes shocklike symptoms, and is lethal in rabbits.[2,5] Other biological effects include lymphocyte mitogenicity,[2,5,6] release of interleukin-1 from monocytes[7,8] and potentiation of chick embryo lethality due to gram-negative lipopolysaccharide (LPS).[9]

The enhancement of chick embryo lethality can be used as a bioassay for TSST-1 in the presence of a fixed sublethal amount of LPS. The chick embryo system is quick, reproducible, and simple to perform.

Preparation and Injection of Eggs. Fertile White Leghorn hens' eggs (or other white eggs) are set in a humidified, self-turning incubator main-

[1] M. S. Bergdoll, B. A. Crass, R. F. Reiser, R. N. Robbins, and J. P. Davis, *Lancet* **1,** 1017 (1981).

[2] P. M. Schlievert, K. N. Shands, B. B. Dan, G. P. Schmid, and R. D. Nishimura, *J. Infect. Dis.* **143,** 509 (1981).

[3] M. S. Bergdoll and P. M. Schlievert, *Lancet* **2,** 691 (1984).

[4] J. P. Davis, J. Chesney, P. J. Wand, and M. La Venture, *N. Engl. J. Med.* **303,** 1429 (1980).

[5] J. C. de Azavedo, P. J. Hartigan, and J. P. Arbuthnott, *in Proc. FEMS Workshop Conf. Bact. Protein Toxins, Seillac, France,* (J. E. Alouf, F. Fehrenbach, J. H. Freer, and J. Jeljaszewicz, eds.), pp. 331–338. Academic Press, London, 1984.

[6] N. J. Poindexter and P. Schlievert, *J. Infect. Dis.* **151,** 65 (1985).

[7] T. Ikbejima, C. A. Dinarello, M. Gill, and S. Wolff, *J. Clin. Invest.* **73,** 1312 (1984).

[8] J. Parsonnet, R. K. Hickman, D. D. Eardley, and G. B. Pier, *J. Infect. Dis.* **151,** 514 (1985).

[9] J. C. S. de Azavedo, R. N. Lucken, and J. P. Arbuthnott, *Infect. Immun.* **47,** 710 (1985).

tained at 37°. After 11 days of incubation, each egg is checked for viability by placing over a candling box in a dark room and examining for the presence of blood vessels. A candling box consists of a small closed wooden box with an oval cut out on one side over which the egg can rest, a light-bulb is fitted within the box so that light is transmitted through the egg, thus making the blood vessels visible. A large vein is selected and marked about 1–2 cm from the air sac. The selected vein should be firmly attached to the membrane. This can be determined by rocking the egg on the candling box and ensuring that no movement of the vein occurs. A small window (≈5 mm square) is made in the shell, just over the marked vein, by gently scoring all four sides of the window using a portable dental drill fitted with a flexible shaft and a 2-cm-diameter diamond-edged cutting wheel (available from any dental supplier). The shell is lifted off using a suitable implement, e.g., the edge of a scalpel blade, taking care not to damage the underlying membrane.

Immediately prior to injection, a small drop of heavy grade liquid paraffin is placed on the membrane, thereby rendering it transparent and making the blood vessel visible. Test samples (100 μl) are injected into the vein using a 1-ml syringe and a 26-gauge (or smaller) needle. Injections should be made slowly in order to prevent separation of the vessel from the membrane and excessive bleeding when the needle is withdrawn. A total of 10 eggs is injected with the same dose. Finally, the windows are sealed with wax and the eggs are incubated at 37°.

Embryos are examined for viability 18–24 hr after injection; dead embryos appear to have lost all the blood vessels from the membrane. A small proportion of embryos may die as a result of excessive bleeding after injection; these can be identified by the presence of a large blood clot around the injection site.

Sample Preparation. TSST-1 and LPS are diluted in pyrogen-free saline using pyrogen-free glassware. Preparations of purified TSST-1 should be free of endotoxin activity; this may be checked by the *Limulus* lysate gelation assay.

Comments. If LPS from *Escherichia coli* (*E. coli* 0111 LPS; Mallinckrodt; St. Louis, MO) is used, the expected dose response of TSST-1 in the presence of 0.1 μg/ml LPS is shown in Table I. The dose–response curve is complex. At low TSST-1 concentrations (<0.1 μg/ml) lethality is directly proportional to dose but a plateau effect occurs at concentrations of 0.5–20.0 μg/ml. At higher concentrations (>20 μg/ml) toxicity decreases with increasing dose. In order to assay TSST-1 samples of unknown concentration, it is essential to establish a dose–response curve at concentrations less than 0.1 μg/ml and to dilute the test sample.

The LD_{50} of LPS alone is approximately 0.33 μg/ml. Table II[9] shows

TABLE I
DOSE RESPONSE OF TSST-1 IN CHICK EMBRYOS IN THE PRESENCE OF 0.1 μg/ml LPS

TSST-1 (μg/ml)	Embryos killed/embryos challenged
0.01	4/9
0.10	8/9
0.50	9/10
1.00	8/10
2.50	8/10
5.00	8/10
7.50	8/10
10.00	8/10
20.00	10/10
30.00	6/10
40.00	6/10
50.00	5/10
100.00	2/10
250.00	0/9

TABLE II
EFFECT ON CHICK EMBRYOS OF VARYING SUBLETHAL DOSES OF LPS WITH FIXED DOSES OF TSST-1[a]

TSST-1 (μg/ml)	Embryos killed/embryos challenged with indicated dose of LPS (μg/ml)				
	0	0.025	0.05	0.075	0.1
0	0/9	0/10	0/10	0/10	0/10
1	0/9	3/10	5/10	7/10	7/10
10	0/9	1/10	5/10	6/10	6/10
100	1/10	0/9	1/9	2/8	1/8

[a] Data from Ref. 9.

the effect on chick embryos of varying sublethal doses of LPS with fixed doses of TSST-1. It is clear from Table II that TSST-1 in the absence of LPS has no effect on 11-day-old embryos.

Assays of TSST-1 in rabbits are insensitive, expensive, and subject to variation since lethality is influenced by age[5] and strain[10] of rabbit. In

[10] J. C. S. de Azavedo and J. P. Arbuthnott, *Infect. Immun.* **46,** 314 (1984).

addition, levels of colonization by gram-negative bacteria also appear to influence the outcome in rabbits.[11,12] Assays based on the mitogenic effect[2,6] and the ability to induce release of interleukin-1 from monocytes,[7,8] although sensitive, are fairly complex to perform and require the use of tissue culture facilities. The chick embryo assay is easily executed, sensitive, and reproducible. However, it should be noted that the assay may not be suitable for crude culture supernatants from TSST-1^{+} strains since staphylococcal products other than TSST-1 may be lethal in chick embryos.

[11] P. M. Schlievert, *Infect. Immun.* **36,** 123 (1982).
[12] P. M. Schlievert, *J. Infect. Dis.* **147,** 391 (1983).

[49] Photoreactive Lipids for the Study of Membrane-Penetrating Toxins

By CESARE MONTECUCCO

Microbial protein toxins exert their action at the plasma membrane level or at intracellular targets. In most cases, at some stage of the intoxication process, they must insert into the lipid bilayer[1] and this membrane penetration step is a still poorly understood aspect of their mechanism of action. A gross conformational rearrangement of the toxin molecule may be necessary to account for a change of its solubility from the hydrophilic solvent–water to the hydrophobic solvent–hydrocarbon chains of membrane lipids. This structural transition may involve a limited number of the total population of toxin molecules. Moreover, the interaction of the toxin with lipids may be a transient one and last only for a short period of time.

Hydrophobic photolabeling with photoactivatable lipids has been recently introduced to study lipid–protein interactions.[2-4] The method is highly sensitive, technically simple, and does not require expensive equipment. It is based on the reagents shown in Fig. 1 characterized by

[1] S. Olsnes and K. Sandvig, *in* "Endocytosis" (I. Pastan and M. Willingham, eds.), p. 195. Plenum, New York, 1985.
[2] J. Brunner, *Trends Biochem. Sci.* **6,** 44 (1981).
[3] H. Bayley, "Photogenerated Reagents in Biochemistry and Molecular Biology." Elsevier, Amsterdam, 1983.
[4] R. Bisson and C. Montecucco, *in* "Progress in Lipid Protein Interaction" (A. Watts and J. J. H. H. M. de Pont, eds.), p. 259. Elsevier, Amsterdam, 1985.

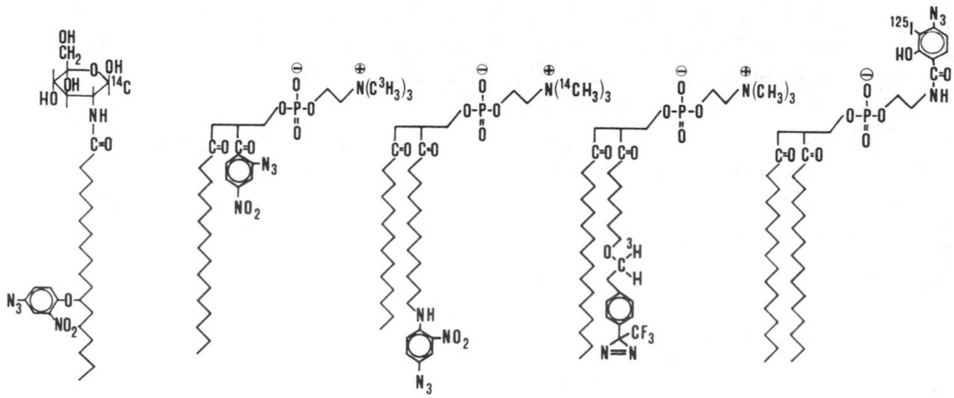

12APS-GlcN PC I PC II PC III ASA-PE

FIG. 1. Structural formulas of the photoactivatable lipid analogs that have been used in studying the lipid interaction of bacterial protein toxins.

the presence of the nitroarylazido or the trifluoromethylaryldiazirine photoreactive groups, which are stable in the dark where they can be manipulated in any of the usual biological operations. On illumination with long wave ultraviolet radiation, they are converted to a nitroarylnitrene (12APS-GlcN, PC I, and PC II) or to a trifluoromethylarylcarbene (PC III). These highly unstable electrophilic intermediates are able to react rather unspecifically with neighboring molecules to yield covalent derivatives. Since the reagents of Fig. 1 are radioactive and the photogenerated intermediate is membrane localized, the distribution of radioactivity among the different components of the system will provide information on their membrane localization.

It should be emphasized that hydrophobic photolabeling of the membrane-inserted toxin transforms a noncovalent lipid–protein interaction (even a short-lived one) into a long-lasting stable bond that can be later studied by analyzing the cross-linked and labeled species.

This method not only allows one to identify those toxin fragments interacting with the membrane but also, in favorable cases, to trace the lipid cross-linking at the amino acid level.

A second major advantage of the method is the high sensitivity, consequent to the use of radioactively labeled reagents, that will result in high signals even with minute amounts of toxin.

Hydrophobic photolabeling has been recently applied to the study of the lipid interaction of several microbial toxins.[5–19]

[5] B. J. Wisnieski and J. S. Bramhill, *Biochem. Biophys. Res. Commun.* **87,** 308 (1979).

Synthesis of Reagents and Properties

The preparations of 12-(4-azido-2-nitrophenoxy)stearoyl[1-^{14}C]glucosamine (12APS-GlcN),[20] 1-palmitoyl-2-(2-azido-4-nitrobenzyol)-*sn*-glycero-3-phospho[^{3}H]choline[21] (PC I), and 1-myristoyl-2-[12-(4-azido-2-nitrophenyl)amino]dodecanoyl-*sn*-glycero-3-phospho[^{14}C]choline[21] (PC II) are rather simple and offer the great advantage of introducing the radioactivity only at the last step of the synthetic scheme. Hence their synthesis can be attempted in a standard biochemical laboratory and has been already performed by different groups.[22–26] Also the preparation of dipalmitoyl(3-[^{125}I]-4-azido-salicylamido) phosphatidylethanolamine (ASA–PE) can be easily accomplished with the possibility of preparing a stock amount of nonradioactive precursor that can be made radioactive when needed; a further advantage of this reagent is the high speed and sensitivity of [^{125}I] detection.[27] The synthesis of 1-palmitoyl-2-[10-[4-[(trifluoromethyl)diazirinyl]phenyl]-8-oxa-9-[^{3}H]decanoyl-*sn*-glycero-3-phospho-

---

[6] B. J. Wisnieski and J. S. Bramhall, *Nature (London)* **289,** 319 (1981).
[7] M. Tomasi and C. Montecucco, *J. Biol. Chem.* **256,** 11177 (1981).
[8] M. Tomasi, G. D'Agnolo, and C. Montecucco, *Biochim. Biophys. Acta* **692,** 339 (1982).
[9] L. S. Zalman and B. J. Wisnieski, *Proc. Natl. Acad. Sci. U.S.A.* **81,** 3341 (1984).
[10] V. W. Hu and R. K. Holmes, *J. Biol. Chem.* **259,** 12226 (1984).
[11] C. Montecucco, G. Schiavo, and M. Tomasi, *Biochem. J.* **231,** 123 (1985).
[12] C. Montecucco, M. Tomasi, G. Schiavo, and R. Rappuoli, *FEBS Lett.* **194,** 301 (1986).
[13] C. Montecucco, G. Schiavo, J. Brunner, E. Duflot, P. Boquet, and M. Roa, *Biochemistry* **25,** 919 (1986).
[14] V. Escuyer, P. Boquet, C. Montecucco, and M. Mock, *J. Biol. Chem.* **261,** 10891 (1986).
[15] Z. T. Farahbaksh and B. J. Wisnieski, *J. Biol. Chem.* **261,** 11404 (1986).
[16] E. Papini, R. Colonna, G. Schiavo, F. Cusinato, M. Tomasi, R. Rappuoli, and C. Montecucco, *FEBS Lett.* **215,** 73 (1987).
[17] V. W. Hu and R. K. Holmes, *Biochim. Biophys. Acta* **902,** 24 (1987).
[18] E. Papini, G. Schiavo, M. Tomasi, M. Colombatti, R. Rappuoli, and C. Montecucco, *Eur. J. Biochem.* **169,** 636 (1987).
[19] C. Montecucco, G. Schiavo, Z. Gao, E. Bauerlein, P. Boquet, and B. R. DasGupta, *Biochem. J.* **251,** 379 (1988).
[20] J. S. Bramhall, M. A. Shiflett, and B. J. Wisnieski, *Biochem. J.* **177,** 765 (1979).
[21] R. Bisson and C. Montecucco, *Biochem. J.* **193,** 757 (1981).
[22] P. Moonen, H. P. Haagman, L. L. M. van Deenen, and K. W. A. Wirtz, *Eur. J. Biochem.* **99,** 439 (1979).
[23] F. G. P. Farley and C. I. Ragan, *FEBS Lett.* **127,** 45 (1981).
[24] J. G. Molotkovsky, Y. M. Manevich, E. L. Vodozova, A. G. Bukrinskaya, and L. D. Bergelson, *Biol. Membr.* **6,** 566 (1985).
[25] M. S. El Kebbaj, J. M. Berrez, T. Lakhlifi, C. Morpain, and N. Latruffe, *FEBS Lett.* **182,** 176 (1985).
[26] R. George, R. N. A. H. Lewis, and R. N. McElhaney, *Biochim. Biophys. Acta* **821,** 253 (1985).
[27] Z. Gao and E. Bauerlein, *FEBS Lett.* **223,** 366 (1987).

choline[28] (PC III) involves many steps and demands a specific chemical background. The generally low yield of coupling the photoreactive fatty acid to lysolecithin, involved in the preparations of PC I, PC II, and PC III, has been recently greatly improved.[29]

12APS-GlcN in its ^{14}C form has a low specific radioactivity. In general terms the lower the specific activity of a labeling reagent the higher is the amount needed to obtain a reasonable signal-to-noise ratio and concentrations of probe may be needed that will perturb the biological system under study. By using [^{3}H]glucosamine a glycolipid of much higher specific radioactivity can be obtained.[30] The photoreactive phosphatidylcholines PC I, PC II, and PC III have a high specific radioactivity and have never been used at a ratio above 1% of the total lipids. Recently a procedure has been introduced to prepare phosphatidylcholine analogs of specific radioactivity higher than 200 Ci/mmol.[29]

Although the carbenes and nitrenes generated by photoactivation of the reagents of Fig. 1 are highly reactive, they do show different reactivities versus the various lateral residues of proteins. From the little information available it can be inferred that the SH group is by far the most reactive one and the primary C–H bond the least reactive one.[4] Hence the distribution of radioactivity among the labeled polypeptides reflects more the nature of the residues exposed to lipids than the actual amount of protein surface inserted in the membrane. The extreme situation can be envisaged of a polypeptide chain in contact with lipids that is poorly or not labeled because it is composed of unreactive aliphatic amino acids. This is less likely with the more reactive PC III than with the other reagents of Fig. 1. In this respect it would be useful to have photoactivatable groups that are so reactive as to be completely unspecific; however, this would be of limited usefulness in studying the membrane interaction of bacterial toxins because in this system an enormously large excess of lipids with respect to the toxin is always present and nearly all the probe radioactivity would be bound to lipids rather than to proteins.

PC I and PC II bear their photoactive group at two different levels of one fatty acid chain to probe different regions of the lipid bilayer. ASA–PE is designed to probe lipid protein interactions occurring at the most external part of the lipid bilayer. Only the protein surface interacting with the polar head groups of phospholipids will be labeled by the tritiated PC I and by ASA–PE while protein regions intercalated into the deeper hydrophobic core of the membrane will be cross-linked to the ^{14}C-labeled probe PC II. Hence by following the $^{3}H/^{14}C$ labeling ratio it is possible to obtain

[28] J. Brunner, M. Spiess, R. Aggeler, P. Huber, and G. Semenza, *Biochemistry* **22,** 3812 (1983).

[29] C. Montecucco and G. Schiavo, *Biochem. J.*, **23,** 309 (1986).

[30] C. Montecucco, R. Bisson, C. Gache, and A. Johannsson, *FEBS Lett.* **128,** 17 (1981).

information on the varying depths of membrane penetration of different polypeptide chains or of different parts of the same protein molecule. This approach has already been very informative in various systems.[7,12,16,31,43]

Although at a superficial look 12APS-GlcN and PC I may resemble detergent molecules they do not behave as such and hence they do not label proteins from the water phase as shown in several different systems.[5,6,9,11,13,32–37,43] For 12APS-GlcN this is due to the long hydrocarbon chain and for PC I to the esterification of position two that makes it a phosphatidylcholine rather than a lysolecithin.

Preparation of Photoreactive Liposomes

All the following operations until the illumination step are performed in a red-lighted room. Amounts are those needed for 10 samples: 10^7 dpm of PC I and 2×10^6 dpm of PC II (taken from stock solutions in absolute ethanol, containing two drops of toluene/ml) are mixed with 15 mg of purified[38] asolectin (from a stock solution in chloroform/methanol, 1 : 1 by volume) in a glass vial (any thick-walled glass vial with a flat bottom and a plastic cap as those used for scintillation counting). In the same way any other lipid mixture can be used instead of asolectin. The solvent is removed with a gentle stream of nitrogen, diethyl ether is added, and the operation repeated. If possible the residual traces of organic solvent are removed under vacuum; otherwise the ether step is repeated at 30° and the vial is kept under the nitrogen flux for several minutes. Deacrated buffer (5.1 ml) is added, flushed with nitrogen, and tightly capped. The vial is placed in a water-sonicating bath (Laboratories Supplies, Hicksville, NY) and irradiated until optical clarity (at least 10 min). The liposomal suspension is divided among 10 different smaller vials in such a way that the 500-μl vol has a thickness of 1–2 mm.

Toxin-Liposome Interaction

The toxin is added and incubated at the desired temperature; protein amounts sufficient to give clear Coomassie blue bands are generally used

[31] J. Girdlestone, R. Bisson, and R. A. Capaldi, *Biochemistry* **20,** 152 (1981).
[32] R. Bisson, C. Montecucco, and R. A. Capaldi, *FEBS Lett.* **106,** 317 (1979).
[33] V. W. Hu and B. J. Wisnieski, *Proc. Natl. Acad. Sci. U.S.A.* **76,** 5460 (1979).
[34] C. Montecucco, R. Bisson, F. Dabbeni-Sala, A. Pitotti, and H. Gutweniger, *J. Biol. Chem.* **255,** 10040 (1980).
[35] L. J. Prochaska, R. Bisson, and R. A. Capaldi, *Biochemistry* **19,** 3174 (1980).
[36] R. Bisson, G. C. M. Steffens, and G. Buse, *J. Biol. Chem.* **257,** 6716 (1982).
[37] J. Hoppe, C. Montecucco, and P. Friedl, *J. Biol. Chem.* **258,** 2882 (1983).
[38] Y. Kagawa and E. Racker, *J. Biol. Chem.* **246,** 5477 (1971).

but the method is sensitive enough that meaningful results can be obtained with tens of nanograms.

Several different parameters can be studied by choosing the appropriate incubation conditions. For example the effect of pH can be tested by adding an appropriate amount of acidic buffer[39] or a dilute solution of acid[11,13]; other parameters such as ionic strength, temperature, and the presence of reducing agents can be tested as well. It should be recalled that membrane-soluble, SH-containing reducing agents cause a conversion of the photoreactive group to a nonphotoactivatable derivative.[40] Hence membrane-impermeant agents such as glutathione or cysteine, rather than the lipid-soluble dithiothreitol or 2-mercaptoethanol, should be used to test for the lipid interaction of reduced toxins with photoactive probes. Even water-dissolved reductant should be removed before illumination when a reagent such as PC I, with the reactive group near the water phase, is used.

Ultraviolet Illumination

Photoconversion of the lipid reagent can be performed with a variety of light sources under different conditions. This crucial step can be operatively monitored by determining the amount of radioactivity associated with the toxin as a function of the following different parameters. (1) Light spectrum: different ultraviolet light sources (mercury arc lamps, xenon lamps, mineralamps) have been used with or without filters or monochromators. Although the absorbance of both carbene and nitrene precursors is maximal below 300 nm these radiations must be avoided because of the damage they may induce on the toxin molecule. Short-wave radiations are removed by inserting between the light source and the sample a glass/water filter, which will also act as a heat filter. The photoconversion of the yellow-colored nitrene precursors 12APS-GlcN and PC II can be conveniently obtained also with the light source of a common slide projector. (2) Power of the source: a whole range of sources has been used from the 2000 W to the microwatts. The higher the power the shorter the time needed to accomplish the photolysis of the reagents. For most experiments a low-wattage lamp such as the UVSL 58 mineralight (Ultraviolet Products, San Gabriel, CA) can be used. (3) Irradiation time: this is highly dependent on the power of the light source, on the extinction coefficient of the photoreactive group at the wavelength used, on the transparency of

[39] P. Boquet and E. Duflot, *Proc. Natl. Acad. Sci. U.S.A.* **79,** 7514 (1982).

[40] J. V. Staros, H. Bayley, D. N. Standring, and J. R. Knowles, *Biochem. Biophys. Res. Commun.* **80,** 568 (1980).

the sample, on the presence of other light-absorbing species such as hemes, and on the geometry of the system. (4) Distance from the light source: the closer the sample is to the light source, the shorter will be the irradiation time. Also important is the thickness of the sample, which should be kept to a minimum.

A simple and inexpensive photolysis apparatus is shown in Fig. 2. The light source is held upside down and the samples contained in scintillation vials are immersed in a water bath, which serves both the function of temperature control and of filtering the protein-damaging radiations below 310 nm. With this set-up a few minutes are sufficient to completely photolyse PC I, PC II, and ASA–PE.

Whatever apparatus is used it is important to check that the conditions used have led to complete photolysis of the reagent and that there is no incorporation of radioactivity due to reactions occurring after the irradiation step. This is assessed by plotting the amount of radioactivity associated to the protein (determined as described below) as a function of varying the irradiation conditions.

Recovery of Protein and Separation from Unbound Lipids

The illuminated toxin–liposome mixture is transferred into conical plastic tubes (1.5 ml) together with 200-μl washings. Precipitation of the protein is obtained either by adding cold acetone to a final concentration of 80% and incubation at −20° for 30 min or by acid precipitation with TCA (final concentration 5–10%) and incubation at room temperature for 30 min. The samples are centrifuged on a swing-out table centrifuge for 10

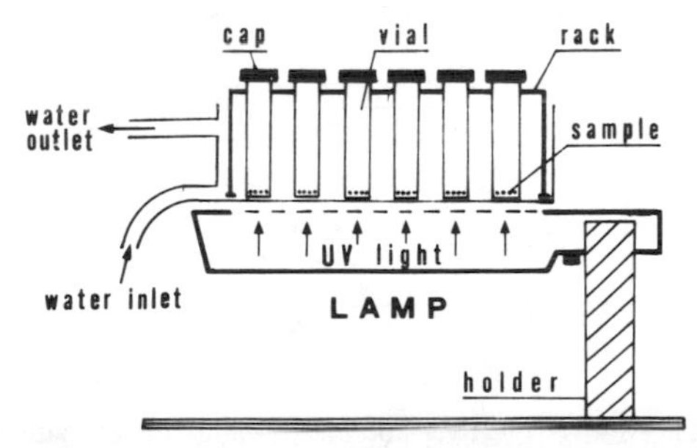

FIG. 2. A simple and effective ultraviolet irradiation apparatus. The lamp is a mineralight UVSL 58 mounted upside down on its holder. Radiations are directed upward versus the samples contained in glass scintillation vials. The sample volume is such that its thickness is smaller than 2 mm. Vials are kept vertical by a rack inside a large Pyrex glass beaker connected to a water-circulating thermostatted bath. During operation the apparatus is covered with aluminum foil. In another set-up, useful for large volumes, the sample is contained in a Petri dish and illuminated from the top.

min and the supernatant is carefully removed with a plastic-tip pipet. Pellets are solubilized with 20–30 μl SDS electrophoresis sample buffer containing Bromphenol blue following the conditions best suited for the particular system under study. Boiling for 2–3 min does not lead to an appreciable release of radioactivity from the protein.

SDS-polyacrylamide gel electrophoresis is a most powerful method to separate all noncovalently bound radioactive lipids from the labeled protein. Different electrophoretic systems can be used with similar effectiveness. At the end of the electrophoretic run the gel is fixed, stained with Coomassie blue or Stains-all, destained, and scanned following usual protocols. The acidic conditions used for fixation do not cause loss of radioactivity covalently bound to the protein.

Figure 3 shows the electrophoresis of diphtheria toxin after hydrophobic photolabeling in the presence of asolectin liposomes at low pH: all the

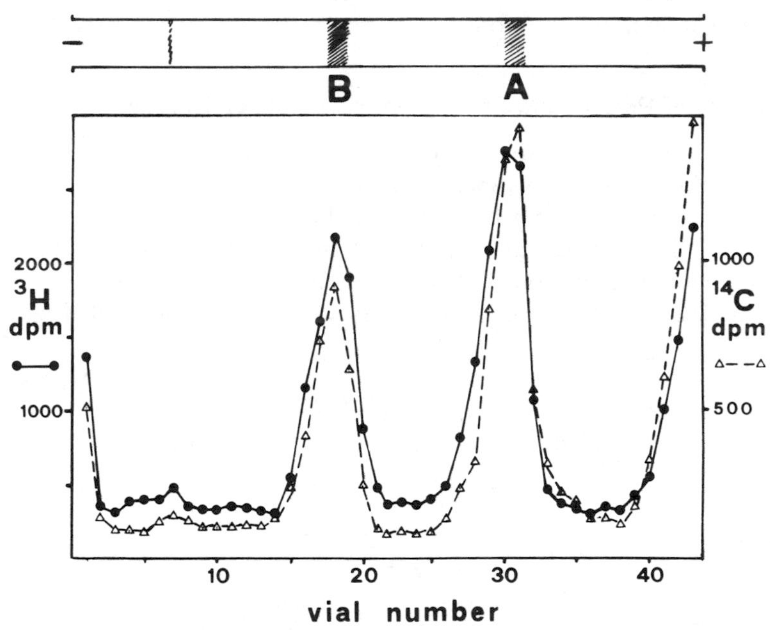

FIG. 3. The top panel shows the two Coomassie blue-stained bands of diphtheria toxin photolabeled at pH 4.5 in the presence of asolectin liposomes containing PC I and PC II, recovered by TCA precipitation, resuspended in an SDS buffer containing 3% 2-mercaptoethanol, and electrophoresed on a 12% Laemmli gel. The lower panel shows the profile of radioactivity associated to the gel as determined by slicing and counting. Both PC I (●) and PC II (△) label the two toxin fragments, indicating that at low pH both chains become able to insert into the lipid bilayer. This result suggests that the membrane-crossing form of the toxin involves both the enzymatic (A) and the receptor-binding (B) toxin fragments.

lipid bound radioactivity is effectively separated from the labeled protein bands and it is found at the gel front without trailing.

Detection of Protein Labeling

The amount of radioactivity associated with the protein can be determined in two different ways.

1. Gel slicing and counting: the gel is sliced in 1-mm-thick slices with a manual slicer (Hoefer Scientific, San Francisco, CA) and the slices distributed in different glass or plastic scintillation vials. If the resolution is high or there is only one protein band on the gel two or three slices can be put in the same vial. Tissue solubilizer (300–400 μl; Soluene 350, Packard) is added per vial and the tightly stoppered samples are incubated overnight. After addition of 3–4 ml of a scintillation fluid and a further incubation for 8 hr at room temperature, the vials are counted for radioactivity in a scintillation counter. A dpm output gives a higher and more reproducible result in view of some possible quenching effect of Coomassie blue and of impurities sometimes present in tissue solubilizers. This method is very laborious but provides a quantitative result and it is the only one that can be used in a double-labeling experiment such as that employing both PC I and PC II at the same time, shown in Fig. 3.
2. Fluorography: when one isotope only is present it may be advantageous to use the much simpler fluorographic method.[41] It provides a higher resolution with respect to gel slicing and counting and hence should be preferred after polypeptide chains migrate close to each other. This method can be rendered semiquantitative by preflashing the film. Moreover when the fluorographic image is obtained the gel can still be cut in slices and the whetted slices can be treated as above for a quantitative determination of bound radioactivity.

Since the labeled protein represents a minority of the total molecules present on the gel (see below) and since the radioactively labeled protein has an increased molecular weight because of the covalent addition of one probe molecule, there may be a shift of the radioactive protein peak with respect to the stained protein band; in SDS-polyacrylamide gels this shift is practically evident only for polypeptides of molecular weight lower than 15,000. However, it constitutes a major problem in HPLC.[42]

[41] R. A. Laskey, "Radioisotope Detection by Fluorography and Intensifying Screens," Rev. 23. Amersham International, Amersham, England, 1984.

[42] J. Brunner, A. J. Franzusoff, B. Luscher, C. Zugliani, and G. Semenza, *Biochemistry* **24,** 5422 (1985).

Yield of Protein Labeling

The amount of radioactivity covalently bound to the protein after photoactivation is low with any integral membrane protein as well as with all toxins so far tested.[4] It is due to the competing reaction pathways that can be followed by the photogenerated intermediate. (1) Protein cross-linking: it depends on the reactivity of the lateral residues exposed to lipids and on the concentration of the protein in the membrane; the higher these two factors the higher is the yield of protein labeling. (2) Lipid cross-linking: it increases with the reactivity of the reagent, with the degree of lipid unsaturation (because double bonds are more reactive than single bonds versus nitrenes and carbenes), and with increasing lipid to protein ratios. (3) Self-quenching: it results from the reaction of the photogenerated intermediate with the still unactivated precursor and it is a function of the amount of probe with respect to total lipids; this is an additional reason for the need of reagents of the highest available specific radioactivity. (4) Decay of the photogenerated intermediate to unreactive species.

For the above-mentioned reasons the yield of protein labeling in addition to being low is also unpredictable. It was around 0.5% for PC I and PC II and 0.008% for PC III with tetanus toxin at pH 4.5 under the conditions described above; the difference between the nitrene- (PC I and PC II) and carbene (PC III)-generating reagents is due to their different reactivities, leading to high lipid labeling by the more reactive PC III; as a consequence a much higher amount of radioactive PC III must be used in studying the lipid insertion of toxins because of the very large lipid-to-toxin ratio always present.

As a consequence of its low and variable yield of labeling, this method cannot provide an estimate of the percentage of toxin molecules actually involved in the process of membrane insertion. However, the incorporation of radioactivity following hydrophobic photolabeling cannot be considered an artifactual result related to that minority of toxin molecules that accidentally come into contact with the lipid reagent since low yields are also commonly found with integral proteins, which are known to possess a hydrophobic domain with a surface in stable and permanent contact with lipids.

Toxin Labeling in Intact Biological Membranes

All the reagents of Fig. 1 have been shown to partition and disperse with high efficiency in the lipid bilayer of cell organelles, viruses, and cells.[5,6,9,19,27,28,43] 12APS-GlcN can be added as an ethanolic solution. Al-

[43] J. Giraudat, C. Montecucco, R. Bisson, and J. P. Changeux, *Biochemistry* **24,** 3121 (1985).

ternatively, it can be handled as the phosphatidylcholine reagents. These probes are dried on the bottom of a glass vial, vacuum pumped, and the appropriate buffer is added. After sonication as above for 10 or more minutes, the suspension is added to the biological material and incubated at the appropriate temperature. The time course of incorporation can be followed by centrifuging small aliquots on a sucrose gradient and counting the pellets.

The high partition efficiency of photoactive phosphatidylcholine analogs with respect to the unmodified phospholipid is very likely related to their higher monomer concentration.

As compared to the liposomal model system, the additional presence of integral membrane proteins poses some problems. In addition to the toxin the integral proteins of the cell membrane will be labeled and hence much more radioactivity must be used. Moreover, some labeled membrane proteins may overlap the toxin bands. This possibility may be difficult to eliminate since certain proteins do not stain and yet are labeled. If highly specific antisera are available the toxin may be recovered after labeling by immunoprecipitation and hence freed of most if not all contaminating labeled proteins.

Acknowledgments

I thank Dr. R. Bisson for several useful discussions.

[50] Electron Microscopy: Assays Involving Negative Staining

By JØRGEN TRANUM-JENSEN

Introduction

Electron microscopes can visualize the quaternary structure of proteins and the spatial organization of supramolecular assemblies in projected two-dimensional views. The prime virtue of the techniques of electron microscopy is that they permit imaging of *individual* molecules and assemblies at a dimensional scale which is otherwise accessible only with difficulty by X-ray and neutron diffraction analysis on crystallized material.

Though a modern electron microscope is capable of imaging objects to a resolution of 0.1–0.2 nm, it is seldomly possible to obtain images of biological structures with a reliable preservation of details below a 2- to 2.5-nm resolution limit. This means that the domain of biological electron microscopy seldomly reaches structures with a molecular mass below 100,000 Da.

Several of the bacterial toxins that damage cell membranes through formation of transmembrane protein channels form oligomeric structures of a size that can be resolved and studied in the electron microscope. This has been shown for several members of the large group of thiol-activated, oxygen-labile toxins that attack cholesterol-containing membranes, e.g., cereolysin,[1] perfringolysin,[2] tetanolysin,[3] listeriolysin,[4] and streptolysin O,[5–7] the last being the most extensively studied in this class of toxins. All these toxins form various-sized oligomeric structures producing large membrane lesions, approaching 30 nm. Staphylococcal α-toxin is another pore-forming protein which generates relatively small, 2- to 3-nm pores, formed by uniform, hexameric assemblies of the toxin.[8,9]

Electron microscopy has proved rewarding in the study of the lesions produced by these toxins, as well as those induced by complement[10,11] and the fungal polyenic cytolysin, filipin.[12,13] The electron microscope can detect the presence of a particular type of structural lesion and its frequently inhomogeneous distribution within a target cell population,[14] providing important structural information to complement biochemical and functional data.

Electron microscopy is generally also very useful as an assay for the purity and homogeneity of preparations of oligomeric proteins because it permits one to distinguish between different assembly forms and nonspecific aggregates that all may be present within the same preparation, and which may be difficult to discriminate by other means.

[1] J. L. Cowell, K.-S. Kim, and A. W. Bernheimer, *Biochim. Biophys. Acta* **507,** 230 (1978).
[2] C. J. Smyth, J. H. Freer, and J. P. Arbuthnott, *Biochim. Biophys. Acta* **382,** 479 (1975).
[3] S. Rottem, R. M. Cole, W. H. Habig, M. F. Barile, and M. C. Hardegree, *J. Bacteriol.* **152,** 888 (1982).
[4] J. Parrisius, S. Bhakdi, M. Roth, J. Tranum-Jensen, W. Goebel, and H. P. R. Seeliger, *Infect. Immun.* **51,** 314 (1986).
[5] J. L. Duncan and R. Schlegel, *J. Cell Biol.* **67,** 160 (1975).
[6] J. E. Alouf, *Pharmacol. Ther.* **11,** 661 (1980).
[7] S. Bhakdi, J. Tranum-Jensen, and A. Sziegoleit, *Infect. Immun.* **47,** 52 (1985).
[8] J. H. Freer, J. P. Arbuthnott, and B. Billcliffe, *J. Gen. Microbiol.* **75,** 321 (1973).
[9] R. Füssle, S. Bhakdi, A. Sziegoleit, J. Tranum-Jensen, T. Kranz, and H.-J. Wellensiek, *J. Cell Biol.* **91,** 83 (1981).
[10] J. Tranum-Jensen, S. Bhakdi, B. Bhakdi-Lehnen, O. J. Bjerrum, and V. Speth, *Scand. J. Immunol.* **7,** 45 (1978).
[11] S. Bhakdi and J. Tranum-Jensen, *Biochim, Biophys. Acta* **737,** 343 (1983).
[12] O. Behnke, J. Tranum-Jensen, and B. van Deurs, *Eur. J. Cell Biol.* **35,** 189 (1984).
[13] O. Behnke, J. Tranum-Jensen, and B. van Deurs, *Eur. J. Cell Biol.* **35,** 200 (1984).
[14] S. Bhakdi and J. Tranum-Jensen, *J. Immunol.* **133,** 1453 (1984).

Rationale of Negative Staining

There are three major and general problems in visualizing biological structures by electron microscopy. First, the generally low atomic number of constituent atoms (C, N, O, H); second, the susceptibility of the structures to radiation damage; and third, the problems of maintaining the structures in high vacuum.

The low atomic numbers (*Z*) encountered in biological molecules are actually quite unfavorable for imaging by transmission electron microscopy because they give little electron scatter, and because the scatter they give is dominated by inelastic collision events between the incident electrons and the specimen. Only elastically scattered electrons that have their energies preserved (but are changed in direction and phase) contribute to formation of a sharp (focused) image. The inelastically scattered electrons that lost energy in the specimen (typically some 25 eV) cause image blurring because they are focused differently (more deflected) by the electron lenses. At the same time the inelastic collisions are the ones that cause profound chemical alterations, mass loss, and structural changes in the specimen, significant for high-resolution details already at specimen irradiation doses of 50 e/nm^2 of specimen. For comparison, it takes a minimum of some 3000 e/nm^2 of specimen to record an image with a signal-to-noise ratio that permits resolution of 2-nm details. At this dose, about 20% of the mass of a protein can be assumed lost.[15,16]

The third problem mentioned above relates to the structural changes that accompany the loss of structural water which will take place in the microscope vacuum, if not before, during the preparation.

The rationale of negative staining is to circumvent the above-mentioned problems through embedment of the biological structures in a substance, a negative stain, that gives a high level of elastic electron scatter and which produces a sufficiently rigid and electron irradiation-stable mold to reliably preserve the surface contours, crevices, and projections of the biological structures during and after their cremation in the electron beam. Evidently, the negative stains themselves should not induce structural changes during the preparation of the specimen. Their distribution should be ruled solely by the space accessible around the molecular contours and should not accumulate specifically on particular structural details (positive staining). Finally, the stain should solidify into a stable mold at a time where the biological molecules still retain their structural water.

Section I gives a number of preparatory procedures for negative staining which we have found useful for ultrastructural studies on pore-form-

[15] K. Stenn and G. F. Bahr, *J. Ultrastruct. Res.* **31,** 526 (1970).
[16] R. E. Thach and S. S. Thach, *Biophys. J.* **11,** 204 (1971).

ing proteins in native target membranes, in liposomal membranes, and in detergent-solubilized form. Section II deals with the preparation of specimen supports and their adsorptive properties. Section III describes the practical handling of specimen application and staining. Section IV gives some physical and chemical properties of the commonly used negative stains, necessary for a critical appraisal of results, and Section V deals with some practical aspects of microscope operation.

Emphasis is placed on the rationales, the prospects, and the cautions of the techniques, the interpretation of results, and recognized artifacts.

I. Preparing Material for Negative Staining

A. Preparing Native Target Membranes for Negative Staining

Erythrocyte ghosts and membrane isolates of other cells can be studied in negative staining without pretreatments. The results are generally unsatisfactory, however, because during the process of negative staining most membranes become grossly distorted by the air–water interface that settles on the grid when the stain dries. Further, ghosts and vesicles are seen as two overlying layers of membrane, seriously reducing the resolution and impeding the image interpretation. These problems may be overcome by fixation of the membranes in osmium tetroxide, which makes the membranes brittle, permitting fragmentation into rigid flakes. The procedure used is simple, as follows.

1. A small volume (e.g., 25 μl) of sedimented membranes is suspended for 15–30 min at room temperature in 10 vol of 0.25–0.5% (10–20 m*M*) OsO_4 dissolved in a phosphate-buffered medium, e.g., 25 m*M* phosphate, at pH 7–7.5. The medium of the membrane preparation should be free of protein solutes and should not contain Tris or other compounds that react with OsO_4.

2. The membranes thus fixed are washed twice in 5–10 m*M* phosphate buffer and resuspended to 5–10 times the original volume. A small aliquot is kept, and the rest is fragmented by shear forces. This is most easily accomplished by forcefully passing the sample 5–10 times through a thin hypodermic needle. A 100- to 250-μl Hamilton syringe (series 700) with a 50-mm-long needle of 110- to 150-μm i.d. is excellent for the purpose. The size of the resulting fragments can be quite accurately chosen by the number of passages and the force applied to the syringe, the proper handling of a given preparation being learned by experience. The unfragmented and the fragmented membranes are then negatively stained as detailed in Section III. The membrane flakes are used for study of the

structures projected "face on" onto the membrane (Figs. 1A and 2A); the unfragmented aliquots serve for study of the structures projected "in profile" along the sharply bent edges of the membrane (Fig. 1B).

Osmium tetroxide reacts with and cross-links unsaturated lipids, proteins, and many polysaccharides. The reactions are very complex and are incompletely known; for recent reviews, see Refs. 17 and 18. It should be kept in mind that OsO_4 may also cleave peptides,[19] and it may cause disassembly of some polymeric protein structures like actin.[20] We have not, within the obtainable limits of resolution, been able to detect structural changes induced by OsO_4 in the pore-forming toxins studied to date. Caution should be exerted on this point in other systems, however.

Glutaraldehyde fixation is considered an excellent stabilizer of protein quaternary structure,[17] documented for some crystalline proteins by no or minimal change of crystallographic performance.[21] It is strongly recommended that glutaraldehyde-fixed membranes be examined in negative staining although such membranes do not break into flakes by shear; rather they go to tatters. Glutaraldehyde is incompatible with Tris and other amine-containing buffers. Free proteins should not be present as these may become attached to the membranes.

B. Proteolytic Stripping of Native Target Membranes

The membrane-inserted oligomeric structures of streptolysin O (SLO) and *Staphylococcus aureus* α-toxin as well as the C5b-9 membrane complex of complement have proved remarkably resistant to proteolytic degradation[7,10] allowing extensive removal of other externally exposed proteins, and leaving the toxin oligomers ultrastructurally unchanged (Fig. 1B). Such stripping is particularly useful for visualizing lesions in profile along the bent edges of membranes in negative staining because other proteins that may obscure the surface presentation of the lesions are removed. For such stripping we have mostly used trypsin (Sigma type III) and α-chymotrypsin (Sigma type II) (Sigma Chemical Co., St. Louis, MO) in combination at individual enzyme concentrations of 0.1–0.2 mg/ml in phosphate-buffered saline, pH 7.8, for 30 min at room temperature, followed by two washes in the buffer. The period may be extended to several hours with no microscopically detected alteration of the toxin structures,

[17] M. A. Hayat, "Fixation for Electron Microscopy." Academic Press, New York, 1981.

[18] E. J. Behrman, *In* "The Science of Biological Specimen Preparation," p. 1. AMF O'Hare, Chicago, 1984.

[19] M. Emerman and E. J. Behrman, *J. Histochem. Cytochem.* **30,** 395 (1982).

[20] P. Maupin-Szamier and T. D. Pollard, *J. Cell Biol.* **77,** 837 (1978).

[21] F. A. Quiocho and F. M. Richards, *Biochemistry* **52,** 833 (1964).

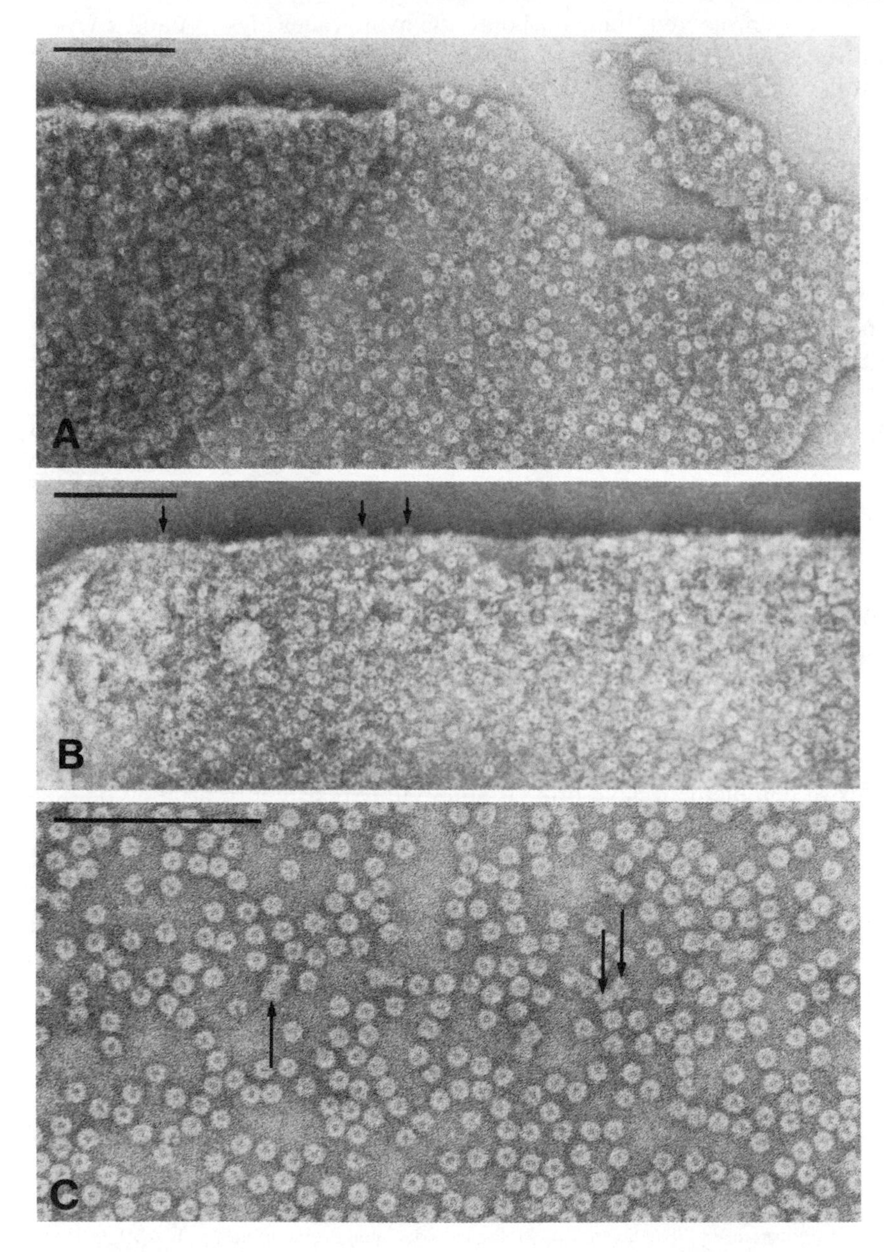

FIG. 1. (A) Rabbit erythrocyte membrane, lysed with *S. aureus* α-toxin, fixed in OsO_4, fragmented by hydrodynamic shear, and negatively stained. Two overlying layers of membrane are seen to the left; a single layer with correspondingly improved resolution of the annular lesions, formed by hexamers of the toxin, is seen to the right. (B) Rabbit erythrocyte

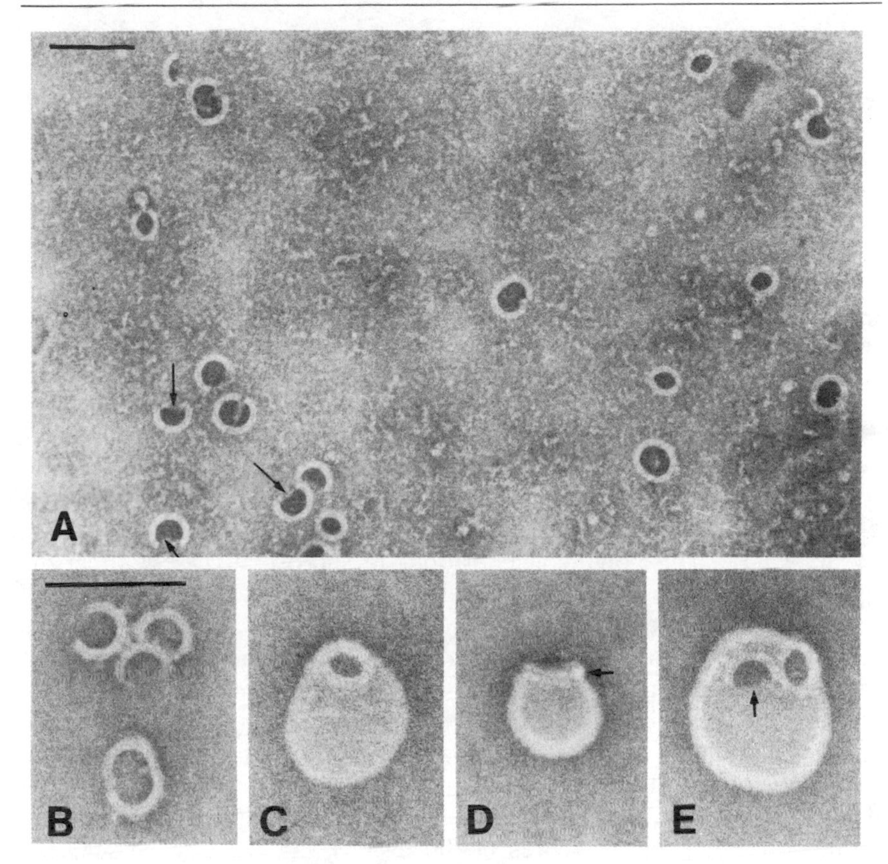

FIG. 2. (A) Fragment of negatively stained human erythrocyte membrane, lysed with streptolysin O. The arc- and ring-shaped oligomers of the toxin surround large membrane channels in which the negative stain can accumulate. Part of the circumference of the lesions may be formed by a free edge of membrane (arrows). (B) Oligomers of SLO, isolated from a primary target membrane. (C–E) SLO oligomers, reincorporated into lecithin liposomes. The light rim at the margin of the liposomes represent the wall of the liposomes that have taken up the stain. Sodium silicotungstate used for negative staining. Scale bars indicate 100 nm.

---

lysed with α-toxin and proteolytically stripped with trypsin + α-chymotrypsin, followed by OsO_4 fixation and negative staining. The α-toxin hexamers are seen as small cylindrical stubs along the sharply bent edge of the ghost. (C) Preparation of negatively stained, purified α-toxin hexamers. The majority of the cylindrical hexamers are preferentially oriented on the grid to be seen in axial projection, exhibiting their 2.5- to 3-nm central pit/channel. A few profile views are marked by arrows. Sodium silicotungstate used for negative staining. Scale bars indicate 100 nm.

but the membranes become very fragile and washings and resuspensions should be performed very gently. Given the same period of proteolytic exposure, SLO-lysed erythrocytes become more degraded than α-toxin-lysed cells, as judged from the electron microscopic appearance of the membranes. Also, lateral aggregation within the membrane becomes very marked for the former lesions compared to α-toxin lesions. These effects are probably due to facilitated entrance of enzymes via the large SLO pores, so the membranes are also digested from the inside. Subsequent OsO_4 fixation is of particular benefit for preventing gross distorsions of digested membranes during negative staining.

C. Preparing Liposomes for Negative Staining

Small (50–200 nm) lecithin liposomes treated with toxin or carrying reincorporated toxin oligomers may be negatively stained without pretreatment. If the liposome preparation has been fractionated on a sucrose gradient, the sucrose concentration should be lowered to below 1%, e.g., by dialysis, because high concentrations of sucrose adversely affect the quality of negative stainings. Generally, an appropriate dilution of a liposome preparation for negative staining corresponds to 2–5 mg of lipid/ml.

Negatively stained liposomes carrying SLO or *S. aureus* α-toxin oligomers as well as complement complexes exhibit a distinctive feature compared to plain, lesion-free liposomes. The former take up the negative stain to exhibit a central stain deposit, whereas the latter are devoid of stain[7,9,22] (Fig. 2C–E). This difference is very consistent and reproducible if the stain layer on the grid is thin. It is interpreted as the result of the entrance of stain via the pores of the lesions and can also be seen for relatively large stain molecules such as silicotungstate (M_r ~3000). If the layer of stain is thick, plain liposomes also tend to exhibit a central stain deposit. This observation is interpreted as an expression of how the liposomes flatten during drying of the stain solution. In thick stain layers the liposome vesicles acquire cup forms with a stain deposit in the concavity, being supported on the convex side by the surrounding stain. The formation of cup forms of intact vesicles is explained by the osmotic drag of water volume from their interior due to the increasing osmolality of the drying stain solution.

When liposomes carrying lesions are thinly negatively stained and contain a central stain deposit, the light rim, representing the bent edge of a vesicle, flattened during drying of the stain, is very often of uneven width, narrow at one pole and smoothly broadening toward the opposite

[22] S. Bhakdi and J. Tranum-Jensen, *Proc. Natl. Acad. Sci. U.S.A.* **75,** 5655 (1978).

pole (Fig. 2C–E). The width of the narrow part corresponds with the expected thickness of the bilayer. The phenomenon is interpreted as the effect of flattening a sphere which forms a sharp edge only at one site while the rest of the edge is rolled over itself toward the opposite pole, but an accompanying reorganization of the lipid structure cannot be ruled out. It is a consistent finding that the membrane-incorporated proteins, which can be assumed to move rather freely in a lateral direction, tend to crowd along the edge at the pole where the lipid rim is thinnest, are only rarely present over the flattened surface of the liposome vesicles, and are almost never present along the thicker parts of the lipid rim. The crowding of protein at the edge is probably due to its escape from the fluid–air interface of the drying stain, as well as the adsorption of the protein to the supporting surface.

D. Preparing Detergent-Solubilized Toxin Oligomers (and Other Proteins) for Negative Staining

Detergent-solubilized preparations of isolated toxin oligomers are usually negatively stained without pretreatment. Concentrations of 0.1–0.3% of sodium deoxycholate and Triton X-100 in the solubilized preparations do not disturb the staining. High concentrations of sucrose should be lowered to below 1%. The appropriate dilution of protein for negative staining must be determined, starting at, e.g., 0.5 mg/ml, and will generally be found between 0.02 and 0.2 mg/ml.

For studies of proteins whose ultrastructure has not been established, it is advisable to fix some specimens in glutaraldehyde prior to negative staining, because structural changes induced by the stain are always a possibility, though generally not a problem. Fixation can be achieved by addition of 1% glutaraldehyde to a dilute suspension, or after adsorption of the protein to the grid by application of a drop of 1% glutaraldehyde for 5–10 min.

II. Preparation of Specimen Supports for Negative Staining

For negative staining of native membranes and membrane fragments we prefer thin Formvar (polyvinyl formal) films, carried on 400-mesh copper grids and reinforced by 3–6 nm of evaporated carbon. We prepare Formvar films on a glass substrate by the technique of Revell and Agar as given by Bradley,[23] who describes several other techniques for thin film preparation. It is important that the Formvar films are prepared very thin,

[23] D. E. Bradley, *in* "Techniques for Electron Microscopy" (D. Kay, ed.), p. 58. Blackwell, London, 1965.

free of interference colors, to exhibit only a grayish, not shining, tone in reflected light when floated on water. The necessary strength of the film is adjusted by the thickness of carbon evaporated on the film after mounting on grids. Good quality, reproducible carbon evaporation is most easily achieved by evaporation of spun carbon wire (obtainable from Balzers Union) in a vacuum greater than 10^{-4} torr.

The electron optical imaging conditions are substantially improved by the use of thin, 3- to 8-nm pure carbon films as supports. These films are exceptionally stable in the electron beam, but mechanically fragile and must be carried on a second film support with holes, so that only those areas of the carbon film that are suspended over the holes are used. These films are of limited use for examination of larger objects or if many copies of membrane fragments are to be examined on each grid, but for the study of proteins and other small particles in solution their virtues are worth utilizing. We prepare such carbon supports by the classical technique of Brenner and Horne.[24] The carbon films are deposited on freshly cleaved mica and floated on a water surface. Grids with holey carbon-reinforced Formvar films are placed on top of the floating carbon film and picked up. Details on the preparation of holey films can be found in Ref. 23.

Surfaces of evaporated carbon are generally hydrophobic and slightly negatively charged, and they give poor adhesion to most specimens. Further, the surface repels the negative stain, and when the fluid layer is thinned during drying of the stain, it breaks up to collect in microdroplets that pile up on some of the adsorbed structures and leave others all too thinly stained. Further, the interfaces of the retracting fluid droplets that sweep over the grid may cause massive distortion of the adsorbed structures. These problems are overcome either by addition of suitable wetting agents to the specimen suspension[25] or by preconditioning of the carbon surface. Preconditioning is achieved by exposure of the grid to agents like bacitracin[25] and polylysine,[26] or by purely physical means. To avoid the addition of foreign substances to the specimen, the grids are exposed to ion bombardment, so-called glow discharge, effected by placing the grids in a high-voltage (3 kV) discharge in a gas, usually atmospheric air, at about 10^{-2} torr for 20–40 sec.[27,28] Commercial attachments for this procedure are available for most vacuum evaporation apparatuses. We use an Edwards unit (Edwards High Vacuum, Crawley, Sussex, England). The

[24] S. Brenner and R. W. Horne, *Biochim. Biophys. Acta* **34,** 103 (1959).

[25] D. W. Gregory and B. J. S. Pirie, *J. Microsc.* **99,** 251 (1973).

[26] R. C. Williams, *Proc. Natl. Acad. Sci. U.S.A.* **74,** 2311 (1977).

[27] M. Reissig and S. A. Orrell, *J. Ultrastruct. Res.* **32,** 107 (1970).

[28] J. Dubochet, M. Groom, and S. Mueller-Neuteboom, *Adv. Optical Electron Microsc.* **8,** 107 (1982).

carbon surface that issues from glow discharge in atmospheric air is negatively charged and very hydrophilic, giving good adherence to most specimens and an excellent, uniform spread of the negative stain that dries down in a continuous layer. It is also possible to control the surface charge and hydrophilicity of the carbon film by various additions to the glow-discharge atmosphere.[28] The carbon surface appears to be best suited for negative staining 15–20 min after a 40-sec glow-discharge treatment at which time the thickness of the negative stain fluid film, left to dry on the grid, can be quite precisely chosen. Immediately after glow discharge it is more difficult to obtain very thin fluid layers by suction with absorptive paper. The high charge density of the carbon surface right after glow discharge may also increase the risk of electrostatic deformation of the specimen. The even hydrophilicity of the surface decays after a few hours as judged from the appearance of progressively more irregular spread of the stain.

The process of adsorption of proteins and membranes on the grid surface involves complex electrostatic interactions.[29] One should be aware that the surface properties of the carbon support may influence the gross conformation of some structures, notably DNA,[30] and though it does not appear to be a common problem, the possibility of molecular distorsions induced by the adsorption process should be borne in mind. Such phenomena of distorsion are likely to occur especially at very low or no salt concentrations because the electrostatic interaction between the protein molecules and the fixed surface charges are thereby increased.

It is common experience that the amount (surface density) of protein molecules, lipid vesicles, or membrane fragments adsorbed to a carbon support is determined by the conditioning of the surface and the concentration of the specimen material as well as the concentration of salt in the suspension applied. The time of contact is less important. This allows for easy adjustment of the area density of specimen by dilution. Poor adhesion of negatively charged particles to carbon surfaces, glow discharged in air, i.e., negatively charged, may sometimes be overcome quite easily by increasing the NaCl concentration in the suspension medium to 50–100 m*M*.[29]

It is frequently observed that protein molecules show up in negative staining with a preferential orientation, due to their geometry and/or the nature of the adsorptive surface relative to properties of the protein, such as polarity of charge. Such effects can sometimes be very marked as

[29] C. L. F. Woodcock, L.-L. Y. Frado, G. R. Green, and L. Einck, *J. Microsc.* **121,** 211 (1981).

[30] J. M. Sogo, R. Portmann, P. Kaufmann, and T. Koller, *J. Microsc.* **104,** 187 (1975).

with *S. aureus* α-toxin adsorbed to carbon films glow discharged in air (Fig. 1C).

III. Staining Procedure

The standard procedure for preparation of a negatively stained specimen is very simple. A drop of the specimen suspension is applied with a pipet to the appropriately conditioned grid which is held by a fine watchmaker's tweezer attached to a fixed stand. The drop is left on the grid for 10–15 sec, but, as mentioned, the time of contact is generally not critical for the amount of material being adsorbed. The drop is then touched with the edge of a filter paper and 10–15 drops of the negative stain solution is flushed over the grid with the filter paper serving as a drain. The last drop is sucked off to leave a thin fluid film that is allowed to dry. If the room atmosphere is humid a gentle stream of air should be used to speed drying because slow drying favors the formation of disturbing crystalline structures in the stain. This is particularly advisable for silicotungstate stains. A thin Teflon coating on the branches of the tweezer is helpful in preventing accumulation of fluid due to the capillary effect between the branches, which may be quite disturbing, because this fluid flows back on the grid when the final thin fluid film is formed.

The important point in the staining procedure is that the thickness of the final dry layer of negative stain be correct relative to the objects examined. For small objects that have a height of less than some 15 nm above the supporting surface it is possible to obtain full embedding in the stain layer without compromising the electron optical imaging by a too high mass thickness of the stain. In this case the stain solidifies before the air–fluid interface reaches the object. Larger objects are incompatible with a useful thickness of the stain, and effects of distorsion (flattening) caused by the interface must be considered in the image interpretation.

If the specimens are thinly shadowed with platinum, evaporated in a vacuum at a shallow angle (e.g., 20°) subsequent to the negative staining, platinum will accumulate on structures raised above the surrounding level of the stain. Much useful information on the degree of specimen embedment and distorsions can be gathered, although, evidently, the shadowed specimen itself is not useful for high-resolution observations.

Whenever quantitative analysis of negatively stained preparations of inhomogeneous material is attempted, it is important to rule out the possibility that different populations of structures in the suspension are differentially adsorbed. The safest way to eliminate such effects is to apply microdroplets of the suspension appropriately diluted with negative stain and leave the droplets to dry on the grid. This can be achieved by spraying[24] or more easily by application from a finely graded Hamilton syringe, utilizing the excellent hydrophilicity of glow-discharged grids that allows formation of a uniform fluid layer spontaneously or aided by a gentle sweep with the side of the needle over the grid. For quantitative sampling of microdroplets, the suspensions should be made very dilute to avoid overlay and clumping of the material. A very reliable, but technically difficult method for quantitative sampling of the particulate contents in a suspension is the so-called agar filtration method.[31] The principle of this method is to filter a small volume of the suspension—to which a known quantity of latex spheres has been added—through a specially prepared collodion film layered on a partially dehydrated agar plate. The pore size of the film is not precisely known and fairly large proteins (IgG) may slip through. Also, collodion films are not suited for high-resolution microscopy because of drift. A prime virtue of the technique is that it permits quite accurate absolute counts of particles in the suspension to be made which are very difficult to achieve by other means.

IV. Negative Stains

A large number of substances, mostly heavy metal salts, have been tried as negative stains and relatively few have been found useful. None of these is ideal, and no single stain is universally applicable. There are good reasons to continue the search for better stains. For a critical appraisal of negative staining electron micrographs it is important to consider some chemical and physical properties of the stain used.

Uranyl (UO_2^{2+}) salts of the organic anions acetate and formate are widely used, the acetate probably being the most commonly used negative stain. Uranyl formate was introduced because its mass density is somewhat higher than the acetate.[32] Uranyl sulfate has also been advocated mostly because of its higher radiation stability.[33] The uranyl salts used are 0.5–2% (12–15 m*M*) solutions in water. The pH of these solutions is low, about 4.4 for the acetate, 4.1 for the formate, and 2.9 for the sulfate. During drying of the specimen at room temperature, the pH decreases as the solutions become more saturated, reaching a pH of 4.1 at about 250 m*M* acetate, pH 4.0 at about 200 m*M* formate, and pH 1.1 at about 500 m*M* sulfate. The pH values of acetate, formate, and sulfate may be adjusted to about pH 5.0 with ammonia. At higher pH the solutions become

[31] E. Kellenberger and D. Bitterli, *Microsc. Acta* **78,** 131 (1976).

[32] R. Leberman, *J. Mol. Biol.* **13,** 606 (1965).

[33] L. F. Estis, R. H. Haschemeyer, and J. S. Wall, *J. Microsc.* **124,** 313 (1981).

unstable and precipitate as polyuranates. Addition of oxalate permits titration to pH 6.5 without formation of precipitates on a short time scale (hours).[34] A similar pH tolerance may be obtained through addition of EDTA.[35] All the uranyl salts precipitate with low (<1 m*M*) concentrations of phosphate, which accordingly should be washed off the grid with saline before addition of uranyl salts. It is possible, however, that traces of phosphate retained in the specimen may improve staining and specimen preservation in some cases.[36]

The uranyl ion is a very competent complex former, and solutions of the uranyl salts with organic anions will contain a mixture of several cationic, neutral, and anionic species of uranyl complexes whose relative representation varies with pH, the naked UO_2^{2+} ion prevailing only below pH ~2.5.[37,38] The uranyl salts of (noncomplexing) inorganic anions like sulfate and nitrate will similarly exist as various hydrolyzed ionic species, i.e., in complex with hydroxyl groups.[39] This implies that both anionic and cationic groups in the specimen will bind the stain in a uranyl salt solution at pH 4–5, i.e., become positively stained. Negative staining is complicated by positive staining because it causes accumulation of stain at locations that would otherwise collect little or no stain. The micrographs can then no longer be interpreted purely in terms of stain exclusion by bulk. Uranyl salts are well-known agents for protein precipitation. Marked binding of uranyl to various proteins at concentrations and pH values relevant to negative staining has been demonstrated.[40] Detailed studies on crystalline proteins negatively stained with uranyl acetate have identified positive staining as a significant complication that in itself limits the resolution of the protein structure to about 2 nm.[41,42]

The stability toward electron irradiation of the uranyl salts with organic anions is not good. At higher doses (on the order of several hundred $e/\text{Å}^2$) as required for direct observations at ×30,000–60,000 magnification on the microscope screen, the stain becomes extensively converted to

[34] J. E. Mellema, E. F. J. van Bruggen, and M. Gruber, *Biochim. Biophys. Acta* **140,** 180 (1967).

[35] D. E. Bradley, *J. Gen. Microbiol.* **29,** 503 (1962).

[36] R. Craig, A. G. Szent-Györgyi, L. Beese, P. Flicker, P. Vibert, and C. Cohen, *J. Mol. Biol.* **140,** 35 (1980).

[37] G. H. Tishkoff, *in* "Pharmacology and Toxicology of Uranium Compounds" (C. Voegtlin and H. C. Hodge, eds.), Natl. Nuclear Energy Ser., Div. 6, Vol. 1, p. 125. McGraw-Hill, New York, 1949.

[38] M. Tzaphlidou, J. A. Chapman, and M. H. Al-Samman, *Micron* **13,** 133 (1982).

[39] S. Ahrland, *Acta Chim. Scand.* **3,** 374 (1949).

[40] J. E. Mellema, E. F. J. van Bruggen, and M. Gruber, *J. Mol. Biol.* **31,** 75 (1968).

[41] P. N. T. Unwin, *J. Mol. Biol.* **98,** 235 (1975).

[42] A. C. Steven and M. A. Navia, *J. Microsc.* **128,** 145 (1982).

uranium dioxide, accompanied by marked shrinkage and granulation of the stain.[43] These effects are also present, though less pronounced, at a minimum dose of about 20–25 $e/Å^2$ required for recording a micrograph at ×30,000–50,000. Uranyl sulfate is much more stable,[33] and we prefer this compound for uranyl staining. It is usually titrated with ammonia to pH 4.

The heteropolyanions phosphotungstate and silicotungstate, used as their sodium or potassium salts, are also widely employed as negative stains. They are generally used as 1–2% solutions in water. Sodium phosphotungstate is commonly prepared by titration of phosphotungstic acid ($H_3PW_{12}O_{40}$) with 1 *N* NaOH to pH 6.5–7. This titration is somewhat peculiar because much more base is consumed than expected from the formula for phosphotungstic acid. This is due to the conversion of $PW_{12}O_{40}{}^{3-}$ ion to $PW_{11}O_{39}{}^{7-}$, which takes place above pH 2, accompanied by formation of meta- and/or paratungstates that may further transform to other macroionic species upon prolonged standing.[44] These conversions are slow, and the titration takes many hours to approach equilibrium. We titrate to pH 6.5–6.8. If the titration is carried beyond pH 7, the $PW_{11}O_{39}{}^{7-}$ ion begins to decompose to wolframate ($WO_4{}^{2-}$) and phosphate; full decomposition occurs at about pH 8.[44]

Sodium silicotungstate can be purchased directly, but not all brands are equally suited for negative staining. We have found the compound obtained from TAAB Laboratories Equipment, Ltd. (Reading, Berkshire, England) superior for our work. It dissolves quite slowly in distilled water and should be aged for some days prior to use. The spontaneous pH of a 1–2% solution lies in the range of 6.5–7.5. In this range the $SiW_{11}O_{39}{}^{8-}$ ion is prevalent. If titrated beyond pH 8 this ion decomposes to $WO_4{}^{2-}$ and $SiO_4{}^{-}$.[44]

The complex chemistry of phospho- and silicotungstates is not commonly recognized in negative staining methodologies, and it is likely that different opinions on the usefulness of these stains for particular purposes may derive from the imprecise definition of the stains obtained after titration. Thus, the pH is often not given or it is unclear if a stated value refers to the pH right after addition of base or after standing for 12–24 hr when the initial pH may have dropped by one unit.

The phospho- and silicotungstate heteropolyanions are large, roughly spherical ions carrying only negative charge. When used at a pH close to neutrality, a limited number of positively charged residues in most specimens reduces the problem of positive staining relative to the situation

[43] P. N. T. Unwin, *J. Mol. Biol.* **87,** 657 (1974).

[44] D. L. Kepert, *in* "Comprehensive Inorganic Chemistry" (J. C. Bailar, H. J. Emeléus, R. Nyholm, and A. F. Trotman-Dickenson, eds.), p. 607. Pergamon, London, 1973.

with uranyl salts. The large diameter of the heteropolyanions, about 1.1 nm,[45] restricts their penetration into narrow crevices, but the "gross" surface contours of the structures are reliably outlined. We consider the phospho- and silicotungstates particularly suitable for estimation of molecular volumes/weights from measured dimensions in electron micrographs. The apparently more elaborate substructural details often seen with uranyl salt staining are more difficult to interpret because of the unknown contribution of positive staining which may reach deep into the structures because of the smaller size of the uranyl ionic species.

The heteropolyanionic stains do not undergo visible granulation in the electron beam even after heavy exposure, but the chemical stability of these stains upon electron irradiation has not been studied in detail.

The isopolyanions heptamolybdate and meta- or paratungstates are also used as negative stains. The heptamolybdate ($Mo_7O_{24}^{6-}$) is commonly employed as 1–4% solutions of the ammonium salt in water. This dissolves with a spontaneous pH of about 4.5 and is usually titrated to pH 7–7.4. This stain has been advocated particularly for the higher osmolality needed to reduce hypotonic damage to closed membranous structures.[46] The electron beam stability of this stain is satisfactory by visible criteria, but the image contrast is rather low.

Sodium tungstate (2% solution in water) has a spontaneous pH of 7.5 and contains only WO_4^{2-} ion. When titrated to lower pH values paratungstate A ($HW_6O_{21}^{5-}$) is formed. This conversion gains momentum at about pH 7.4 and is complete at about pH 6. With aging (weeks) or boiling, the paratungstate A transforms progressively to the larger paratungstate Z($W_{12}O_{41}^{10-}$) species.[44] Sodium tungstate is inferior as a negative stain above pH 7.4–7.5 because it tends to crystallize when the specimen is dried. Titrated to pH 6.0–6.5 with tungstic acid and aged, the stain is comparable to phosphotungstate. Sodium tungstate and its isopolyanionic derivatives are not in common use as negative stains probably because of a lack of recognition of the critical importance of pH and age for its performance as a negative stain. The stain can work well and its properties should be explored further. Possibly the size heterogeneity of ionic species in a solution aged 3–4 weeks at pH 6.5 is favorable for packing the stain.

Nonionic or weakly ionic negative stains have been sought to avoid unwanted ionic interactions with the specimen. Methyl phosphotungstates and methylamine tungstate have been used,[47] but have not been

[45] A. J. Bradley and J. W. Illingworth, *Proc. R. Soc. London, Ser. A* **157,** 113 (1936).

[46] U. Muscatello and R. W. Horne, *J. Ultrastruct. Res.* **25,** 73 (1968).

[47] R. M. Oliver, this series, Vol. 27, p. 616.

widely adopted. Glucose,[48] aurothioglucose,[49] and metrizamide, an iodinated sugar,[50] have also been used as negative stains. They give little electron scatter, close to that of the specimen, and image formation is based almost exclusively on phase contrast. These compounds are as irradiation sensitive as the specimen and can be applied only for studies on crystalline specimens, utilizing the redundancy of information for subsequent image processing to overcome the low signal-to-noise ratio inherent in very low dose exposures.

V. Microscope Operation

It is beyond the scope of this chapter to give general guidelines for high-resolution microscope operation. An excellent treatise covering theoretical and instrumental aspects of high-resolution microscopy can be found in Ref. 51. Only a few points pertinent to microscopy of negatively stained biological specimens at the semi-high resolution level of 2–2.5 nm will be mentioned below.

First, it is essential to keep the beam exposure of the specimen to the minimum required for recording the image. This is best achieved by literally taking the photographs in the blind, carrying out all focusing and astigmatism corrections at a field just next to the field to be recorded.[52] Convenient low-dose recording kits for presetting and timing of a sequence of condensor lens, beam deflector, and shutter operations are available from the major microscope manufacturers. If searching is necessary, e.g., for membrane fragments that are not uniformly spread over the grid, it should be performed at low (e.g., ×5000) magnification and at low illumination intensity, just sufficient for distinguishing gross structures. The magnification for image recording should be selected based on the fact that the specimen dosage increases proportional to the square of the linear magnification. The beam damage on biological specimens soon equals the gain in resolution at increased magnification, and there is hardly ever any point in operating over ×60,000 linear magnification. Most photographic emulsions for electron recording are sufficiently fine grain to tolerate ×10 secondary magnification. Detailed attention should be given to the photographic process. For example, expensive gains achieved up to the point of photographic exposure can easily be lost in the

[48] P. N. T. Unwin and R. Henderson, *J. Mol. Biol.* **94,** 425 (1975).

[49] W. Kühlbrandt and P. N. T. Unwin, *J. Mol. Biol.* **156,** 431 (1982).

[50] S. D. Fuller, R. A. Capaldi, and R. Henderson, *J. Mol. Biol.* **134,** 305 (1979).

[51] J. C. H. Spence, "Experimental High-Resolution Electron Microscopy" (C. E. H. Bawn, H. Fröhlich, P. B. Hirsch, and N. F. Mott, eds.). Clarendon, Oxford, England, 1981.

[52] R. C. Williams and H. W. Fisher, *J. Mol. Biol.* **52,** 121 (1970).

subsequent inexpensive procedures of development and photographic copying. Reviews of the properties of electron emulsions and principles for their handling may be found in Refs. 51 and 53. We have found the Agfa Scientia 23 D 56 plates developed for 5 min at 20° with nitrogen burst agitation to a speed of 0.8 $\mu m^2/e$ in Kodak D-19 developer very satisfactory for most purposes.

For high-resolution image recording and interpretation it is essential to be familiar with the amplitude and phase contrast in electron image formation. Mapping the focus dependence of the contrast transfer function of the particular microscope one is using is helpful. A detailed guide for this procedure is found in Ref. 54.

[53] R. C. Valentine, *Adv. Optical Electron Microsc.* **1,** 180 (1966).

[54] D. Misell, *in* "Image Analysis, Enhancement and Interpretation" (A. M. Glauert, ed.), Vol. 7. North-Holland, Amsterdam, 1979.

[51] Electron Microscopy: Assays Involving Freeze-Fracture and Freeze-Etching

By JØRGEN TRANUM-JENSEN

The freeze-fracture replication technique, developed by Moor and Mühlethaler,[1] has two unique features. First, it exposes structures residing within the lipid bilayer matrix for electron microscopic examination, and second, it permits the study of biological structures "fixed" only by freezing, thus circumventing a number of artifacts linked to other preparation methods for electron microscopy. Though the latter advantage was the main impetus for development of the technique, it is rarely fully exploited because of the major difficulties encountered in freezing native biological fluids without introducing ice-crystal artifacts. Thus, the transformation of fluid water and the usual physiological solutions into an amorphous solid phase requires cooling rates in the order of 10^6 K/sec.[2,3]

[1] K. Mühlethaler, *in* "Freeze-etching: Techniques and Applications" (E. L. Benedetti and P. Favard, eds.), p. 1. Soc. Fr. Microsc. Elect., Paris, 1973.

[2] H. Moor, *in* "Freeze-etching: Techniques and Applications" (E. L. Benedetti and P. Favard, eds.), p. 11. Soc. Fr. Microsc. Elect., Paris, 1973.

[3] U. Riehle and M. Hoechli, *in* "Freeze-etching: Techniques and Applications" (E. L. Benedetti and P. Favard, eds.), p. 31. Soc. Fr. Microsc. Elect., Paris, 1973.

This can be achieved only to a depth of 2–3 μm because of the constraints set by the conduction of heat through ice and its removal from the surface. In practice, however, microcrystallites of ice up to 3–4 nm are acceptable because they fall below the resolution limit of the technique, determined by the thickness and the grain size of the replica made of the fracture face. By standard platinum/carbon replication the point-to-point resolution will seldom be better than 5 nm. This extends the maximum obtainable useful freezing depth to about 25 μm, i.e., about the thickness of two cell layers. For many applications this imposes serious difficulties. Useful freezing of deeper layers requires the addition of cryoprotectants, like glycerol in high concentration, which in turn for most specimens demands a preceding chemical fixation to avoid artifacts induced by the cryoprotectants. Therefore, by far most freeze-fracture studies are performed on glutaraldehyde-fixed and cryoprotected material, utilizing only the first feature of the technique—the visualization of the interior of a membrane with its integral proteins through cleavage of the membrane bilayer into its constituent two monolayers. For freeze-etching, i.e., sublimation of ice from the fracture face to reveal the true outer and inner surface of the membranes, the freezing must be performed in pure or close to pure water, which in general necessitates a preceding chemical fixation.

The rather low resolution obtainable with standard freeze-fracture techniques limits the potential for resolution of molecular details of the intramembranous organization of small pore-forming proteins like *Staphlyococcus aureus* α-toxin. Also, the possibility of plastic deformation of proteins during the process of fracture should raise caution against subtle interpretations of small details.[4,5] Another limitation stems from the lack of simple means for direct determination of the chemical nature of the structures observed in the replica of the fracture. Thus, inverted lipid micelles residing within the lipid bilayer could erroneously be taken for large integral membrane proteins.[6,7] Various labeling techniques for individual intramembrane particles have been developed,[8] but they are troublesome and the resolution is generally rather poor.

Given these limitations, the technique is still very useful and for proteins forming channels of more than 8-nm diameter, like the C5b-9(m)

[4] U. B. Sleytr and A. W. Robards, *J. Microsc.* **110,** 1 (1977).

[5] J. Lepault and J. Dubochet, *J. Ultrastruct. Res.* **72,** 223 (1980).

[6] A. J. Verkleij, C. J. A. van Echteld, W. J. Gerritsen, P. R. Cullis, and B. de Kruijff, *Biochim. Biophys. Acta* **600,** 620 (1980).

[7] R. G. Miller, *Nature (London)* **287,** 166 (1980).

[8] J. E. Rash, T. J. A. Johnson, C. S. Hudson, F. D. Giddings, W. F. Graham, and M. E. Eldefrawi, *J. Microsc.* **128,** 121 (1982).

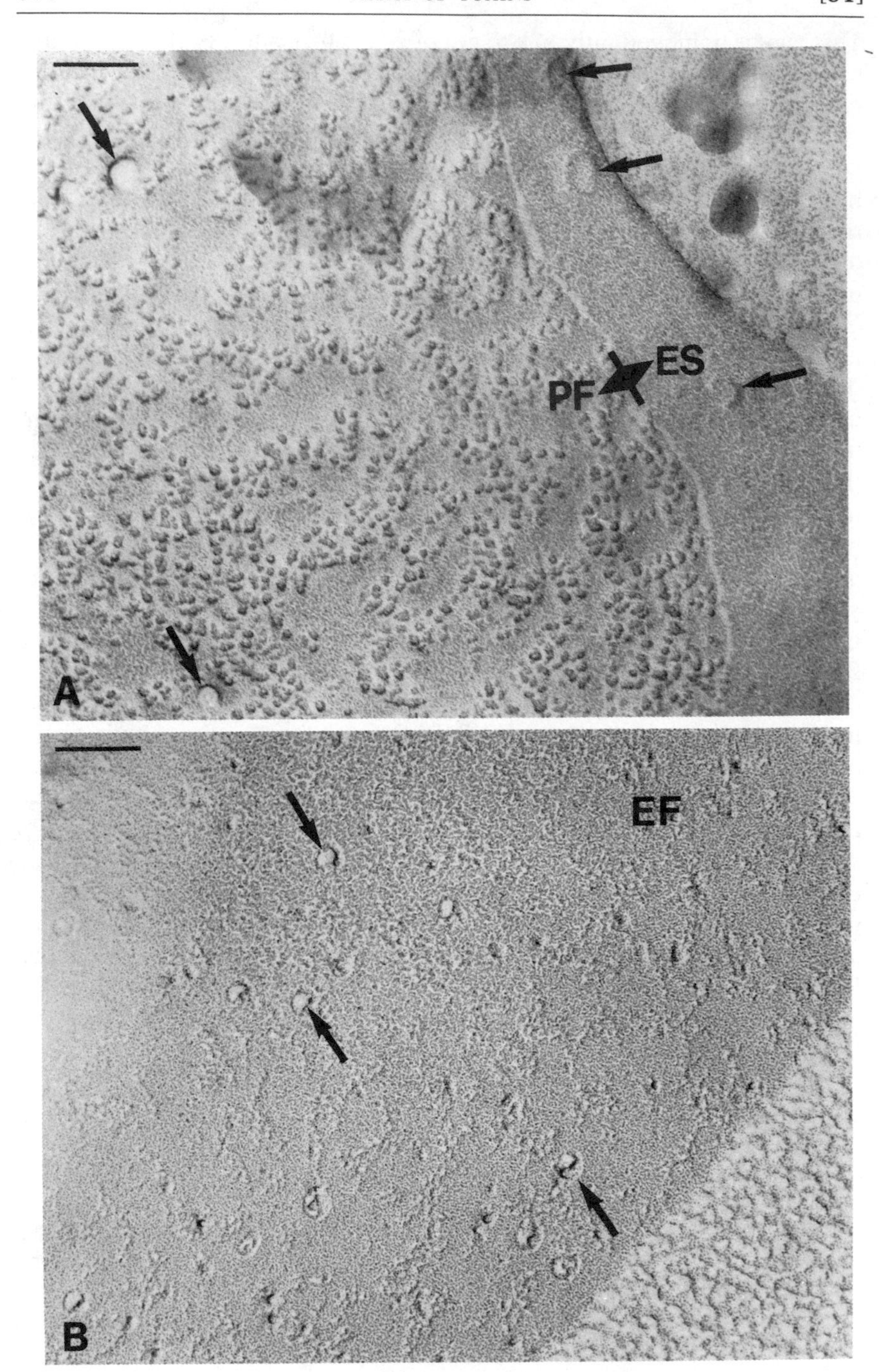
A
PF
ES
B
EF

complex of complement[9] and the oligomers of streptolysin O[10] (Fig. 1) and similar SH-activated toxins, the technique may yield valuable supplementary information on the nature of the lesions. The preparation of replicas of the two complementary fracture faces of the same membrane is a powerful but laborious extension of the technique that may give very conclusive interpretations in return.[9]

The following sections give short accounts of the various preparative steps in the standard freeze-fracture and freeze-etch techniques with an emphasis on the rationales, the problems, and the major precautions to be taken in the various steps. Extensive accounts may be found in Ref. 11. Technical operation guides should be obtained from the manual for the particular apparatus used. In our laboratory we are using a Balzers BAF 301 unit equipped with a liquid nitrogen-cooled baffle over the diffusion pump, a rotary specimen stage, electron guns for Pt/C and C evaporation, and a quartz crystal thin-film monitor (Balzers, Liechtenstein).

I. Chemical Fixation of Cells and Membranes for Freeze-Fracture

Whole cells in suspension, membrane isolates, and liposomes can be directly frozen in thin layers without significant ice-crystal formation by special techniques, detailed in Sections III and IV. These techniques are sufficiently laborious, however, to justify a general recommendation that initial experiments should be performed on chemically fixed and cryo-

[9] J. Tranum-Jensen and S. Bhakdi, *J. Cell Biol.* **97,** 618 (1983).

[10] S. Bhakdi, J. Tranum-Jensen, and A. Sziegoleit, *Infect. Immun.* **47,** 52 (1985).

[11] J. H. M. Willison and A. J. Rowe, *in* "Replica, Shadowing and Freeze-etching Techniques: Practical Methods in Electron Microscopy" (A. M. Glauert, ed.), Vol. 8. North-Holland, Amsterdam, 1980.

FIG. 1. Replicas of erythrocytes, lysed with streptolysin O and rotary shadowed with Pt/C at an angle of 25° (see Chapter [50] in this volume for negatively stained analogous specimens). (A) is from a specimen frozen in 5 m*M* phosphate buffer. The fracture was etched before replication whereby the outer surface of the membrane (ES) has been uncovered. The toxin-induced lesions are distinguished on this surface as rings (arrows), but the contours are somewhat blurred by a contaminating layer formed by the buffer salts contained in the volume of ice sublimated. The surrounding (retracted) ice surface is seen at upper right. At the PF face (i.e., the inner membrane leaflet seen from outside) ring-shaped toxin-induced holes are seen in the lipid plateau between integral membrane particles. The bottom of the holes is free of platinum. (B) is from a specimen frozen in 30% glycerol and shows the EF face (i.e., the outer membrane leaflet seen from the inside) with protruding annular structures corresponding dimensionally to the negative stain image of SLO lesions. Surrounding ice is seen at the lower right, and exhibits a protruding network of a eutectic mixture of glycerol/water, and depressions corresponding to crystals of pure ice that have been etched. Scale bars indicate 100 nm. The membrane faces are labeled according to the generally adopted nomenclature defined in Ref. 42.

protected material, which is fairly easy to prepare and which can give a fast survey of a large number of membranes. If the study includes freeze-etching, whole cells will need to be chemically fixed because they do not stand the environment of water with no or minimal salt or buffer required for etching.

The standard preparation involves fixation for 30–60 min in 2–4% (200–400 m*M*) glutaraldehyde buffered to pH 7–7.2 in, e.g., 50–100 m*M* phosphate. It is advisable to perform the fixation, as well as preceding and following procedures up to the step of freezing, close to the normal physiological temperature of the specimen to prevent possible temperature-dependent phase transitions of the membrane lipids that may cause rearrangement of the structures of interest. We have noted such effects, particularly with bacteria, and in some instances keeping and fixing material for freeze-fracture in the cold has deleterious effects. Depending on the method of freezing to be used, the fixed material may be kept in suspension or sedimented to a pellet. If sedimented right after addition of glutaraldehyde most membrane preparations and whole cells will form pellets sufficiently coherent to be handled as small blocks. If the pellet is too loose it may help to add 0.5–2% albumin to the suspension and to apply glutaraldehyde initially to a final concentration of only 1%, immediately followed by centrifugation. This usually leaves sufficient time for sedimentation before the albumin gels. Subsequently, the glutaraldehyde concentration above the pellet, which should be small (~1 μl), is raised. A Beckman microfuge is well suited for these procedures. Using polyethylene tubes the pellets are easily recovered by cutting off the tips of the tubes. Osmium tetroxide-fixed material is generally not used for freeze-fracture of membranes because the membranes will not split properly in two monolayers, unless the membrane lipids are fully saturated.[12]

II. Cryoprotection

The essential effects of cryoprotectants, occasionally termed "cryofixatives,"[13] as used for freeze-fracture are as follows:

1. to lower the temperature of homogenous nucleation, i.e., the temperature at which crystal nuclei are formed by the water molecules themselves. The region of maximally obtainable supercooling is hereby extended from close to −40° for pure water to about −55° for the case of the most commonly used cryoprotectant, 30% glycerol[14]

[12] R. James and D. Branton, *Biochim. Biophys. Acta* **233,** 504 (1971).

[13] F. Franks, *J. Microsc.* **111,** 3 (1977).

[14] D. H. Rasmussen and A. P. MacKenzie, *in* "Water Structure at the Water–Polymer Interface" (H. H. G. Jellinek, ed.), p. 126. Plenum, New York, 1972.

2. to inhibit the activity of natural heterogeneous nucleators

3. to reduce the growth rate of ice crystals which has an appreciable maximum at about −20°, (~10^8 nm/sec measured perpendicular to the crystal face) but is lowered with decreasing temperature[3,15]

4. to raise the temperature at which recrystallization of ice takes place, i.e., the ice structure becomes unstable with the formation of larger crystals at the expense of smaller ones. For pure water, recrystallization starts at −120° and reaches a significant speed at −100°. The commonly applied cryoprotectant, 30% glycerol, raises the recrystallization temperature to about −60°,[16] which is well above the normal working range of −90 to −110° for freeze-fracturing and freeze-etching.

Glycerol may be applied directly to a fixed specimen suspension or to small blocks of pellets without preceding washings for removal of fixative. If the specimen is osmotically fragile, like erythrocyte ghosts, it should be added in small increments.

Using 30% (w/v) (3 *M*) glycerol it is possible to avoid significant ice-crystal formation to depths of about 100 μm with conventional freezing techniques. In deeper layers larger crystals will form. Glycerol and water forms a eutectic mixture at 67% (w/w) glycerol.[17] This mixture can be frozen as a crystal-free eutectic solid in thick layers. By freezing 30% glycerol, the lower cooling rates in deeper layers leave time for phase separation. Crystals of pure water can grow till glycerol is concentrated up to about 67% (w/w) in the residual fluid. This mixture then passes into solid phase as an entity at about −45°. A well-illustrated account of freeze-fractured ice-crystal structures and their interpretation is found in Ref. 18.

Certain high-molecular-weight substances like polyvinylpyrrolidone (PVP) and hydroxyethyl-starch (HES) with polymeric weights of about 40,000 may be used as cryoprotectants in concentrations of 25% (w/w).[19–21] They do not penetrate intracellularly, and they probably act by suppressing heterogeneous nucleators, which prevail extracellularly,[22] thereby permitting a higher degree of subcooling before intracellular nucleators become active or homogeneous nucleation gains momentum.

[15] P. V. Hobbs, "Ice Physics." Clarendon, Oxford, England, 1974.

[16] B. J. Luyet, *Ann. N.Y. Acad. Sci.* **125,** 502 (1965).

[17] M. L. Shepard, C. S. Goldston, and F. H. Cocks, *Cryobiology* **13,** 9 (1976).

[18] L. A. Staehelin and W. S. Bertaud, *J. Ultrastruct. Res.* **37,** 146 (1971).

[19] F. Franks, M. H. Asquith, C. C. Hammond, H. L. B. Skaer, and P. Echlin, *J. Microsc.* **110,** 223 (1977).

[20] F. Franks, *Scanning Electron Microsc.* **2,** 349 (1980).

[21] A. P. MacKenzie and D. H. Rasmussen, *in* "Water Structure at the Water–Polymer Interface" (H. H. G. Jellinek, ed.), p. 146. Plenum, New York, 1972.

[22] D. H. Rasmussen, M. N. Macaulay, and A. P. MacKenzie, *Cryobiology* **12,** 328 (1975).

These substances are not used often because they are generally less effective. Their main attraction is that they can be applied in high concentrations to living, unfixed cells.

A short review of the principles and problems of cryoprotection can be found in Ref. 23.

III. Mounting the Specimen for Freezing

Specimen holders are usually made of copper or gold because of the excellent thermal properties of these materials. Their thermal conductivity is about 200 times that of ice and about 1000 times that of water, while their heat capacities per unit volume are in the order of those of ice and water within the relevant temperature range. This is the reason why the cooling rate of the specimen is not critically dependent on the mass of the holder accompanying it during freezing. It is even possible to cover the surface of the specimen with a thin plate of copper without significantly sacrificing the cooling rate in the specimen.[24] Sandwich holders are used in which a suspension of specimen is sandwiched together with one or two electron microscopy gold grids between two copper plates, 100 μm thick. The fracture is produced by pulling the plates apart after freezing. These holders are used for propane-jet freezing (see Section IV).

Another type of holder, particularly suited for preparation of complementary replicas in the Balzers freeze-fracture units, consists of two plates, each with a central drilling, aligned to form a tube inside which the specimen is frozen. Two complementary fracture faces result from separation of the plates after freezing.

The simplest type of support consists of a small plate on which the specimen is frozen. Such holders are used for fracture by cutting with a razor blade.

Special holder assemblies for freeze-fracture of cultured cell monolayers may be found in Refs. 25 and 26.

IV. Freezing

The simplest way of freezing the mounted specimen is by immersion in a small bath of a cryogen that itself is cooled in a dewar with liquid nitrogen. The essential requirements for a good cryogen are (1) a low

[23] H. Skaer, *J. Microsc.* **125,** 137 (1982).

[24] A. Elgsaeter, T. Espevik, and G. Kopstad, *Proc. Annu. Meet. Electron. Microsc. Soc. Am.* **38,** 752 (1980).

[25] P. Pscheid, C. Schudt, and H. Plattner, *J. Microsc.* **121,** 149 (1981).

[26] T. Espevik and A. Elgsaeter, *J. Microsc.* **123,** 105 (1981).

melting point, close to the temperature of boiling liquid nitrogen (−196°), (2) a broad temperature interval between the melting point and the boiling point, (3) a high heat capacity and thermal conductivity, and (4) a low viscosity and surface tension. Of the many candidates tested, propane (melting point −189°, boiling point −42°) probably gives the highest cooling rates,[27] possibly surpassed by ethane,[28] but the accuracy of the measuring methods is difficult to assess. An admixture of 20–50% isopentane to propane lowers the melting point to just below −196°, which may be convenient because this cryogen bath does not solidify upon standing in the liquid nitrogen.[29] Propane is highly flammable and many laboratories prefer using Freon 22 ($CHClF_2$, melting point −160°, boiling point −41°), which gives cooling rates comparable to those reached with propane.[27] Liquid nitrogen itself is useless as a cryogen because an insulating layer of gas is formed by boiling around the specimen. Nitrogen slush at the melting point of nitrogen (−210°)[30] is also a poor medium because the interval to the boiling point (−196°) is too narrow to prevent boiling around the specimen. A nitrogen slush, kept above the critical pressure (33.5 atm), theoretically should be an excellent cryogen.[31]

A bath of propane or Freon 22 is easily formed by maintaining a gentle stream of the gas into a metal container held in liquid nitrogen until the bath is collected due to condensation. When the cryogen starts to solidify at the walls of the container, the bath is ready. If the bath solidifies on standing, a sufficient fluid volume is formed by melting with a metal rod. The mounted specimen, held between the tips of a watchmaker's forceps with thin branches to minimize the flow of heat, is introduced into the cryogen with a smooth, quick movement (about 1 m/sec) that should not be discontinued until the specimen is submerged 2–3 cm into the fluid. This is important because convection of the cryogen around the specimen improves the cooling rate. The cooling rate obtainable with this method of freezing lies in the range of 1000–5000°/sec[27] and yields fine quality freezing to a depth of about 30 μm in cryoprotected specimens.

An excellent freezing method which requires a special apparatus is propane-jet freezing, where an instantaneous jet of propane at about −185° is directed toward both sides of the specimen held in a sandwich holder.[32] The apparatus is commercially available from Balzers Union. The constant jet of cryogen against the sandwich yields cooling rates

[27] M. J. Costello and J. M. Corless, *J. Microsc.* **112,** 17 (1978).

[28] N. R. Silvester, S. Marchese-Ragona, and D. N. Johnston, *J. Microsc.* **128,** 175 (1982).

[29] B. Jehl, R. Bauer, A. Dörge, and R. Rick, *J. Microsc.* **123,** 307 (1981).

[30] A. P. MacKenzie, *Biodynamica* **10,** 341 (1969).

[31] W. B. Bald, *J. Microsc.* **134,** 261 (1984).

[32] M. Müller, N. Meister, and H. Moor, *Mikroskopie* **36,** 129 (1980).

better than 10^4°/sec for thin (~40 μm) specimens which thereby may be frozen without cryoprotection. The method has proved effective also for preservation of temperature-labile lipid phases.[33,34] A number of further refined freezing methods exist designed to yield even higher cooling rates. These include slam-freezing, where the specimen is placed against a metal block cooled by liquid nitrogen or helium,[35,36] and spray-freezing, where a specimen suspension is frozen as aerosol droplets.[37] Another sophisticated method utilizes freezing point depression and lowering of the temperature by homogeneous nucleation of water at high (~2000 atm) pressures.[3,38]

To avoid overinterpretation of structures observed in the deeper layers of a frozen specimen where ice crystals have grown to appreciable size due to lower cooling rates, a number of accompanying phenomena that may possibly affect protein structures should be mentioned. First, during growth of the ice crystals, dissolved substances become excluded from the ice volume to concentrate in the residual fluid phase until a eutectic point of the residual fluid is reached. When an NaCl solution is frozen, salt may be concentrated up to 4.6 *M* (23%, w/w) until the eutectic point is reached at −21°.[39] The temperature will tend to stay at this level as long as the ice crystals grow and the residual fluid will be concentrated because of the heat of crystallization liberated from the forming ice.[2] The specimen, e.g., membranes, are excluded from the growing ice crystals and become exposed to high ionic strengths before the residual fluid settles. The concentrations reached in the residual fluid depend on composition and cooling rate. For a mixture of glycerol, NaCl, and water cooled at a modest rate, the ternary eutectic point is reached at −80° at a concentration of 63% (w/w) glycerol and 1.6 *M* NaCl.[17] In Ca^{2+}-containing solutions, high concentrations of this ion will also be reached.[39] A second phenomenon which may be of relevance is the formation of an electrical potential between the growing ice crystals and the fluid phase because of the charge separation due to differences in the ease with which different ions are included (as contaminants) in the ice crystals. In certain dilute

[33] R. van Venetië, W. J. Hage, J. G. Bluemink, and A. J. Verkleij, *J. Microsc.* **123,** 287 (1981).

[34] M. J. Costello, R. Fetter, and M. Höchli, *J. Microsc.* **125,** 125 (1982).

[35] G. P. Dempsey and S. Bullivant, *J. Microsc.* **106,** 251 (1976).

[36] J. E. Heuser, T. S. Reese, M. J. Dennis, Y. Jan, L. Jan, and L. Evans, *J. Cell. Biol.* **81,** 275 (1979).

[37] L. Bachmann and W. W. Schmitt, *Proc. Natl. Acad. Sci. U.S.A.* **68,** 2149 (1971).

[38] M. Müller and H. Moor, *in* "Science of Biological Specimen Preparation" (J.-P. Revel, T. Barnard, and G. H. Haggis, eds.), p. 131. AMF O'Hare, Chicago, 1984.

[39] J. E. Tanner, *Cryobiology* **12,** 353 (1975).

electrolyte solutions, potentials of several hundred volts may arise,[15] in physiological fluids probably only a few volts, but precise data relevant to freeze-fracture preparation are not available. A third factor to consider is the possibility of lipid phase transitions upon cooling. Thus the hexagonal type II phases that may be of relevance for biological membrane structure will be preserved only by very high cooling rates.[33]

V. Fracture

During transfer of the specimen to the cooling stage of the freeze-fracture apparatus care should be taken that the specimen is not warmed above its recrystallization temperature. Initially the stage should be maximally cooled and broad thermal contact with the specimen holder should be ensured. This is essential for the temperature measurements to be reliable when the measuring probe is located in the stage. A vacuum greater than 10^{-5} torr is required before fracturing (see Section VI).

Fractures are produced either by cutting or by breaking. For cutting, the edge of a razor blade, mounted in a holder perfused with liquid nitrogen, is advanced through the specimen, much like in a microtome, cutting off 10–20 μm/stroke. The material removed is in the form of ice-powder and only the fracture face of the block is usable. A large part of this face is not useful because the razor blade has scraped over it, causing local heating and micromelting artifacts. The main advantage of the cutting method is that the fracture can be quite precisely directed through the optimally frozen superficial layer of the specimen block. When the fracture is produced by breaking the specimen block, two intact and complementary fracture faces issue. By means of special mechanical devices these two fracture faces can be positioned for simultaneous replication under identical conditions to yield two complementary replicas.[9,40]

No matter how the fracture is produced, the fracture plane follows the path of least resistance. The two lipid monolayers of a frozen bilayer are held together only by weak van der Waals interactions because the hydrophobic interaction (''bond'') is an entropic phenomenon existing only in an environment of fluid water. The bilayer therefore represents a cleavage plane of much less resistance than the surrounding ice whose mechanical strength is ruled by abundant hydrogen bonds. When the fracture advances through the ice to meet a frozen bilayer membrane it tends to follow this membrane until it eventually deviates too far from the average

[40] K. Mühlethaler, W. Havenstein, and H. Moor, *in* ''Freeze-etching: Techniques and Applications'' (E. L. Benedetti and P. Favard, eds.), p. 101. Soc. Fr. Microsc. Elect., Paris, 1973.

fracture plane in the surroundings. At this point the fracture again enters the ice. A general consensus on this interpretation of fracture and a corresponding nomenclature system for membrane fracture faces has been determined.[41,42] Transmembrane (integral) proteins represent an additional type of "bond" between the monolayers of a biological membrane. With the possible exception of very protein-rich membranes, like mitochondrial inner membranes,[43] these proteins do not have sufficient strength to prevent the cleavage of the bilayer, but their presence may increase the frequency of pure cross-fracture events.[44]

In the process of fracture, the intramembrane domain of integral membrane proteins becomes exposed. A particular species of protein generally follows one of the fracture faces and appears as a little knob elevated above the lipid plateau. It leaves a little complementary pit in the other monolayer. The polypeptide strand of some proteins may break in the process.[45] Carbohydrate moieties residing exterior to the membrane, i.e., embedded in ice, surely will be broken off. It is likely that proteins with only a single peptide strand spanning the membrane will cross-fracture in or close to the cleavage plane and will be difficult to visualize by subsequent shadowing. Possibly, only proteins with several intramembrane strands become regularly visible. It is possible that proteins may become plastically deformed by the fracture.[4,5,46] Such deformations are likely to decrease with decreasing temperature,[47] but fractures at temperatures below −120° require vacuum conditions that are not easily met, as detailed in the following section.

VI. Vacuum, Contamination, and Etching

The dominant constituent of the residual atmosphere in a conventional freeze-fracture apparatus is water.[47,48] As soon as the fracture faces are exposed to the vacuum, condensation or sublimation processes begin at the surface. If the vapor pressure of ice at the specimen temperature is

[41] D. Branton, *Proc. Natl. Acad. Sci. U.S.A.* **55,** 1048 (1966).

[42] D. Branton, S. Bullivant, N. B. Gilula, M. J. Karnovsky, H. Moor, K. Mühlethaler, D. H. Northcote, L. Packer, B. Satir, P. Satir, V. Speth, L. A. Staehlin, R. L. Steere, and R. S. Weinstein, *Science* **190,** 54 (1975).

[43] F. S. Sjöstrand, *J. Ultrastruct. Res.* **69,** 378 (1979).

[44] S. Kirchanski and D. Branton, *Proc. Annu. Meet. Electron. Microsc. Soc. Am.* **38,** 756 (1980).

[45] H. H. Edwards, T. J. Mueller, and M. Morrison, *Science* **203,** 1343 (1979).

[46] A. J. Verkleij and P. H. J. T. Ververgaert, *Biochim. Biophys. Acta* **515,** 303 (1978).

[47] H. Gross, E. Bas, and H. Moor, *J. Cell Biol.* **76,** 712 (1978).

[48] H. Moor, *in* "Freeze-etching: Techniques and Applications" (E. L. Benedetti and P. Favard, eds.), p. 21. Soc. Fr. Microsc. Elect., Paris, 1973.

lower than the partial pressure of water in the vacuum, water will settle as contaminating ice on the fracture face. Assuming that every water molecule impinging on the surface is immobilized, the rate of contamination mounts to about one monomolecular layer of water per second at a pressure of 10^{-6} torr. Improving the vacuum to 10^{-7} torr reduces this contamination rate by an order of magnitude. In addition to the contamination, which is ruled by the balance between partial pressure of water and the temperature of the specimen, another type of contamination takes place due to specific adsorption of water molecules to particular structures in the system. Such adsorption sites may serve as nucleators for deposition of ice. Thus, certain structural details may become "specifically" decorated with ice,[49,50] providing misinformation regarding the dimensions of small details. For this reason one should work well away from the region of temperature and pressure where contamination will take place. A conventional freeze-fracture apparatus in good condition should easily reach 10^{-6} torr and the vacuum can be lowered to 10^{-7} torr by pumping for 2–4 hr. Lowering the vacuum significantly below 10^{-7} torr requires the use of special pump systems in combination with heating arrangements for the vacuum chamber for the desorption of gases.[47] Assuming that the residual pressure is largely due to water, the approximate temperature limits for contamination lie at $-100°$ for 10^{-5} torr, at $-110°$ for 10^{-6} torr, and at $-120°$ for 10^{-7} torr. For this reason conventional freeze-fracture apparatuses equipped with oil-diffusion pumps should not generally be operated at specimen temperatures below $-120°$. Significant reduction of contamination can be achieved, however, by using a baffle cooled with liquid nitrogen located above the diffusion pump as well as by a liquid nitrogen-cooled shield placed closely around the specimen.[51,52]

Contrary to contamination conditions, if the vapor pressure of ice at the specimen temperature exceeds the partial pressure of water in the vacuum, ice will sublimate from the fracture face. This process is termed "etching" and may be utilized to uncover the true inner and outer surfaces of membranes hidden in the ice below the fracture face. A prerequisite for this technique is that the specimen be frozen in pure or almost pure water. All solutes contained in the volume of ice sublimated to uncover a surface will be deposited onto that surface as a contamination layer that blurs details (Fig. 1A). Crisp and reliable definitions of struc-

[49] H. Gross, O. Kuebler, E. Bas, and H. Moor, *J. Cell Biol.* **79,** 646 (1978).
[50] D. Walzthöny, H. Moor, and H. Gross, *Ultramicroscopy* **6,** 259 (1981).
[51] R. L. Steere, *J. Microsc.* **128,** 157 (1982).
[52] S. Bullivant, *in* "Science of Biological Specimen Preparation" (J.-P. Revel, T. Barnard, and G. H. Haggis, eds.), p. 175. AMF O'Hare, Chicago, 1984.

tures revealed by deep etching will be obtained only by sublimation of pure ice.[53] The rate of sublimation is critically dependent on the temperature. At $-90°$ it is in the order of 10^2 nm/sec, at $-100°$ it has already declined to about 2 nm/sec, and at $-110°$ it is as low as 0.1 nm/sec.[54] Because the temperature at the specimen surface is not known with high precision, etching rates are in practice a bit difficult to control. Usually, a specimen temperature of $-100°$ is sought. It is important that the specimen be shielded by a liquid nitrogen-cooled cold trap during the relatively long period of etching ($\frac{1}{2}$–3 min) to diminish contamination, e.g., with hydrocarbons originating in the oil-diffusion pump. This is easily achieved when fractures are produced by cutting if the blade is simply left over the specimen.

In general it is helpful to allow time for some etching in standard fractures also, because the edges of the fractured membranes will then stand out clearly. In some cases it is essential to rule out etching, e.g., in the comparison with etched specimens to demonstrate water-filled channels.[9] The best way to avoid etching is by fracture through breaking of the specimen when the evaporation source for replication is already activated whereby the replication starts virtually at the very moment the fracture is produced.

VII. Replication

The first step in the replication of the fracture face is shadowing with a heavy metal, deposited to an average thickness of 0.5–2.5 nm. The second step is to reinforce this very thin surface replica with a layer of pure carbon, 10–25 nm thick. Ideally, the distribution of the metal should be ruled solely by the geometrical relations between a point-shaped evaporation source and the topographical details of the fracture face. This implies that the evaporated metal atoms move in linear paths through the vacuum and remain on the very site of collision with the fracture face. Only metals with very high melting points and correspondingly extremely low vapor pressures can be used. Further, their atomic mass must be high to give sufficient electron scatter for a contrasting image formation in the microscope, and they must stand treatment with harsh chemicals used for cleaning the replica. No material fully meets these requirements. Thus, some movement from the site of collision is witnessed by micrograininess of the deposited film and the demonstration of crystalline structure by

[53] K. R. Miller, C. S. Prescott, T. L. Jacobs, and N. L. Lassignal, *J. Ultrastruct. Res.* **82,** 123 (1983).

[54] J. G. Davy and D. Branton, *Science* **168,** 1216 (1970).

electron diffraction. A mixture of tantalum and wolfram evaporated at about 3000° produce the finest, nongrainy films,[55] but the replicas are difficult to prepare and clean, and they may oxidize. The most widely used and an excellent shadowing material is platinum coevaporated with carbon in about equimolar ratio (95% Pt, 5% C by weight).[55] The coevaporation of carbon suppresses the formation of Pt microcrystals, and a grain size of only 15–30 Å in the replica is easily achieved in a conventional apparatus. The grain size can be significantly reduced by lowering the specimen temperature. With Pt/C shadowing it appears that there is often an element of specific "decoration" of particular specimen details.[55] Such decoration phenomena are well documented for some other metals.[56]

The evaporation source may be as simple as two aligned carbon rods, the one pointed and wrapped with a few centimeters of thin Pt wire. A dc current or full-wave rectified ac current of 20–50 A is passed through the rods and a spring load mechanism ensures contact between the rods as the pointed tip evaporates. Various designs for this type of source may be found in Ref. 11. The pointed rod is connected to the positive pole of the current source, the other is grounded. This reduces a harmful flow of electrons between the evaporation source and the specimen, which is at ground potential. A grounded diaphragm of the smallest practical size (~5 mm) should be inserted between the source and the specimen to reduce the thermal load, and the source-to-specimen distance should be about 10 cm. It is important that the tips of the carbon rods are thoroughly degassed in the vacuum prior to fracture and shadowing by passing a current just insufficient for evaporation for 1–2 min, otherwise the vacuum will seriously degenerate during evaporation. Excellent shadowing may be obtained by this method if thoroughly standardized.[51,57] Many investigators find that the reproducibility of the method is unsatisfactory, mostly because the ratio of Pt to C in the deposited film is somewhat difficult to control. The carbon component is often too high, reducing the image contrast. Also, the deposition rate is difficult to control. Rates should be set for 1–2 Å/sec. Rates that are too high may result in harmful heating of the fracture face when the kinetic energy of the incident atoms dissipate on top of the inescapable thermal irradiation from the white-glowing source. A quartz crystal thin-film monitor is very useful for controlling layer thickness and deposition rates.

Improvement of the quality as well as the reproducibility of Pt/C

[55] H. Moor, *Philos. Trans. R. Soc. London, Ser. B* **261,** 121 (1971).
[56] D. Studer, H. Moor, and H. Gross, *J. Cell Biol.* **90,** 153 (1981).
[57] R. L. Steere, E. F. Erbe, and J. M. Moseley, *J. Microsc.* **111,** 313 (1977).

evaporation is achieved by so-called electron guns in which a small pellet of Pt–C alloy is heated by bombardment with accelerated electrons. The glowing surface that the specimen "sees" is small, and harmful charged particles that "etch" the specimen are sorted out by deflection in a high-voltage field.[55,58]

The standard geometry of shadowing is unidirectional at an angle of 45° relative to the average fracture plane. The inclination of a particular small area of the fracture face is not known, however, and it can be quite different from the average plane. Therefore, calculations of heights based on the lengths of shadows are very approximate. This can be overcome by bidirectional shadowing from two opposite sources,[11] but only if the layer of deposited metal is thin relative to the object height. Unidirectional shadowing gives a very plastic image that is generally fairly easy to interpret in terms of elevations and depressions. The information content is limited, however, by the fact that part of the circumference of a structure, e.g., a membrane protein protruding from the membrane fracture face, lies in shadow. This is overcome by rotary shadowing where the specimen is set to rotate during shadowing[59] (Fig. 1). The full contour of a particle or a hole is thereby outlined by the deposited metal. It is sometimes difficult to distinguish knobs and holes/depressions in rotary-shadowed replicas because both are similarly outlined. If the depression is deep enough relative to the shadowing angle, metal will not reach the bottom and the depression will appear entirely blank. Knobs always collect some metal on the top, but the top layer is thinner than the layer collected on the sides of the knob, and should not be misinterpreted as a central pore in the knobs. A theoretical treatise on the layer geometry in rotary-shadowed replicas is found in Ref. 60. We have found rotary shadowing at an angle of 25° relative to the average fracture plane to be generally very useful. At more shallow angles smaller structures become visible, but larger parts of the fracture face lie in shadow during some part of the rotation. In general, the interpretation of rotary-shadowed replicas is facilitated by comparison with unidirectionally shadowed analogous specimens.

The replica obtained by the heavy metal film alone is much too fragile to be handled and cleaned. Therefore, an additional layer of 10–25 nm pure carbon is evaporated from a separate source perpendicularly onto the specimen. The thickness of this layer is not critical. The pure carbon is almost invisible in the electron microscope when viewed in focus because of the low electron scatter and the absence of phase contrast in exact focus.

[58] H. P. Zingsheim, R. Abermann, and L. Bachmann, *J. Phys. E.* **3,** 39 (1970).
[59] L. H. Margaritis, A. Elgsaeter, and D. Branton, *J. Cell Biol.* **72,** 47 (1977).
[60] L. Landmann and J. Roth, *J. Microsc.* **139,** 221 (1985).

VIII. Cleaning of the Replica

Following replication the specimen is thawed. A replica of a frozen suspension will float on a water surface and a replica attached to a block of tissue will often spontaneously float at the surface, carrying the block submerged in the water due to the hydrophobicity of the pure carbon surface of the replica. In sandwich holders, the replica is lifted off together with the intercalated gold grid. When two grids are intercalated the fracture will often run between them and the two complementary replicas can be recovered. Gold grids are convenient because they can be placed directly on the cleaning fluids. All traces of the original specimen must be thoroughly removed when cleaning a replica. The most effective cleaning method for Pt/C replicas is to float them on concentrated chrome–sulfuric acid or a 1 : 1 mixture of 15% sodium hypochlorite and 1 *N* NaOH for 1–4 hr followed by washing in several changes of distilled water. A small blunt glass rod is well suited for lifting the replicas from the baths.

The cleaned replica is finally floated on a very clean water surface. With a forceps a 200–400 mesh electron microscope grid is then brought into contact with the replica and pressed into the water. The grid is turned, replica side up, lifted out of the water, and air dried. The adhesion of the replica to the grid is improved if the grid is made sticky beforehand, e.g., by dipping it into chloroform in which glue from a piece of cellophane tape has been dissolved (3–4 cm cellophane tape/50 ml chloroform).

[52] Two-Dimensional Crystals of Tetanus Toxin

By J. REIDLER and J. P. ROBINSON

Introduction

The mechanism by which toxin molecules traverse cell membranes remains unclear. This is due both to the enzymatic nature of the active subunit and the difficulty of biochemical isolation of intermediates of proteins intimately associated with membranes. Three-dimensional crystals of some toxins have been obtained but the mechanism of toxin entry has not yet been elucidated. Although two-dimensional crystals are currently limited in resolution, the application of novel two-dimensional crystallization techniques[1,2] to toxins can provide further insight into the

[1] E. Uzgiris and R. Kornberg, *Nature (London)* **301,** 125 (1983).

[2] J. Reidler, E. Uzgiris, and R. Kornberg, *In "Handb. Exp. Immunol,"* (D.M. Weir, ed.), Blackwell Scientific, Oxford.

nature of the interactions between membranes and toxins. Tetanus toxin forms two-dimensional crystalline arrays at the air–water interface in the presence of a monolayer film containing a small percentage of the natural receptor, ganglioside G_{T1}.[3] Such crystalline arrays form in a time scale of hours, require approximately 1 μg purified toxin per experiment, and yield a resolution in projection of 20–30 Å in uranyl acetate negative stain. Here, we describe methods which detail procedures for the formation of these two-dimensional crystals.

Experimental

Tetanus toxin is prepared as described elsewhere in this volume,[4] then dialyzed into citrate–phosphate buffer at pH 5.6 and ionic strength of 0.07.[5] Triple-distilled deionized water is used for all dilutions. Dialyzed toxin is diluted to 50 μg/ml in the same buffer just prior to use, and 20-μl volumes of the diluted material are placed in each microtiter well (Lux #5200) using a Pipetman P20 (see Fig. 1).

Comments. Protein concentrations are often varied between 25 and 500 μg/ml during the initial search for crystalline areas. Buffer conditions and pH also play a substantial role in the crystallization process and a variety of conditions should be explored. For example, varying the pH over two units is not unusual. Concentrated protein (1–5 mg/ml) diluted into standard buffers provides a rapid screening for crystallization conditions. The protein can be diluted into the buffer either before or after the formation of the lipid monolayer. The latter method may minimize protein denaturation by lipid solvents, but mixing can disturb the monolayer. Protein purity is quite important, and repeated freeze/thaw cycles should be avoided.

Lipids

Ganglioside G_{T1} is purchased from Supelco, Inc. (Cat. #4-6035) and used without further purification. Egg phosphatidylcholine (Cat. #P-5388) is purchased from Sigma and should be >99% pure. Ganglioside (G_{T1}) and phosphatidylcholine (PC) are dissolved in chloroform at a weight ratio of 1 : 20 G_{T1} : PC, which corresponds to an approximate molar ratio of 1 : 50.

[3] J. Reidler and J. Robinson, *in* "Proceedings of the Electron Microscopy Society of America" (G. W. Bailey, ed.), pp. 528–529. San Francisco Press, San Francisco, California, 1985.

[4] J. Robinson, this volume [13].

[5] G. Lee, E. Grollman, S. Dyer, F. Beguinot, D. Kohn, W. Habig, and M. Hardegree, *J. Biol. Chem.* **254,** 3826 (1979).

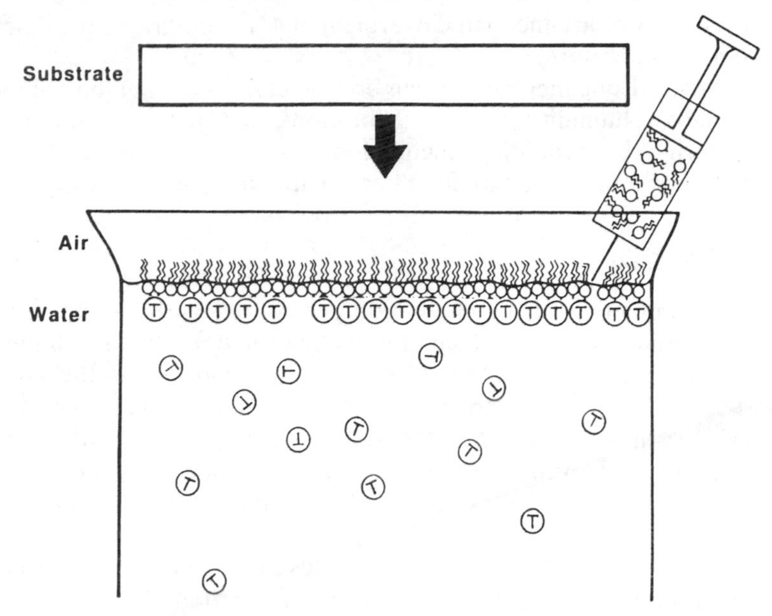

FIG. 1. Schematic diagram of crystallization technique for the formation of two-dimensional crystals on water. Lipid : ganglioside mixtures are dissolved in volatile solvent and added to the surface of the toxin solution (T's in circles) using a syringe. After crystal formation, a substrate (electron microscope grid) absorbs the crystalline layer for viewing.

This lipid mixture is added carefully to the surface of the microtiter wells at a concentration of 1 mg/ml and a total volume of 0.25–0.5 μl from a 10-μl Hamilton syringe, thus allowing the formation of a monolayer of lipid to form on the surface of the tetanus toxin solution.[6]

Comments. Lipids are generally stored under argon at 1 mg/ml at −70°. Material is withdrawn with a 10-μl Hamilton syringe and a very small drop is touched carefully to the surface of each well. A slight excess of lipid produces better results. Hexane is preferable as solvent where possible instead of chloroform as it reacts less with the plastic plates. Teflon plates can also be used.

Incubation

In order to avoid evaporation from the wells, water is added to any unused wells and the corners of the microtiter dish. Preparations are incubated for 4 hr or more at room temperature (24°) before staining.

[6] G. L. Gaines, "Insoluble Monolayers at Liquid–Gas Interfaces." Wiley (Interscience), New York, 1966.

Other preparations are incubated overnight at 4° in addition to the initial 4 hr at room temperature.

Comments. Long incubations can best be carried out in a humid incubator (relative humidity >95%). Vibrations and handling should be avoided during the incubation. Incubation time is as short as 2 hr for high concentrations and as long as 24–48 hr for the lower concentrations.

Grids and Staining

At the completion of the incubation period, dry carbon-coated grids are placed directly on the surface of monolayer film for at least 30 min to allow strong association between the lipid monolayer and the carbon support film. The grids are then washed and stained quickly with a 1% solution of uranium acetate in water. Excess stain is removed at the edge with a Kimwipe, then the grids are air dried and viewed in the electron microscope. Organized areas are readily visible at magnifications of ×25,000.

Comments. Nitrocellulose-coated 200 mesh grids are coated with carbon and stored in a desiccator until use. The formation of the crystalline arrays prior to absorption onto the grid eliminates the need for diffusion oil during carbon coating.[2] Copper, nickel, and silver grids have all been used. The grids are raised carefully from the wells with tweezers and turned over for staining. Glass pipets are used during staining. The first drop can be either distilled water or stain, and rinses the surface free of buffers and unbound protein. Dry Kimwipe areas should be used to avoid too rapid removal of the liquid from the grid surface.

Image Processing

Negatives are examined in an optical diffractometer[7] to determine the degree of organization. Areas containing many diffraction spots are marked and scanned at 50-μm resolution on a Perkin-Elmer flat-bed densitometer. Digitized data are processed on a VAX-780 using image-processing programs developed by N. Unwin and co-workers at Stanford University.[8]

Comments. Electron diffraction can replace optical diffraction but is time consuming and degrades crystal quality. An extremely simple (but cumbersome) optical diffractometer consists of a monochromatic beam

[7] E. Salmon and D. DeRosier, *J. Microsc.* **123,** 239 (1981).

[8] L. Amos, R. Henderson, and P. N. T. Unwin, *Prog. Biophys. Mol. Biol.* **39,** 183 (1982).

and a long focus lens (about 400 mm). Negatives are placed between the beam and the lens and viewed by an observer at the beam focus behind a small beam stop. Lower magnification negatives will create larger diffracted images. Computer processing of crystals can be carried out at numerous facilities.

Results

Tetanus toxin forms two-dimensional crystals with mixtures of G_{T1} and phosphatidylcholine (PC) under conditions similar to the natural receptor. These crystals are hexagonal in projection, with a center-to-center distance of 75 Å, and a typical field is shown in Fig. 2. An optical diffraction pattern of a selected area is shown in Fig. 3. Three orders of diffraction spots are readily visible, indicating a resolution in negative stain of about 25 Å. Negative stain limits resolution to typically around 20 Å.

No crystals are formed and no binding observed in the absence of G_{T1}, indicating a specific interaction between the G_{T1} and the toxin. High ratios of G_{T1} to PC do not produce crystals, and this could be due to any number

FIG. 2. Transmission electron micrograph of tetanus toxin crystals with 1% uranyl acetate negative stain. ×200,000.

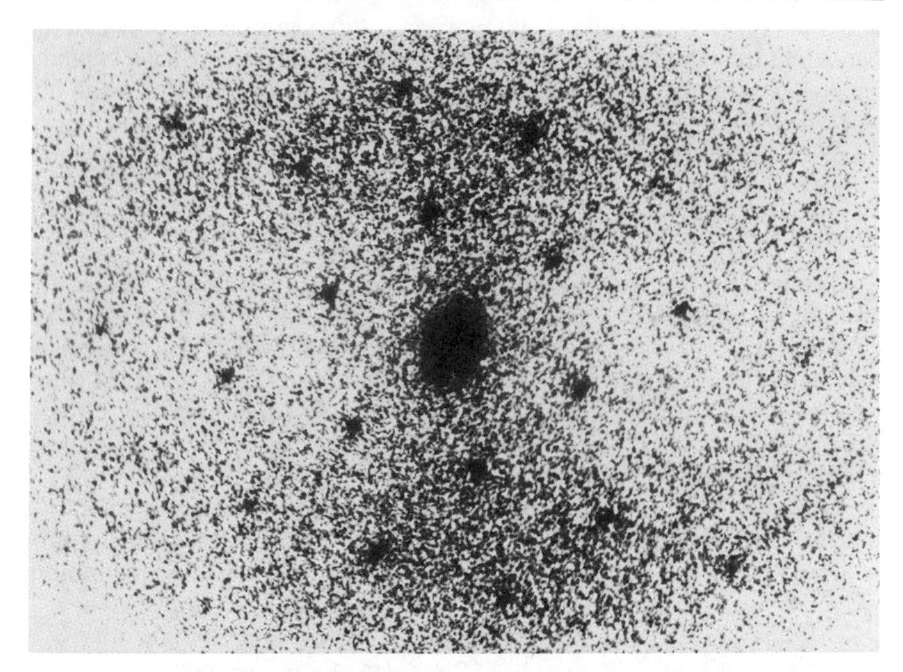

FIG. 3. Optical diffraction pattern of a selected crystalline area of tetanus toxin.

of factors including micelle formation, surface charge alterations, local pH changes, and steric hindrance. A density of 2 mol% of G_{T1} in the monolayer is sufficient to form a uniform layer of toxin, due to the difference in the relative surface area occupied by G_{T1} and toxin. In addition, no crystals are formed with ganglioside mixtures, or with gangliosides G_{M1} or G_{D1b}, which bind tetanus toxin with lower affinity. It is possible that crystals can form under different incubation conditions with these gangliosides, but these have not yet been found.

Direct proof of toxin in these crystals will require cocrystallization with monoclonal antibodies against specific regions of the toxin molecule. Monoclonal antibody cocrystals will also aid the interpretation of processed images. An unsymmetrized processed image is shown in Fig. 4. Monoclonals will allow direct visualization of additional protein density to different parts of the molecular structure. Alternatively, better resolution with different stains or incubation conditions may allow accurate determination of the toxin subunits without monoclonal antibodies. Better resolution will be required in order to begin the reconstruction of the three-dimensional image by tilting the specimen crystals in the electron microscope coupled with image processing.

FIG. 4. Unsymmetrized processed image of tetanus toxin crystals. Resolution limit for this image is about 30 Å. Center-to-center distance is 75 Å.

Discussion

Tetanus toxin forms two-dimensional crystals at monolayer surfaces analogous to the natural receptor and under conditions which mimic toxin entry. High-resolution toxin crystals formed at the membrane interface could provide a direct structural solution to the mechanism of toxin entry across the membrane. Examination of three-dimensional images of high-resolution toxin crystals under differing conditions could lead to direct observation of toxin structural conformations at the membrane related to toxin entry.

Image processing with gold glucose and vitreous ice for increased resolution are planned. Monoclonal antibodies against different active portions of the toxin molecule will allow further mapping of crystalline arrays of toxin. Preliminary observation of crystalline arrays of the structurally homologous botulinum toxin indicate a similar size and structure to the tetanus toxin.

Acknowledgment

The authors wish to thank Dr. John Murray for densitometer scanning tetanus toxin negatives.

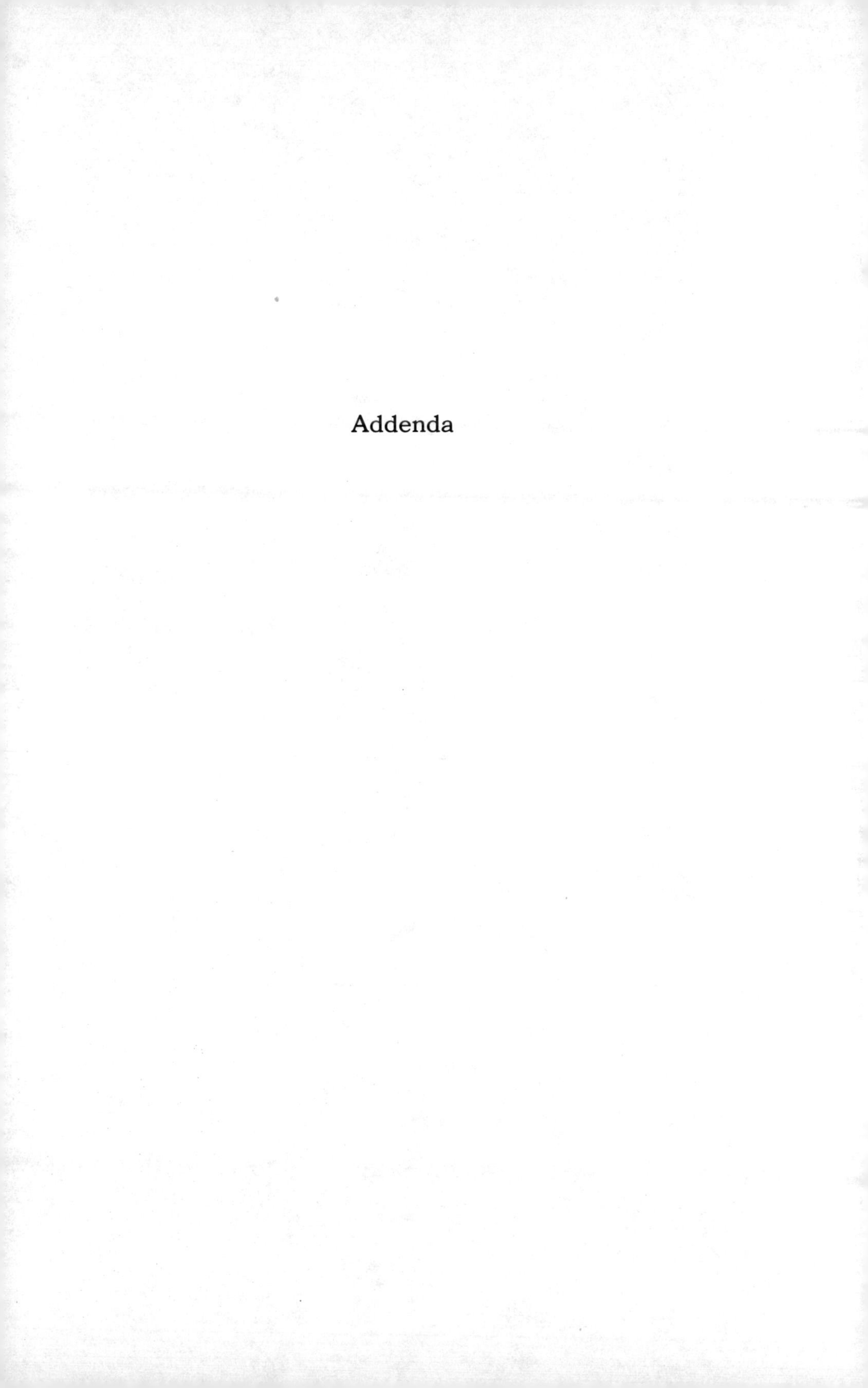

Addenda

Addendum to Article [20]

By G. A. Bohach, S. J. Cavalieri, and I. S. Snyder

Polyethylene Glycol Precipitated Hemolysin

Bhakdi *et al.*[15] reported purification of the hemolysin from Todd–Hewitt broth using polyethylene glycol (PEG 4000). The supernatant (4°) from 1000 ml of media is brought to 3% (v/v) in glycerol. Solid PEG (200 gm) is dissolved in the supernatant by stirring and the preparation incubated at 4° for 60 min. The precipitate is collected by centrifugation (16,000 *g* for 20 min) and dissolved in approximately 8 ml saline. The hemolysin preparation showed a single protein band (107,000 Da) by PAGE (stained with Coomassie blue) and by immunoblotting. The typical hemolytic titer of the dissolved precipitate was reported as 8000–16000 HU as compared with 64–128 HU in the culture supernatant. The specific activity of the hemolysin preparation and the presence or absence of endotoxin was not reported.

[15] S. Bhakdi, N. Mackman, J-M Nicaud, and I. B. Holland, *Infect. Immun.* **52,** 63 (1986).

Addendum to Article [22]

By Gerald T. Keusch, Arthur Donohue-Rolfe, Mary Jacewicz, and Anne V. Kane

Scale-up

In order to obtain larger amounts of toxin, production can be scaled-up to 20-liter batches by growing the organisms in a fermentor. A starter culture is prepared by inoculating several single colonies from a fresh plate of *S. dysenteriae* 1, strain 60R, into 25 ml complete MSB medium and incubating at 37° for 6–7 hours. Twenty liters of MSB, sterilized *in situ* in a New Brunswick MF 128S fermenter equipped with a dissolved oxygen probe and microprocessor control, is inoculated in the late afternoon with the starter culture. Overnight incubation at 37° proceeds under vigorous aeration to maintain a high set point (90%) for dissolved O_2 content.

The organisms are harvested on a Pellicon tangential flow membrane filtration system (Millipore, Bedford, MA), with centrifugation of the re-

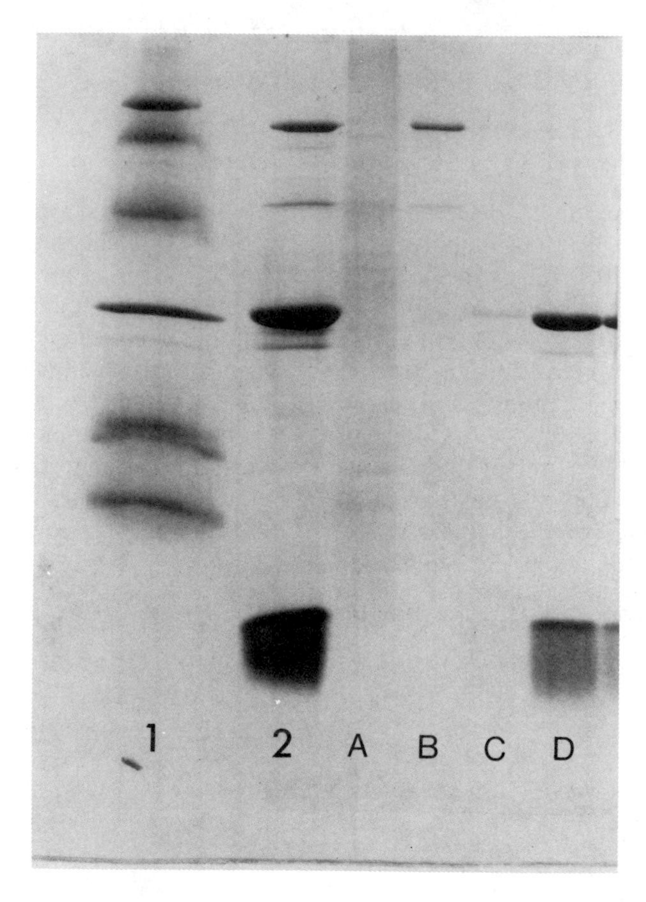
1
2
A
B
C
D

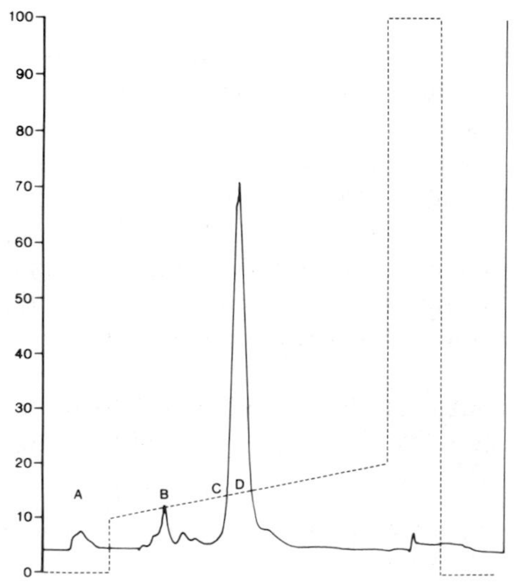
100
90
80
70
60
50
40
30
20
10
0
A
B
C
D

sulting cell slurry at 5,000 *g* to yield the final pellet (generally ~14 gm wet weight). The pellet is washed and reconstituted as for smaller batches, then lysed by passage through a French pressure cell at 30,000 psi. The lysate is cleared by an initial spin at 5,000 *g*, followed by centrifugation at 37,000 *g*. The supernatant is further treated with ammonium sulfate to obtain the fraction precipitating at 30–60% saturation. These proteins are reconstituted in one one-hundredth of the original culture volume and are dialyzed extensively against 10 m*M* Tris–HCl, pH 7.4. The crude protein preparation is further purified by passage over Blue Sepharose and Polybuffer 96, as described on page 156, except for an increase in the column sizes to 400-ml and 45-ml bed volumes, respectively.

The mean yield of toxin antigen in the crude preparations from the fermentor lysates has been 0.64 mg/l of culture or 12.8 mg/20-liter batch. The percent yield from the succeeding purification steps has been comparable to that obtained with the smaller volume columns (Table 1, p. 158). The final yield following Bio-Gel P-60 is ~42% of the initial toxin antigen, or 0.27 mg/liter initial culture volume, which is not significantly different from the yield of small volume shaker cultures. However, the scale-up does result in the frequent contamination of the final toxin with two minor protein bands (Fig. 2, upper panel). These bands may be successfully removed from the toxin preparation by fast-protein liquid chromatography (FPLC). The Bio-Gel eluate is passed over an anion exchange column (MonoQ, Pharmacia/LKB, Piscataway, NJ) equilibrated with 20 m*M* triethanolamine–HCl, pH 8.0, and eluted with a gradient formed with 0.5 *M* NaCl in the same buffer. This results in pure toxin containing the two polypeptide subunits in peak D (Fig. 2). Toxin recovery from this additional step is virtually complete.

FIG. 2. Anion exchange FPLC (MonoQ column) purification of 20 liter fermentor production of Shiga toxin. Upper Panel: Toxin samples from the Bio-Gel P-60 step before and after application to the MonoQ column were loaded onto a 15% polyacrylamide gel, subjected to electrophoresis as in Fig. 1, and stained with Coomassie blue. Lane 1, molecular weight standards; lane 2, Shiga toxin prior to FPLC separation; lane A, fraction A from the MonoQ column; lane B, fraction B from the MonoQ column; lane C, fraction C from the MonoQ column; lane D, fraction D from the MonoQ column. Lowel Panel: Elution profile of Shiga toxin from a MonoQ column, eluted with a gradient formed by 0.5 *M* NaCl in 20 m*M* triethanolamine-HCl, pH 8.0. The dotted line indicates the percent of salt solution in the gradient. Fractions A, B, C, and D were separately collected and subjected to polyacrylamide gel electrophoresis. Bioactivity was present primarily in fraction D.

Author Index

Numbers in parentheses are footnote reference numbers and indicate that an author's work is referred to although the name is not cited in the text.

A

B

C

D

E

F

G

H

I

J

L

Q

R

S

V

W

Y

Z

Subject Index

A

B

C

E

G

H

I

O

P

R

S

T

Y